# Geography of the Physical Environment

*About this Series*

The *Geography of the Physical Environment* book series provides a platform for scientific contributions in the field of Physical Geography and its subdisciplines. It publishes a broad portfolio of scientific books covering case studies, theoretical and applied approaches as well as novel developments and techniques in the field. The scope is not limited to a certain spatial scale and can cover local and regional to continental and global facets. Books with strong regional focus should be well illustrated including significant maps and meaningful figures to be potentially used as field guides and standard references for the respective area.

The series appeals to scientists and students in the field of geography as well as regional scientists, landscape planners, policy makers, and everyone interested in wide-ranging aspects of modern Physical Geography. Peer-reviewed research monographs, edited volumes, advance and undergraduate level textbooks, and conference proceedings covering the major topics in Physical Geography are included in the series. Submissions to the Book Series are also invited on the theme 'The Physical Geography of…', with a relevant subtitle of the author's/editor's choice.

More information about this series at http://www.springer.com/series/15117

Gábor Mezősi

# The Physical Geography of Hungary

With contributions by Timea Kiss

Springer

*Author*
Gábor Mezősi
Department of Physical Geography
and Geoinformatics
University of Szeged
Szeged
Hungary

*Contributor*
Timea Kiss
Department of Physical Geography
University of Szeged
Szeged
Hungary

ISSN 2366-8865 ISSN 2366-8873 (electronic)
Geography of the Physical Environment
ISBN 978-3-319-45182-4 ISBN 978-3-319-45183-1 (eBook)
DOI 10.1007/978-3-319-45183-1

Library of Congress Control Number: 2016949564

Translation from the Hungarian language edition: *Magyarország természetföldrajza* by Gábor Mezősi, © Akedemiai Kiado 2009. All Rights Reserved. ISBN 978-963-05-8976-5.
Timea Kiss has contributed to Chapter 1 of this Work.


Cover image by Sonja Weber, München

Printed on acid-free paper

This Springer imprint is published by Springer Nature
The registered company is Springer International Publishing AG
The registered company address is: Gewerbestrasse 11, 6330 Cham, Switzerland

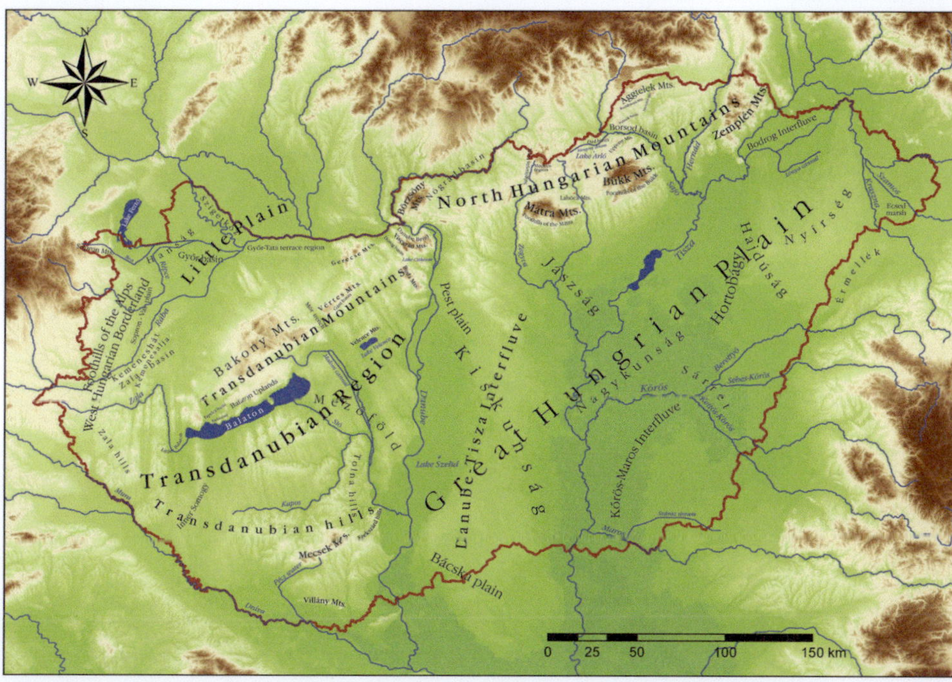




The landscapes of Hungary

# Preface

It is a challenge to publish a comprehensive work on the physical geography of Hungary. Several excellent works have been written with a similar content, from various points of view. Some of them aimed to analyse the palaeo-geographic changes of Hungary (Martonné 1995, Schweitzer 2002), and others highlighted the issues of relief development (Lovász 1994, Pécsi 1996) or concentrated on the different landscapes and regions (Kogutowicz 1930, *Magyarország tájföldrajzi sorozata 1967–1988*).

The objective of physical geography is to analyse and explain the geographical patterns created by the interaction of soil, hydrology and surface forms. In fact, the integration of these disciplines constitutes the "core knowledge" of geography and not the thorough examination of them one by one. It is evident that various specialized disciplines are known in physical geography; e.g., geomorphology describes and explains surface forms, whereas soil geography defines the origin, character and potential use of soils. It is definitely true that a grounded, comprehensive analysis of natural conditions is difficult without this classical knowledge. Each discipline may motivate and require the analysis of several geographical questions.

The integrative interpretation and approach of our natural environment is not a simple task, especially if we take into consideration that the socio-economic factors also play a role in it. One of the pathways of an integrated approach must lead into the geographical landscapes; however, now a little bit different direction was taken and aimed at describing the correlations of some natural conditions in a regional sense.

It is "one-sided", since it examines only the effect relationships of a few natural processes. It also undertakes the presentation of the most important physical geographical processes of the natural units, because the analysis of impact chains would be far from complete without them. Certain elements of the present volume attempt to explore some topics in a novel way. It extends beyond classical knowledge and addresses a few integrated issues in some of its segments.

Nevertheless, it does not aim at the complete synthesis of the geographical environment, but rather at describing some of the correlations of the soil–water–air relief system. It is also hard to undertake a "synthesis" because the different rate and scale of the factors active in the geographical systems is an especially difficult issue—just to mention one of the most evident reasons. For example, it is complicated to describe the effects of the climate on the relief or the soil types and processes, because we know the tendencies and the rate of change on the global scale, but the answers are difficult to formulate on a regional scale, even if the discussion required this. The challenge is not only in the above-mentioned magnitude differences, but also in the need to measure the overall impact of a lot of parameters. Therefore, the present work can only undertake to highlight the importance of some effect relationships and to demonstrate through some examples how a partial analysis may lead to an impact analysis.

The use of a certain surface affects the whole geo-system, not only the individual parameters. This systemic interrelatedness is not always conspicuous when using the geographical elements, because the relationship of the factors is sometimes loose, and it is hard to recognize that the processes influence the system by their combined impact. It is becoming acknowledged more clearly with the more and more intense exploitation of the resources and assets, because accessing them is becoming more and more costly and hazardous. It is difficult to follow exactly how and which factor is changed by the environmental use without the above-mentioned approach.

The present book is organized around two major frames of thought, adopting a traditional structure in this respect. The first part discusses the general physical geographical issues of Hungary; however, it does not aim at a thorough chronological and spatial analysis of each physical geographical factor (e.g. hydrography, soil or vegetation) due to space constraints. Instead, we provide an overview of the evolution of the main landforms, their general state, the more important processes and the rhythm of their natural changes, that is, in our opinion, all the information that physical geography can use in complex, multifactorial analyses. Undoubtedly, we pay more attention to relief evolution, because this is one of the scientific tasks of physical geography.

The second part with a regional content provides a physical geographical overview on a macro-regional scale. The meso- or micro-regional analysis and a monographic completeness are not regarded as an aim here, either, since we know of excellent summaries, at both meso-regional (see the Landscape Geography of Hungary) and micro-regional levels (*Magyarország kistájainak katasztere I–II*). Nevertheless, through a few examples on this scale, this part undertakes to demonstrate the environmental conflicts, values and hazards in an area and the specific landscape features, which require a comprehensive geographical knowledge.

Finally, the author wishes to express his gratitude to the reviewers for their detailed assessment; their advice and suggestions have enriched the present work. Gratitude is also due to the colleagues who participated in the compilation of the present volume.

Szeged, Hungary
2011

Gábor Mezősi

## References

Lovász Gy, Majoros Gy (1994) Magyarország természeti földrajza. JPTE, Pécs
Martonné Erdős K (1995) Magyarország természeti földrajza. KLTE, Debrecen
Pécsi M (1996) Geomorphological Regions of Hungary. MTAFKI, Budapest
Schweitzer F (2002) Pleisztocén. In: Karátson D (szerk.) Pannon enciklopédia, Kertek, Budapest, p 130–135

# Acknowledgements

My greatest debt is towards Tímea Kiss, who not only wrote the Sects. 1.4, 1.5.2 and 1.8 of Chapter 1 but reviewed the other parts of the book. I owe special thanks to Katalin Erdős, Gergely Horváth and György Lovász for their detailed, thorough and encouraging reviews. I am also very grateful to Teodóra Bata for research assistantship, to Balázs Benyhe for creating figures and to György Sípos for his assistance in proofreading.

# Contents

# Part I

# General Characteristics of the Physical Geography of Hungary

# 1 Relief Evolution

**Abstract**

In this chapter, the palaeo-geographical changes of the central part of the Carpathian Basin are described, since the Palaeozoic until nowadays. It analyses the evolution of the landscape based on plate tectonics, besides the main influencing factors of palaeo-geographical forms and processes are also studied in connection with climate and vegetation changes. The Quaternary landscape development processes and their consequences are described in detail. In Hungary, fluvial processes are dominant; therefore, the development of the fluvial network, the formation of terrace systems and their controlling factors (e.g. tectonic movements) are emphasized too. The other main form assemblage of the region is connected to blown sand and loess formation. In the mountainous areas, the dominant processes are the different development ways of karstic and non-karstic surfaces and the periglacial forms and processes. The summary of geomorphological, climatic, vegetational and hydrological processes follows a temporal order; thus, the chapter gives a detailed and complex analysis on the Pleistocene, Late Glacial and Holocene evolution of the region.

## 1.1 Paleozoic and Mesozoic Relief Evolution and Palaeo-Geography

Paleozoic and Mesozoic palaeo-geographical history of the Carpathian Basin is quite complex, because the crust pieces currently located side by side and representing an integral unit may have been as far as thousands of kilometres from each other at the time of their evolution. Even though the climate was warmer and more uniform at that time, the different ecological conditions resulted in distinct surface evolution and sediment formation. The crust pieces forming the Carpathian Basin once were located at different places on Earth; thus, they will not be discussed in detail. The description follows the evolution of the micro-plates called ALCAPA (Pelso) and Tisia-Dacia, which came about as a result of subsequent evolution and fusion. However, it must be considered that the two micro-plates are separated along the Middle Hungarian structural line (lineament), which was active until the Miocene and only got into their current positions in the Mid-Tertiary.



G. Mezősi, *The Physical Geography of Hungary*, Geography of the Physical Environment,
DOI 10.1007/978-3-319-45183-1_1

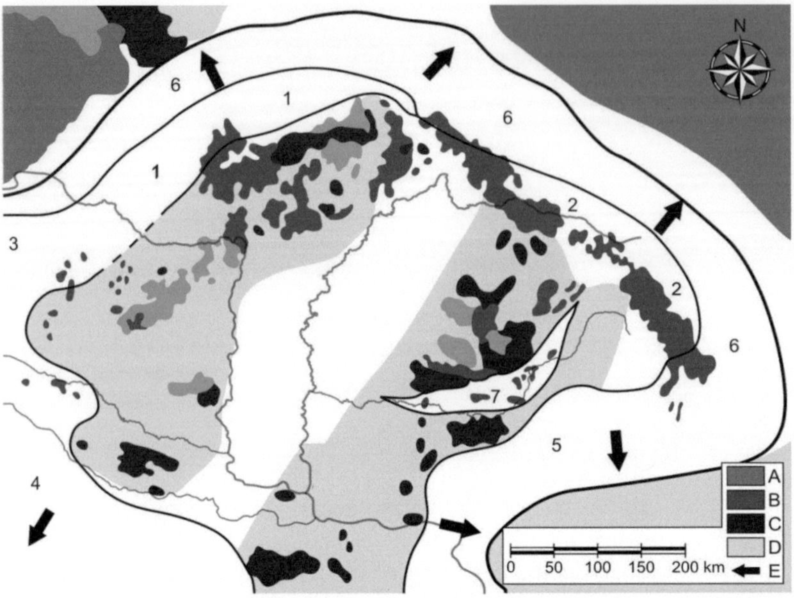


**Fig. 1.1** The structure of the Carpathian Basin (after Prinz 1936). *A*—Pre-Carboniferous massif, *B*—eruptive rocks, *C*—carboniferous blocks, *D*—carboniferous folded structures, *1*—mountainous belt of the Felvidék (Northern Hungary), *2*—mountainous belt of Transylvania, *3*—folded strata of the Alps shifted to the north, *4*—folded strata of the Dinarids, *5*—folded strata of the Transylavanian Alps (Erdélyi-havasok), *6*—the youngest flysch folds connecting the Alps and the Transylvanian Alps, *7*—minor folded strata of the Metalliferous Mountains in Romania

There have been several evolutionary models on the structural formation of the basin, which were first comprehensively systematized by Prinz (1936). In his view, the basement of the basin was formed by an "intermediate mass" called the Tisia massif. It is a stable micro-plate which consists of old rocks, and it is partly responsible for the formation of the arc of the Carpathians (Fig. 1.1). The modern plate tectonics identified this unit as a piece of a micro-continent existing in the basement of the basin.

The researches done before 1950 could hardly rely on drilling results; therefore, Schmidt's idea (1961 in: Bulla 1964) on the thinning of the crust beneath the basin and the existence of the Paleozoic and Mesozoic basement sediments in a narrow NE-SW strip should be highly appreciated. The geological map of the basement of the basin made in the 1980s reflects this structure (MNA 1989, Fig. 1.2); however, the explanation today differs. The reason for the existence of the trips is not the existence of special geosynclinals here, since they can be explained by specific plate movements. Two important factors have to be highlighted when interpreting the evolution of the basin.

A. One of the key factors in the development of the basin is the Central Hungarian Lineament stretching along the Zagreb-Kulcs-Zemplén Mts. line. It was discovered ca. 50 years ago, and it became soon obvious that the basin can be divided into two very distinct parts along this line. The SE part shows similarities with the sequence of sediments of the (Paleo-) European continent, while the NW part is connected to the African continent (which lies to the south), or to the micro-continents detached from it, and it could be compared to the sediment sequences of the Eastern Alps (Stegena et al. 1975; Wein 1978). Due to the scarcity of experience in plate tectonics, when the theory appeared, some researchers interpreted this main line as a subduction line and associated several phenomenon with it, for example, the thinner crust, the resulting

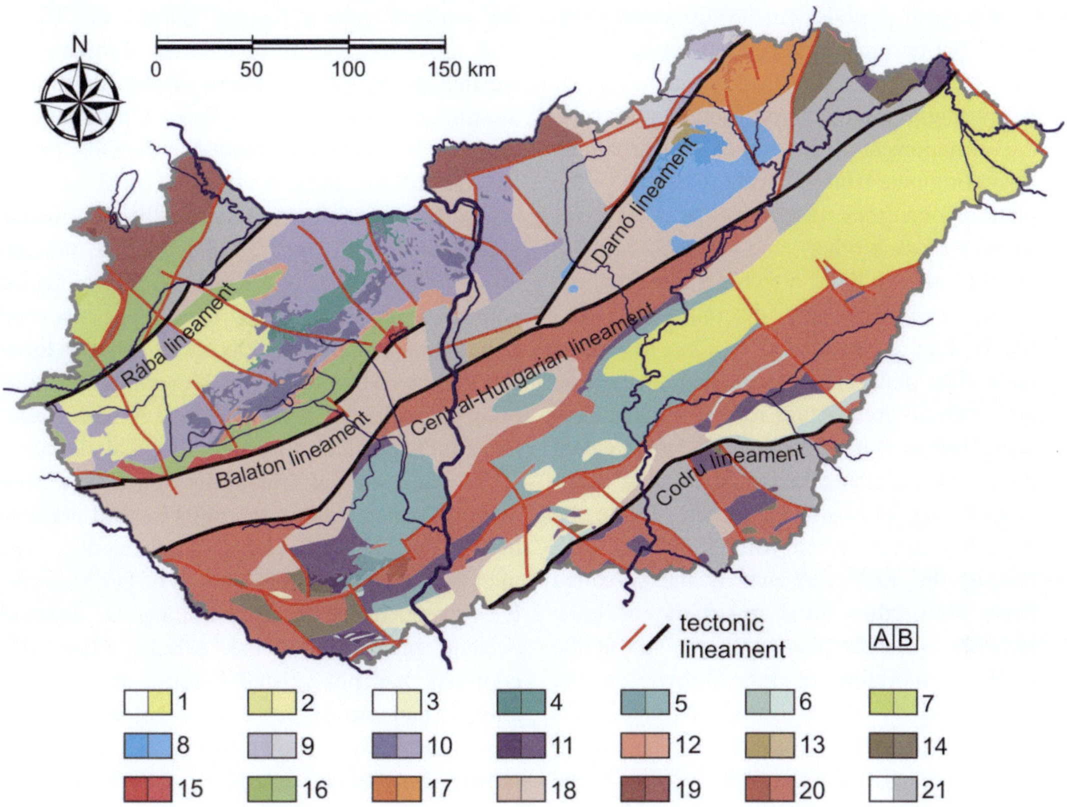


**Fig. 1.2** Map of the Tertiary basement of the Carpathian Basin, Hungary. *A*—surface formations, *B*—subsurface formations; *1*—Cretaceous-Eocene flysch, *2*—Upper Cretaceous (Bakony type) sedimentary rocks, *3*—Upper Cretaceous (Lowland type) sedimentary rocks, *4*—Jurassic and Lower Cretaceous (Bakony type) sedimentary rocks, *5*—Jurassic and Lower Cretaceous (Mecsek type) sedimentary rocks, *6*—Jurassic and Lower Cretaceous (Villány type) limestone, *7*—Jurassic and Cretaceous (Alpine type) metamorphic rocks, *8*—Triassic and Jurassic (Bükk type) limestone and slate, *9*—Triassic and Jurassic (Gömör type) limestone, dolomite and slate, *10*—Triassic (Bakony type) limestone and dolomite, *11*—Triassic (Mecsek type) limestone and dolomite, *12*—Permian (Bakony type) limestone, *13*—Carboniferous-Permian (Bükk type) limestone and slate, *14*—Carboniferous-Permian (Villány type) sedimentary and volcanic rocks, *15*—Carboniferous granite, *16*—Paleozoic slate, *17*—Paleozoic crystalline limestone and slate, *18*—other Paleozoic and Mesozoic formations, *19*—old crystalline slate, *20*—old granite and crystalline slate, *21*—unknown basement

positive heat flux, the changes in the subsidence rate of plains, or the occurrence of basaltic volcanism subsequent to andesite volcanism (Stegena et al. 1975).

Other ideas have also occurred when interpreting the role of the Central Hungarian Lineament. For example, Wein (1978) recognized the similarity of the sediments in the Mecsek Mts. and the Carpathians and also the occurrence of flysch between Szolnok and Debrecen outside of the range of the Carpathians. He considered the line as a fault line along which a part of the inner and outer belt of the Carpathians turned into the basin with a rotational motion. From the mid-1980s, the intrusion of the Transdanubian Mountains from the Drauzug into their current position (during the Eocene and Miocene) was interpreted in a similar way, as the members of the Transdanubian Mountains fitted into the environment of the Eastern Alps. The evolution of the structure will be introduced later from a plate tectonic aspect. The palaeo-geographical characteristics and the surface processes of the main geological periods will be discussed separately for the two main micro-continents.

B. According to the other well-known approach, the formations at the bottom of the Carpathian Basin originate from different locations of the globe (sometimes thousands of kilometres away from each other), and they shifted close to each other without a tectonic deformation. Their movement was discontinuous, and finally, they amalgamated creating new plate units. The plate unit formed and evolved in the southern area of the former Pangea, and later, Tethys Bay or Tethys Ocean is called ALCAPA (referring to the location it originates from). This unit is located NW to the Zagreb-Kulcs-Zemplén Mts. line (Fig. 1.2). Part of the ALCAPA unit is the so-called Pelso Unit (after the Latin name of Lake Balaton). Another plate unit evolved at the northern shoreline of the Tethys, on the southern parts of the European plate. This unit is developed from the amalgamated subunits of Tisza (see Prinz's Tisia) and Dacia micro-plates, which are located SE to the Central Hungarian Lineament. The above-mentioned two large units only became adjacent at the end of the Cretaceous. The plate movements and the resulted landforms will be discussed later.

### 1.1.1 Palaeo-Geography of the Paleozoic

In the Carpathian Basin, there are evidences on orogeny (Kadomian and Caledonian) from the Precambrian and the early Paleozoic, but they were metamorphosed several times; thus, the analysis of these sediments has hardly provided any useful information on palaeo-geography (Péró 2000).

The oldest formations are located in the Sopron Mountains, and they belong to the ALCAPA unit and evolved of sandy sediments. The oldest are micas, which are related to the Variscan orogeny in the Cambrian period, while the younger gneiss formed during the Upper Carboniferous. The SE boundaries of these rocks are indicated by the (tectonic) Rába Lineament. Southeast of this line, the Pelso Unit could be found (it became amalgamated in the Oligocene). On the surface, these rocks form small and flat hills. The oldest formations are the Ordovician quartz phyllite along the Lake Balaton, the Silurian shale at Lovas and Alsóörs (it is a protected geological site), but below the surface a several hundred metres thick rock body stretches as far as the Velence Mountain. This Devonian limestone body is presented near Polgárdi, its coral remnants indicate shallow sea and tropical climate.

The unit around Bükk Mts. must have been located farthest among the units located north of the Central Hungarian Lineament. The micro-plate carrying the Bükk is of Dinaric type. It evolved on the ocean rim along the southern oceanic platforms of the African plate (this continent had not existed before), and presumably, it had moved from there before the Carboniferous as it is shown in Fig. 1.3. Due to its presumed situation, several hundred metres of a sandy, clayey and molassic type sediments accumulated during the Upper Carboniferous, similar to those in the Southern Alps. The "graben" between Igal and the Bükk is tectonically unique because of its significant movements and was considered to be a single folded unit, though nowadays it is believed to consist of different parts. The graben stretches from NE to SW today, and it was filled during the Carboniferous and the Jurassic periods. Here, the sedimentation started by Carboniferous shales and limestone lenses (Bükk Mts.). There is an early Permian sediment hiatus, but later a shallow sea environment may have prevailed resulting in the development of sandstones, shales and carbonate lagoon sediment facies (Fig. 1.4).

It is difficult to find an analogy for the Uppony and Szendrő Mts. in the Carpathians (Péró 2000). The Ordovician and Silurian sandstones and shales contained iron and manganese ores; however, the former had been mined and processed for hundreds of years in the smelters of Újmassa. However, the manganese nodules have only useless amount (although their exploitation

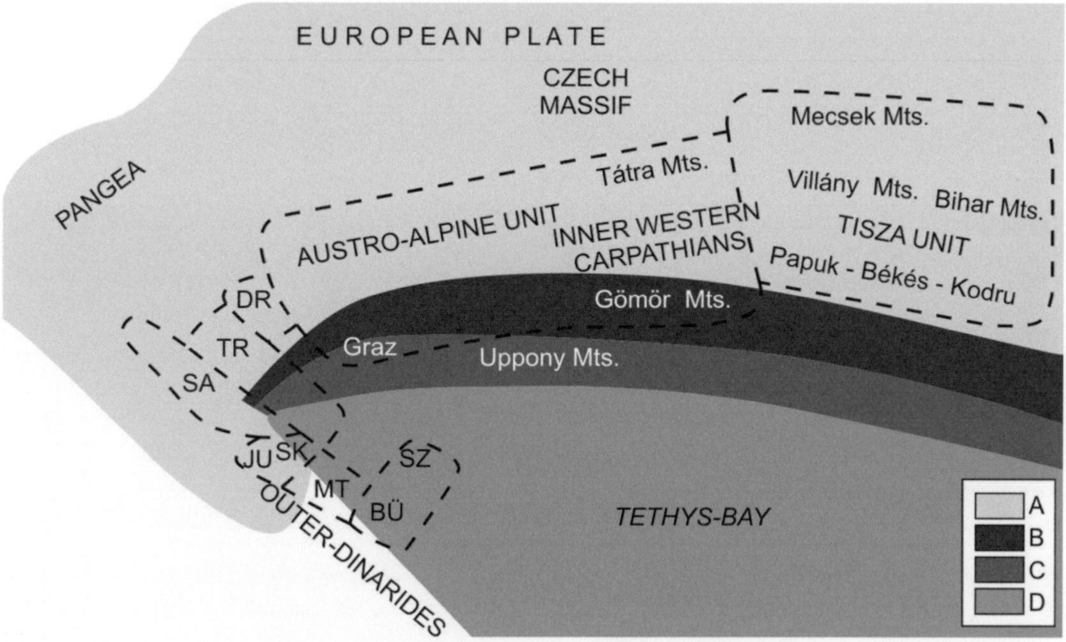


**Fig. 1.3** Location of geological units of the Carpathian Basin in the Middle Carboniferous (after Haas et al. 1999). *A*—terrestrial units, *B*—shallow marine clay, *C*—shallow marine carbonate, *D*—flysch, *DR*—Drauzug, *TR*—Transdanubian Hills, *SA*—Southern Alps, *JU*—Julian Alps, *MT*—Middle Transdanubia, *BÜ*—Bükk Mts., *SZ*—Southern Karawanks (the denotations remain the same in the following figures)

has been attempted). Their setting can be studied in the vicinity of the water reservoir of Lázbérc. The most intensive sedimentation started with Devonian carbonate sediments. A sequence representing almost the whole Devonian could be found in the Szendrő Mt: shales were covered by sandy sediments, indicating the movement of the micro-plate into another environment. (Later, the limestone crystallized and it got the name of "Rakaca marble". It was used as ornamental stone as long as its fragmentation allowed it.)

The tiny Vilyvitány Block (few square kilometres) is a geological curiosity in Hungary, as its early Paleozoic mica is probably the oldest stone in the basin.

Between Lake Velenceand Lake Balaton, granite batholith is the characteristic formation, which developed in connection with the Variscan orogeny, intruding into the older, Upper Carboniferous slates. Nowadays, just its small part is on the surface (Velence Mt.), but it had probably repeatedly gone through planation. Wobbly stones (or "woollen bags") remained on the surface as the evidence of tropical weathering, though they probably got to the surface during the Quaternary, embedding in granite gravel and sand (Image 1.1). According to the interpretations of some palaeo-magnetic studies, the granite crystallized in the southern hemisphere along the tropic of Capricorn and postvolcanic processes also played role in its evolution. It resulted in some lead and zinc ores, which were mined in Pátka until the beginning of the 1970s.

Limestone and dolomite were formed during the Permian in the ALCAPA area (joining Tethys). At the end of the period, gypsum-like sediments deposited in the lagoons, which could be found in the vicinity of Perkupa and Alsótelekes at the foothills of the Rudabánya Mts. The terrestrial areas were dominated by arid or semi-arid environment, though the term semi-dry could be the most suitable description (Majoros and Péró 2000). On this environment, thick weathering crust and red soils developed, and planation took place on the semi-arid environment. However, some palaeontologists refer to

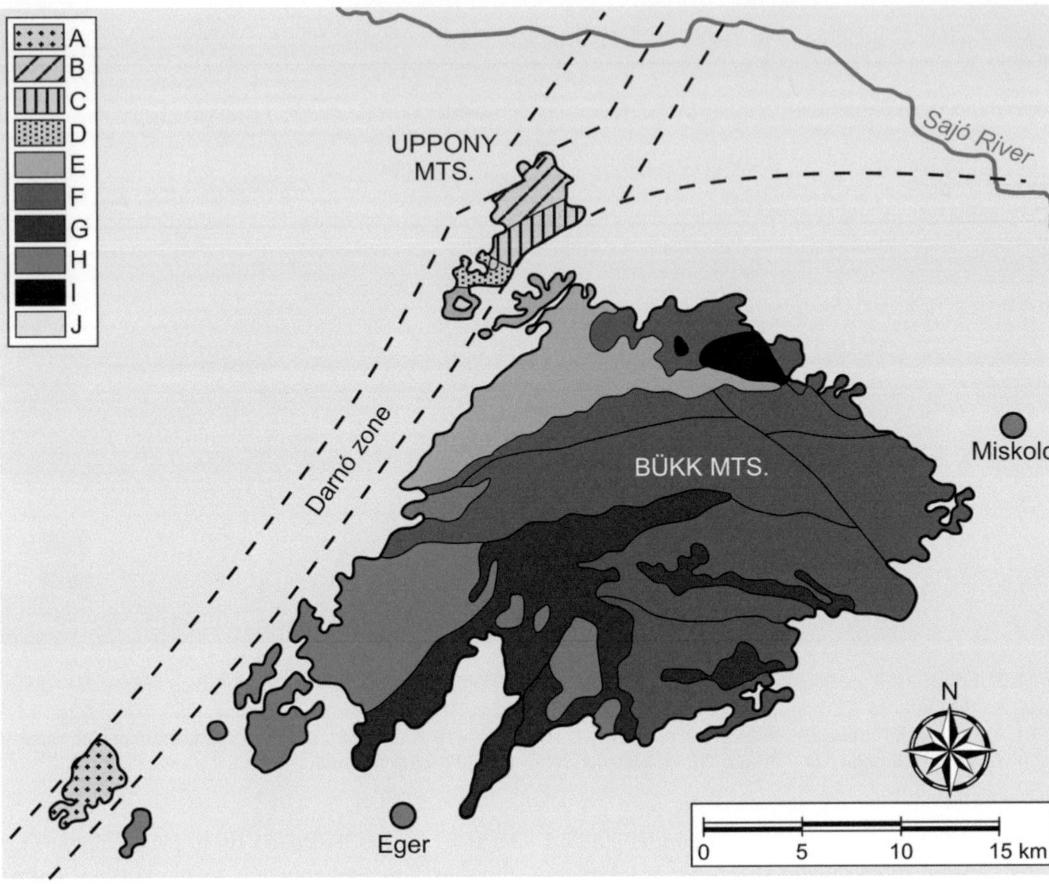


**Fig. 1.4** Geological sketch of the Bükk and the Uppony Mountains (after Kovács 1989, in Haas 2001). *A*—mellitic, *B*—Uppony Mts., *C*—Tepličany Unit, *D*—Nekézseny Conglomerate, *E*—Upper Paleozoic, *F*—Triassic, *G*—Jurassic, *H*—Szarvaskő–Mónosbél strata, *I*—Kis-fennsík strata *J*—Cenozoic formations

warm and humid environment based on dendrological analysis of tree ring-less tropical trees. At the end of the Permian, about 250 million years ago, there was mass fauna extinction; thus, ca. 54 % of existing families disappeared, including ancient crabs and insects.

Assuming that landscape evolution took place in a dry and warm climate, the red colour of the sandstones could be explained. This rock is easily carved; therefore, it is the characteristic building material near its surface occurrences (Révfülöp and Balatonalmádi). Under the surface, it stretches towards the Lake Velence, but here it is already replaced by dolomite deposited in a lagoon.

The Paleozoic rocks of the Tisza and the Dacian structural units (later they merged) constitute the basin basement SE of the Central Hungarian Lineament, evolved uniformly during the Alpine orogeny. The NE-SW ridges of the units developed in the Devonian as a result of convergence of continental plates. The oldest terrestrial formations of the basement form three strips from the Mecsekalja across the Danube-Tisza Interfluve and the Körös Plain (Fig. 1.5). At the bottom of the Upper Carboniferous and Permian sedimentary strata, sandstone and shale (with traces of anthracite) can be found. They are covered by other Permian layers of deltaic and fluvial origin, which

**Image 1.1** Wobbly stones in the Velence Mountain (photo by Kiss, T.)

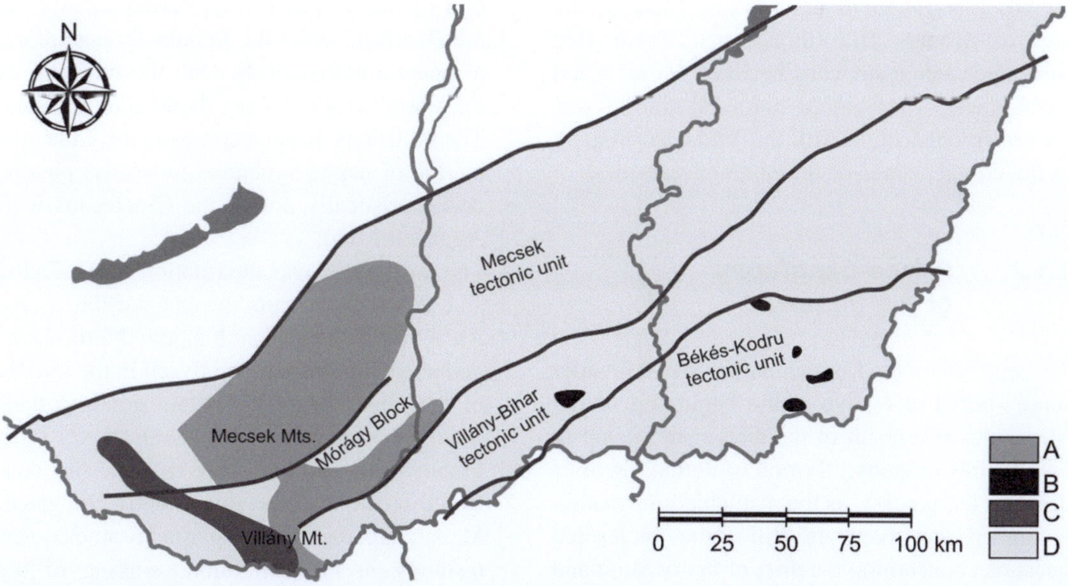


**Fig. 1.5** Upper Carboniferous and Permian formations of the Tisza Unit refer to dry terrestrial environment (after Majoros and Péró 2000). *A*—Mecsek, Lower and Upper Permian rocks, *B*—Mecsek, Lower Permian rhyolite in the basement of the Great Hungarian Plain, *C*—Upper Carboniferous and Lower Permian formations, *D*—supposed erosional area during the late Paleozoic

gradually became red upwards, indicating a warm and dry climate.

Grey and green sandstones (1000 m thick) of fluvial origin deposited on top of the Upper Permian sequence, containing uranium. The formations near Kővágószőlős and Kővágótöttös were analysed in detail to support uranium extraction. The mining started in the 1950s and terminated in the mid-1990s, and today, these ores are considered as strategic reserves. Formerly, the storage of radioactive waste of low and medium radiations was planned in the mines, but finally, it was realized in the granite of Bátaapáti. The above-mentioned sediment sequence, which was formerly regarded as Permian, was completely covered by a clastic red sandstone formed in a deltaic and nearshore environment (Barabás and Barabásné Stuhl 2005). The sequence of sandstone and aleurite layers were cemented by siliceous geyser eruptions, which later affected by selective erosion. These towers (called "babás szerkövek") are important protected geological features on the steep cliffs of the southern slope of the Mecsek Mts.

One of the most common rock of the Tisza Unit is the 330-350-million-year-old Lower Carboniferous granite, which is on the surface at the SE foreground of the Mecsek Mts., in the area of Mórágy Hill (though its 20-km-wide subsurface range stretches between Mórágy and Kecskemét). These large mass of granite was formed in connection with the Variscan orogeny in the contact zones of the plate movements.

### 1.1.2 Palaeo-Geography of the Mesozoic

The structure of the Carpathian Basin in a broader sense started to evolve at the beginning of the Mesozoic, as a result of the tectonism related to the Alpine orogeny. Several different theories describe the process, as the available information is limited and there are still some unclarified questions concerning the drift of the oceanic and continental plates. Besides, the point-like information fragments could be fitted differently, resulting in different evolutionary models.

However, the key elements of the Mesozoic evolutionary history are known, based on the facts of plate tectonics. Its main features are the formation and evolution of the Tethys Ocean, the opening and later the subduction and disappearance of its bays (Vardar and Pennin), and the mosaic-like fragmentation of the coasts. As a result of this series of processes, the fragmented crust pieces of the basin evolved. Originally, these units were in a coastal position and attained their present position due to the intricate movements caused by the colliding African and European plates. As a result of the plate movements the units of the Bükk and Transdanubian Mts. moved onto the African plate, while the units of the Aggtelek-Rudabánya Mts., the Mecsek and the Tisza Unit were positioned on the shelf area of the European plate.

According to Csontos and Vörös (2000a, b), four tectonic phases could be distinguished in the geological timescale:

- The first phase is the opening of the Tethys and the formation of the Vardar and Pennin oceanic bays (Figs. 1.6 and 1.7). The formation of the Vardar Ocean started as early as the Permian, while the Pennin Ocean started to open simultaneously with the collision of the Vardar Ocean during the Middle Jurassic;
- The collisions were accompanied by the formation of nappe structures in several phases, characteristically during the Cretaceous (e.g. Aggtelek Mts.);
- The third phase was the rotation of ALCAPA and Tisza-Dacia units in opposite directions. As a result, Transdanubia, the North Hungarian Mountains and the flysch in the middle of the Great Hungarian Plain got into their present position (see Fig. 1.9 and Sect. 1.2);
- Finally, due to these movements, the coalesced blocks became significantly elongated. Meanwhile, the asthenosphere became cooler, resulting in the continuous sinking of the Pannonian Basin.

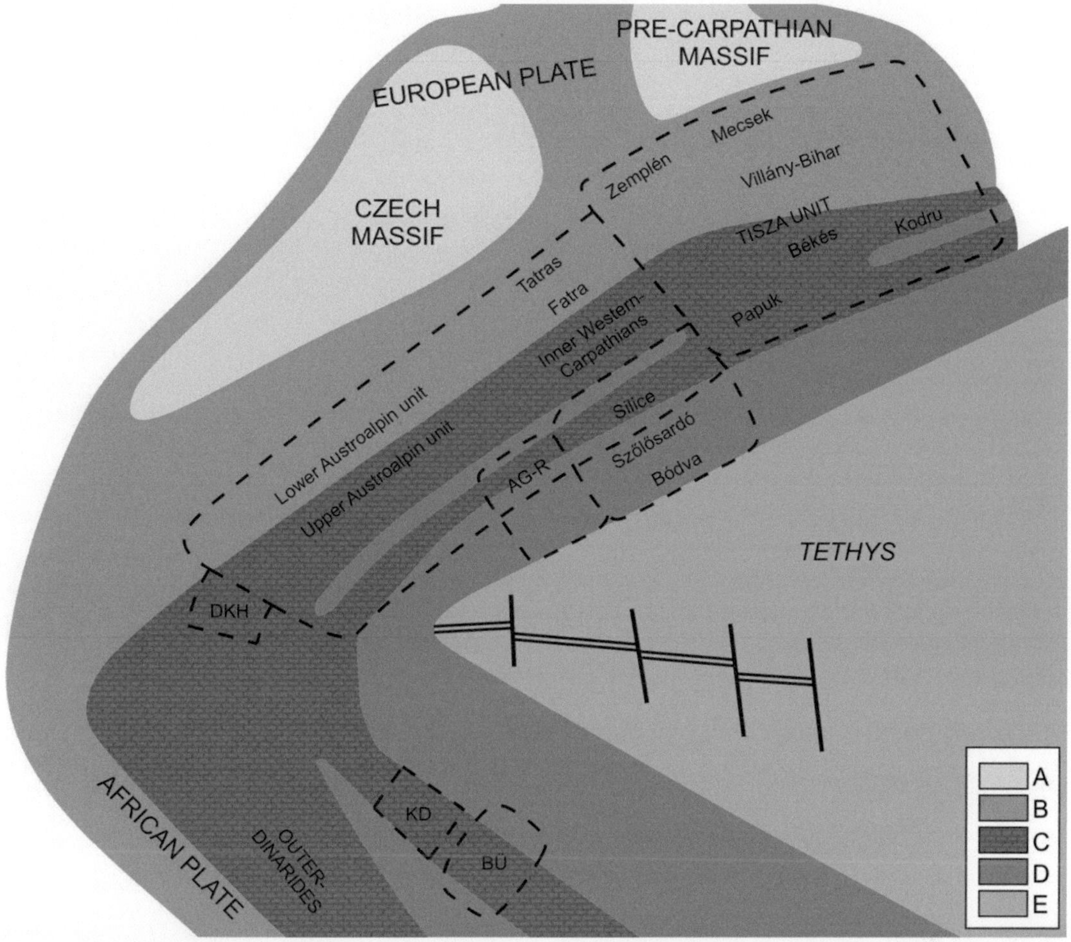


**Fig. 1.6** Location of the main structural units in the Upper Triassic (after Haas et al. 1999). *A*—terrestrial area, *B*—terrestrial sediments, *C*—shallow sea carbonate platform, *D*—pelagic carbonates, *E*—ocean floor, *DKH*—Transdanubian Mountains Unit, *KD*—Middle Transdanubian (Száva) Unit, *AG–R*—Aggtelek–Rudabánya Unit, *BÜ*—Bükk

These tectonic activities considerably determined the Mesozoic palaeo-geographical conditions and sedimentary processes.

During the Permian and Triassic periods, the landscape development on the Tisza Unit and ALCAPA (which were remote to each other) was rather different. On the terrestrial surfaces of the Tisza Unit, significant erosion took place, which produced early Triassic sand and red sandstone in certain locations.

However, the pieces of the ALCAPA Unit were situated at the edge of the Vardar Ocean; thus, a significant thickness of maritime sediments developed on the subunits of the Bükk, Dinarides and Transdanubian Mountains. At the beginning, the sediments were accumulated in the coastal areas or in a shallow sea, as it was indicated by the gypsum formation at Perkupa lasting from the Permian until the beginning of the Mesozoic. The vegetation was mainly characterized by sparse pine forests and the fauna by some large ancient reptile species (Vörös 2000). Soon after the large-scale extinction between the Permian and Triassic, a rapid ecological recovery happened, and thick-bedded limestone deposited of the calcite shells of protozoa. The sedimentation included only the smaller parts of the ALCAPA Unit (e.g. Bükk, Transdanubian Mts.,

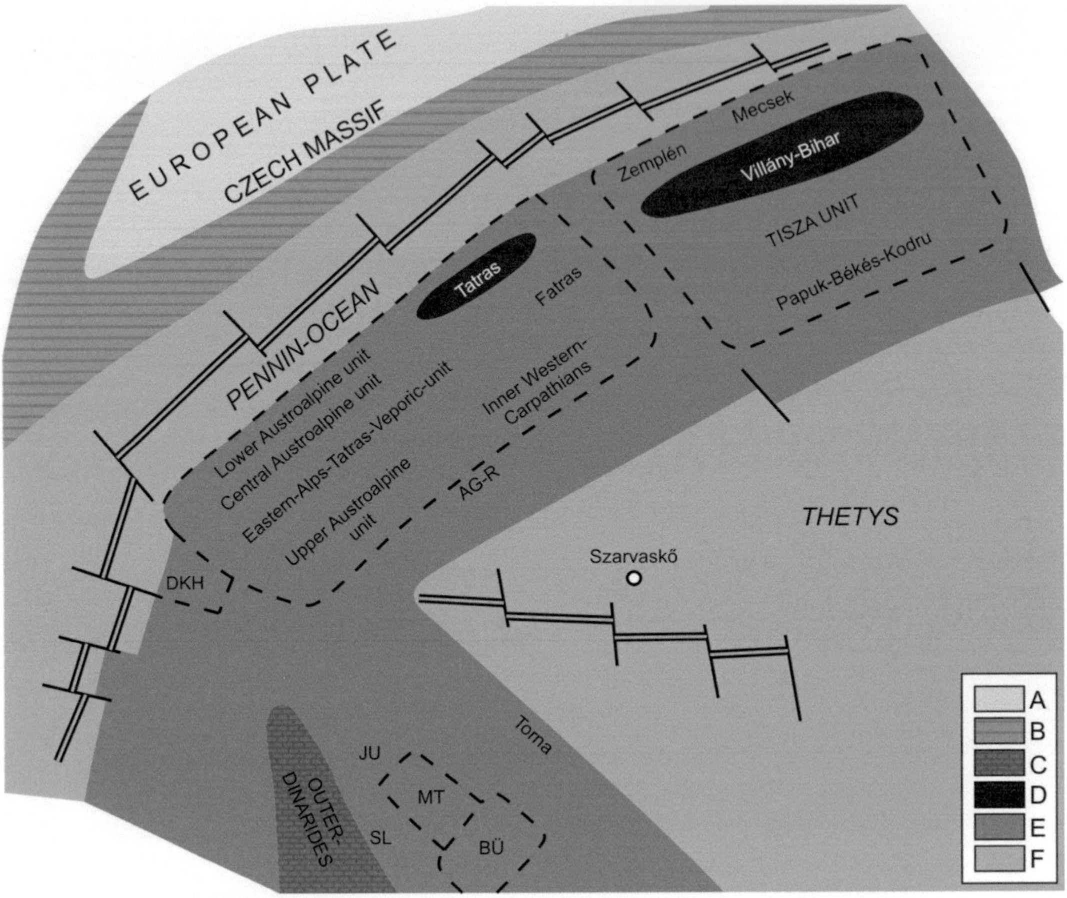


**Fig. 1.7** The structural geological reconstruction of the Middle Jurassic (after Haas et al. 1999). *A*—terrestrial area, *B*—shallow and deep-sea marl and sediment, *C*—shallow sea carbonate platform, *D*—shallow sea carbonates, *E*—pelagic carbonates, *F*—ocean floor, *DKH*—Transdanubian Mts. Unit, *AG–R*—Aggtelek–Rudabánya Unit, *JU*—Julian Alps, *MT*—Middle Transdanubia, *BÜ*—the Bükk

Middle Transdanubian and Száva units), though the regression of sea level was about 80–100 m. Serpentinite and talc were formed in the Lower Alps as a result of metamorphosis of volcanic rocks, in connection with the separation of the Pennin Ocean during the Middle Jurassic. This talc was mined in the neighbourhood of Felsőcsatár.

During the Triassic, limestone and dolomites were formed in the sinking sea on the southern edges of the Bükk and Transdanubian Mts. units (Fig. 1.6). In the Transdanubian unit, the sandy and marly sediments were covered by about 200-m-thick dolomite (Megyehegy Formation), which accumulated on a carbonate platform, and finally, in the Late Triassic, 1000-m-thick dolomite and (Dachstein) limestone were deposited in the quickly subsiding sea. Similar conditions characterized the Bükk unit: the carbonate foreground mainly consists of very thick limestone layers, referring to atoll and lagoon environments, but also to deep-sea conditions. The slow subsidence of the seafloor was accompanied by tectonism connected to the collision of the Vardar Ocean, which finished in the Middle Jurassic. The process resulted in the decomposition of the uniform ocean basin. The tectonism resulted in subwater volcanism, which produced, for example, pillow lavas (Bükk Mts. at Szarvaskő).

The Aggtelek-Rudabánya unit structurally fits the Eastern Alps, especially because it was situated in the northern foreground of the Vardar oceanic bay of the Tethys. At the beginning of the Triassic, sandstone and shale were accumulated on the shelf. At the beginning of the middle Triassic, this shallow part of the sea was separated from the open ocean, creating anaerobe sedimentary environment; thus, bituminous Gutenstein limestone was deposited rich in organic material. After a phase characterized by reef and lagoon environment, the sea started to deepen and the carbonate platform became dissected. In the deep water, Wetterstein limestone evolved in 1000 m thickness, and this limestone characterizes the Aggtelek-Rudabánya Mts. today. This palaeo-geographical phase continued in the beginning of the Jurassic too, when limestone and shale accumulated in a small area (Bérczi and Jámbor 1998).

In the beginning of the Jurassic, the limestones indicate changing water depth too. In the Transdanubian Mts., limestone developed, and manganese ores (at Úrkút, Eplény) refer to short oxygen-deficient environment. The manganese mining ceased due to the depletion of the ore resources. On the walls of the mines, different forms of manganese precipitations and crusts could be observed, as well as subsequent tropical karst forms (Image 1.2). Some researchers (Hartai 2005) dated the marine manganese formation to the Middle Jurassic (Dogger age).

By the end of the Cretaceous, the Vardar and Pennin oceans disappeared due to subduction, and simultaneously, the Tisza and ALPACA plates collided, forcing the Tisza plate to shift towards SE direction, and on the other hand, the ALPACA moved in a NE direction, as the European and African plates collided (Fig. 1.8 and see detailed explanation in Sect. 1.2).

In the first half of the Cretaceous, significant carbonate sedimentation took place for a considerable long period in the Transdanubian Mts. The process quickly terminated, and the sediments were uplifted by compressive forces to the surface. This terrestrial surface became covered by the eroded products of metamorphical and igneous rocks originating from the surroundings (of that time). The limestone landscape had moderate topography, and it probably looked like an eroded (planation) plain. The planation process took place on tropical climate and lasted for 50 million years. (In the following geological periods, planation was also supported, though they did not comply with these two requirements; thus, it should be considered as the last planation period.) The tropical planation involved the carbonate surfaces, and the weathering resulted in laterite and cone karst forms (the tropical karst forms of the former bauxite mine of Iharkút are now under nature protection; Image 1.3). The bauxite formed during the Cretaceous accumulated (with other, above-mentioned materials) in tropical karst forms in the Bakony and Vértes Mts.. These formations were covered by Eocene sediments, protecting them from subsequent erosion.

The first half of the long Cretaceous was mainly characterized by regressional and transgressional processes under the homogeneous tropical climate. In the second half of the Cretaceous, herbaceous plants appeared in the flora, and this was the time when dinosaurs spread in high variability. The landform development was dominated by tropical karstification and planation. At the end of the Cretaceous, the ancient reptiles became extinct when a large-scale environmental change took place, which was explained by a meteorite impact or a climate crisis caused by volcanism.

In the Lower Jurassic (Lias Age), the Mecsek Mts. was located on the coastal edge of the Tethys (Vardar) Ocean. The climate became humid; thus, in the coastal bogs and marshes, rich vegetation developed, consisting of horsetail, cycad and pine species (Vörös 2000). The organic sediments became repeatedly covered by sandy and clayey sediments, and after a while, they got to great depth due to the large thickness

**Image 1.2** Palaeo-karst forms covered by iron and manganese crust in the former mine of Úrkút (nowadays a protected geological site, own photograph)

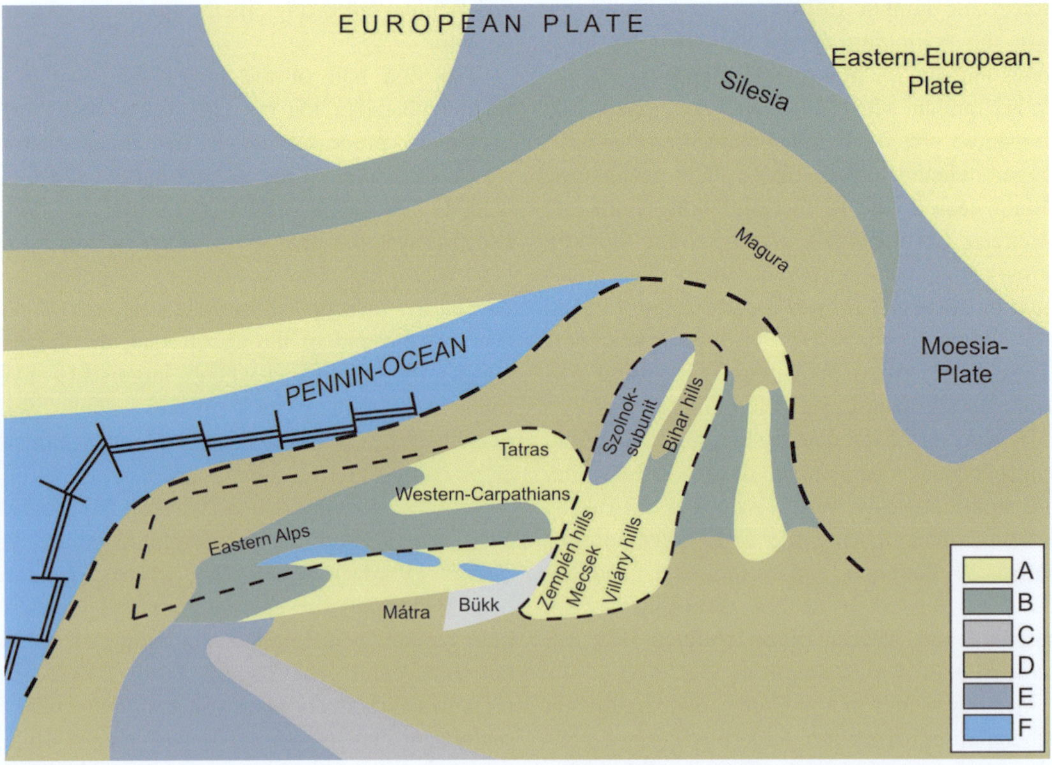


**Fig. 1.8** The geological situation at the end of the Cretaceous (after Haas et al. 1999). *A*—dry land, *B*—shallow and deep-sea marl and sediments, *C*—shallow sea carbonated platform, *D*—flysch, *E*—pelagic carbonates, *F*—ocean floor, *DR*—Drávazug, *JU*—Julian Alps, *MT*—Middle Transdanubia, *BÜ*—the Bükk

**Image 1.3** Tropical cone karst in the Darvastó Mine (own photograph)

of sediments, where good-quality coal was formed. Later, these coal layers were dissected by tectonic movements; thus, they can be exploited with difficulties. The exploitation of formations thicker than 1000 m has a 200 years' tradition (Pécs, Komló and Szászvár), but the mining has been stopped by now. However, there are still several environmental issues, which will be discussed in details in Chap. 11. The Middle and Upper Jurassic periods were accompanied by deep-marine (pelagic) formations (e.g. marl). In the Villány Mt., during the Jurassic, a shallow sea dominated or sediment hiatus existed, but its ammonite-bearing sediments are the most extensive occurrences in the Carpathian Basin.

The Villány Mts. became a terrestrial area for a while in the Lower Cretaceous; thus, its surface was karstified on the tropical climate. The bauxite formed at this time was also mined at Nagyharsány. The archipelago character of the Mecsek and Villány Mts., the collision of the Vardar Ocean, and the movement of the micro-plates of ALPACA effected significant basaltic volcanism (Vörös 2000).

## 1.2 Palaeo-Geography of the Tertiary

### 1.2.1 Palaeo-Geography and Landforms of the Eocene

The geological structure and the palaeogeography of the region had fundamentally changed by the end of the Cretaceous. By the Tertiary, only two oceanic basins remained, the Magura Ocean (later Carpathian Mountains) and the Dinarid-Slovenian Basin. They were separated by two plate fragments (which constitute the basement today), approximately along the Central Hungarian Lineament: the ALCAPA situated to the NE and the coalesced Tisza-Dacia units to the SW (Fig. 1.9).

In the Eocene, especially in the beginning of the period, terrestrial environment prevailed, and the marine sedimentation characterized the area just between the so-called Paleogene Line and the Rába Line. The sedimentary border of the

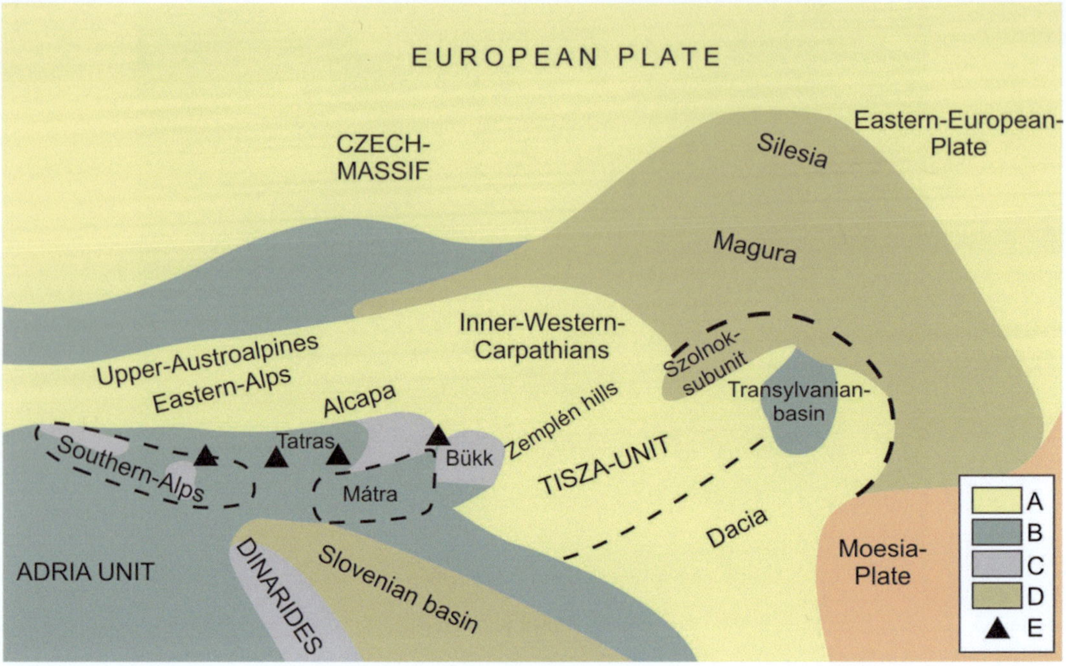


**Fig. 1.9** Location of the main structural units in the Upper Eocene (after Haas et al. 1999; Nagymarosy 2000a). *A*—terrestrial environment, *B*—shallow and deep-sea marl and sediments, *C*—shallow sea carbonate platform, *D*—flysch, *E*—andesite volcanic archipelago, *TR*—Transdanubian Hills, *JU*—Julian Alps, *MT*—Middle Transdanubia

Paleogene stretches along the southern part of the present-day mountain ranges, and it indicates that Paleogenic non-terrestrial sediments could be found just north of it, and to the south of the Rába Line (Table 1.1). The marine transgression from the SW started in the middle of the Eocene, and by the end of the period and at the beginning of the Oligocene, it extended to the southern edge of the North Hungarian Mountains. (To be more precise, it extended to the pieces of the ALCAPA Unit, e.g. Bakony and Vértes Mts., and then, the deposited material shifted towards NE until the Miocene.) At the beginning of the sedimentation clay, marl and limestone deposited, reaching 1000 m thickness in some places (e.g. in the Zala Hills), while it was only 10 m thick in the Gánt and Dorog basins. In pores of the limestone, a significant amount of crude oil was stored in the Zala region. Upper Eocene large-scale marine sedimentation and transgression mainly affected areas east to the Danube (those to the north of the Paleogene Line; Fig. 1.10). Along the shallow coasts, bogs and marshes developed, and the arboreal vegetation fitted to the tropical conditions (e.g. swampy and eucalyptus forests), providing favourable conditions for brown coal development (Tatabánya, Oroszlány and Nagyegyháza). The maximum thickness of the formation is 100 m (the total thickness of coal layers is about 10 m). Brown coal was often mined together with the underlying Upper Cretaceous–Lower Eocene bauxite in the 1970s and 1980s ("Eocene Programme"). The mining was maintained by water extraction, which constituted a serious environmental risk, for example, the springs of Tata dried, and the discharge of wells in Budapest also decreased (Alföldi and Kapolyi 2007). There were times when the quantity of water extracted was 40-fold of the coal quantity (Hubai 1992)! At the same time, coal mining was strongly hindered by the danger of water intrusion due to the karstic

**Table 1.1** Chronological division of the tertiary (after Haas 2001)

| (million years) age | | | stage |
|---|---|---|---|
| 3, 4 | PLIOCENE | | Romanian (Piacenzian) |
| 5 | | | Dacian (Zanclean) |
| 6, 7 | MIOCENE | upper | Pontian (Messinian) |
| 8, 9, 10, 11 | | | Pannonian (Tortonian) |
| 12, 13, 14, 15, 16 | | middle | Sarmatian (Serravallian) |
| 20 | | lower | Badenian (Langhian) |
| | | | Karpatian (Burdigalian) |
| | | | Ottnangian (Aquitanian) |
| 24 | OLIGOCENE | upper | Egerian (Chattian) |
| 30 | | lower | Kiscellian (Rupelian) |
| 37, 50 | EOCENE | | |
| 57, 67 | PALEOCENE | | |

environment, which was one of the reasons for the termination of mining in the Dorog Basin. Thus, the brown coal mines were shut in 2004 because of the disadvantageous site features, the low efficiency of mining and the resulted environmental problems.

The other typical raw material of the region was the previously introduced bauxite, which formed during the Cretaceous and the Eocene. However, in the first half of the Eocene when terrestrial conditions prevailed, this bauxite was eroded and transported (there are no Paleocene formations in Hungary) and deposited in the former tropical dolines between the cone karst on the southern and northern edges of the Bakony and Vértes Mts. The cone karsts were preserved in their original, vertical position despite of their horizontal shift by 600–700 km towards NE (Image 1.3).

The flysch belt built of Middle and Upper Eocene marine sediments is located between Szolnok and Debrecen (Fig. 1.10). Its material consists of mainly marls and sandstones originated from the intensive erosion of the uplifting Carpathian Mountains. These sediments should be located in the outer arc of the Carpathians, though the Szolnok flysch basin is located in its inner side. According to Wein (1978), this flysch formation rotated into the basin from NE (from the outer side of the arc of the Carpathians), along the Central Hungarian Lineament, which served as a transform fault. (Others claim that the Mecsek Mts. shifted to their present location from the carbonate cliff belts of the Carpathian Mountains.)

During the Eocene, volcanism and hydrothermal activity took place from the Zala Hills to the Mátra Mts. This is in connection with tectonism at the beginning of a subduction (Martonosi 1994). Thus, the Magura Ocean in the northern part of the Tethys collided as a result of a subduction due to SW-NE forces. As a result, in the middle of the Eocene in the area of the ALCAPA Unit, a line of andesite-dacite volcanoes developed. These formations can be found on the surface in the Velence Mts. (at Pátka, in an intensively eroded form) and in the Mátra Mts. (Lahóca-hegy). The large amount of deeply located copper ores in Recsk are also a product of this volcanic activity, though nowadays the andesite of Recsk has been dated to the Lower Oligocene.

The Eocene juxtaposition of the ALCAPA and Tisza-Dacia units is an important issue of the tectonic studies. Two models were developed to

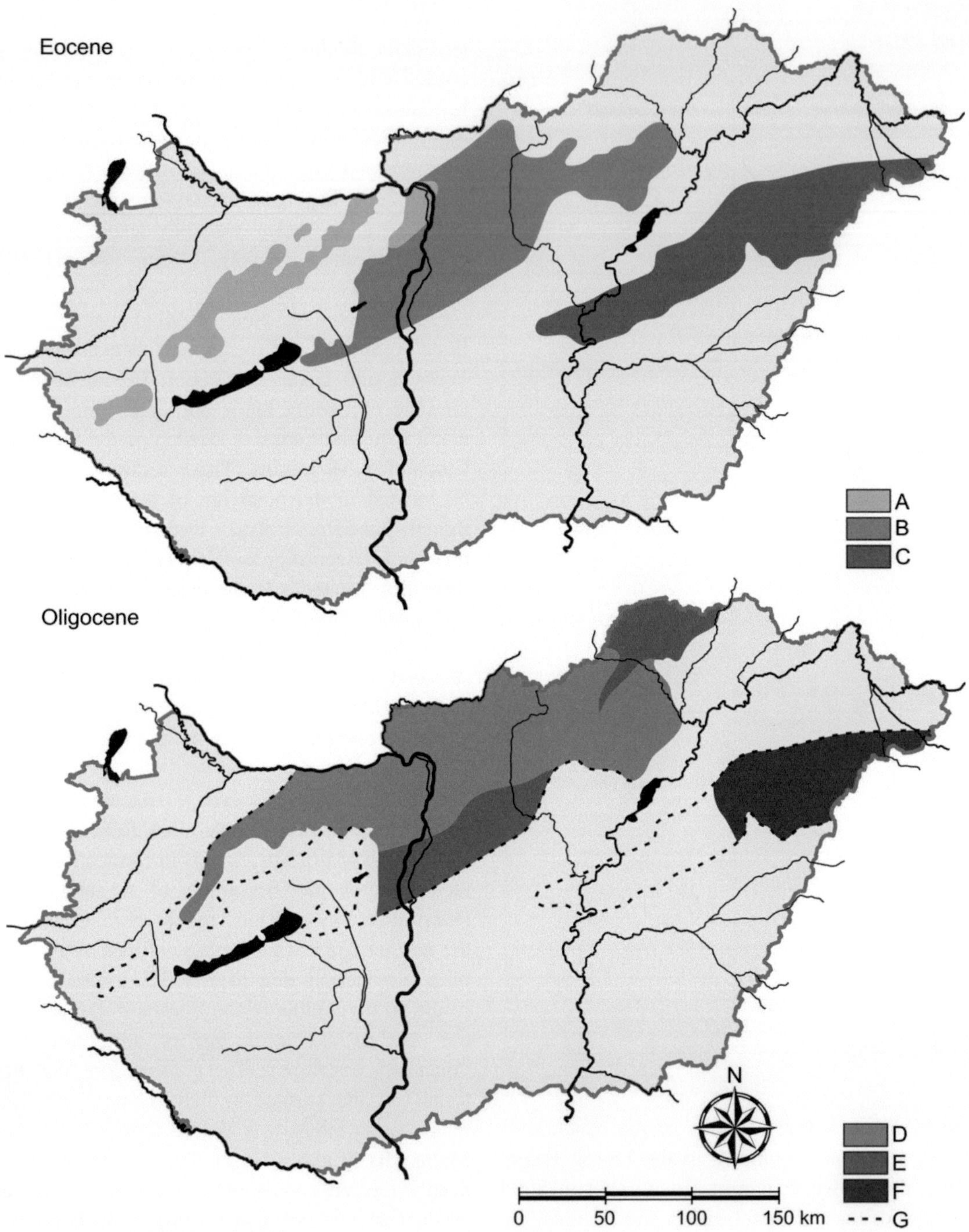


**Fig. 1.10** The territorial extension of Eocene and Oligocene formations (after Haas 2001). *A*—Eocene sequence with Miocene transgression, *B*—Eocene sequence with Upper Eocene transgression, *C*—Szolnok flysch belt with Eocene sediments, *D*—Oligocene formations in the southern territories with a sediment hiatus, *E*—Early Miocene formations, *F*—flysch covered by Oligocene sediments, *G*—border of the Paleogenic formations

explain this geological event: the first explain the juxtaposition by lateral shifts, while the other by rotational movements. As a result, these units that have highly different Mesozoic and Paleogenic sediment structures got close to each other (Csontos and Vörös 2000). It must be noted that the two units were dislocated by the movements of the African and European plates towards different directions and by their subsequent collision. During the collision, the units of the Transdanubian Mts. and Bükk were ejected, and they shifted to the E-NE and squeezed into the basement, which had a thinned crust. The ejection has been known since the 1980s, as during the survey of the quantity of bauxite, volcanic material was found in the bauxite. Consideration the subsequent, maybe multiple redepositions, the origin of these materials could be quested. They were probably originated from the Eastern and Southern Alps, where volcanoes were active at that time. The debate on the rate of ejection, or on the distance of the plate shifts (100 or 500 km) until the Lower Miocene, has not ended yet. Probably, this shift could have just started in the Upper Eocene.

In the first half of the Eocene marine, transgression did not affect the region of the Carpathian Basin; thus, terrestrial landscape evolution could have been significant in the humid tropical and climate. There is no evidence of a significant cooling in this period (37–57 million years), and it started gradually just in the Oligocene (Kordos 1979). Theoretically, the environmental conditions made the continuation of bauxite formation possible. In the Middle Eocene, a marine transgression started; the temperature of the sea was 23–24 °C, providing good conditions for coral reefs. As the sea gradually became enclosed from the Paratethys and its salty sea water was replaced by freshwater, an endemic flora and fauna developed in it (Hartai 2005; Nagymarosy 2000a).

At the end of the Eocene, marine marl sediments were deposited in the transgressional areas, under similar ecological conditions. These sediments cover larger areas; thus, they covered the area of the Buda Mts. and the southern edge of the North Hungarian Mountains too (Nagymarosy 2000a). This initially shallow and warm tropical sea had normal salt content, and corals lived in it. Limestone deposited in the sea, while in the deeper waters clay and marl sedimented. At this time deposited the material of the Castle Hill at Budapest, and the Buda Marl used for paint production.

### 1.2.2 Palaeo-Geography and Landforms of the Oligocene

During the Oligocene, the marine transgression affected almost the same area that of the Eocene, with the significant difference that the area of the Transdanubian Mts. emerged from the sea; thus, terrestrial (erosional) processes became dominant. Majority of today's Carpathian Basin became terrestrial area in the Oligocene, and the sea reached the foothills of the North Hungarian Mountains just in the second half of the period.

In the Eocene, tropical climate dominated, and it continued at the beginning of the Oligocene; however, later the climate shifted to subtropical. Thus, while at the beginning of the Oligocene, the mean temperature of each month was above 20 °C, and at the end of the period, it was only valid for half of the months, and the amount of precipitation also decreased. This transitional character of the climate also influenced the flora and the fauna; however, it is difficult to reconstruct this palaeo-geographical environment due to the low number of fossils, the shortness of the period and the simultaneous appearance of pine, deciduous and monsoon forests, which do not coexist today (Báldi 1976; Kretzoi 1987).

In the first half of the Oligocene, significant erosion characterized the relatively elevated Bakony, Vértes and Buda Mts. (This is also called intra- or infraoligocene denudation.) Part of the Eocene sediment cover disappeared in the Buda Mts., and the planated horsts became exhumed (Martonné 1995). The progressive climate change affected the geomorphical processes too. Besides the tropical areal erosion, in the Upper Oligocene linear erosion became the most significant process. It is characterized by the

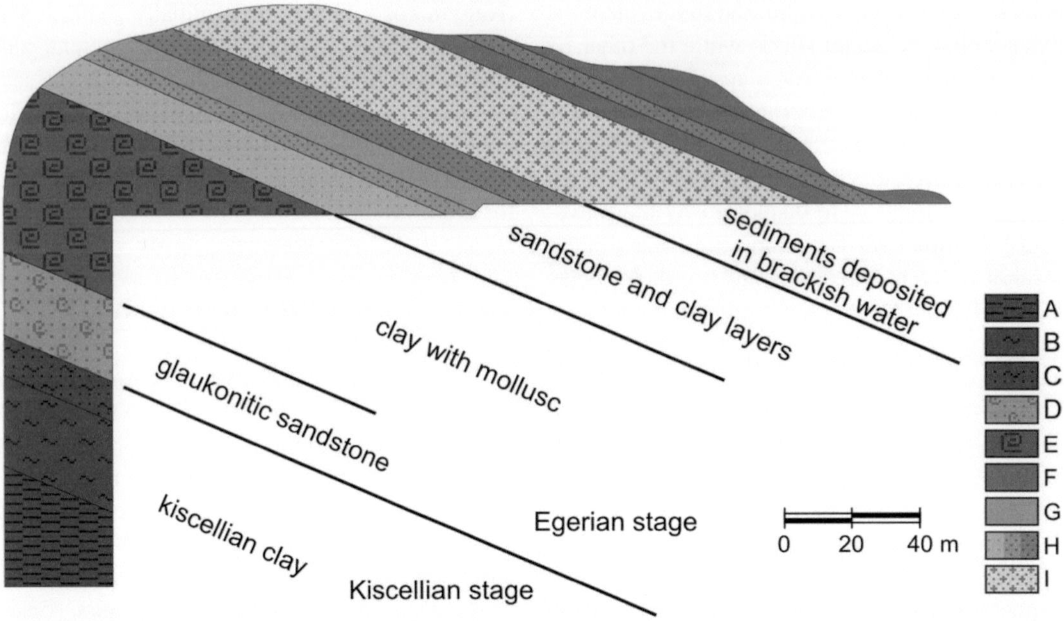


**Fig. 1.11** Cross section of the Wind Brick Factory of Eger, which exhibits the Lower Oligocene and the Upper Oligocene in a regressional sequence (after Haas 2001). *A*—tuff marl, *B*—clayey marl, *C*—sandy clayey marl, *D*—tuff with glauconitic sandstone, *E*—clay with mollusc, *F*—silt, *G*—clay, *H*—sand and sandstone, *I*—gravelly sand

Csatka Formation, a layer of multicoloured clay, fluvial pebble and sand deposited on the eroded surface in a few hundred metres thickness. It has a typical occurrence in the Keszthely Mts. (Nagymarosy 2000a).

In contrast, the strip NE of the Vértes and the Danube was covered by sea (Fig. 1.10). At the beginning of the Oligocene, some bays with poor oxygen supply (similar to the present-day Black Sea) developed in the vicinity of Budapest and Bükk Mts. The deeper seas became anaerobe, probably because of their separation from the oceans, and consequently, they had poor wildlife. There was no vertical sea water exchange; thus, either oxygen could not get into deeper layers; therefore, the large amount of organic material accumulated on the seafloor could not decompose. These sediments are poor in fossils and consist of clay and marl (Tard Clay Formation), and in some places, they are the source and parent material of some hydrocarbon occurrences in the region of Bükkalja and Bükkszék (MT 2002).

From the Middle Oligocene, the connection between the sea and the ocean was re-established. Simultaneously, the climate deteriorated, and it also influenced the water temperature, as it was indicated by the nanoplanktons and the macro-fauna during the Egerian Age (Báldi-Beke and Báldi 1974). The process may have been enhanced by the sea–ocean connection towards northern seas. At the same time, the anaerobic state terminated; thus, the black-slated Tard Clay Formation was replaced by the Kiscell Clay Formation (Fig. 1.11). This sediment cover extended to larger regions: Buda Mts., to the surroundings of Esztergom (also involved in Oligocene denudation) and to the area of North Hungary (until the Bükk Mts.), where its thickness is up to 600–800 m. The Kiscell Clay was formed in a basin on an open sea, it consists of homogeneous clay, and it is the most common Oligocene sediment. It was used as early as the time of the Roman Empire, 2000 years ago, when it was a raw material in brick manufacturing in the construction works of Aquincum (KOZ 2009).

The Hárshegy Sandstone, also formed in the Early Oligocene, covers the Kiscell Clay Formation, and it is the coastal or shelf equivalent of the latter. Hárshegy Sandstone could be found in the area of Esztergom, and in the Buda Mts., it is poor in fossils and its thickness is only a few dozen metres. The sandstone cemented by silica probably during subsequent volcanism (Nagymarosy 2000a); thus, it became a hard and resistant rock. It was mined near Pilisborosjenő and Esztergom for a building material and also in the territory of the Romhány for fireproof bricks (Hartai 2005).

At the end of the Oligocene, the deepest basin was situated along the Buda Mts.—foothills of the Bükk Mts. line. Its typical sediment is schlier (marl), which deposited in a 200-m-deep sea in calm water, and it is several hundred metre thick and mainly consists of clay and fine sandy silt.

### 1.2.3 Palaeo-Geography of the Miocene

The Hungarian chronology of the Miocene reflects the characteristics and changes of the western Paratethys compared to the Mediterranean region (Table 1.2). The beginning of the Miocene was characterized by terrestrial processes and later by the formation of the Pannonian Basin, archipelagic conditions and an extensive volcanic activity.

Lower Miocene (Eggenburgian Period) started by a moderate transgression from SE and SW. Its sediments were deposited shallow water, and they consist of sand and schlier (marl) in the Buda Mts. (basically, the Oligocene processes continued). Other Lower Miocene basins also developed, but these served as sedimentary areas just by the end of the period. In the region, the Styrian Basin, the East Slovakian Basin and the Vienna Basin have similar sediment sequences. These basins may have been the preliminary form of the Pannonian Basin. In the area of today's Hungary, terrestrial processes dominated (Fig. 1.12), but the sediments also refer to freshwater marshlands and limnetic environment. Thus, for example, in the Ottnang Period, 10–15-m-thick coal lenses and brown coal layers were deposited on shale at Brennberg in the Sopron Mts. They are covered by a thick fluvial sequence, indicating their terrestrial origin. The coal at Brennberg was one of the most important sites of the Hungarian coal mining: the exploitation started in 1789 and lasted for 200 years. Four or five productive coal layers of 100–200 m were found in the Sajó Valley, the Ózd and the Salgótarján Basins from the same period, but they were deposited in such freshwater marshes that later had brackish water. These wetlands were surrounded by bald cypress forest and by mangrove later, in accordance with climate and salt content; thus, the depressions transformed into lagoons. As a result of the transgression of the sea, sand and marl were deposited on this sequence, preserving the coal layers underneath. The Hungarian brown coal mining has practically ceased to exist by now.

A smaller limnic basin in the western foreground of the Mecsek Mts. was filled by terrestrial sediments, containing coaly layers too. Two main fluvial sediment transport directions (routes of palaeo-rivers) have to be mentioned: the first flowed into the basin connected it to the Mecsek Mts. from the SE, and the second flowed towards the Buda Basin towards an E-NE direction from the Alps (Fig. 1.12). One of the most important sites from this period is at Ipolytarnóc, which has a special palaeontological significance. The sediment of sand and multicoloured clay accumulated on a riverbank preserved several leaves and footprints of the period (Image 1.4), which were protected and cemented by the silica and a pyroclastic cover (Lower Rhyolite Tuff Formation).

In the Carpathian period, the SW-NE transgression from the Mediterranean region became strong (Fig. 1.12). It continued in the Badenian period too, and the subtropical sea filled the graben-like depressions (Nagymarosy 2000b). Sandstone, schlier and clayey marl accumulated between the terrestrial stripes (Little Hungarian

**Table 1.2** Chronological division of the Miocene (after Hámor 2001)

| age (million y) | orogenic cycles | main regional events | age | | Western Parathetys regional stages | sub-stage | recommended main stages | |
|---|---|---|---|---|---|---|---|---|
| 2 | rhodanian | basalt 3-4 m. y | PLIOC. | up. | Romanian | | | up. |
| | rhodanian | | PLIOC. | low. | Dacian | | | low. |
| 5, 7 | rhodanian, leithanian | basalt 9-10 m. y | MIOCENE | upper | Pannonian | | KASZPI | upper |
| 12 | leithanian | rhyolite 12-14 m. y | MIOCENE | middle | Sarmatian | | KASZPI | upper |
| 13 | styrian | 13-14 m. y | MIOCENE | middle | Badenian | up. | KASZPI | upper |
| 14 | styrian | andesite 14-15 m. y | MIOCENE | middle | Badenian | mid. | MEDITERRANEAN | middle |
| 16 | styrian | | MIOCENE | middle | Badenian | low. | MEDITERRANEAN | middle |
| | styrian | rhyolite, dacite 15-17 m. y | MIOCENE | lower | Karpatian | | MEDITERRANEAN | middle |
| 17 | styrian | | MIOCENE | lower | Ottnangian | | ATLANTIC | lower |
| 18 | styrian | rhyolite 18-20 m. y | MIOCENE | lower | Eggenburgian | | ATLANTIC | lower |
| 22 | savic | | MIOCENE | lower | Egerian | up. | ATLANTIC | lower |
| 25 | savic | | OLIGOCENE | | Egerian | low. | | |

Plain, Outer Somogy, South Great Hungarian Plain, South Transdanubia; Hámor 1984). The nearshore location is indicated by abrasional gravels (Börzsöny Mts.) and coal (Ózd Basin, Sajó Valley), and even abrasion terraces were described the Mecsek Mts. (Lovász and Majoros 1994).

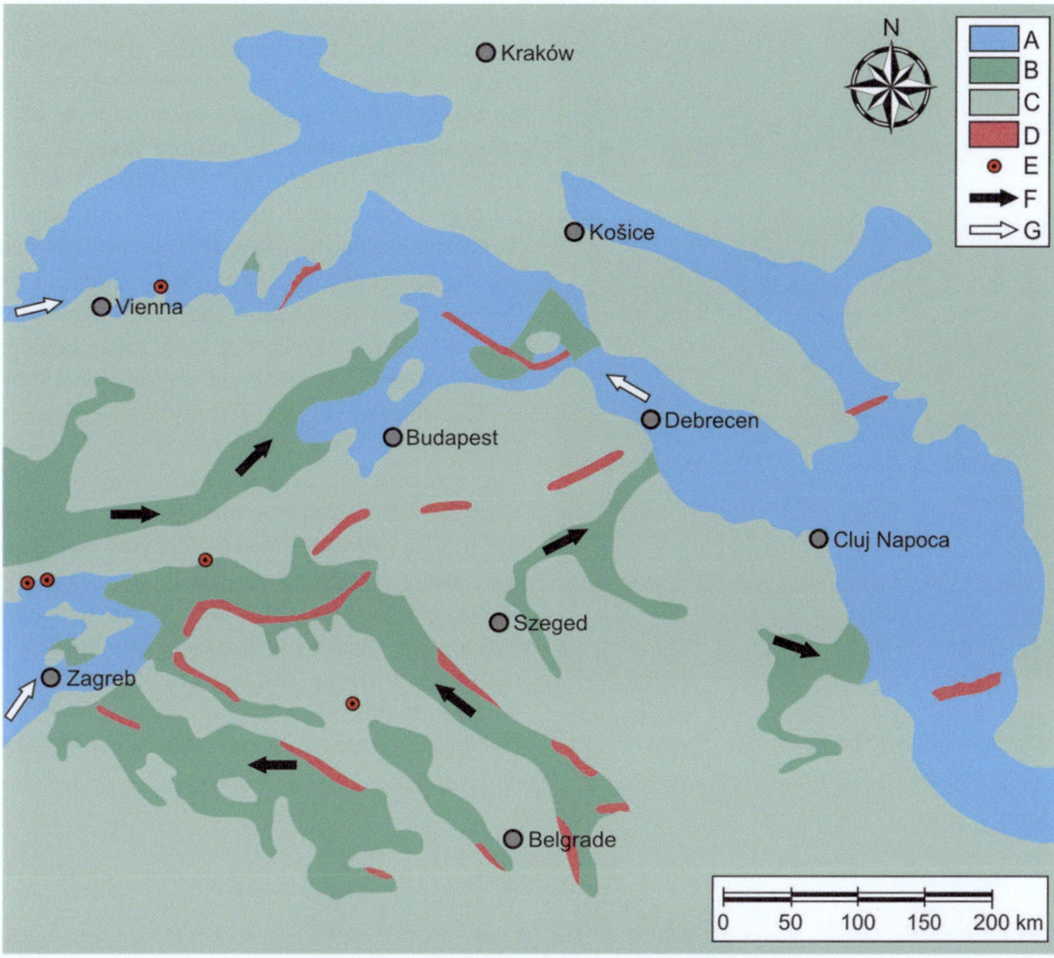


**Fig. 1.12** Main elements of the palaeo-geography of the Carpathian Basin in the Lower Miocene (after Hámor 2001). *A*—shallow or deep water cover (mollusc, marl), *B*—fluvial and floodplain formations, *C*—terrestrial areas, *D*—probable volcanic eruption ridges, *E*—andesite bodies, *F*—direction of fluvial sediment transport, *F*—sea transgression

The most characteristic Miocene landforms in the Carpathian Basin are connected to volcanism. Though the largest volcano on Earth was probably not in this region (although Székely 1997—supposed it was the Mátra Mts.), it was the most active volcanic region in the Miocene in Europe. The volcanism was caused by the subduction of the oceanic crust of the Outer Carpathians, the related thinning of the continental crust and the creation of a "back-arc basin", which was the beginning of the formation of the Pannonian Basin. The thinner intrusive lithospheric plate, rich in water, reduced the melting temperature of the mantle, which led to magma formation (Müller 2000). The volcanic activity was undoubtedly the most intense in the Badenian Period, though it had already started in the Lower Miocene (Fig. 1.13).

In the Miocene, three main volcanic periods were identified, which used to be called "lower", "middle" and "upper rhyolite tuff" levels referring to the connected tuff production. These volcanic levels used to be regarded as main stratigraphical levels in North Hungary. The exact locations of eruptive centres are not well known, though it is acknowledged that the

**Image 1.4** The layers of Ipolytarnóc preserved petrified tree (photograph by Kiss, T.)

high-energy eruptions covered the surface by volcanic material in large areas. The growing number of radiometric dating does not prove the probability of two large cessation periods, contrarily to the traditional threefold division. The traces of the oldest, ca. 20-million-year-old ("lower") rhyo-dacite tuffs indicate eruption in Transdanubia, in the Great Hungarian Plain and the North Hungarian Mountains (Image 1.5). The sediment that covers the 17–17.5-million-year-old sandy sediments of Ipolytarnóc belongs to this sequence too (Márton and Márton 1996).

The zones of volcanic activity can be typically found at 150–200 km from the subduction line. The volcanoes consist of mostly andesite, they were formed in a marine environment, and they are similar in size to Mount Vesuvius and Etna. The main evolutionary period of Börzsöny, Mátra and Visegrád Mts. was in the Badenian Period. The volcano of Börzsöny produced tuff of "Lower Rhyolite Tuff" age, and afterwards, it became an (effusive) type producing lava. Volcanic cones and domes emerged, which defined

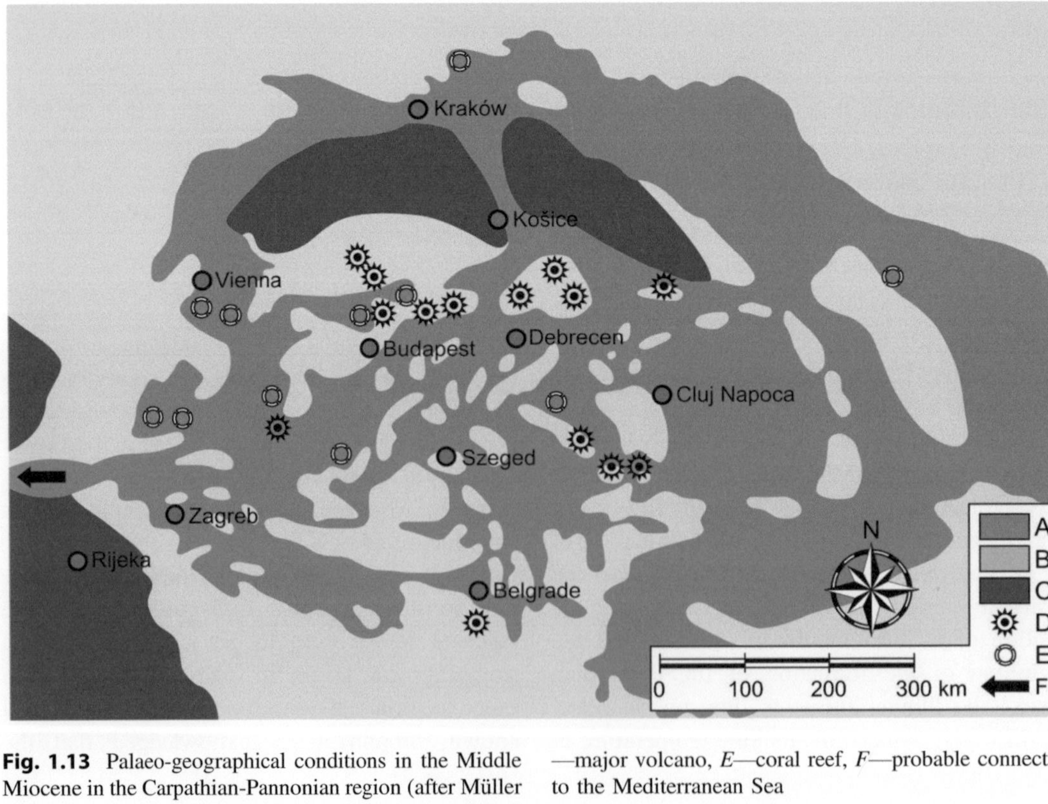


**Fig. 1.13** Palaeo-geographical conditions in the Middle Miocene in the Carpathian-Pannonian region (after Müller 2000). *A*—transgression, *B*—flat land, *C*—mountains, *D*—major volcano, *E*—coral reef, *F*—probable connection to the Mediterranean Sea

**Image 1.5** The "Lower Rhyolite Tuff" was covered by Leitha limestone; in between them, Badenian andezite variations could be found at the geological trail of Sámsonháza (own photograph)

the character of this stratovolcano. The forms differ in the eastern and western wings (see the regional section of the present book), which could be explained by the different quantity of gas in the magma. If the amount of dissolved gas was low, it appeared as lava on the surface, and if its amount or the water content was high, it typically appeared as pyroclast.

The morphology of Börzsöny Mts. definitely shows the volcanic form with its central caldera with supposedly more eruption centres (Fig. 1.14). The classical morphological analysis applies two methods: the gradient (measured on the volcanic slopes) and the river network patterns (radial pattern indicates a volcanic cone; Székely 1997). However, the volcanic explosions and lava flows did not appear exclusively, but alternately. Similar characteristics describe the Visegrád Mts., where the volcanic activity took place in a marshy coastal area mainly during the Badenian.

Period is ca. 14–15 million years. The volcanic processes forming the Visegrád Mts. were similar to those of Börzsöny, but fierce explosive activity took place several times due to contact with karst waters (Korpás 2000). Due to the large amount of pyroclast, the volcanic slopes provide favourable conditions for mass movements. The relatively consistent bedrock became less stable because of the burden of volcanic mass and became affected by landslide hazard. Mass movements are significant not only in the Visegrád Mountains (mainly in the eastern side, e.g. Sztaravoda Valley), but also at the foot of Mátra (Szabó 1995).

The Mátra Mts. are the most thoroughly analysed members of the inner Carpathian volcanic arc due to their rich mineral ores. They belong to the inner arc of the Carpathian Mountains, though they still have several unclarified issues. The volcano belongs to the same arc as the Visegrád and Börzsöny Mts., that is, the volcanic activity started in a shallow

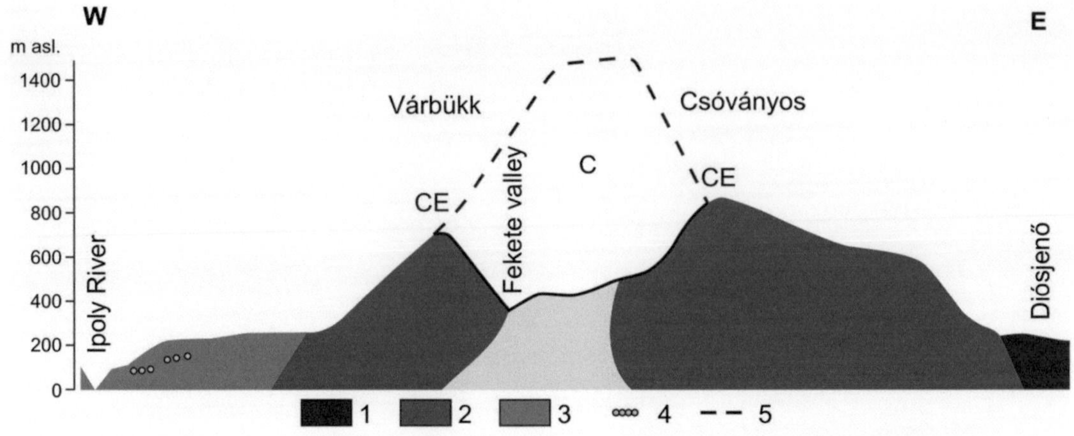


**Fig. 1.14** Reconstruction of the palaeo-volcano of Börzsöny Mts. (after Székely 1997). *1*—Carpathian schlier, *2*—Lower Badenian stratovolcanic formations, *3*—Pannonian sand and clay, *4*—Pleistocene terrace gravel, *5*—supposed palaeo-volcano, *C*—caldera, *CM*—caldera rim

water. The sediments of the Eocene and Oligocene deposited on the Triassic limestone and shale, and the products of the volcanic activity in the Miocene covered them all: first the lower rhyolite tuffs and then Badenian volcanic material, whose majority was andezite, dacite and rhyolite (in tuff forms; Image 1.6). The debate over the periods of eruption, the structure of the volcano and the location of the eruption centres has not been closed yet (these are discussed in details in the chapter of the North Hungarian Mountains).

The Zemplén Mts. are also members of the inner Carpathian volcanic arc, developed in the Badenian Period. The model outlined above is also valid here: the volcanoes developed over an 18–20-million-year-old "Lower Rhyolite Tuff" level. However, the volcanoes have a particular pattern, which is visible looking at a geological or geomorphological map: the volcanoes do not have one eruption centre, and they are situated in the intersection of a nearly S-N fault lines and other fault lines perpendicular to them.

The precious non-ferrous metal ores can principally be linked much more to Miocene than to Eocene volcanism. One group of the ores was formed of hydrothermal solutions: they precipitated from the vapours and solutions of hydrothermal waters in the vent or in the joining dike systems. For example, gold, silver and copper ores cemented the porous layers of lava flows and tuffs at the supposed eruption centre at Recsk (Lahóca Hill). Here, the contact metamorphism connected to Eocene volcanism had already resulted in zinc and copper ores. The subvolcanic intrusion in the Upper Eocene created one hundred million tons of copper filling the cracks, but its exploration is difficult as is situated at a great depth and has an average concentration of 1 %.

The dike systems containing gold, silver and lead ores were formed along the dacite body intruding into the andesite at Nagybörzsöny. The dike network (rich in lead, zinc and copper) at Gyöngyösoroszi was formed by the subvolcanic body intruding into the pyroxene andesite subsequently. A subvolcanic meta-somatized body intruded into the caldera created after the collapse of the Sarmatian andesite volcano at Telkibánya, forming the argentiferous dike system (Zelenka 2000). Kaolin and perlite formations were the result of weathering of the Sarmatian volcanic material in the Zemplén Mts. The significance and the extent of mining were also expressed by the fact that Telkibánya (gold) and Rudabánya received mining town rights, and other settlements were also established with the purpose of mining (e.g. Nagybörzsöny for gold and silver). At Rudabánya, iron, copper and silver ores were mined, resulted by a

**Image 1.6** Basalt columns in Somoskő, in North Hungary (photograph by Kiss, T.)

meta-somatism, which is supposedly connected to Oligocene movements (Szemán 2009).

From the Badenian period, besides volcanic materials, Leitha limestone was formed in shallow water, under a warm climate (indicated by coral reefs). Its known occurrences are in the Fertő-mellék Hills, Bakony Mts., Mecsek Mts. and the Tétény Plateau (Image 1.7). The fossils of the sea segmented by islands refer to warm, subtropical climate and a weak connection with the Mediterranean (Fig. 1.13; Müller 2000). In the deeper areas of the sea, clay and marl were formed, and brown coal and lignite deposited in the joining marshes of lagoons (at Hidas, Várpalota). Aleurite (alginite) was formed at Pula, but it is not extracted any more.

The closed littoral connection to the Mediterranean had a thorough environmental impact on the Upper Miocene and the Sarmatian period. The sea with a reduced salt content extended to the Aral Sea, and its effective sinking resulted in significant thickness of sediment accumulation (Fig. 1.15), as it is reflected by the widespread Sarmatian limestone and sandstone layers. The Vardar Ridge divided the territory of the Carpathian Basin into two parts: one of them was sinking, while the other was elevating (Hámor 2001), constituting the basis of the Pannonian Sea.

Several separate data on the climate of the Miocene exist; however, to unify them is not an easy task. The data prove gradual cooling during the Miocene. Thus, the majority of the vegetation used to consist of subtropical species, and tropical species just had a low proportion; but the proportion of temperate zone species increased from 10 to 40–50 % during the Miocene (Kordos 1979). The beginning of the Miocene was characterized by a 17–18 °C annual mean temperature, with 6 °C in January and 30 °C in July and

**Image 1.7** The quarry of Fertőrákos has been a UNESCO World Heritage Site since 2001 (own photograph)

by about 1400 mm annual precipitation. The significantly higher number of data from the later periods shows warm winters, cooler summers and a decreasing mean temperature (14 °C), which all refer to an oceanic climate influence. The Sequoia forests from the Badenian and Sarmatian periods are thermophilous, and the vegetation with several conifers evokes a Mediterranean environment.

### 1.2.4 Palaeo-Geography of the Late Miocene

The sinking of the basin mainly created by plate movements was relatively quick during the Pannonian Period, which started about 12 million years ago. (The Outer Carpathian oceanic crust subducted, resulting in the thinning of the continental crust; thus, a depression could be created behind the arc.) The marine conditions started to disappear due to filling up by sediments and the loss of connection to the Mediterranean Sea. The sea had quite different depths: it was up to 900 m deep at the deepest points and much shallower near the coast. The centre of the basin had gradually shifted towards SE (Fig. 1.16). The sea became more and more shallow and resembled gradually a lake, as its salt content had decreased to 8–12 ‰. The water of lake should have increasing salt content under these circumstances (it may have happened for a short period at the end of the Sarmatian), but the water household was influenced by a significant outflow of the salty water through the Iron Gates Gorge and the inflow of freshwater from precipitation (Müller 2000). The changing water level is shown by the abrasion terraces along the southern and south-eastern edges of the Mecsek and Transdanubian Mts. and by the existence of pure

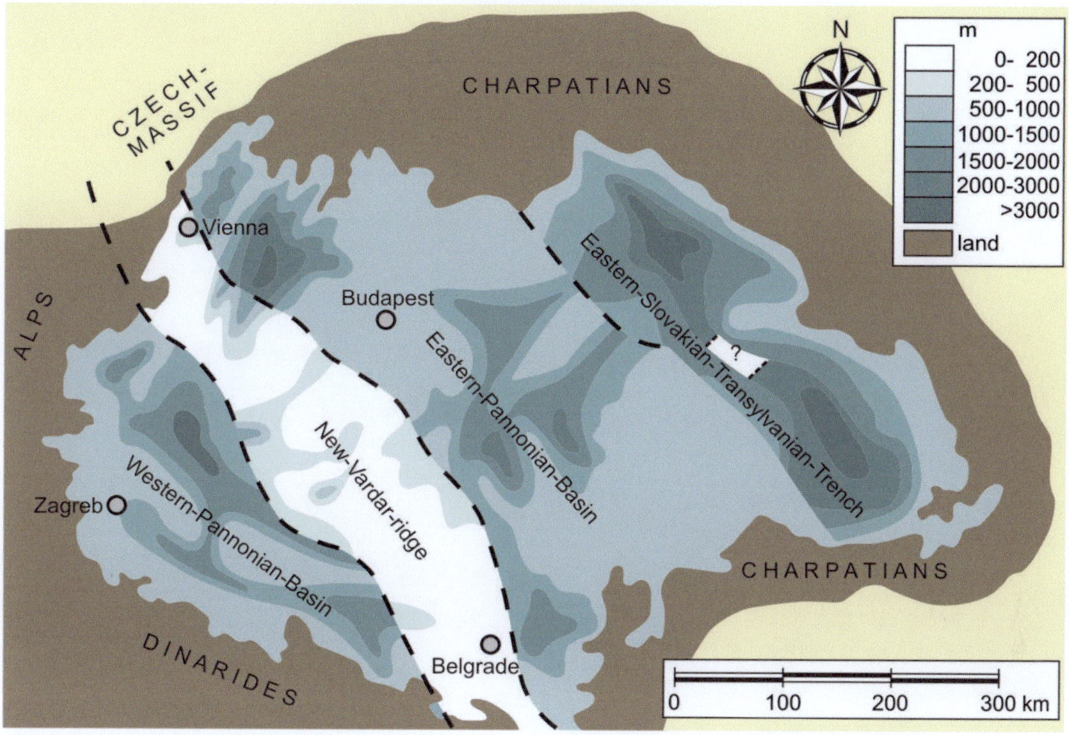


**Fig. 1.15** The thickness of Late Miocene sediments of the Carpathian Basin (after Hámor 2001)

quartz sand (the raw material for glass manufacturing) in the coastal zones.

The simultaneous uplift of the mountains and sinking of the basin and the resulted increased erosion produced a large amount of sediment. The deltas filled up the sinking basin gradually intruding into the lake (Juhász et al. 2006; Fig. 1.17), and the terrestrial areas were filled up by rivers transporting material from the intensively eroding territories. Clayey marl and calcareous marl deposited in the shallow areas and aleurolite and clay in the deeper waters, while in the coastal zone sand, gravel and conglomerates created several thousand metre thick deposits (Hartai 2005). The deltas intruded the basin form NW and NE, and they were also growing laterally. The subsurface topographical differences are clearly indicated by the approx. 6–8-million-year-old canyons (Juhász et al. 2007). The gradual and intensive sinking of the Pannonian Basin is reflected by the 3000–4000-m-thick sediment sequence formed during the 7 million years of existence of the basin. The lake became filled up and receded by the end of the Miocene, due to the effects of the slower sinking and short regional climate change. The climate becomes arid for a short time, although some scientists refer to the existence of traces of the Lake Pannon (Fig. 1.18). This climate change affected mainly the western part of the Mediterranean Sea, where salt and gypsum layers deposited, giving the name to the period: Messinian Salinity Crisis. This 5.4-million-year-old arid period is also reflected by the desert varnish of pebbles (Schweitzer and Szöőr 1992). The cessation of the salinity crisis and the moderation of the arid conditions took place at the end of the Pannonian Period, and many researchers link the following Pliocene Period to the opening of the Strait of Gibraltar.

It could be concluded that during the Pannonian Period, the terrestrial, marine-lacustrine environments created special ecological conditions; thus, their automatic connection to the end

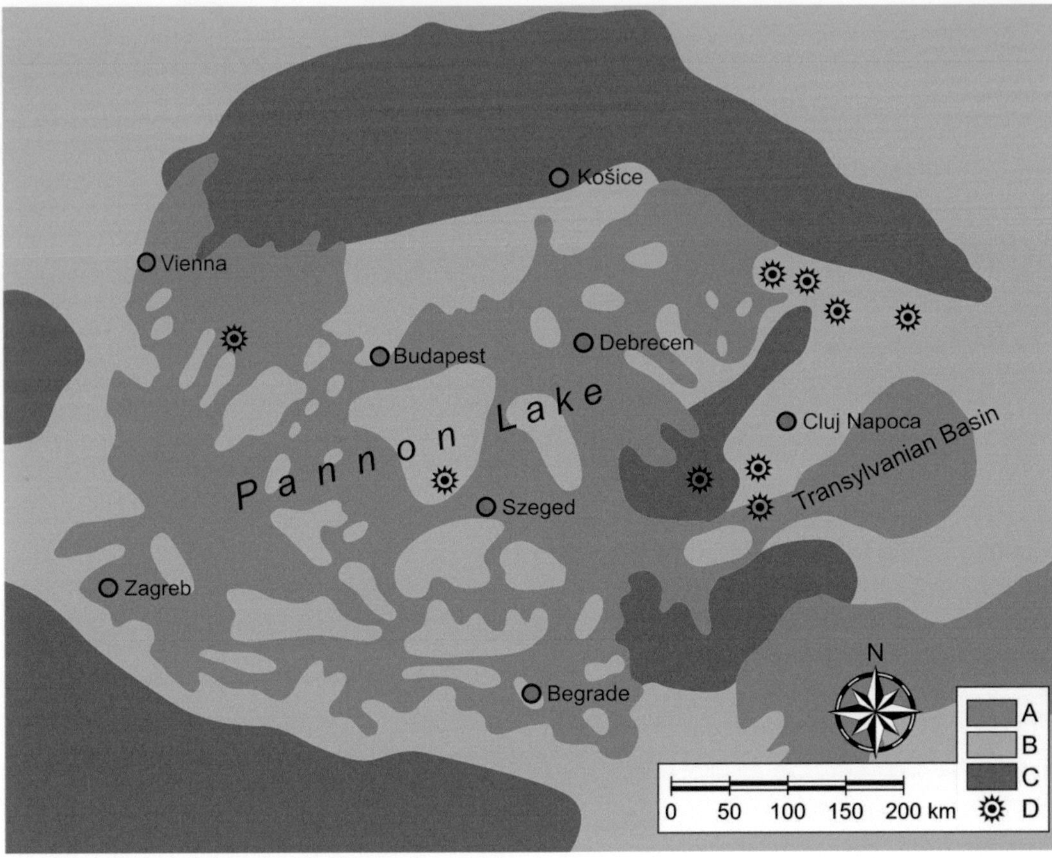


**Fig. 1.16** The approximate extension of Lake Pannon in the Early Pannonian Period (after Hámor 2001; Müller 2000). *A*—water cover, *B*—flat land, *C*—high mountains, *D*—volcano

of the Miocene and the application of the geological analogies valid in the Mediterranean are not grounded. The annual mean temperature was 13–15 °C, and the amount of precipitation was slightly over 1000 mm, referring to the joint effects of subtropical and temperate climates which are nowadays present in the northern zone of the Mediterranean Sea. In the climatic history of the Pannonian, there are no references to significant fluctuation; thus, the transition from sea to lake and its changes may be attributed to just tectonic factors and not climatic ones (Kordos 1979).

The period between 12 and 5.4 million years was special regarding its geological history and also the wildlife of its seas and lands. The marine or lacustrine flora and fauna was unusually rich, not typical of lakes, because the endemic evolution—which started in a marine environment and terminated in a closed freshwater lake—was extraordinarily fast in geological history (Hartai 2005). Extensive terrestrial areas existed in the Pannonian (such as today's Danube-Tisza Interfluve; Fig. 1.16), and the subtropical fauna was characterized by ancient horses, antelopes, rhinoceroses and elephants.

The main economic importance of the Pannonian sediments is in its lignite deposits and the fact that Pannonian sediments are the reservoir rocks of Hungary's major hydrocarbon and (thermal) water resources. Lignite is located in the foothills of the Mátra and Bükk Mts., and it refers to periodic shoreline changes typical in deltaic environment of the Pannonian Sea (Lake).

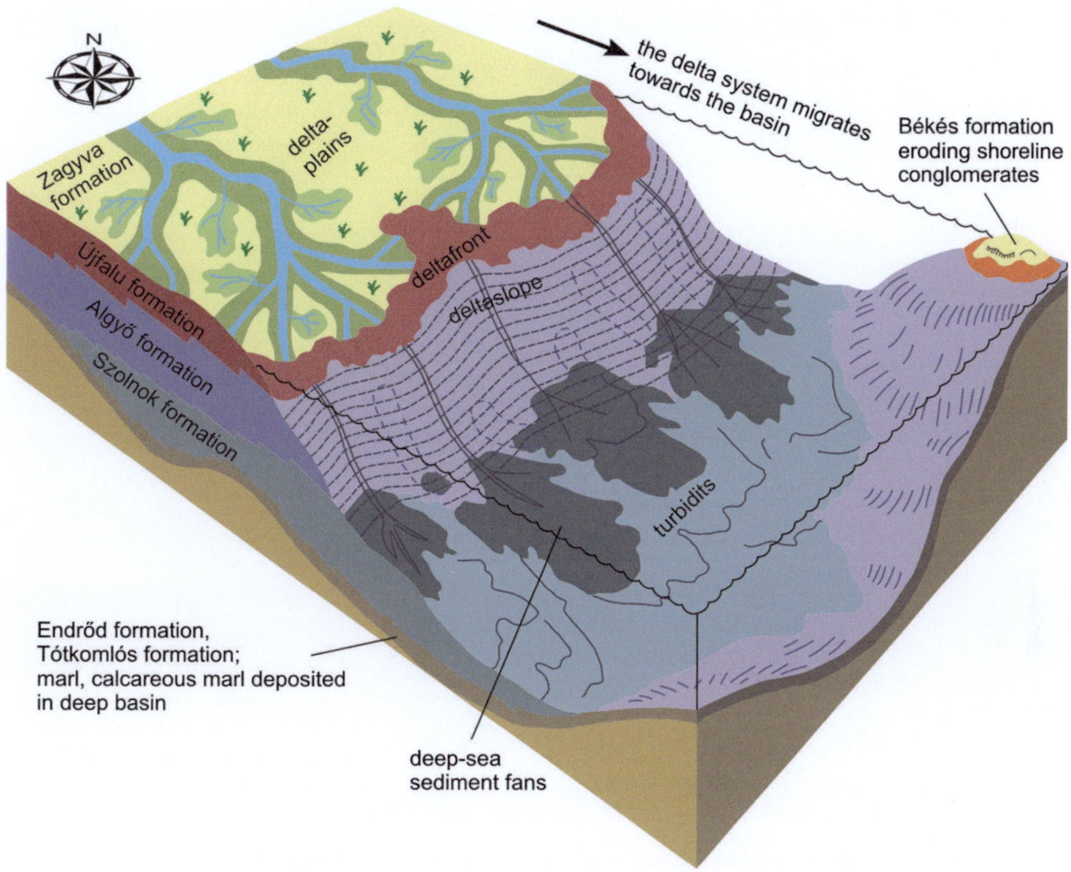


**Fig. 1.17** Theoretical sketch of deltas at the rim of the Pannonian Basin (after Juhász et al. 2006)

Several lignite layers were identified at Bükkábrány and, but only one or two reach more than 0.5–1 m thickness out of the more than three dozen layers (Fig. 1.19). Bald cypresses fossilized in a vertical position, and they indicate subtropical boggy coastline environment (Image 1.8).

Apart from the foreground of the present-day mountain range, in other areas of the delta plain, lignite was also deposited, for example in the Great Hungarian Plain at Algyő, but these layers are in a much larger depth; thus, their significance is much smaller. The lignite occurrences around Szombathely (Torony) and in South Somogy have similar character to those of the 7-million-year-old sediments, but they have smaller economic importance and serve just professional interest. Apart from Pleistocene loess, the Pannonian clay constitutes an important basic material in brick manufacturing.

The significance of the Pannonian sediments in water and hydrocarbon geology will be discussed later (Sect. 3.2).

### 1.2.5 Palaeo-Geography of the Pliocene

The Pannonian formations have been broadly studied owing to their valuable resources, unlike the 2.5–5.4-million-year-old formations of the Pliocene. The research of this period is difficult, because the basin had filled up (though by slower rate than before), and the lacustrine sedimentation was replaced by fluvial processes. The temperature and the amount of precipitation decreased,

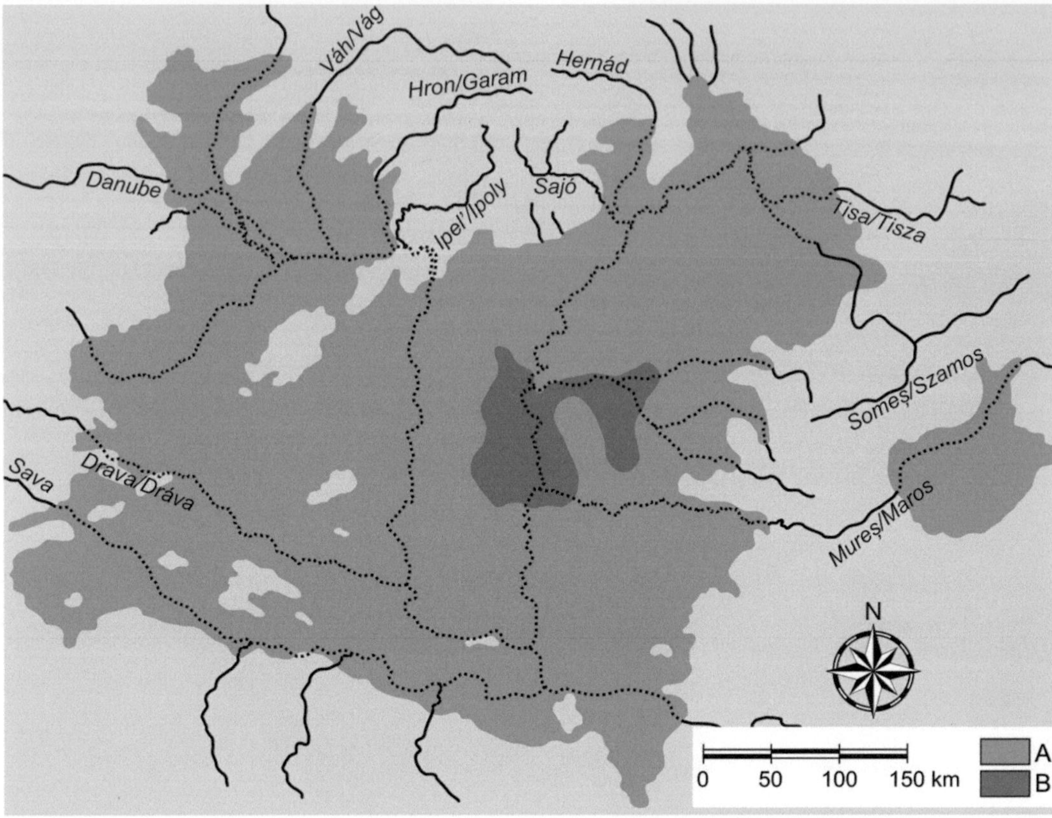


**Fig. 1.18** The Lake Pannon during its largest (*A*) and latest (*B*) extension (after Borsy 1991)

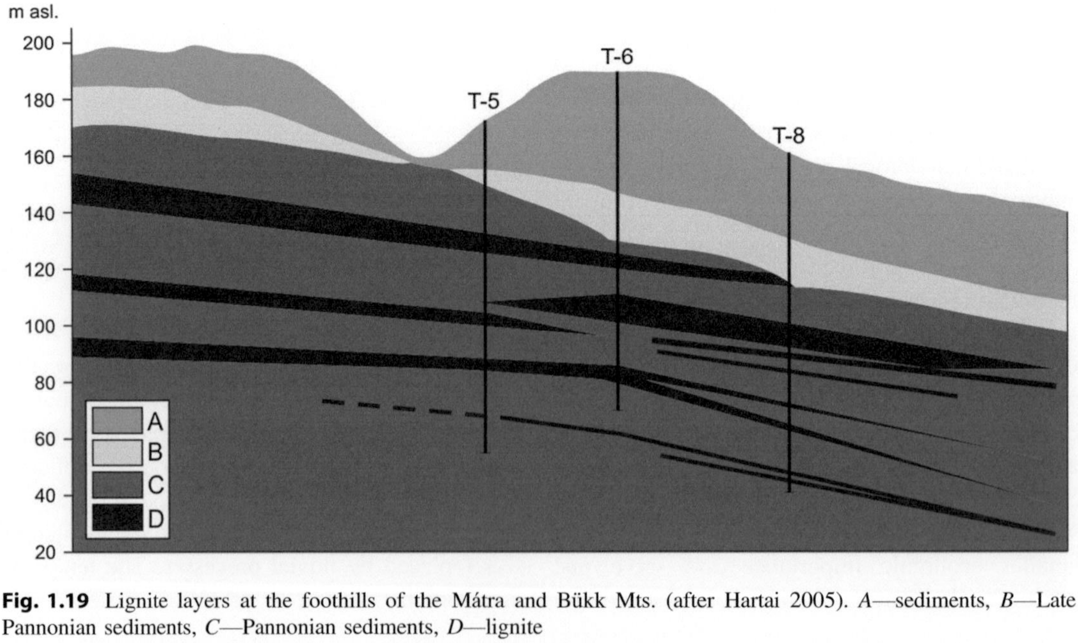


**Fig. 1.19** Lignite layers at the foothills of the Mátra and Bükk Mts. (after Hartai 2005). *A*—sediments, *B*—Late Pannonian sediments, *C*—Pannonian sediments, *D*—lignite

**Image 1.8** The stems of fossilized bald cypress preserved in the lignite mine of Bükkábrány (photograph by Hevesi, A.)

referring to continental-like climate, which reduced the intensity of surface processes.

The key issue in surface evolution was the assumption of fluvio-lacustrine sedimentation which had a long scientific tradition. It supposed the existence of an aggrading river network, which connected lakes and deposited cross-bedded sand. The idea has been debated by many (quoted by Martonné 1995). One of the weaknesses of the theory is that the sinking areas (Jászság and Körös basins, Lower Tisza Valley) were not active simultaneously during the Pliocene; perhaps, only the latter two were really sinking, and these movements were not parallel (the Lower Tisza Valley might have been the relic of the Pannonian Lake). Transdanubia did not fit in this picture either, because the sinking of the Dráva and Győr basins was not coincident. The fluvio-lacustrine state might only prevail in the end of the Pannonian; moreover, this sediment type is missing in the majority of the Great Hungarian Plain, expect for the Gödöllő Hills, where this sediment type is covered by the desert varnished pebbles indicating arid climate at the end of the Pannonian (Schweitzer 1993). Ádám (1975) tried to prove that the cross-bedded sand could be formed rivers. The Pliocene should be regarded as the first main period of alluvial fan formation. Some of the alluvial fans connected closely to mountains (e.g. those joining the Bakony and Kőszeg Mts.) are covered by basalt, providing a possibility to determine their age. The best conditions for alluvial fan development were present only later, in the Pleistocene, especially in

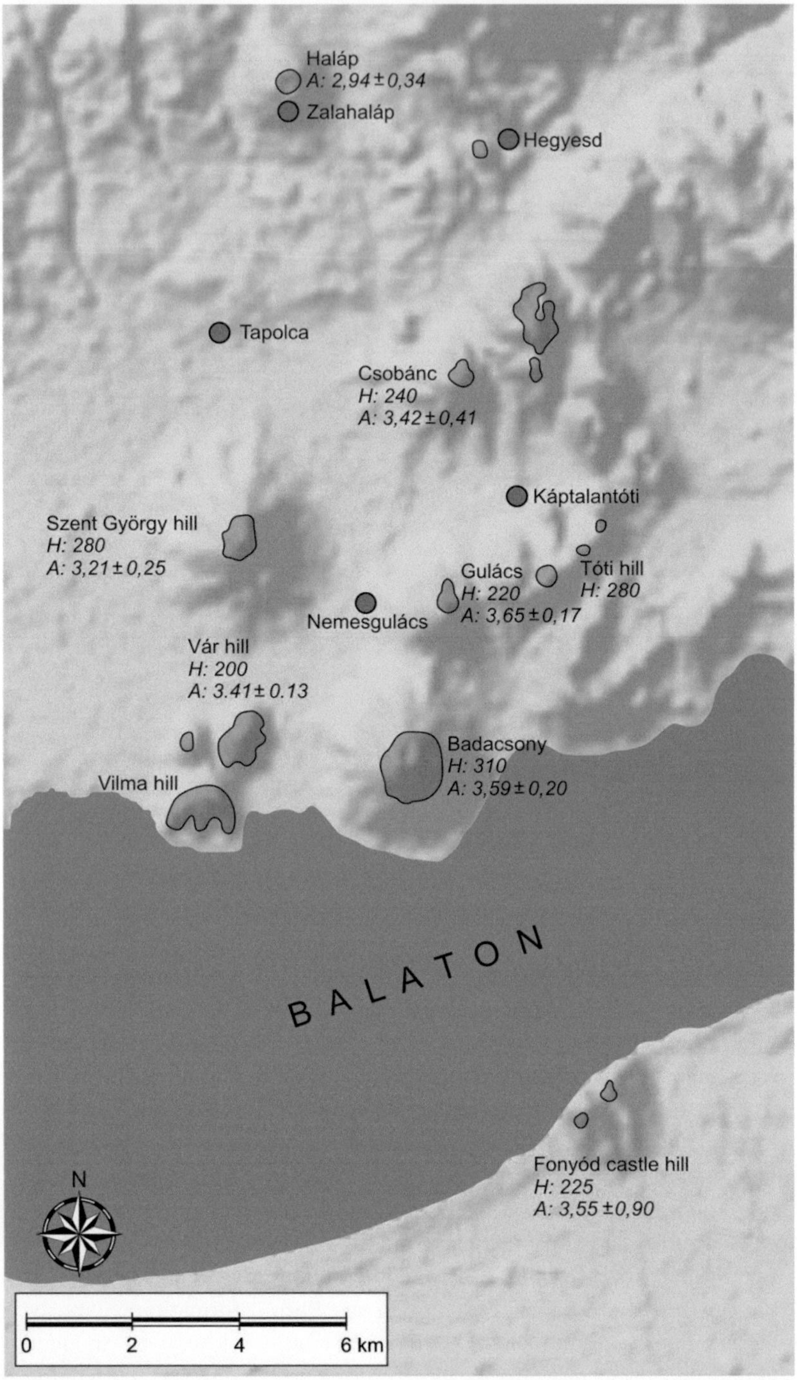


**Fig. 1.20** Basalt buttes in the Tapolca Basin and their radiometric ages (after Borsy et al. 1987). *H*—altitude, *A*—K/Ar average age expressed in million years

cases of the largest alluvial fans built by the Ancient Danube, Körös, Maros and Tisza rivers and by their tributaries (see Fig. 1.27).

The formation of basaltic volcanoes in the Pannonian, and to a greater extent in the Pliocene, is closely related to the evolution of the Pannonian Basin. The continental plate crust became thinner behind the subduction along the arc of the Carpathians (its thickness is approximately half of the thickness of those under the

**Image 1.9** The former basalt mine of Hegyestű, close to Zánka (photograph by Kiss, T.)

Alps or the Carpathians), and the crust of the back-arc basin elevated; thus, the hot basaltic magma reached the surface (Harangi 2000). The thinning of the crust is dated to 16–18 million years, and it is one of the major explanations for the formation of the Pannonian Basin. The basalt volcanism itself started later, about 10 million years ago, first in the western part of the basin and in the area of Burgenland. Afterwards, ca. 3–4 million years ago, smaller volcanic fields developed in the Tapolca Basin and the Balaton Uplands (Fig. 1.20; Image 1.9). The material of the multiphase volcanism broke through Pannonian sediments, and the basalt cover preserved the former sediments from denudation; thus, volcanic buttes developed. According to some researchers, fluvial erosion also played a role in the formation of buttes. Oil shale (alginite) was created from the algae that had established in the craters; it was mined in Pula as a soil fertilizer (Hartai 2005). The basalts situated in the area of Medves Region, and the Little Hungarian Plain can be linked to the Late Pliocene. Those around Salgótarján (Medves Region) are 1.5–2.3 million years old, and the one next to the settlement of Bár is about 2 million years old, being one of the latest volcanoes in Hungary. The majority of the basalt mines are nature reserves (e.g. Badacsony Mt.); thus, mining has stopped in them.

The tectonic dissection of the surface started already in the Pliocene at some places: it is indicated by the 3–4 million years old travertines found on Kőpite Hill at Dunaalmás at 290 m elevation and lower (Schweitzer 2002). Similar occurrence is known (and exploited) in Tardosbánya.

## 1.3 Chronology and Main Palaeo-Geographical Features of the Quaternary

The processes of the Quaternary played an outstanding role in the evolution of the present-day surface and landforms. Thus, the question is rather which areas and forms were not influenced by active Quaternary processes (e.g. the accumulation of loess or sand and periglacial events). The Pleistocene is characterized by significantly cooling climate, though the chronology is much more diverse. The mean temperature was lower by 10 °C in some periods, while in others it was a few degrees higher than today. The surface-forming processes, the flora and the fauna adapted to these drastic climate changes. At the beginning of the Pleistocene, these transformations were not rapid; therefore, it is difficult to define the starting date of the period. It can be determined from various points of view resulting in significantly different results. For example, the evolution of vegetation and species, or the development of special landforms (e.g. terraces, travertine levels), or palaeosol sequences could be considered. Thus, matching these approaches and determining the beginning of the Pleistocene are not easy tasks.

The solution accepted by the most important professional organization (INQUA) is to define the beginning of the Pleistocene on the basis of palaeo-magnetic dating connected to geomagnetic reversals (Walker 2005). The magnetizable particles (e.g. in volcanic rocks) preserve the direction of the Earth's magnetic field when the rocks cools, according to the actual, but ever-changing north-south direction. As a result, there are segments that show the current north-south pole direction, and there are reverse segments too, when today's southern direction was the northern one, that is, they indicate a period of reversed magnetic polarity. These palaeo-magnetic periods, for example on the basis of the ages of volcanic rocks, could be determined. The reversed Gilbert period started 5.5 million years ago (at the beginning of the Pliocene), it was followed by the Gauss Chron with the same polarity as the present, then the reversed Matuyama Chron started 2.5–2.6 million years ago, and the present is the Brunhes, which lasted for 700 thousand years. The ages may be different according to the applied different measurement techniques (e.g. potassium-argon method, or refined astronomically), but the Matuyama-Gauss reversal means the beginning of the Pleistocene within this framework (Table 1.3).

The palaeo-magnetic periods are divided by reversal events, or subchrons. The Olduvai Event (approx. 1.8–1.9 million years ago) must be highlighted, which can be connected to the appearance of the prehistoric man; therefore, many consider this age as the beginning of the Quaternary. The other well-defined event is the Jaramillo Event about 0.9 million years ago, which is regarded as the end of the Quaternary by many, because according to them, it is also the end of a sedimentation period, since, for example, the loess and alluvial accumulation processes also changed at that time (Pécsi 1984, 1996). Geomorphological points of view could be considered too, when determining the beginning of the Pleistocene. The appearance of ice sheets ca. 10 million years ago, and the advances of younger glaciers caused regional environmental changes. These ice-related processes can be synchronized for about 2.5 million years. The Alpine glaciation (Biber, Donau, Günz, Mindel, Riss and Würm) and that of Northern Europe (Tegel, Cromer, Elster, Saale and Vistula) refer to a climate-based chronology (Table 1.3). The timescale that applies fluvial terraces is said to be geomorphological, but in fact, it is essentially influenced by climate changes, if those of tectonic origin were not taken into consideration. Seven terraces fit in the 2.5-million-year-long period in the Carpathian Basin; however, more terraces can be identified, whose development can be dated back to 3.5 million years (Schweitzer 2002). The number of terraces along the Danube is uncertain due to the duplication of terraces (their detailed analysis is in Sect. 1.4).

The biostratigraphical data provide a different chronology. Though the younger periods (e.g.

**Table 1.3** Chronology of the quaternary (own compilation after Pécsi and Kretzoi 1985; Schweitzer 2002)

| magneto-statigraphy | age | litostatigraphy | biostatigraphy | | terraces and travertine-springs | | geomorphological levels | Eastern-Europe norhern part | Alps |
|---|---|---|---|---|---|---|---|---|---|
| Brunhes 0 | QUATERNARY | MENDE-BASAHARC | | | travertine I. | I. floodplain | | visztula ice age | würm ice age |
| | | D-T* $Th_1$ | Szántórium | Ságvár | | | | eem interglacial | riss-würm interglacial |
| | | $Dh_2$ | | Pilisszántó | | II/a terrace | | | |
| | | MF | | Istállóskő | | | | saale ice age: warta glacial | riss II glacial |
| | | BD | | Tokod | | | | saale ice age: trene interglacial | |
| | | | | Szeleta | | | | saale ice age: drente glacial | riss I glacial |
| | | BA | | Érd | travertine II. | | | | |
| | | | Solymárium | Subalyuk | | | | | |
| | | | | Varbó | | | | holstein interglacial | mindel-riss interglacial |
| | | MB | | Süttő | | II/b terrace | | | |
| | | Phe | | Solymár | travertine III. | | | elster ice age | mindel ice age |
| | | | | Uppony | | III. terrace | | | |
| | | Mtp | | | | | | cromer: (W-Europe) Netherlands IV interglacial, C glacial, III interglacial, A glacial, II interglacial, B glacial, I interglacial | günz-mindel interglacial |
| | | PAKS | Biharium | | travertine IV. | | | | |
| | | | | Vértesszőlős | | IV. terrace normal | | | |
| | | PD | | Tarkő | | | | | |
| 0,73 | | PDk | | Templomhegy | | | | cromer, bavel: II glacial, b interglacial, I glacial, a interglacial | |
| | | | | | travertine V. reversed? | | | | |
| 0,90<br>0,97 | | DUNAFÖLDVÁR | | Nagyharsány | | V. terrace reversed | | | |
| 1,15<br>1,17 | | | | | | | | menap ice age | günz ice age |
| | | | | Montepaglia 2 | travertine VI. | | | | |
| | | $Dfr_1$ | | Montepaglia 1 | reversed? | VI. terrace | glacis formation in larger valleys and on the edge of intermountain basins | waal interglacial | duna-günz interglacial |
| 1,55 | | $Df_1$ | Villányium | Betfia | travertine VII/a | | | eburon ice age | duna ice age |
| Matuyama 1,67 | | | | | | | | | |
| 1,87 | | | | Kislángium | travertine VII. | VII. terrace (Kemeneshát) oldest alluvial sediment | early alluvial fan development | tegelen interglacial: in Netherlands beerse glacial | duna-biber interglacial |
| 2,01<br>2,04<br>2,12<br>2,14 | | | | | | | | | |
| | | | | Beremendium | | | Dömsöd pediment | ice age before tegelen (brüggen) | biber ice age |
| 2,48 | PLIOCENE | | Csarnotanum | | | | | | |
| | | | | | travertine VIII. | pediment formation | | | |
| Gauss 2,92<br>3,01<br>3,05<br>3,15 | | Kulcs | | Węzě | | | lower pediment formation on Mátraalja | | |
| | | | | | travertine XI. | | | | |
| 3,40 | | | Ruf inium | | | | | | |
| Gilbert | | * Dunaújváros-Tápiósüly | | | | | climacteric period of pediment-formation | | |

Biharium and Villányium) became older by half a million year due to modern dating techniques (Pécsi and Kretzoi 1985), much more information exists on the biosphere of these two periods. The Villányium fauna, characterized by ostrich and camel, indicates dry and warm conditions (at least at the beginning, ca. 3 million years ago), which remained the same until 1.8 million years ago. Thus, the border of the Pleistocene is set at 1.8 million years, from this point of view (Schweitzer 2002).

Whichever point of view is considered, pediment surfaces exceeding the chronological border could be found, although they have different origin (they could be 3–4 million years old). The fluvial terraces that are 3 million years old or older, or the older loesses provide some unclarified scientific issues.

### 1.3.1 Tectonic Changes in the Pleistocene

The subsidence of the crust pieces that had already started to sink in the Miocene continued. Thus, the areas of the Little Hungarian Plain, the Great Hungarian Plain and the Dráva Valley were filled up by Pleistocene alluvial sediments, and only small lowland surfaces became covered by loess (Fig. 1.21). The subsidence reached about 300–400 m in average. At the same time, the hills and mountains elevated by ca. 200–300 m during this 2.5 million years (the uplift and the consequent intensive erosion had already started in the Late Pliocene). Altogether 1000-m relief difference was created compared to the surface of the Pliocene sediments (Rónai 1974). The spatial and temporal differences in the

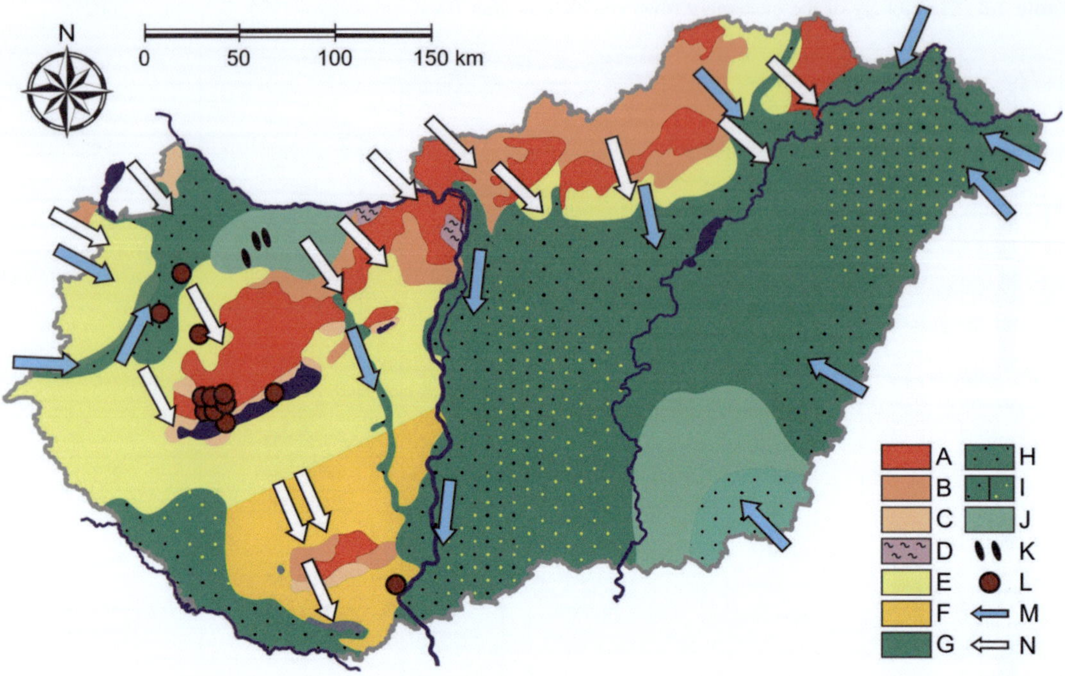


**Fig. 1.21** Quaternary palaeo-geographical map of Hungary (after Jámbor 2001). *A*—rapidly uplifting erosional area (by 100–250 m at the end of the Pleistocene), *B*—slowly elevating erosional area (by 50–100 m, respectively), *C*—area of wind erosion, *D*—travertine on the edge of karst mountains, *E*—loess area, *F*—red clay accumulation, *G*—fluvial accumulation, *H*—alluvial fan, *I*—alternating fluvial and blown sand sediments, *J*—alternating alluvium and loess, *K*—buttes shaped by wind, *L*—early Pleistocene basalt volcanoes, *M*—direction of fluvial transport, *N*—wind direction

subsidence and uplift influenced the changes of the river network, the rate of periglacial and aeolian processes.

The sediments in the subsiding areas are nearly horizontally bedded, and they become wedged towards the mountains. This suggests that the Late Pliocene and Pleistocene sedimentation affected large areas, which is also proven by the great thickness (up to 1000 m) of sediments, though the young sediments on the elevated areas eroded too (Bada and Horváth 1998). The sedimentological analyses did not refer to bias between the Upper Pannonian and Pleistocene layers; thus, they cannot be structurally separated. However, based on palaeo-magnetic measurements, it is possible to separate them; thus, the 2.6-million-year-old Matuyama-Gauss reversal is at 440 m depth at Vésztő, or at 420 m at Dévaványa. Based on these measurements, several areas could be identified, which subsided more intensively than the average of the basin (Rónai 1985). Such area is the South Tisza Valley, which is also tectonically preformed (indicated by the 660-m-thick Pleistocene sedimentary strata at Mindszent), the territory of the Körös Basin (respectively 442 m at Komádi), and the Lower Zagyva Valley (respectively 432 m at Jászladány). According to some researchers, the thickness of the alluvial sediment sequence is ca. 1000 m in the Makó Graben (MT 2002). In the area of Transdanubia, approx. 200–300 m fluvial aggradation was measured in the Dráva Graben. The spatial differentiation of the subsidence had a diverse pattern in the Quaternary. In general, fluvial sediments deposited continuously on the older formations. The areas mentioned above were sinking faster than the average, but while this sinking was continuous during the Quaternary in the Körös Basin, the Lower Tisza Valley and the Little Hungarian

**Table 1.4** The rate of sedimentation at Vésztő, based on palaeo-magnetic measurements (after Rónai 1985)

| Name of palaeo-magnetic turn | Depth (m) | Absolute time (million years) | Development time of 1-m-thick sediment (years) | **Aggradation rate** (mm/y) |
|---|---|---|---|---|
| Brunhes–Matuyama turn | 148 | 0.72 | 4900 | 0.2 |
| End of Jaramillo | 190 | 0.89 | 4000 | 0.24 |
| Beginning of Jaramillo | 210 | 0.94 | 2500 | 0.4 |
| End of Olduvai | 328 | 1.76 | 7000 | 0.15 |
| Beginning of Olduvai | 365 | 1.91 | 4100 | 0.25 |
| Matuyama–Gauss turn | 480 | 2.47 | 4900 | 0.2 |
| Gauss–Gilbert turn | 620 | 3.41 | 6700 | 0.15 |
| Cochiti–Nunivak | 683 | 3.82 | 6500 | 0.15 |
| Sidufjall | 755 | 4.25 | 5600 | 0.16 |
| Gilbert Epoch–5 turn | 940 | 5.44 | 6400 | 0.15 |
| Epoch 5–Epoch 6 turn | 1010 | 5.77 | 4700 | 0.21 |

Plain, the process had become active just by the beginning of the Pliocene and the end of the Pleistocene in the Dráva Graben and by the end of the Quaternary in the Lower Zagyva Valley. The sinking was not continuous, 6–7 cycles are distinguishable, which resulted in 30–80-m sediment individually, and they probably correlate with the glacial phases (Jámbor 2001). The average rate of sinking was 0.2 mm/year, and its spatial variations influenced the direction changes of the river network (Table 1.4).

Apart from these large-scale subsidence tendencies, several smaller but more intensively subsiding areas influenced the directions of rivers and the surface evolution tendencies. Later the developed floodplain levels became important locations for the prehistoric man's settlements. For example, at the end of the Würm, the sinking Baja-Kalocsa graben significantly altered the direction of the Danube (Jaskó and Krolopp 1991), or from the middle until the end of the Würm sinking of mountain foreground basins was typical. For example, in the case of Lake Balaton, this altered not just the entire water network, but also the landform evolution (Marosi and Szilárd 1981) or the activation of the similar mountain foreground basins (e.g. Sárrét, Tikacs in Fejér County) in the Holocene made the areas appropriate for human use.

During the Quaternary, the mountain ranges and hills were uplifted. Its rate (50–100 m) was smaller in case of the hills, while the elevation of mountainous areas was higher (250–350 m). The greatest uplift affected the Bükk Mts., which was about 440 m (Hevesi 1986). The rate of uplift could be calculated from the location of well caves of different age and elevation (see Chap. 12 . about the Bükk Mts.), but the travertine levels are similarly informative (see Chap. 10 about the Gerecse Mts.). The fluvial terraces also provide useful information on the tectonic movements. The fluvial terraces of the Danube Bend (with the members of the Dunazug Mts.) were elevated by up to 250 m. For example here, the terrace No. IV. (about 350,000 years old) is about 150 m higher than the same terrace level between Komárom and Esztergom, which was not affected by intensive elevation.

The processes of uplift and subsidence are still going on. The Quaternary rate was 0.2 mm/year (in both directions) and could be characteristic in the long run; however, in the short run, this rate could be considerably different (see Sect. 1.9). In the elevated or elevating areas, the changing uplifting rates are expressed by the altered discharge and sediment transport of rivers and thus by the evolution of terraces. The situation is completely different in the sinking areas, since here the rivers had no sufficient energy to incise; thus, the rate of sinking is proportional to the type and rate of sedimentation. Some researchers explain the different spatial and temporal distributions of uplift and subsidence by long-wavelength lithospheric folding (Nádor et al. 2005; Bada and Horváth 2001).

### 1.3.2 Evolution of the River Network

The most discussed issue of the Hungarian physical geography is the evolution of the river network, partly because fluvial processes are very important factors in landform evolution, but also because paleo-hydrological forms and sediments must be considered in water management. The issue has been studied from various points of view, but there are only few stable elements that may provide help in understanding the evolution of the system. Nowadays, plenty of information was collected on the evolution of the water system, but just an evolutionary model could be drawn due to the scarcity and ambiguity of data. Therefore, some debated issues that are not properly addressed and can modify the model to a smaller extent will be also discussed.

Several scientists had to deal with the evolution of the river network of the Carpathian Basin, and they have tried to answer the scientific questions with the given background available at that time. Thus, the main question in Cholnoky's (1910) monoglacial approach was whether the water regime of the basin had had a positive balance at the time of the evolution of the river network. He gave a positive answer for the second half of the Pliocene; thus, probably, there had always been an outflow through the Iron Gates, and he also regarded the north-south flow direction of the Danube in the Carpathian Basin as a stable direction since the birth of the river. The novelty of Prinz's (1936) approach was that he took into consideration the role of the surrounding mountains, and he explained the evolution of the river network by centripetal and centrifugal forces. Sümeghy's (1944) model on river network evolution became widely accepted, though he created it based on limited amount of geological data and observations (Fig. 1.22). His model had two main new elements: (1) the description of the flow directions of the Danube and the Ipoly rivers at the end of the Pliocene and (2) the identification of a flow direction of the Danube across the Danube-Tisza Interfluve in the Quaternary. Today, most of the experts agree on the main points of his ideas; however, a lot of new data were provided on the age of the river network evolution, the flow direction changes and their causes.

For methodological reasons, it is advisable to analyse the evolution of the water system of the Danube and the Tisza rivers separately, because the important changes are one magnitude older in the case of the Danube than in the case of the Tisza. Their evolutionary history will be integrated in the analysis of the end of the Quaternary.

#### 1.3.2.1 The Evolution of the River Network of the Danube in Hungary

In the Pliocene, the Danube reached the Vienna Basin and filled it up gradually (the Vienna Basin structurally is the counterpart of the Győr Basin). In the Middle Pliocene, the Danube flowed towards the Carpathian Basin through the Ebenfurt Gate (Somogyi 2000b), which is proven by the significant amount of sandy sediment filling the basin. These fluvial sediments cover the 6-million-year-old cross-bedded sand and clay layers (see Sect. 1.2). The development of fluvio-lacustrine depositional environment is the result of the hydrological connection between the depressions in the western and the eastern Pannonian Basins after the regression of the Pannonian Sea and later the Lake Pannon. However,

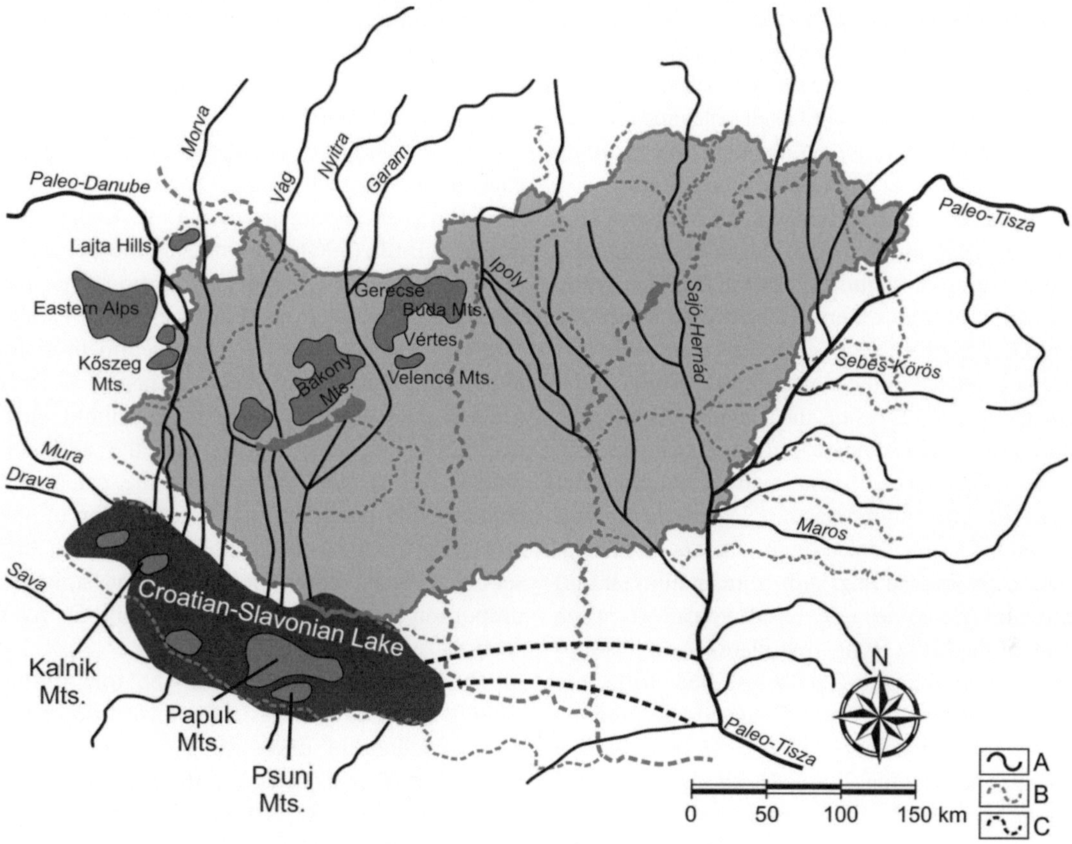


**Fig. 1.22** The water network of the Carpathian Basin at the end of the Pliocene (after Sümeghy 1944, Somogyi 1961, 2000b). *A*—water network at the end of the Pliocene, *B*—present-day direction, *C*—the most likely outlets of the Croatian-Slavonian Lake

recently, serious doubts have emerged about the existence of such a water network. The western Pannonian basins (e.g. Steyr Basin, Dráva Basin and Száva Basin) and the eastern Pannonian basins (e.g. Vienna Basin, Little Hungarian Plain, Danube Valley and Jászság) have subsided by a different rate since the Middle Miocene. Simultaneously, on the Vardar Ridge, which separated these basins, the amount of accumulated sediment was just the tenth (Fig. 1.16; Hámor 2001). The Danube arriving through the Ebenfurt Gate headed towards the western Pannonian basins, which became active at that time, and the Danube and its tributaries covered the entire territory of Transdanubia with cross-bedded sandy sediment of a significant thickness. Szádeczky (1938) described Danube confluence to the Croatian-Slavonian Lake. The first landscape evolution models of the water system emphasized the role of the Danube: some explained the 200-m sediment hiatus between the volcanic buttes of the Tapolca Basin by the erosional activity of the Danube, and they compiled arguments for supporting the hypothesis that the Danube flowed across the Tapolca Basin and the bed of the Kis-Balaton (e.g. Pécsi 1975; Ádám 1975). The erosion of the buttes could be explained by deflation during the arid Pliocene (Cholnoky 1918); thus, according to this model, the erosional activity took place 1–2 million years earlier than the 3–4-million-year-old main period of basaltic volcanism. On the other hand, Bulla (1964) emphasized the role of fluvial erosion. In the past decades, a lot of information has been found about the Billege Gravel Formation, which was intended to prove the presence of the

Danube; however, it turned out to have a non-Danubian origin, but it was transported from eroding Bakony Mts, and the formation was even permeated by conduits of basalt volcanoes. In the mine of Lesencetomaj, an excellent site exhibits the Billege Gravel Formation (Pécsi 1996).

At the end of the Pliocene, the Vardar Ridge became uplifted. The first researches explained the changing flow direction of the Danube by the elevation of the watershed between Keszthely and Gleichenberg (e.g. Zala Hills), and this had been connected to the elevation of the hills in the Quaternary. As a result, the Danube changed its flow direction and started to flow towards the eastern Pannonian basins, while its erosional base was the Győr Basin. Nevertheless, later researches considered other factors too, e.g. geomorphological (e.g. shift from an alluvial fan) and geological ones (e.g. basalt volcanism in the edge of the Little Hungarian Plain). As a result of these changes, the Danube shifted from the Eberfurt Gate to the Bruck Gate and later into the Devín Gate, which is still active (and built a younger alluvial fan; Fig. 8.2). The Danube filled up the Győr Basin by 400 m. In the western part of Transdanubia, gravel deposits covered the former (fluvio-lacustrine) sandy sediments by erosional discordance, which may indicate the border between the Pleistocene and the Pliocene in the Little Hungarian Plain.

From the beginning of the Pleistocene, the Danube flowed along the Bruck/Devín Gate through the Győr Basin, towards the Danube Bend. The Danube Bend was already a strait in the Pannonian Sea, and its sedimentological evidences can be seen on Szent Mihály Mt. The Danube Bend had an epigenetic and an antecedent character at the same time; thus, the former flow direction got inherited, and the Danube appearing in the strait incised by the same rate as the uplift of the area (see Sect. 1.4). (The latest researches found evidence on much younger, Late Pleistocene development of the Visegrád Strait (Ruszkiczai-Rüdiger et al. 2006).) Having passed through this strait, the Danube joined the Tisza in the northern part of the Lower Tisza Region (Fig. 1.23). Thus, the sinking central part of the Carpathian Basin served as an erosional base, attracting the Danube from the NW and the Tisza from the NE. As a consequence, the Danube built an extensive alluvial fan in the Danube-Tisza Interfluve.

The geomorphologists have studied the presence of the Danube in the Danube Bend for a long time, discussing its evolution. Kádár supposed in Bulla (1964) that the Danube Bend had been created by the headward erosion from the direction of Jászság or the Lower Tisza Valley. However, the amount of volcanic sediment in the alluvium of the Pest Plain is just 5 %, but according to the model above, it should have exceeded this value. In general, the alluvium always reflects the lithology of the catchment area; in this case, the upper catchment of the regressing river has a volcanic origin; thus, this sediment type should be represented a larger proportion than in the Győr-Esztergom river section.

The other debated issue is the role of the sinking basin of Jászság in the formation of the flow directions of rivers. As it was mentioned earlier (see Sect. 1.2), during the Pannonian, the entire Carpathian Basin was filled up by deltas from the north. The sinking of the Jászság (Lower Zagyva Valley) became intensive and started to influence the flow directions just later. This contradicts the former ideas, i.e. the Jászság was the direct erosion base of the Danube.

The alluvial fan origin of the Danube-Tisza Interfluve has become commonly accepted. The gravel formations of the foothills and terraces can be traced in the sediment sequence of the Great Hungarian Plain. Pécsi (1959) identified the gravel material of terrace No. IV at Kecskemét at a depth of 400 m. The terrace material of the Pest Plain could be separated from the normal (alluvial) sequence of the Danube-Tisza Interfluve along the southern edge of the Pest Plain (Ócsa-Sári line). The identification of the older and higher terraces is difficult partly because of the urbanization and partly because of the young, Danubian alluvium with normal stratification deposited on the older ones (e.g. south part of the Pest Plain), in contrast to the terrace sequence that has reverse stratification on the Pest Plain. A critical question in the evolution of the alluvial fan is the evolution of the Danubian terrace No. V and the origin of its gravel material. Some scientists claim that its evolution started in the Upper Miocene, which could have continued in the first half of the Pleistocene (Pécsi 1991; Burján 2002). The problem is complicated by the fact that these older forms could be the remains of delta formations (see Sect. 1.2). The Danube flowing through the Danube-Tisza Interfluve may have had a complex, multichannel and anastomosing pattern, and its flow direction had not changed significantly from the beginning of the Pleistocene until the Middle Würm. The sediment sequence is dominated by fluvial sand up to 50–100 m depth, and fluvial sand, blown sand and loess alternate close to the surface, reflecting the temporal and spatial changes of the channel system and the environment (Lóki 1994).

There are several explanations on the transition of the flow direction of the Danube to the (present-day) north-south direction. The date of the flow direction change can be well defined geomorphologically, since the terraces older than the Late Würm (II/a) are missing along the present-day flow direction (Pécsi 1959); thus, at that time, the Danube had a SE direction across the Danube-Tisza Interfluve with a normal stratification. The following geomorphological explanations exist for the channel shift:

- The Danube shifted from its alluvial fan, as the Danube-Tisza Interfluve was higher than its environ;
- The Coriolis force affected the river (Jámbor 2001); this force corresponds to the force involved in the Baer's law affecting the rivers with a north-south flow direction. However, this could not have played a significant role because the north-south section is too short.
- The NW-SE transform faults reactivated in the Würm, still having a modest activity today (e.g. earthquake in Dunaharaszti in 1956). Besides, the subsidence of the Baja-Kalocsa graben started about 30–40 thousand years ago (Jaskó and Krolopp 1991), which may have been the strongest influencing factor, especially because of its erosional effect (see Fig. 1.51).

#### 1.3.2.2 The Transforming Water System in Transdanubia

Before the development of the north-south flow direction of the Danube (Fig. 1.22), the Croatian-Slavonian Lake was the erosional base of the rivers arriving into the Carpathian Basin from Western Alps and from the North-western Carpathians west of the Garam/Hron River. Ipoly River had already joined the water system of the Lower Tisza (bypassing the Börzsöny Mts. from the east until the beginning of the Pleistocene). The present-day flow direction of the Ipoly River probably exists just since the Middle Quaternary (Mike 1991), though Borsy proved that it changed to the present flow direction already at the beginning of the Quaternary, based on terrace morphological researches.

Based on the heavy mineral analysis of a thick, alluvial sediment sequence stretching towards south, several scientists claimed that the river system of the Garam and Nyitra/Nitra rivers passed through the Mór Graben at the beginning of the Pleistocene (Ádám 1959). In the case of the river systems of the Vág and Morva rivers, no real data could prove their course between the members of the Transdanubian Mountains. When the Danube started to flow from west to east, the structure of the water network of Transdanubia changed considerably: the rivers arriving from the NW Carpathians joined the Danube, and a new river system developed in Transdanubia.

The river system of the Rába significantly changed too, which drained the streams of the Eastern Alps into the sinking Győr Basin. Meanwhile, it built an extensive alluvial fan (Kemeneshát and Kemenesalja), which was also influenced by the smaller rivers (e.g. Ikva, Répce, Gyöngyös and Perint) and their alluvial fan sequence (Sopron-Vas Plain; Fig. 1.22). The beginning of the alluvial fan formation of the Rába is reflected by the fluvial gravels, which cover the Late Pliocene–Early Pleistocene basalt volcanoes; thus, the alluvial fan started to develop ca. 2.5–3.5 million years ago. It is another issue

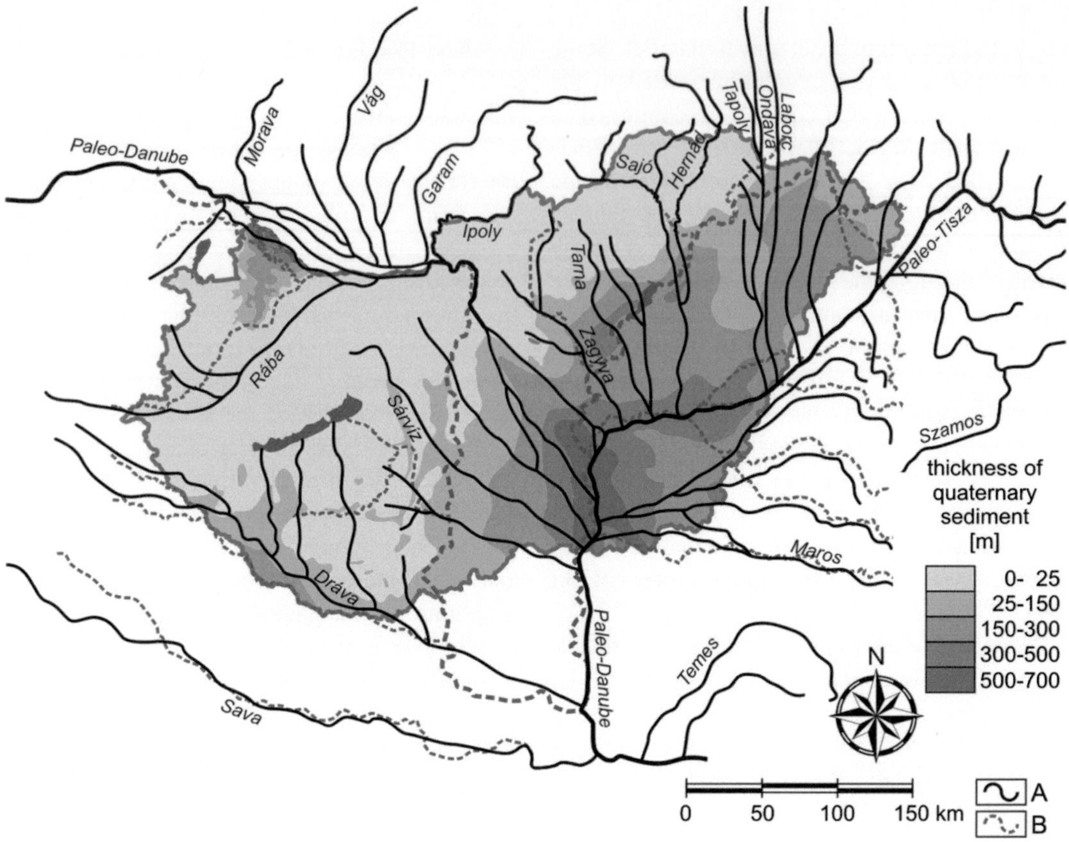


**Fig. 1.23** River network at the beginning and in the middle of the Quaternary (after Borsy 1989; Gábris and Nádor 2007)

that at the time of the evolution of the Danubian terrace No. IV, the Rába shifted from this alluvial fan towards west and formed an asymmetrical terraced valley.

The ancestors of the Zala River reached the Zala Hills in the Middle Pliocene (their sediments were found at Zalaegerszeg), and later, the Zala joined the Marcal River draining water towards the Győr Basin. This water system existed until the beginning of the Würm, when the increasing activity of the subsiding Dráva Graben increased the headward erosion of a stream from the plain of the Rinya towards north; thus, the Zala was captured at Türje. As a result, the Marcal River became independent, still flowing northward, but the Zala turned to south. The second capture of the Zala River is connected to the evolution of the basin of Lake Balaton, which changed the flow direction of the river.

In contrast, the Dráva River, except its lower reach, remained in its wide trench, influenced by the activity of Holocene subsidences (Bodnár et al. 2009).

The rivers that had their source in the south-eastern slopes of the Transdanubian Mountains (e.g. Paleo-Sárrét River) flowed through the Danube-Tisza Interfluve towards the subsiding South Tisza Valley (Fig. 1.23). Their connection to the Danube was hindered by its growing alluvial fan; thus, they probably joined the Danube in the southern part of the basin. The reconstruction of the courses of these streams is difficult due to the scarce data. Major flow direction change was caused by the formation of the Baja-Kalocsa graben, when the Danube changed its flow direction to the present (north-south) location; thus, the streams arriving from the Transdanubian Mts. were captured. The same evolutionary route, that is, the separation from the supplying territory, was followed by the stream arriving from the Eastern Mecsek Mts., which built the basis of the alluvial fan of the Bácska Plain.

#### 1.3.2.3 The Evolution of the Water System of the Tisza River in the Carpathian Basin

The Pannonian Sea covering the Basin gradually had shrunk, and its last remnants probably located around the South Tisza Valley (see Fig. 1.18; Borsy 1989). The subsiding Carpathian Basin was compressed in N-S and NW-SE directions at the Late Miocene, resulting in the uplift of the mountains around the basin and the accelerated sinking of its central part. By the beginning of the Pleistocene, two large, local sedimentary basins had developed (South Tisza Valley and Körös Basin), which were the local erosion bases of the waterflows arriving from the mountains (see Sect. 1.2.5). The Körös Basin used to be the main sedimentary basin of the water system of the Tisza: all rivers east of today's Tarna River joined it (Fig. 1.24). The Körös Basin and the South Tisza Valley were connected; thus, the water of the Körös Basin flowed into the water system of the Tisza along the present-day course of Kurca River.

The most controversial issue in connection with the water system of the Tisza concerned the spatial and temporal changes of the courses of the Tisza, Szamos and Kraszna rivers. During most of the Quaternary, the Tisza and the Szamos reached the Körös sedimentary basin through the

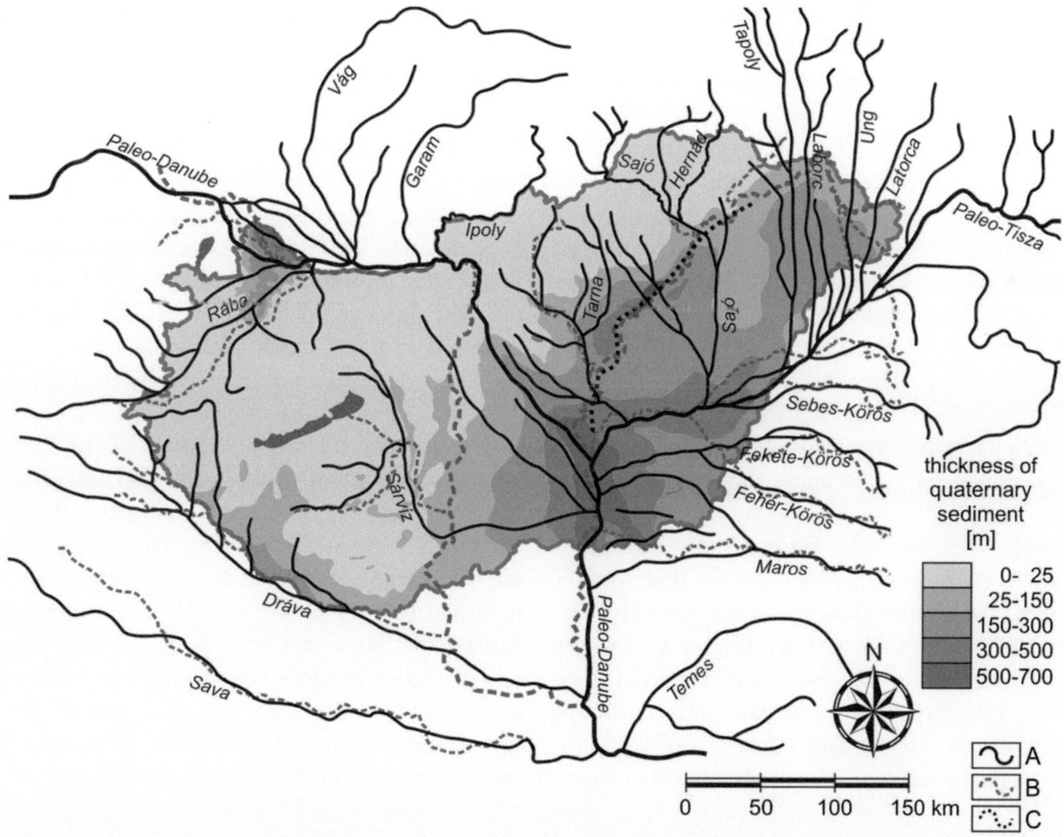


**Fig. 1.24** River network in the Middle Würm (after Borsy 1991; Nagy et al. 2007)

Érmellék, but sometimes they could have built the southern part of the Nyírség alluvial fan too. Nowadays, the hydrographical connections between the Érmellék, Berettyó and Körös basins are studied in detail (Nádor et al. 2005).

The idea that the Tisza originally flowed across the Érmellék, and about 20–22 thousand years ago, it changed its course towards the sinking Szatmár-Bereg Plain was supported by radiocarbon dating (Borsy 1989) and by geomorphological data (Félegyházi et al. 2003). After leaving the Szatmár-Bereg Plain, the Tisza reached the central part of the Great Hungarian Plain, flowing across through numerous subsiding areas (e.g. Bodrogköz and Rétköz). According to this model, the appearance of the Tisza changed the southward courses of the rivers of the Bodrog's system (e.g. Tapoly, Ondava, Laborc, Ung and Latorca Rivers); thus, they could not flow across the Nyírség alluvial fan. These rivers had flown across the Nyírség for a long time: they built the alluvial fan and created still existing valleys ("nyírvízlapos") on the surface. As the central part of the Nyírség was uplifted, in the South Nyírség, these valleys preserved their original southward flow direction, but in the North Nyírség, they have taken the opposite direction towards north. In this approach, the alluvial fan of several hundred metre thickness of the Nyírség was basically built by the tributaries of the Bodrog River, and the alluvial fan formation terminated at the end of the Pleistocene (Borsy 1989). The blown sand forms of the Nyírség developed during the last glaciation (ca. 19–23 thousand years ago); the aeolian processes were enhanced by the uplift, the dry climate and the sparse vegetation. However, the latest studies revealed that the subsidence of the Great Hungarian Plain was uneven both in time and in space, and it influenced the sinking of the Bodrogköz and the Central Tisza Region too, where the driving factor may have been the subsidence of the Jászság Basin. On the basis of this series of subsidences, Gábris and Nádor (2007) presupposed the appearance of the Paleo-Bodrog, which connected the depressions (Fig. 1.25), and this theory may completely change the former model on water network changes (Fig. 1.24), since the river system of the Bodrog may have been an independent unit before the appearance of the Tisza. The coexistence of a river with a relatively high water discharge along the Tokaj-Szolnok line and of the Tisza and Érmellék-Berettyó systems was already raised by Frányó (1992) and Gábris (2003a), and they also proved this by the analysis of high-resolution seismic profiles (Nagy et al. 2007).

According to Gábris and Nádor (2007), the avulsion of the Tisza River from the Érmellék (Image 1.10) to the Szatmár-Bereg-Bodrogköz-Rétköz depressions probably has not happened 20–22 thousand years ago as it was supposed by earlier studies, but later. The subsidence of the Érmellék area may have decreased ca. 14 thousand years ago, while the subsidence of Bodrogköz and the elevation of the Nyírség became more intensive. Their superimposed effect was that as a result of an avulsion, the Tisza flowed towards Bodrogköz, and then, it turned towards south-east and reached the former course of the Paleo-Bodrog. According to the latest analyses, the Szamos River may also have had a different course change as it was believed earlier, since it flowed across the Érmellék until 14–16 thousand years ago (Félegyházi et al. 2003). Somogyi (2000b) also emphasizes that the Szamos appeared in the Szatmár Plain only after filling the Baia Mare Basin, but on the Szatmár Plain, only the sediments of the Tisza and Túr rivers could be found under the sediment of the River Szamos. According to the latest OSL measurements and heavy mineral analyses of the Érmellék, the sediments were transported onto the Great Hungarian Plain from this direction 47 thousand years ago and then 10 thousand years ago by the Tisza and by a large, parallel river (e.g. Szamos). In addition, the analyses of the last ten years show that the rivers took their present flow directions only 10 thousand years ago in this region (Thamó Bozsó et al. 2007a, b).

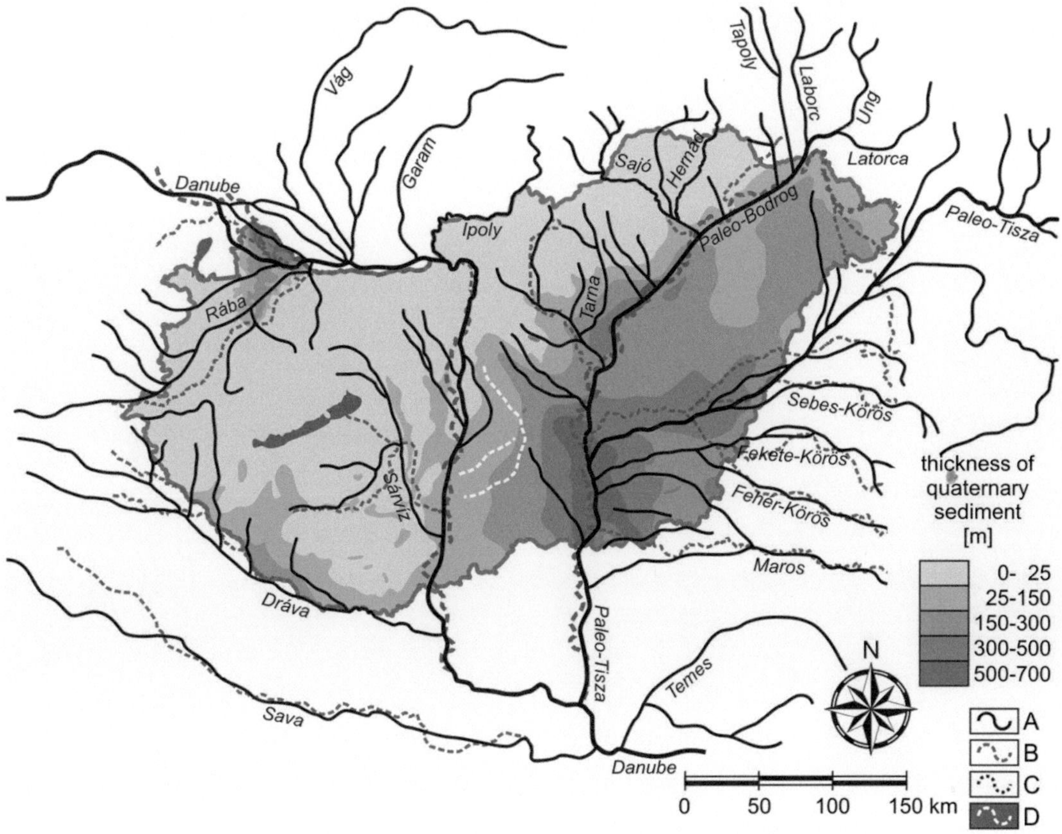


**Fig. 1.25** The river network in the Late Pleistocene (Borsy 1989; Gábris and Nádor 2007)

The Maros River during its evolution filled up the inner lake of the Transylvanian Basin, and then, it reached the Great Hungarian Plain, flowing across an intercolline valley. Here, it built a sprawling alluvial fan of 80–100 km radius, depositing large amount of sediment since the Late Pliocene (this river had the steepest profile among the lowland rivers). The Maros River changed its Quaternary flow direction adapting to the actual slope conditions. It usually transported large amount of sand and gravel, which became valuable aquifers. The Maros flowed towards the Körös Basin during most of the Quaternary, while it built Würm terraces between Déva and Lipova and along the present-day Temes and Béga rivers. At the end of the Pleistocene, the Maros shifted into the South Tisza Valley from the Körös Basin, and the several abandoned palaeo-channel zones (e.g. along the Száraz-ér, Aranka) reflect its continuous southern shift. Simultaneously, infusion loess accumulated between the palaeo-channel zones.

There are still a lot of unsolved questions that should be answered in connection with the evolution of the river network, e.g. the evolution and the age of the Visegrád Strait, and the origin of the cross-bedded sand in Transdanubia. Also, several "critical places" exist which could serve as key points when creating a more accurate Pleistocene landform evolution model (e.g. the Ezüst-hegy at the Hungarian-Slovenian-Austrian border, or the Lower Tisza Region). Some alternative models also exist besides the models discussed above, and one of them is presented by Fig. 1.26. This approach offers a largely different solution for the evolution of the (Danubian) water network (Mike 1991).

**Image 1.10** The Érvölgy and the rim of the Nyírség at Pocsaj (photograph by Kiss, T.)

### 1.3.3 Alluvial Fans in the Carpathian Basin

The alluvial fans played an important role in the aggradation of the basin, apart from the delta formations. A few extensive alluvial fans define the type, the ecological character, the environmental values and problems of the Hungarian plains, together with numerous small debris cones and alluvial fans attached to the mountains and hills (Fig. 1.21). The surfaces of the alluvial fans have a 3–6° gradient in average. However, their apical areas have higher gradient, which modified the fluvial transportational processes. Huge amount of material deposited in the alluvial fans, and its quantity depends on the uplift rate of the catchment and the subsidence rate of the basin. In general, the sediment is getting finer farther from the apex depending on the slope conditions.

In the Great Hungarian Plain, the base of the Danube-Tisza Interfluve, the Nyírség, and the Körös-Maros Interfluve is an alluvial fan, though the largest alluvial fan of the Danube developed in the Little Hungarian Plain in Transdanubia. Smaller alluvial fans built by the Körös River or the streams of the North Hungarian Mountains at their southern edge also exist (Fig. 1.27). In general, the steepest channel is located in the axis of the fan, in conformity with geomorphological laws: it is (was) in the case of the Old Danube in the Szigetköz, of the present-day channel of the Maros, and of the palaeo-channels of the Danube-Tisza Interfluve.

The evolution of each alluvial fan is connected to the Pleistocene, since this was the time when the tectonic, climatic, lithological and hydrological conditions were appropriate. Naturally, on the geological scale, alluvial fans and debris cones were created earlier as well, though

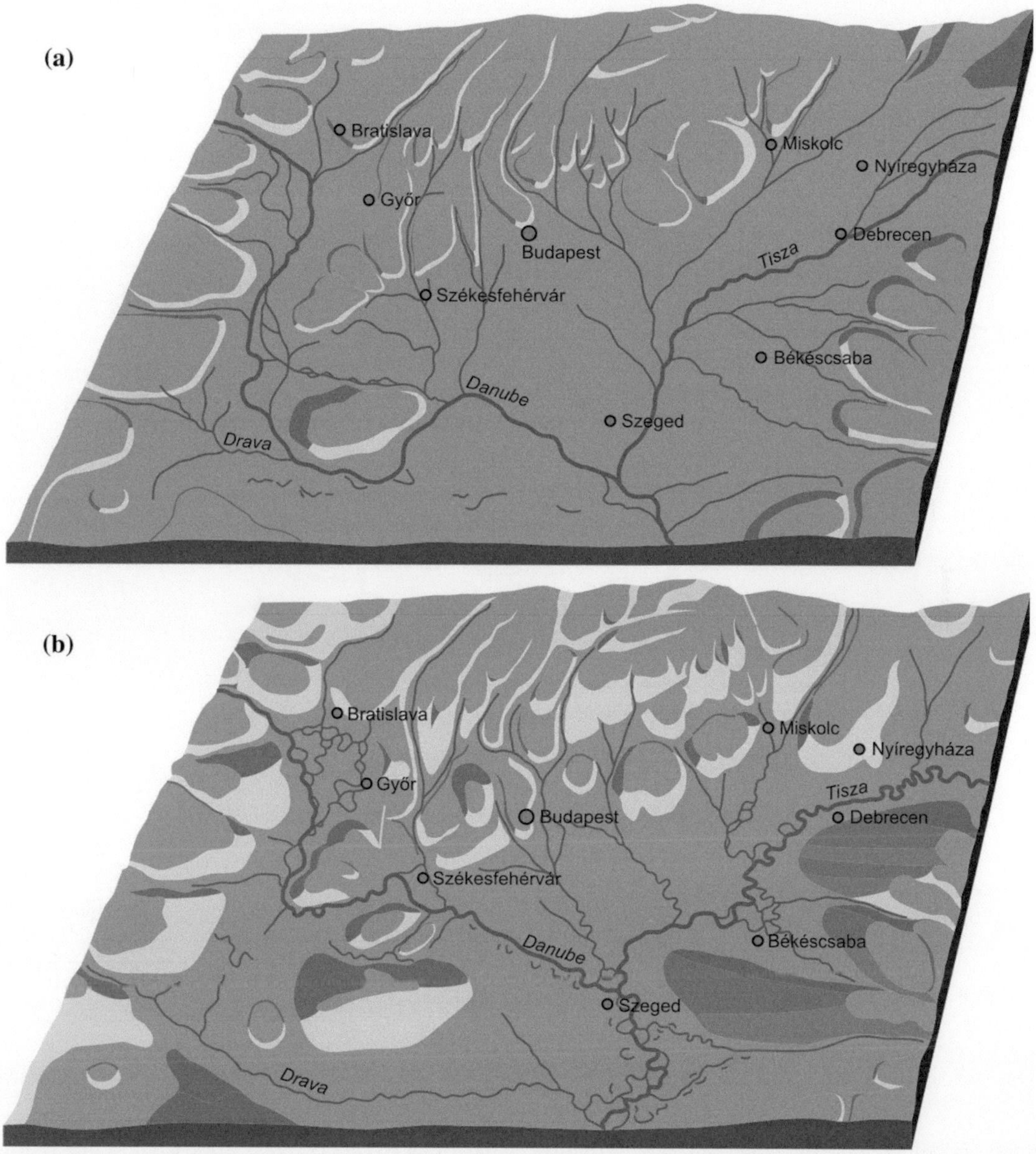


**Fig. 1.26** Alternative palaeo-hydrography of Hungary in the first half (**a**) and at the end (**b**) of the Eopleistocene (after Mike 1991)

under less ideal conditions. Regarding their age, each of the Quaternary alluvial fans has a slightly different evolutionary history.

The formation of the alluvial fan(s) of the Little Hungarian Plain probably started at the beginning of the Pleistocene. The alluvial fan that constitutes the Danube-Tisza Interfluve was built by the Danube, and this process had lasted for the majority of the Quaternary, until the river left its alluvial fan due to tectonic activities described previously. Compared to the Szigetköz, the alluvial fan of the Danube-Tisza Interfluve differs in its structure, as the alluvial fan formation frequently was interrupted in the territory of Kiskunság, and the water supply decreased to such an extent that loess and blown

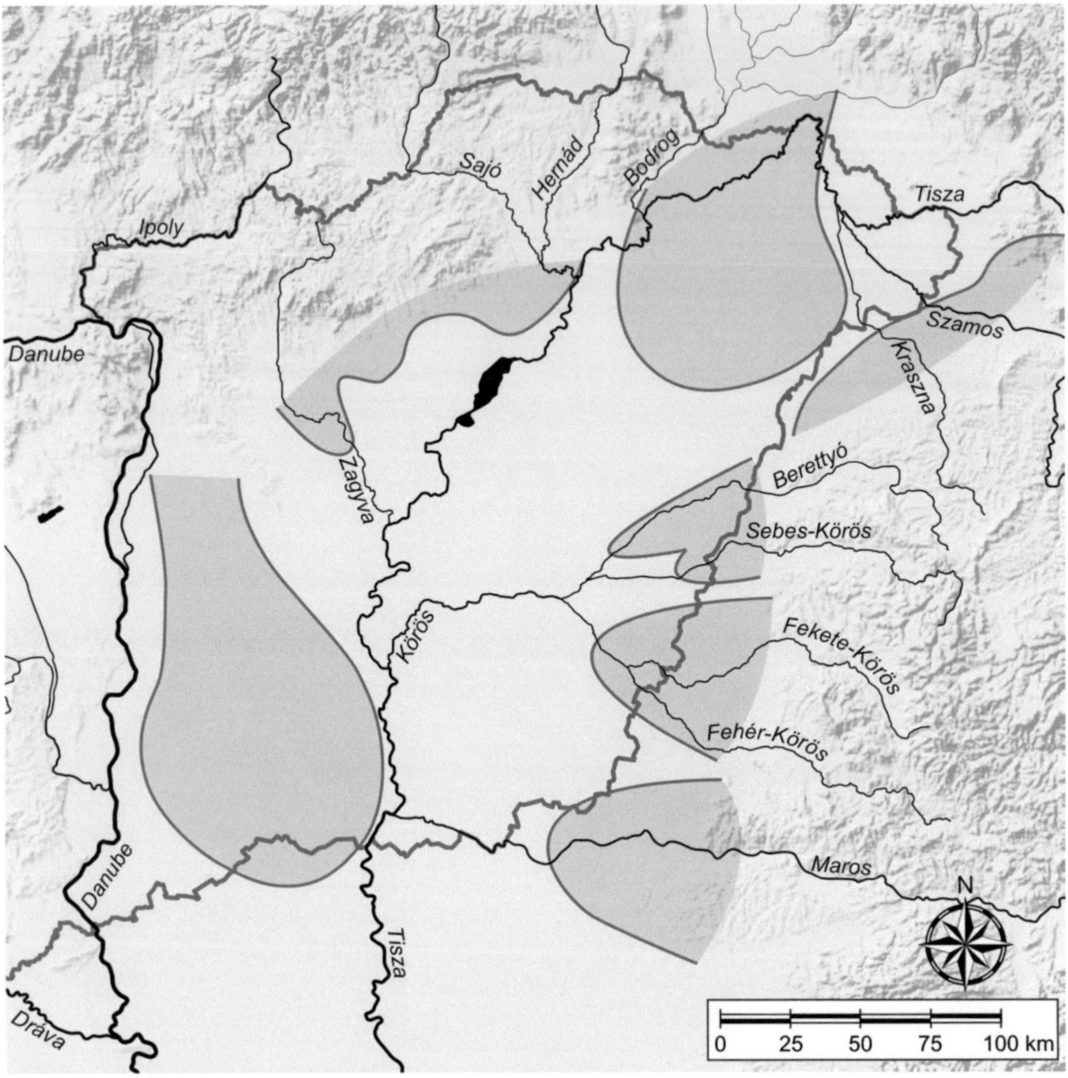


**Fig. 1.27** Alluvial fans in the Great Hungarian Plain (after Hámor 2001; Nádor et al. 2005)

sand accumulated on the surface, creating deposits of several hundred metre thickness, which refers to the alternation of these sedimentary processes.

The Pleistocene history of the large alluvial fan of the Maros River was defined by the channel shifts of the river. The deviation of the course directions from the main NW axis of the alluvial fan determined whether the Paleo-Maros joined to the central part of the Körös Region or to the Lower Tisza Valley. It the latest case terraces developed along the present-day Temes and Béga Rivers (Lovász 2006). The Late Würm and Holocene palaeo-channels are well visible on the present-day surface. The SRTM radar images clearly show the incised valley of the Maros at Semlac-Pecica (Fig. 1.28) and the series of river bed remains (Fig. 1.28) along the Fehér Körös (Timár et al. 2006). The present flow direction and structure of the Maros River started to form only at the end of the Holocene.

The material of the alluvial fan Nyírség has a thickness of 200–300 m, and it consists of sediments of different origins. The alluvial fan was built

**Fig. 1.28** Relief of the alluvial fan of the Maros River and the floodplain of the Lower Tisza based on SRTM data (after Timár et al. 2006)

during the Late Glacial by Carpathian rivers composing the Bodrog River and arriving from north and flowing into the Körös Basin. During some phases of the Pleistocene, the central and southern parts were also built by the Tisza and Szamos rivers; therefore, the alluvial fan is multifaceted.

## 1.4 Evolution of Terraces in the Carpathian Basin (Kiss, T.)

### 1.4.1 The Location of Terraces and the Influencing Factors of Their Evolution in Hungary

Smaller or larger terrace remains can be found along the mountainous and hilly sections of nearly all large Hungarian rivers. The most studied and most extensive terrace system stretches along the Danube, from the terraced areas of Győr and Tata through the Danube Bend to the Pest Plain (Fig. 1.29). (Rivers flowing into the Danube have well-developed terraces too, as these forms join the terraces of the Danube.) Terraces also developed along the Hernád, Sajó, Rába, Mura rivers and along smaller rivers too, but for example, the terrace system of the Dráva River is asymmetric, as it developed just on the southern side of the river (Bodnár et al. 2009). Since the most complete and longest terrace system was formed along the Danube, this will be the focus of the present chapter. We must note that each river has its unique terrace system; thus, to find similarities between them is not an easy task, even if the influencing factors were similar in a given time.

Terrace formation can be triggered by changes in climate, vegetation and tectonics. The most

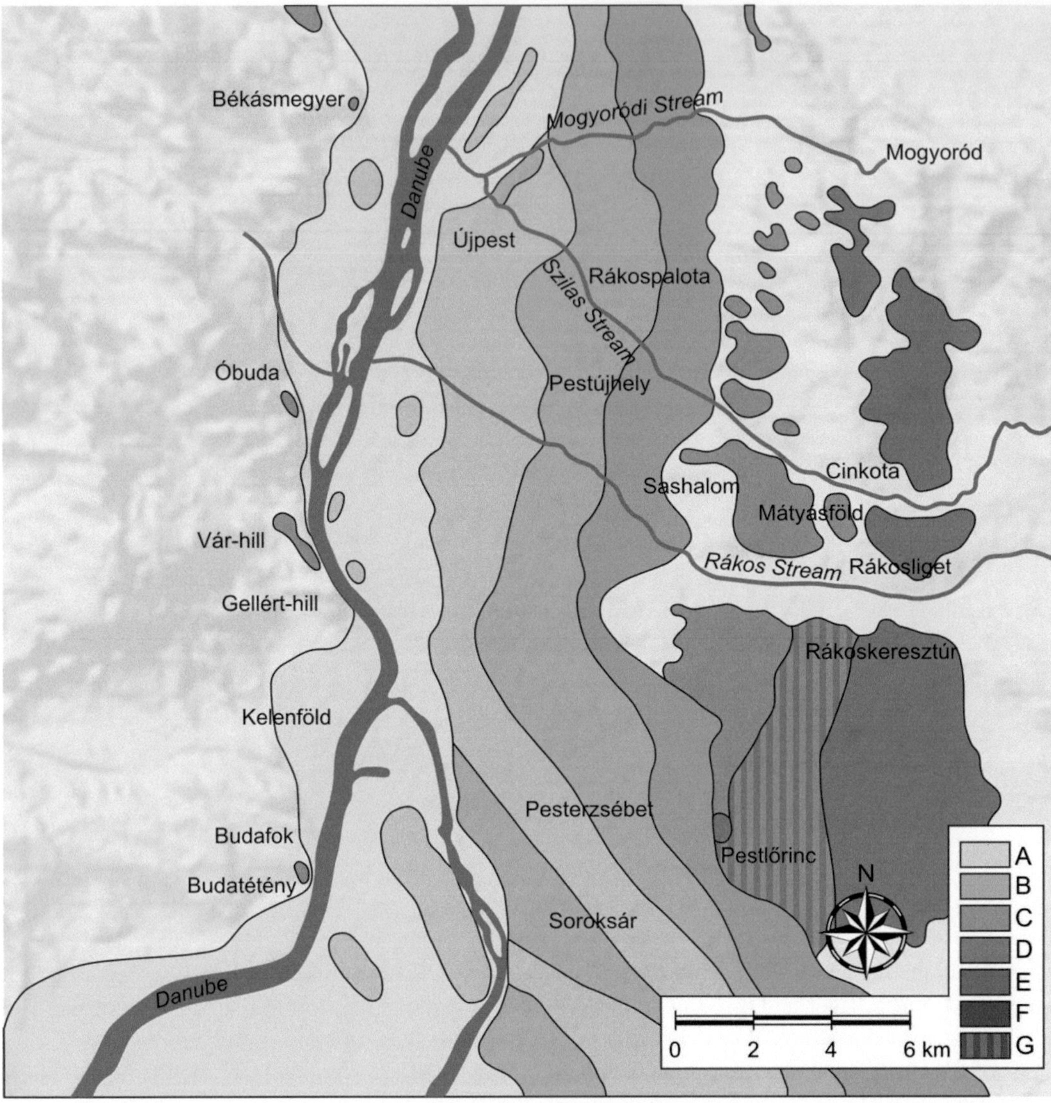


**Fig. 1.29** Terrace levels on the Pest Plain (after Pécsi 1991). *A*—active floodplain, *B*—terrace II/a (Würm), *C*—terrace II/b (Riss–Würm), *D*—terrace III (Riss), *E*—terrace IV, *F*—terrace V, *G*—terraces IV–V

significant geomorphological changes happen suddenly, at the termination of a long even climate phase (Gábris 2006, 2008). Supposing that the amount of precipitation remains constant, run-off increases at the beginning of a cooling period because of the decreased evaporation. The river incises due to the increased run-off (and builds a floodplain in the lower sections of coarse sediment at the same time). When the climate deterioration continues, the vegetation becomes scarcer; thus, the amount of sediments carried into the channel increases. The amount of surplus sediment may even be increased by intensified frost weathering; thus, the river accumulates in its upper sections (becomes braided).

In contrary, if the climate turns warmer and wetter, the run-off increases quickly since the vegetation will be denser only later. As the grassy vegetation is sufficient to inhibit surface erosion, small amount of sediment transported into the river, that is, it will be incised. Later, the run-off decreases due to the uptake and evaporation by the vegetation; therefore, just small amount of sediment washed into the river. This sediment load is fine-grained due to the decline of physical weathering and the intensification of chemical processes; thus, the river pattern may become meandering relatively quickly.

As we have seen, incision can be the result of cooling or warming; at the beginning of the climatic changes, when the system loses its equilibrium. The process of incision is a few hundred or thousand years long, while floodplain evolution may last for tens of thousands of years (Gábris 2006).

In the case of the Danube, the terrace development process was even more complicated due to the glaciation of the catchment during the Pleistocene. Thus, in the high mountains, large amount of sediment was produced during the glacials, and in the beginning of the interglacial periods owing to the increased amount of meltwater, the transported sediment was carried towards the basins, where it accumulated. In the meantime, the rivers intensively eroded their beds by the transported material (incision), and when their discharge dropped, the rivers aggraded their channels (Pécsi 1959, Burján 2002).

Tectonism also plays a fundamental role in terrace development (Pécsi 1959, 1991). The increased gradient in the catchment area leads to incision and intense surface erosion in the upper section of a river and to a coarse-grained floodplain formation in the lower one. However, tectonic movements have never been consistent, neither in time or in space; thus, we cannot directly connect the incision to one subsiding or uplifting period. Moreover, the tectonic movements are slow; thus, their impact can be eliminated or just strengthened by climate change. This is generally true in the case of the catchment area of the Danube; however, the tectonism rather dissected the Hungarian terrace systems instead of facilitating their evolution (Pécsi 1959). The plains in Hungary—within well-defined periods—continuously subsided, while the mountains uplifted; thus, the terraces (and the alluvial sediments of the same age) became dissected, that is why the same terrace is at different elevations (Fig. 1.30).

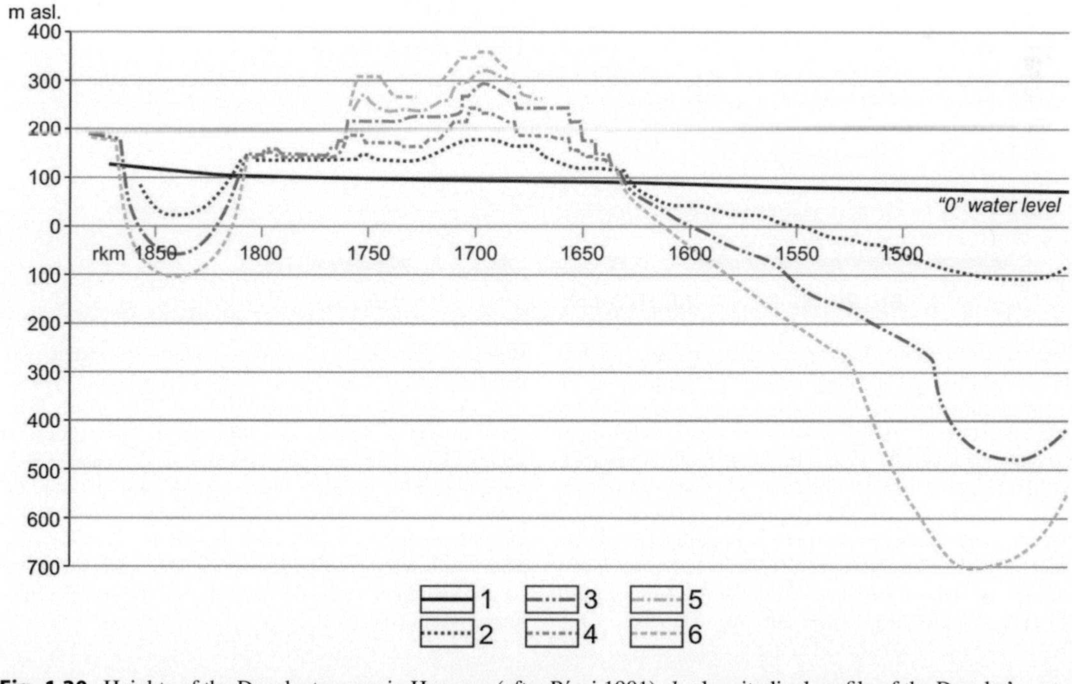


**Fig. 1.30** Heights of the Danube terraces in Hungary (after Pécsi 1991). *1*—longitudinal profile of the Danube's water surface at "0" cm, *2*—terrace III, *3*—terrace IV, *4*—terrace V, *5*—terrace VI, *6*—terrace VII

### 1.4.2 The Influencing Factors of the Development of the Hungarian Terrace Systems

Though numerous scientists have studied the causes (tectonic or climatic changes) and the age of the evolution of the Hungarian terrace systems (e.g. Bulla, Kéz, Pécsi and Gábris), there are still several unsolved scientific problems.

The climatic causes of the terrace development have been studied in connection with the glacial–interglacial changes during the Pleistocene: the aggradation of valley bottoms was connected to glacials and the valley incision to interglacials accordingly. As a result, 4–7 Danubian terraces were distinguished, but it becomes more and more evident that there was a higher number of terraces (Fig. 1.31); thus, for example, two terraces were formed during the Würm (II/a and II/b). The researches made outside the Carpathian Basin revealed that surface formation is not the most intensive during the "peaks" of glacial and interglacial periods, but rather in the transitory periods (Gábris 2006). That is, terrace evolution may have happened during the transition into an interglacial and into a glacial, as well. In addition, the Carpathian Basin is a tectonically very active territory, and for this reason, we must not neglect the role of crust movements (Pécsi 1959). For example, the Danube Bend has elevated at 0.44 mm/year in the past 360 thousand years, while the Danube has incised by at a maximum rate of 1.6 mm/year during 270 thousand years (Karátson et al. 2007).

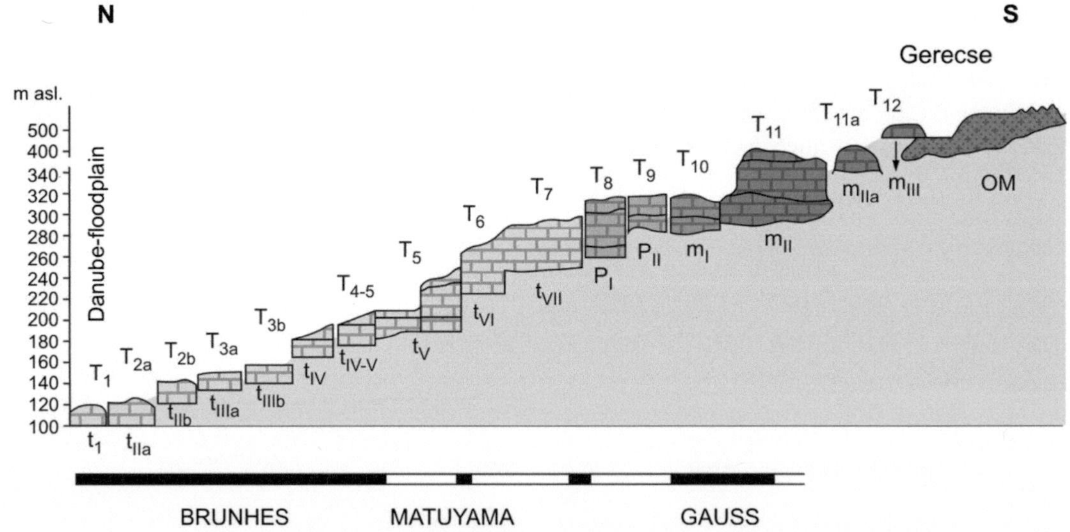


**Fig. 1.31** Geomorphological levels (terraces, foothill surfaces and wave-cut forms in the West Gerecse, along a cross section between Dunaszentmiklós and Dunaalmás (after Pécsi et al. 1988). *tI-VII*—terraces, *T1-12*—geomorphological levels, *PI-II*—Pliocene foothill surfaces, *mI-III*—Upper Miocene abrasion platforms and delta gravel, *OM*—Cretaceous peneplain covered by Oligocene-Miocene gravel sheets, *t1*—Holocene floodplain of the Danube, *tII/A-tVI*—Pleistocene terrace levels and alluvial fans of the Danube and the Által Brook, *tVII*—Upper Pleistocene terrace of the Danube, *T1*—Holocene travertine, *T2a*—Late Würm travertine (14 thousand years), *T2b*—Riss–Würm (127 thousand years), *T3a*—Late Riss (135 thousand years), *T3b*—Early Riss (190 thousand years), *T4*—Mossbach (350 thousand years), *T5*—Cromer (730–960 thousand years), *T6*—travertine with Upper Villány fauna (1.4–1.6 million years), *T7*—travertine with Upper Villány fauna (1.6–1.8 million years), *T8-9*—Pliocene travertines, *T10*—Upper Miocene travertine, *T11-12*—Upper Miocene travertine on abrasion platforms or delta gravel on the shores of the former Pannonian Sea

### 1.4.3 Identification and Dating Methods of Terraces

Erosional and accumulation terraces can be created during terrace development processes. Only lateral erosion took place during the formation of erosional or rock terraces, but floodplain sedimentation did not; thus, their surface consists of bare rock or fluvial gravel patches may cover them. By contrast, the evolution of accumulational or gravel terraces was more complex, as the first period of intensive floodplain aggradation by thick gravel layers was followed by a period of incision, when the former floodplain became elevated to such an extent that even the highest floods were unable to inundate its surface (Pécsi 1959).

Terraces can be identified higher by even 280–300 m than the present mean water level of the Danube (Fig. 1.30). The gravel accumulations in sinking areas (e.g. Little Hungarian Plain and Danubian Plain) cannot be separated from the terrace development, since they were formed at the same time as the gravel material of the terraces in the respective floodplain; thus, these plains have a normal stratification (i.e. the older layers are covered by younger ones), while the gravel terraces have inverse stratification (older material is at higher level).

The terraces are assigned by Roman numbers, starting at the active floodplain of the Danube (terrace No. I.), which could be divided into high and low floodplain levels, which could be covered by water during floods. In contrary, terraces are never covered by floods. If a terrace discovered to have two parts later (e.g. terrace No. II), the parts are marked by II/a and II/b. Only floodplains developed in the Little Hungarian Plain and the Great Hungarian Plain, whereas 7 terraces were formed in the mountainous sections of the Danube (e.g. in the northern foreground of Gerecse Mts. and in the Danube Bend at Visegrád), and 3–5 terraces evolved between the lowland and mountainous sections (Pécsi 1959, 1991).

The identification and dating of Danubian terraces is difficult due to various factors. First of all, the terrace surfaces were dissected by tectonic movements; therefore, the same terrace can be found in different elevations. In this way, terraces of a different age could get into a similar altitude. Secondary, loess, blown sand or travertine may have covered the terraces after their development. The thickness of the travertine sometimes reaches 10–12 m, diminishing the original elevation differences and making the identification and dating of terraces extremely difficult (Fig. 1.32). The travertine was deposited

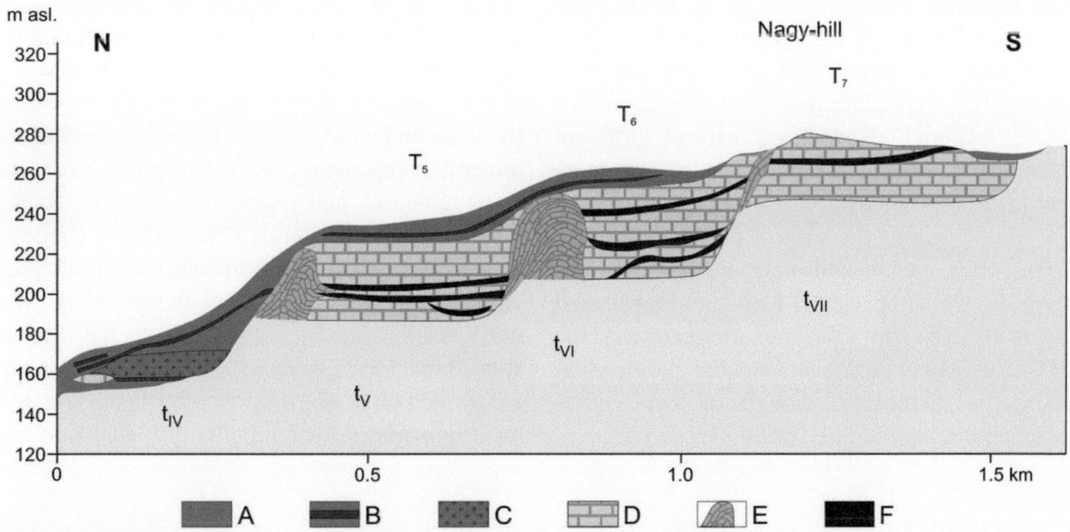


**Fig. 1.32** Danube terraces No. V–VII at Dunaalmás and their sediment cover (after Ádám et al. 1987). *A*—loess, *B*—palaeosol in loess, *C*—terrace gravel, *D*—travertine, *E*—travertine barrier, *F*—palaeosol in travertine, *tIV-VII*—Danubian terraces, *T5-7*—travertine levels

by thermal waters springing at the level of the actual, local erosional base (water level of the Danube). If the mountains uplifted, the level of the erosion base relatively decreased and the karst springs also came to the surface at lower elevations (Pécsi 1991). There are still existing thermal springs on the active floodplain levels (e.g. the Fényes Spring in Tata). A very colourful wildlife was reconstructed from the fossils in travertine; for example, the world's famous Lower Paleolithic remnants of Vértesszőlős were also preserved in it.

We must not forget that the terrace formation may consist of a chronological and spatial sequence of incision and aggradation, and the limit between these processes repeatedly shifted upstream or downstream along the river: the filling of basins took place in the lower section at the time of the terrace formation of the upper section. Therefore, it is not enough to identify the terraces just on the basis of their elevation, but other methods have to be applied. Moreover, the definition of "age" has two approaches: some researchers identify the age of the terrace as equal to the age of fossils found in it (i.e. they determine the date of the floodplain evolution), while others consider the time of the incision as the age of the terrace (Gábris 2006).

In the early researches, the dating of terraces was indirect; thus, an age of a terrace was determined based on the ages of sediments covering them (e.g. blown sand and loess), or periglacial (e.g. cryoturbation phenomena of the gravel material. Nowadays, several different dating techniques are applied supplying more and more accurate data. For example, the age of travertine covering the terraces can be defined by U/Th, ESR, AAR and magnetostratigraphical methods, while the ages of loess and blown sand layers by TL and OSL measurements, if the sediment contains organic material radiocarbon dating, is applicable, though in same cases tephrachronology is also useful (Table 1.5).

### 1.4.4 The Factors Transforming the Terraces

No uniform terrace stretching along the Danube could be identified. One of the reasons is tectonism, since the terraces became dissected by fault lines and elevated to different elevations. Thus, the longest, almost continuous terrace is the youngest, and the more fragmented are the older. The dissection became even more pronounced by the erosional and derasional valleys of the joining streams. These valleys usually connected to the terrace beneath them, and they have terraces themselves (Pécsi 1991). Terraces characterize only the narrowing valleys, and they remain there only when the lateral erosion of the river does not destroy them. Since the terrace system of the Danube is the narrowest in the Danube Bend at Visegrád, here the lack of terraces No. V and VI could be explained by lateral erosion.

## 1.5 Forms and Processes of Loess and Blown Sand Territories in Hungary

### 1.5.1 Some Physical Geographical Issues of the Loess

Loess is coarse silt composed of 0.01–0.05-mm grains, which is a slightly diagenized rock that is characteristically porous, consistent, unstratified, able to compose steep walls and easily degraded by water and wind. Quartz constitutes most of its mineral component (50–70 %), but it contains clay minerals and carbonates too (1–20 %) which cement the grains (Pécsi 1993). Carbonates provide the material for the loess dolls precipitated at the actual groundwater level. We must assign a special importance to the discussion about loess, because excellent fertile soils are formed on it, and it is also one of the few raw material sources for the building industry in the

**Table 1.5** Ages of terraces in Hungary and their relationship with the different chronologies (after Gábris 2008)

| 1000 years | magneto-statigraphy | δ18O | OIS | termination | glacial statigraphy | terraces | age of travertine (1000 years) |
|---|---|---|---|---|---|---|---|
| | | 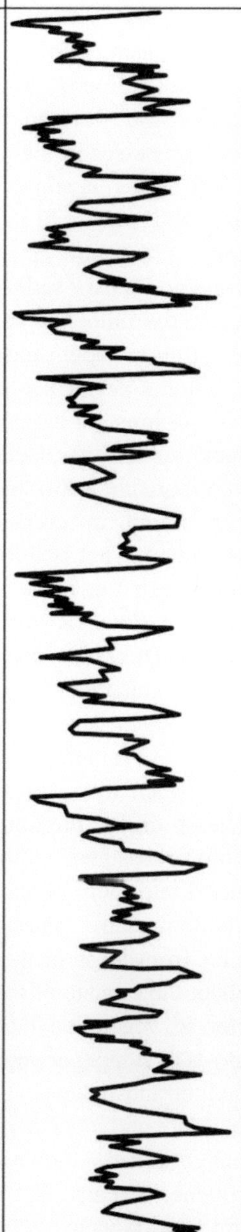 | | T I | holocene | t II/a | 12 |
| | | | 3 | | Würm Weichsel | | 30 |
| | | | | | | | 70 |
| 100 | Blake | | 5/e | T II | R/W Eem | t II/b | 110 |
| | | | | | | | 160 |
| 200 | | | 7 | T III | Riss Saale | t III/a | 200 230 |
| | | | | | | | 290 |
| 300 | | | 9 | T IV | | t III/b | 335 |
| | | | | | | | 370 |
| 400 | | | 11 | T V | M/R Holstein | t IV? | |
| | Emperor | | | | Mindel Elster | | |
| 500 | | | 13 | T VI | | t IV? | |
| 600 | | | 15 | T VII | | | |
| 700 | | | 17 | T VIII | Günz/ Mindel | | |
| 800 | | | 19 | | | | 800 |
| 900 | | | 23 | | | | |
| | | | 25 | | | | |
| 1000 | Jaramillo | | 27 29 | | Günz | | 1000 |
| 1100 | | | 31 33 | ? | | t V? | |
| | | | 35 | | | | |
| 1200 | Cobb Mountain | | 37 | ? | | t V? | |
| 1300 | | | 39 41 43 | | Danube/ Günz | | |

plains, since the Hungarian brick factories obtain their basic materials from loess and often from alluvial clay.

The Hungarian loess can be classified into different types genetically, but the largest categories could be combined. Basically, each type has undergone the loessification described by Berg, that is, the diagenesis of the accumulated material. The classifications mostly focus on the ways of the transportation of the powdery grains and the circumstances of their accumulation. The loess occurrences in the Carpathian Basin occupy a special place not only because of their large extension, but also because they definitely show the environmental changes of the Quaternary. Berg's theory on loess formation can easily be adapted to the circumstances in Hungary: it could be considered as a soil formation, which results in loess or a certain soil type according to the prevailing environmental conditions.

The most characteristic loess types are those accumulated by wind, where the wind transported and deposited the powder material that had originated in the foreground of the ice sheets and glaciers. This material transformed into loess through diagenesis in the presence of steppe vegetation (Fig. 1.33). The spread of typical loess and its great thickness (e.g. 60 m in the SW of Mezőföld) are explained by the fact that the NW winds prevailing in the Pleistocene could accumulate its material in this thickness in the downward (lee) side of the Transdanubian Mountains. The so-called infusional loesses have a similar evolution: in this case, the material of the loess deposited onto a wet surface (e.g. floodplain) and became loess (e.g. in the area of the Körös-Maros Interfluve). This type is called hydroaerolite in geology. There is an almost hundred-year-old debate on the concept of infusional loess: its mineral content is claimed to be fluvial origin by Pécsi (1993) and Sümeghy's opinion; however, it could have an aeolian origin too. Different palaeosols developed between the loess layers, referring to the changing environmental circumstances.

The aeolian sediments in the hilly regions may have been transported to their present location by different solifluction processes and mass movements and by Holocene soil-erosional processes. These loesses are identified as "slope loesses", but some local names exist for them too, e.g. "Palóc Loess". The spread of loess in the Carpathian Basin is limited by the elevation, as they could not be found above 400 m above sea level, because the elevated deposits were eroded by slope processes in the humid environment.

The loess layers are unstratified; however, sand and palaeosol layers are intercalated in them, which are jointly called loess profile. These non-loessy sediments indicate the changes in the environmental conditions (see palaeosols); the question about sand is not only whether it has an alluvial or an aeolian origin, but also what kind of climate it indicates, that is, whether it is a glacial or an interglacial deposit. The picture is even more colourful, because some evidence has been found in connection with blown sand movement. That is, blown sand movements can be connected not only to dry periods, but also to humid periods as well, under certain circumstances. Of course, loess and sand occur in various combinations: e.g. loessy sand or sandy loess, which already show some stratification (e.g. by derasional relocation).

Between loess formation periods soils could develop on the previously deposited losses or it could be weathered when the cold and dry climate turned into more humid (oceanic) or warm (sub-Mediterranean). These palaeosols (Fig. 1.11) differ from loess in their colour. The palaeosols reflect the climate of the given period; thus, they refer to vegetational changes too: existence of forests or steppe, or vegetation like in present-day Mediterranean areas. In Hungary based on the palaeosol and loess sequences, ca. 20–22 climatic changes were indentified based on a loess sequence at Paks. The first descriptions of loess stratification were made a hundred years ago (see Bulla 1937; Cholnoky 1926), which followed the Ice Age chronology of that time. Establishing the ages of soils used to be an important pillar of stratigraphy, also because of the genetic explanations. The classical geomorphological, mineralogical and petrological analyses were

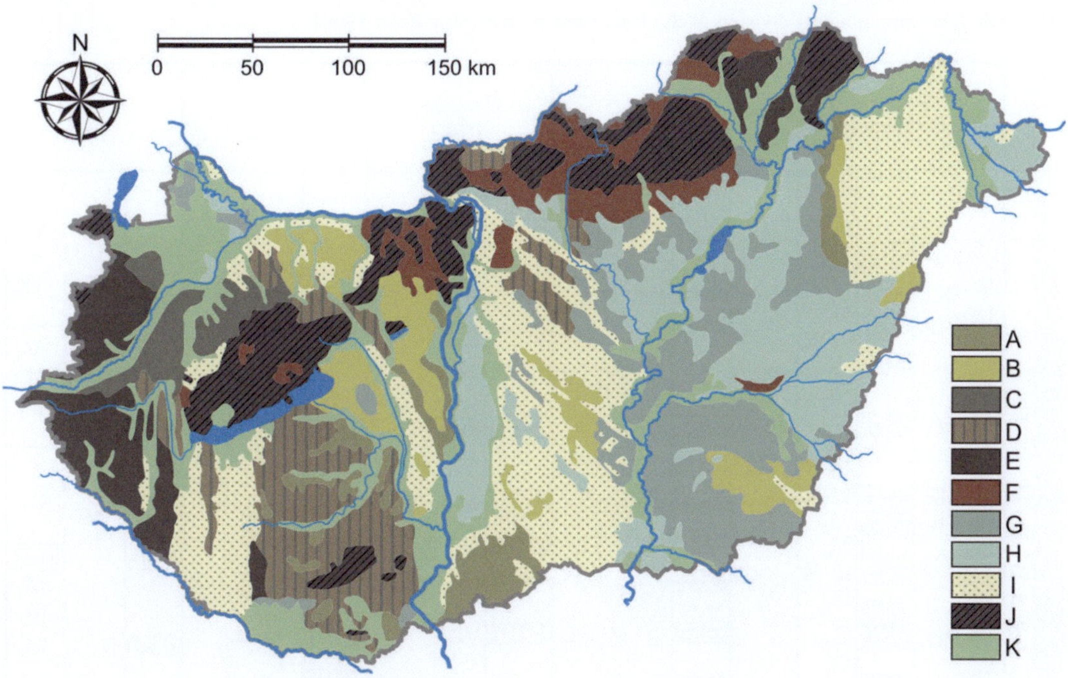


**Fig. 1.33** Loess and blown sand areas in Hungary (after Pécsi 1993). *A*—typical loess, *B*—sandy loess, *C*—derasional loess, *D*—brown loess in patches, *E*—brown loess, *F*—loess derivatives, *G*—alluvial loess-like sediments, *H*—infusional loess, *I*—blown sand, *J*—mountains, *K*—Holocene fluvial sediment

complemented by thermoluminescence dating (TL measurement range approx. 130–150 thousand years), ESR or radiocarbon methods (measurement range approx. 30–40 thousand years). The dating of loess intercalating the soils became also possible with these techniques. Nowadays, modern measurement methods (amino acid stratigraphy—AAR, oxygen isotope stratigraphy—OIS, malacological analysis, optical luminescence—OSL) modified or refined the ages of soils.

The different loess profiles in Hungary represent the Pleistocene comprehensively, though not completely. The loess profiles are grouped based on their ages. The young loesses are rich in carbonate, divided by chernozem-like, dark brown fossilized soils. The older loesses contain less carbonate (although loess dolls are common), compacted sandy layers divide them, and reddish brown forest soils intercalate in them. The material of the Dunaújváros-Tápiósüly and the Mende-Basaharc sequences represent the younger type, as they are up to 350 thousand years old; thus, they formed in the Riss and Würm. The sequence of Paks and the Dunaföldvár Formation are older with several palaeosols (e.g. PD1-2 or Pv1-3).

Pécsi (1993) identified five typical loess formation periods in the Carpathian Basin (though the oldest, the so-called red clay of Kulcs belongs to the series logically, but it is already not a loessy sediment).

The sequence of Paks is the best divided and most studied formation in Central Europe. Approximately 10 buried soils and 9–12 loessy layers were dated back to the Brunhes-Matuyama reversal (Table 1.6), and the loesses are not older than the Jaramillo event, according to Pécsi. The sequence starts with the old loess under the Paks-Dunakömlöd soil (PDk), because underneath is a Pleistocene alluvial sediments (which also contain remnants of Mediterranean forest soils), but they cannot be regarded as loesses (Image 1.11).

**Table 1.6** The chronological division of the Paks loess profile (after Pécsi 1993)

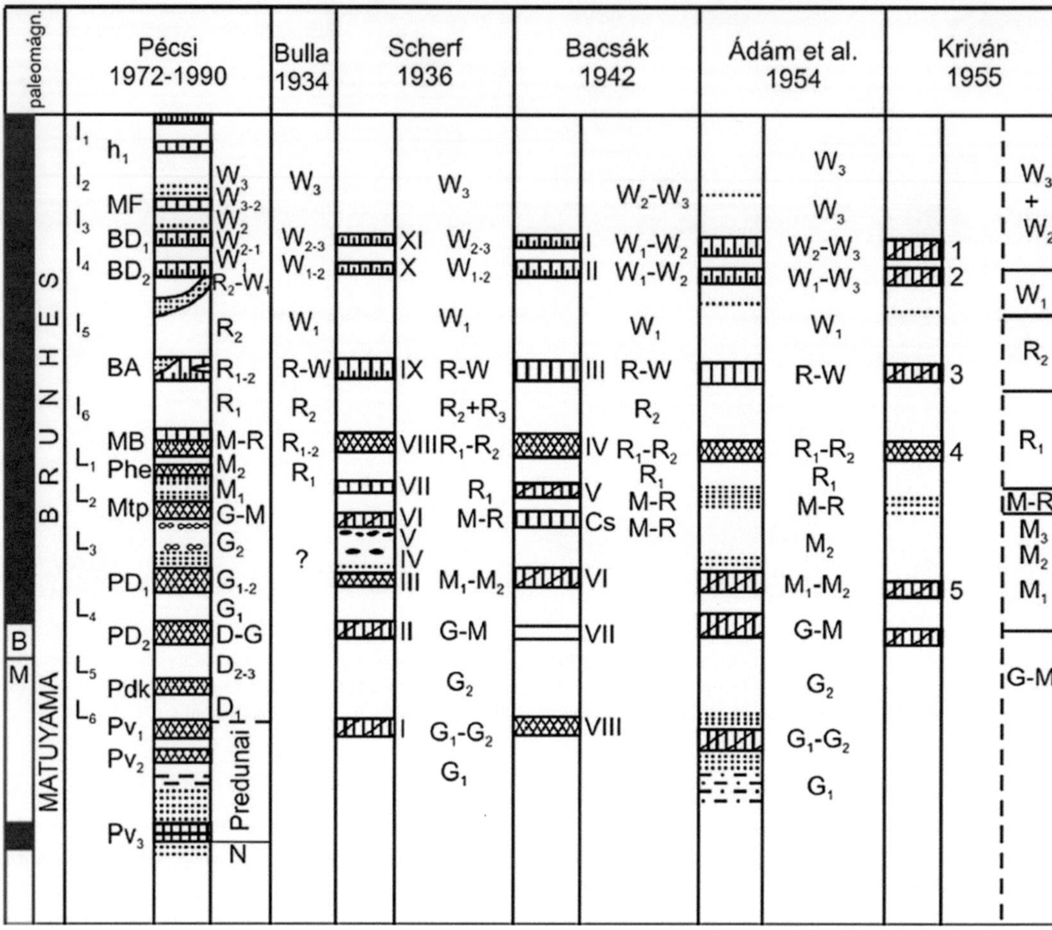


The older loesses are divided by some forest soils formed in various climates and by a well-developed reddish brown palaeosol. Three other loess layers and three reddish brown soil layers can be found under them (all of them younger than the Jaramillo). The double soil complex of Paks (PD) includes two well-identifiable levels, which are separated by 2-m loess layer. The approximately 700-thousand-year-old Brunhes-Matuyama palaeomagnetic reversal can be found in its bottom (Kordos and Schweitzer 2000). Both are well-developed, Mediterranean, dry forest soil remains (maybe, chestnut brown soils—Pécsi 1993), whose level C is also characterized by carbonate accumulation (loess dolls) and large krotovinas (filled-in animal burrows).

Young loesses can be found in the upper 30-m layer of the sediment sequence (Table 1.6). Their typical dividing soils are as follows: MF, BD, BA and MB. The volcanic tuff (the so-called Bag Tephra) above soil MB is estimated to be 300–400 thousand years old, is a precious tracer in the dating of soils and loess profiles (Horváth et al. 1992).

The Mende-Basaharcsequence is divided by four well-identifiable palaeosol levels. The uppermost, the Upper Mende (MF, forest steppe soil), is often a dual facies from the Würm. Under it is a same type palaeosol called Basaharc

**Image 1.11** Double palaeosol developed in the Riss-Würm interglacial, in the young Basaharc Loess sequence (own photograph)

Double (BD) with double soil layers, which can be synchronized with other outcrops as well (Table 1.5). The earlier analyses considered it to be also from the Würm, but the newest measurements show that it is much older, that is, from the Riss-Würm interglacial. The next layer, the Lower Basaharc Alsó (BA) is a chernozem-type palaeosol. The sequence is completed by the palaeosol described in the outcrop in Mende, having double facies (the upper layer is chernozem, and the lower one is forest soil) (Mende Base—MB), which is thought to be from the Riss by many and claimed to be even older by Pécsi (1993).

The youngest Dunaújváros-Tápiósüly loess sequence is dissected by tundra-like palaeosols, referring to a cold and dry climate at the time of their formation. Two, poorly developed soils (humus horizons) were formed in it.

The hottest debate took place about the age of soil BD (Basaharc Double). Several results were obtained by various methods. For example, Butrym and Maruszczak (1984) determined its age as 40–45 thousand years, while Zöller et al. (1994) refer to 135 thousand years. The latest measurements show that this palaeosol is older than the Würm, which is also supported by the results of the latest TL examinations (130 thousand years). Because of these measurements, the former data on soils BD and BA dating back to the second half of the Würm cannot be accepted any more (Table 1.7; Horváth et al. 2006). Today, the BD soils are claimed to belong to the Riss-Würm interglacial, and the Basaharc complex cannot be situated on the Danubian terrace II/b, but rather on terrace IV., since BA is delimited from underneath by the Bag Tephra defined to be 350–360 thousand years old (Ruszkiczay-Rüdiger et al. 2006). The question is even more complicated, because it is plausible that this complex slipped into its present place by slope mass movements.

Considering the geomorphological forms of loessy areas, the same notions are used mostly for the loess formations as for the karst and glacier forms. This is not a genetic explanation, of course, because these territories were not created by the same processes, but undoubtedly, similar erosional forms can be described. Among the small-scale landforms, loess dolines are present in each of the territories covered by typical loess, which are 1–2-m-deep forms without an outlet (Kerényi and Kocsisné 1990). Whereas their formation used to be explained by karst processes (as we have mentioned, carbonated material cemented the loess grains); nowadays, their evolution is explained with subsurface erosion and subsidence caused by suffosion.

The identification of larger forms presents numerous problems. For example, the Öreg-hegy in Dunaújváros is sometimes mentioned as a loess circus although mass movements are responsible for its formation. The frequent, benched landslides of the N-S-oriented, loess-covered high banks of the Danube mean a real environmental hazard (Mezősi 2008). According to Pécsi, the narrow sand layer underneath the

**Table 1.7** The Hungarian (young) loess types and ages in thousand years with the dating methods (after Horváth et al. 2006). *Source* http://www.fsz.bme.hu/mtsz/barlang/5lezar/legm.htm; www.index.hu—2011

| | Pecsi (1995) | Bulla (1938) | Krivari (1955) | Pecsi (1975) | Wintle and Packman (1988) | Oches and McCoy (1955) | Pecsi (1955) | Frechen et al. (1977), Novothny et al. (2002) |
|---|---|---|---|---|---|---|---|---|
| | talajtipus | | | $C^{14}$ | TL | ARR | $C^{14}$, TL, becsles | TL, IRSL |
| talaj | csernozjom | | | | | | 0–11.3 | |
| I1 | | | | | 15.8 ± 1.3 | | | |
| I1 | | | | | 17.6 ± 1.4 | | | (13–20) |
| h1 | humusz h. | | | (16.75 ± 0.4) | | | $W_3$ (16–17) | |
| I1 | | | | | 23.2 ± 1.9 | | | |
| h2 | humusz h. | | | (20–22) | | | $W_2/W_3$ (27–32) | |
| I2 | | | | | 24.0 ± 2.0 | | | (25–35) |
| $mf_1$ | erdossztyepp-talaj | | | (28–29) | 20.9 ± 1.7 | | $w_2/w_3$ (45–60) | |
| I3 | | | | | 43.4 ± 3.8 | | | (50-60) |
| $mf_2$ | erdossztyepp-talaj | | | 32 | R/W (74.7 ± 6.5) | 0IS 5 (<128BP) | $W_1/W_2$ (85–105) | OIS5 (65 ± 10) |
| I4 | | | $w_2$ | | R (79.2 ± 6.1) | | | OIS 6 (>130) |
| $BD_1$ | erdossztyepp-talaj | $W_2$/ $W_3$ | $w_2$ | | | | $R_2/W_1$ (120–140) | |
| $bd_2$ | erdossztyepp-talaj | $W_1$/ $w_2$ | | | | >245 BP | $R_2/W_1$ (150–170) | |
| I5 | | | $W_1$ | | | | | |
| BA | erdossztyepp-talaj | R/W | | | | >339 BP | $R_1/R_2$ (195–230) | |
| I6 | | | $R_2$ | | | | | |
| $mb_1$ | erdossztyepp-talaj | | | | | | $MR_3/MR_1$ (280–310) | |
| $mb_2$ | barna erdotalaj | | | | | <423 BP | $MR_2/MR_3$ (320–360) | |
| LI | | | $R_1$ | | | | | |

loess conducts water into the river from the west; thus, it could lose its consistency under the 60-m-thick loess sequence and may not be able to hold the upper sequence; thus, it promotes the mass movement (Fig. 1.34). The loess creeks belong to the erosional loess forms. They are mainly formed in the lower sections of the erosional-derasional valleys, and they indicate a low resistance to erosion (Lovász and Majoros 1994).

The deep roads that cut through loess are the most typical examples of loess forms of anthropogenic origin. The traffic disrupted the stability of the loess surface, triggering erosion. These forms are quickly developed by intensive erosion and by the fact that loess is able to be permanently stable in high, steep walls. The deep roads may become loess creeks with time, which make the further erosion possible, but they will hinder traffic (Image 1.12).

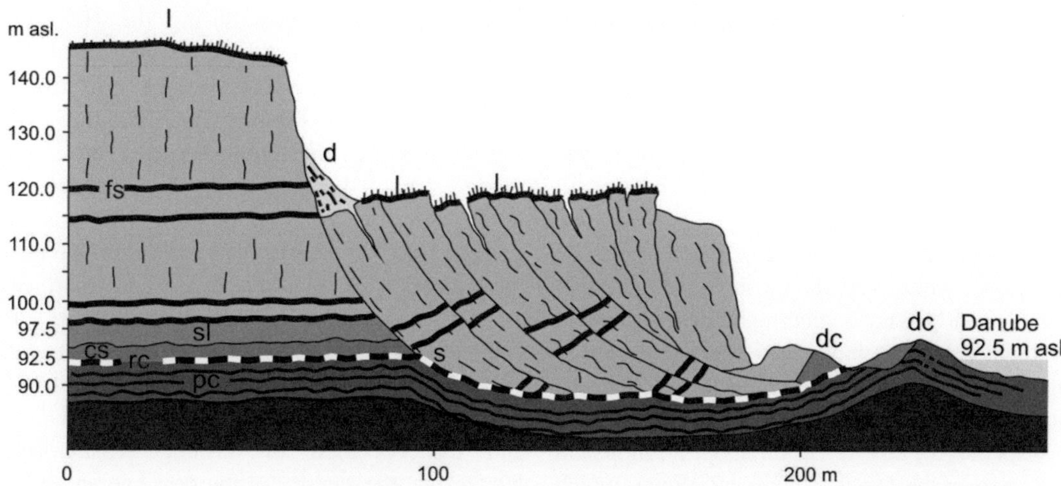


**Fig. 1.34** Landslide at Dunaföldvár causing environmental problems (explanation of the mass movements in the 1970s, after Pécsi 1993). *l*—loess, *o*—debris, *z*—the land mass and Pannonian clay protruding from the Danube river bed, *cs*—slide plate, *ft*—fossil soils, *hl*—sandy loess, *ta*—clayey loamy soil, *va*—red clay, *pa*—Pannonian clay

**Image 1.12** Deep loess road cut through loess in the Szekszárd Hills with indications of linear erosion at the bottom (photograph by Kiss, T.)

## 1.5.2 Blown Sand Forms and Processes (Kiss, T.)

### 1.5.2.1 Location and Evolution of Blown Sand Areas

Nearly 20 % of the territory of Hungary is covered by blown sand which can be found in the largest extension on alluvial fans, e.g. on the Danube-Tisza Interfluve, in the Nyírség and in Inner Somogy (Fig. 1.33). Besides, sandy areas developed along rivers: on terraces (e.g. the Danubian terrace II/b), on islands (e.g. on the Csepel Island), on erosional islands of floodplains (e.g. in Solt Plain or Bodrogköz) and in the floodplains in the form of coastal dunes.

The development history of sand areas was the same in its main features on all alluvial fans. Following the rule of alluvial fan formation, the rivers repeatedly slipped off from the highest areas of the alluvial fan. The high and loose sedimentary areas became drier; thus, the strong, erosive winds could blow it out and accumulate the sand into different forms. The aeolian activity was facilitated by tectonic uplifts, which made the surface of the alluvial fans drier. The process was further intensified by the fact that the majority of the alluvial fans lost their rivers due to avulsions; thus, their whole area became drier. The rivers often incised into a rim of their alluvial fan, capturing other streams running through on the alluvial fans. Since the above-mentioned processes took place in different periods, the aeolian form assemblages were also created in different eras (Borsy 1991).

The Inner Somogy sand area is believed to be the oldest, whose aeolian formation could have started as early as the beginning of the Pleistocene, although evidences have been found only about blown sand movements in the Würm and the Holocene (Marosi 1970; Lóki 1981). The Danube took its present flow direction about 30 thousand years ago in the Danube-Tisza Interfluve, which means that the formation of blown sand could start in the northern part of the region. The Nyírség alluvial fan lost its rivers last among the alluvial fans. It became dry when the River Tisza shifted towards the Bodrogköz about 14–16 thousand years ago (see more in Sect. 1.3), and it captured the tributaries passing through the Nyírség.

The sand forms often deposited on each other and transformed into each other, which raised the possibility of more sand movement periods. Sand movements older than the Würm are known only from the analyses of borings and loess outcrops: for example, 10 blown sand layers were identified in the Pleistocene sediment sequence in the Danube-Tisza Interfluve, and 5.0 and 2.5–2.6-million-year-old blown sand layers were also found (Molnár 1965; Borsy 1991).

The evolution of forms seen today may have been the most intense 22–27 years ago, when the steppe became sparse on the cold and dry climate; thus, the blown sand movement could start (Lóki 2009). Diverse forms developed transforming the surface of the alluvial fans completely. Late glacial sand movements are mainly known from the Nyírség (radiocarbon measurements) and the northern apex of the alluvial fan of the Danube-Tisza Interfluve (OSL measurements); however, the data show a large dispersion (Gábris 2003). The climatic changes of the Late Glacial influenced the evolution of sand forms, as well.

While the Pleistocene sand movements probably involved all the blown sand territories, the intensity and extension of the aeolian activity may have been diverse during the Holocene (Sects. 1.8 and 1.9). At the beginning of the Holocene, sand movements took place in several sand areas (e.g. the Danube-Tisza Interfluve, the Danubian terrace II/a, South Nyírség), but in the second part of the Holocene, the anthropogenic impact may have overlapped by the natural driving forces, which gave momentum to sand movements. Freely moving sand surfaces reached a significant proportion due to the intense land use in the eighteenth and nineteenth centuries. Later, sand movements became rare, because of the intensive forestation of sand areas and the cessation of farming (Kiskunság). Nowadays, the activity of the best-known blown sand surface next to Fülöpháza (Danube-Tisza Interfluve) is artificially maintained.

#### 1.5.2.2 The Aeolian Landscape

The typical aeolian forms belong to the blowout—residual ridge—hummock form assemblage, and to the class of parabolic dunes. The direction of the wind creating the forms was not constant, though they were mostly NW winds. The exception is Inner Somogy, where the erosive winds used to blow from the north (Lóki 1981), and the Nyírség, where the NE winds were also strong (Borsy 1961, 1991).

Blowouts are often found in the original alluvium in an elongated form following the wind direction, and they are separated by residual ridges. Their length is 100–500 m, while their width is 25–200 m. The shallowest (2–3 m) ones are in Inner Somogy where the groundwater level is higher, while in other territories, they are up to 8–10 m deep. The more extensive (up to several kilometres long and wide) deflation flats and basins sometimes include secondary aeolian erosional forms and hummocks (Borsy 1991). The material of blown outs deposited in the form of hummock at the end of the former. In case of strong winds, the material of hummocks was spread it into sand sheets. The blowout dunes have a 10–330-m-long ridge, and they have various shapes (Fig. 1.35).

When the hummocks were transported further from their blowouts, they could form dune fields (typical in Kiskunság) or parabolic dunes. The

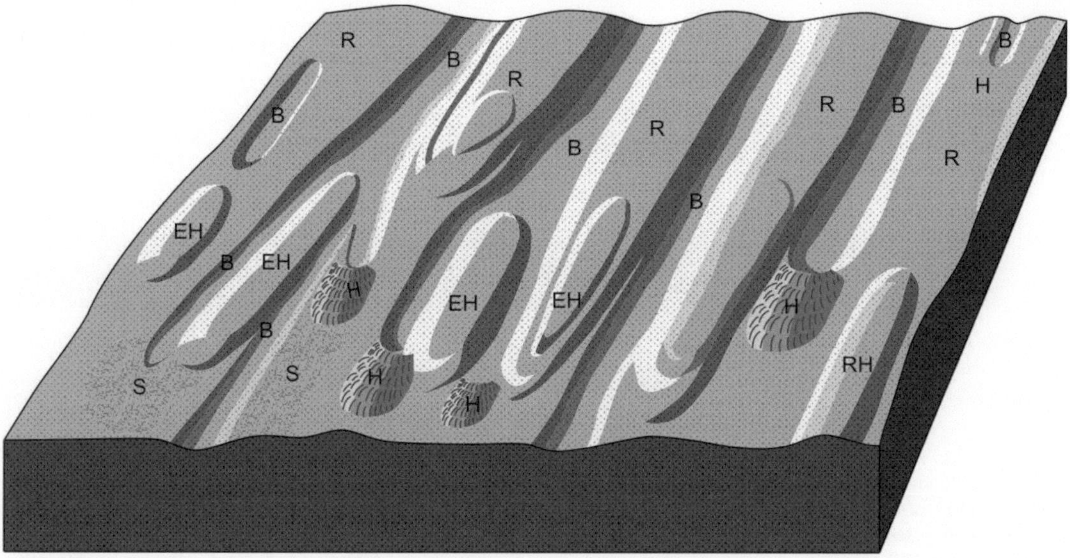


**Fig. 1.35** Spatial structure of blown sand forms in Inner Somogy (after Marosi 1970). B: blow-out depression, M: residual ridge, G: hummock, LG: elongated hummock, L: sand sheet, MHG: elongated hummock deposited on a residual ridge

parabolic dunes are the most common forms in the Southern Nyírség, with an arc length of 0.5–2 km, and they can be up to 15–20 m high. According to Borsy et al. (1969), the majority of the parabolic dunes in the Nyírség are asymmetrical, they have a small western wing, which is explained by the strong NNE winds, and there was not enough room for the development of the western wing, because the dunes are too close to each other. However, according to Kiss et al. (2009), the asymmetric parabolic forms are extremely rare in the South Nyírség.

The Pleistocene aeolian landforms may have transformed significantly during the last thousand years, especially in areas under significant human impact. These dunes were dissected by wind erosion; thus, secondary blowouts were formed on their slopes, and the interdune areas were covered by the blown-out material in the form of sand sheets; thus, the relief of the landscape became more even (Fig. 1.36). In addition, the dunes under intensive agricultural use became smaller (by 0.8–6 m) because of run-off erosion; the process was especially intensive in the humid Inner Somogy (Lóki 1981).

### 1.5.2.3 Main Blown Sand Areas in Hungary

On the alluvial fan of Inner Somogy, divided by the Marcal Ridge, aeolian processes became dominant in the dry periods of the Würm, as a consequence of an avulsion of the Danube, which resulted in the loss of the water system. During the long aeolian periods, the wind eroded the sand from the northern part of the alluvial fan and accumulated it in its southern part, in ca. 5 m thickness. Therefore, the northern part of Inner Somogy could be considered as the aeolian erosional zone of the alluvial fan, which is characterized by blowouts, residual ridges and deflationflats, and where the residual sand had a coarser grain composition due to the deflation (Lóki 1981). However, the deflation flats are smaller and shallower than those in other blown sand areas. The specific landform of the area is the blown hole with a maximum depth of 8 m. The accumulational forms in the southern Inner Somogy are parabolic dunes: they are often superimposed on each other, and based on the latest OSL data, they are very young forms (Fig. 1.37).

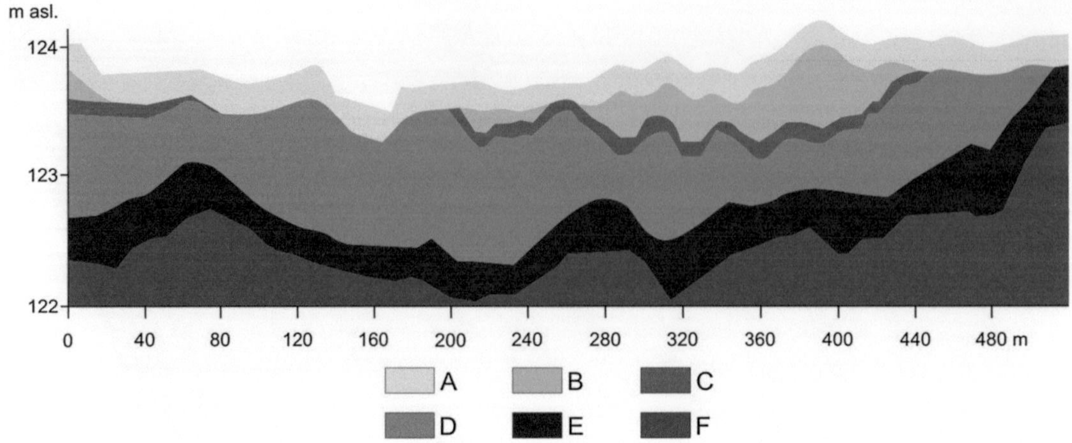


**Fig. 1.36** Blown sand repeatedly covered the original, approx. 2500-year-old surface, while in the intervals of the movements palaeosols were formed. As a result of the process, the original elevation differences of the surface eliminated (after Nyári et al. 2008). *A*—present-day ploughed surface, *B*—blown sand from the fifteenth century, *C*—soil from the Early Middle Ages, *D*—sand layers from the third to ninth centuries AD, *E*—Subboreal/sub-Atlantic palaeosol, *F*—Early Holocene blown sand

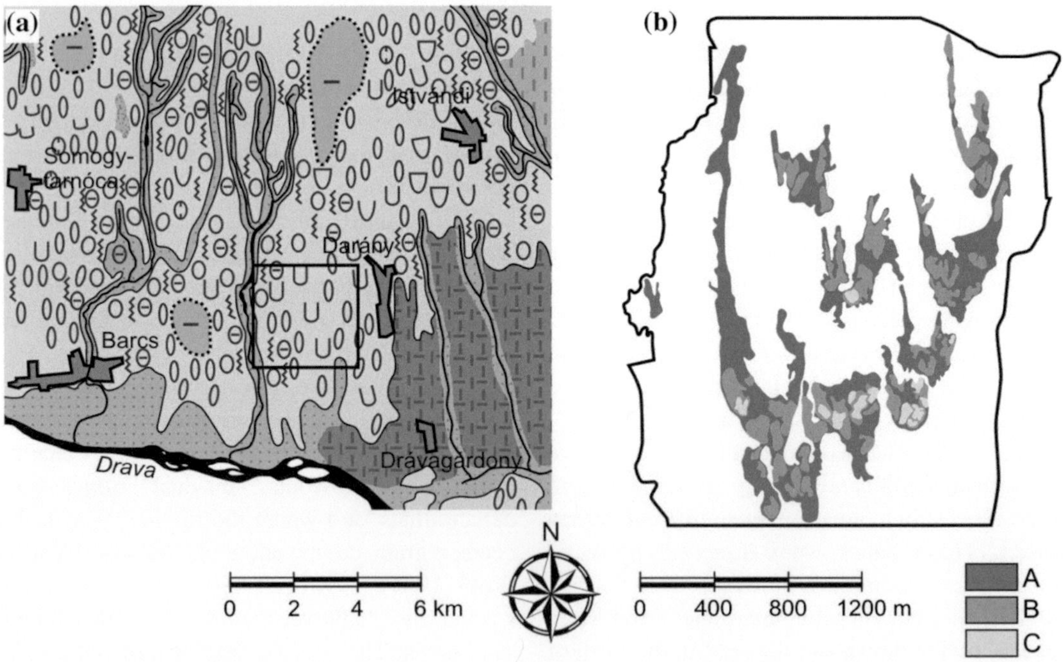


**Fig. 1.37** **a** The geomorphological map of South Inner Somogy shows various aeolian forms and N-S-oriented valleys that dissect the sand area (after Lóki 1981). **b** On the more detailed map, superimposed parabolic dune generations appear west of Darány, which have different ages: *A* 15–16 thousand years, *B* 11–13 thousand years, *C* 100–200 years old (drawn by Györgyövics)

**Image 1.13** Hummocks behind each other at Fülöpháza (photograph by Kiss, T.)

The surface of the alluvial fan of the Danube-Tisza Interfluve became increasingly drier due to the uplift of its central part; thus, an extensive aeolian activity could start in the Middle Würm and in the last and coldest stadial. Special characteristic of the region is the existence of large, flat surfaces, which were wetlands before water regulations. Between these flat areas, sand dune fields developed at elevated surfaces (e.g. dunes of Fülöpháza, Bugac and Orgovány; Image 1.13). The Danube-Tisza Interfluve is mainly characterized by blowouts, residual ridges and hummocks (Fig. 1.38), though hairpin-like parabolic dunes can also be found sporadically.

The aeolian development of the Nyírség region started when the northern parts of its alluvial fan became dry as a result of avulsions. Since its landforms are the youngest (from the Late Glacial) among the sand areas, the forms are integral and easily identifiable. Blowouts and residual ridges prevail in north, while parabolic and valley-marginal sand dunes in the south (Image 1.14). The thickness of the blown sand reaches even 32 m in the accumulation zones, especially in the central areas of the Nyírség, which is the thickest among all the sand areas (Borsy 1961, 1991).

The former valleys used to play an important role in the aeolian formation of the Nyírség, since they were continuously humid areas. Their wetness correlated with the timing of loss of their waterflow (Kiss and Bódis 2000). The valleys in the north Nyírség may have dried out completely in the time of the sand movements at the end of the Pleistocene, and sand dunes invaded them covering their alluvial deposits, whereas the valleys have remained wet until today at the eastern edge of the South Nyírség; thus, they could efficiently stop the migrating parabolic dunes, which accumulated on their edges, and in this way, the highest (up to 18 m) valley-marginal sand dunes of the Nyírség developed (Fig. 1.39).

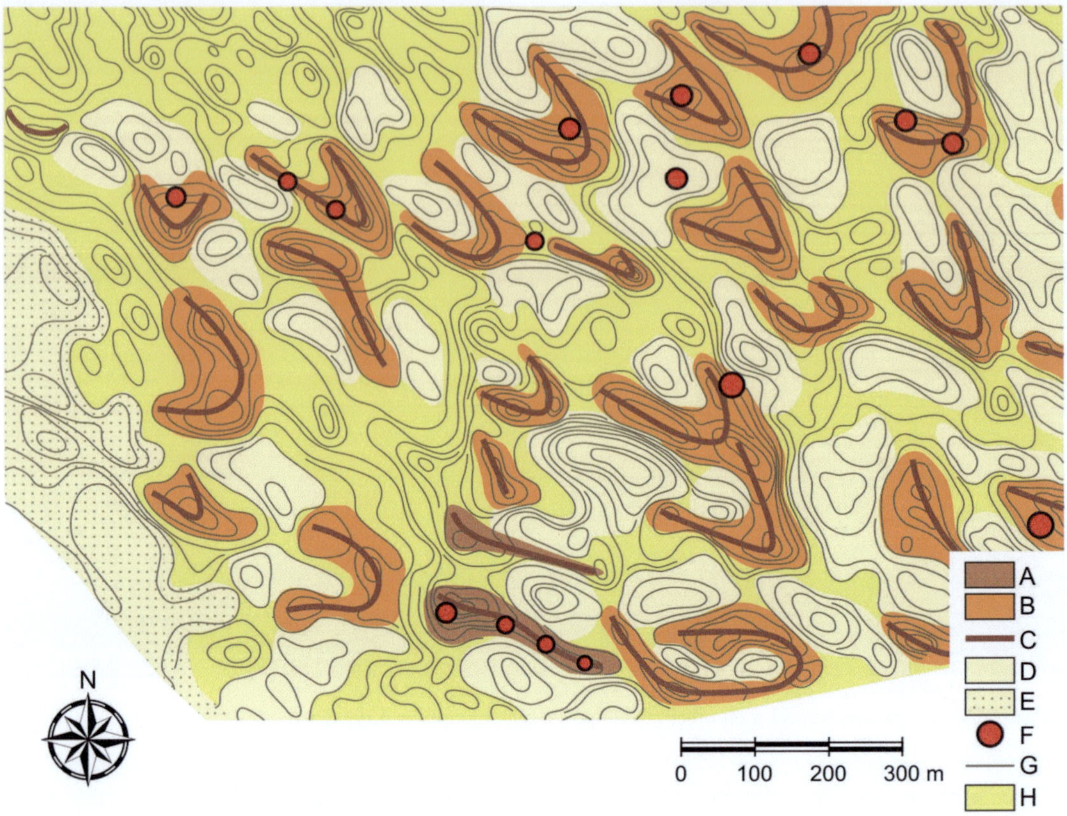


**Fig. 1.38** A geomorphological sketch of the Fülöpháza sand dunes. The hummocks and blowouts were formed by the NW winds, and they are arranged in accumulational zones (Kiss and Tornyánszki 2006). *A*—abandoned wing of a dune, *B*—hummock, *C*—brinkline of a dune, *D*—blowout, *E*—flat surface, *F*—postgenetic deflation, *G*—contour line, *H*—dune field

#### 1.5.2.4 Environmental Issues of Sand Areas

The harmful effect of deflation does not only involve sandy soils, although the blown sand surfaces are the most endangered, especially in the dry spring period. On the basis of erosion pin measurements carried out on experimental plots, it was detected that up to 4–10-cm material could be carried away during a single wind event, which equals 230 tons of material loss in a 50 × 50 m plot (Mezősi and Szatmári 1996). This may significantly damage the young arable crops (by corrosion or burial) and reduces the organic material and nutrient content of soils.

## 1.6 Periglacial Forms and Processes

Periglacial forms constitute the most characteristic landforms in the North Hungarian Mts., Transdanubian Mts., Mecsek Mts., Kőszeg Mts. and Sopron Mts. However, less spectacular periglacial forms could be found hidden and buried in the neighbouring plain and hilly areas. Székely (1977) estimated the rate of periglacial processes (e.g. frost action), and according to his measurements, ca. 40-m-thick correlative sediments were developed in the mountains; thus, these processes played major role in surface

**Image 1.14** Valley-marginal sand dunes in the South Nyírség (own photograph)

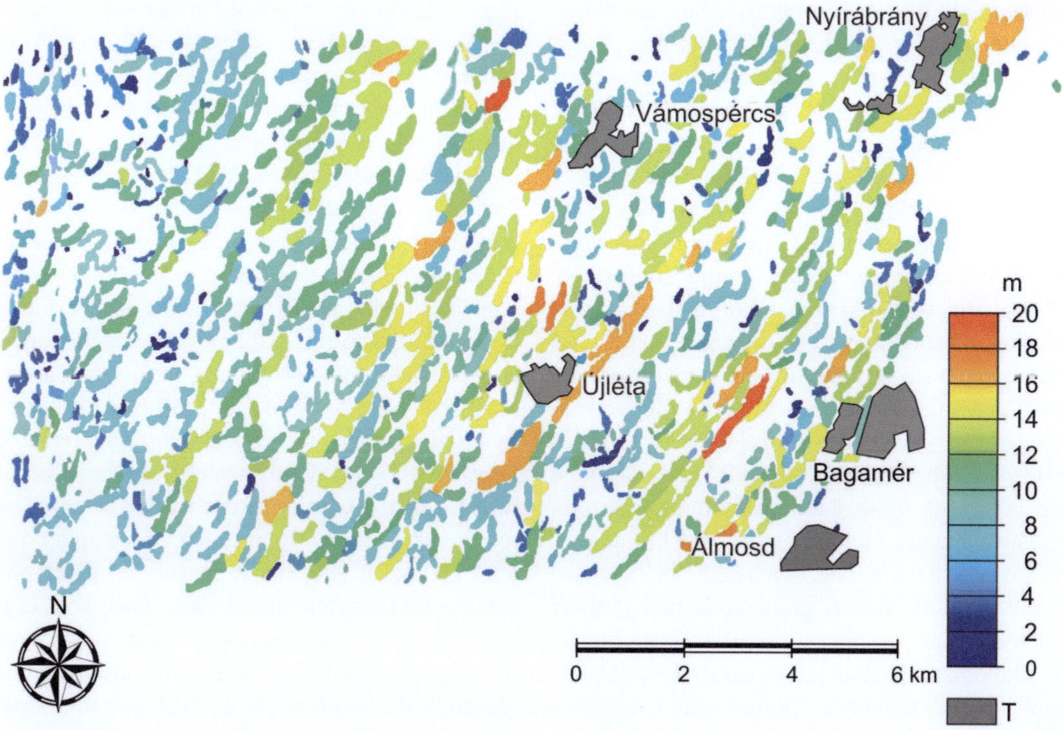


**Fig. 1.39** The relative heights of dunes in the South Nyírség show that they become higher towards south-east and thus towards the former river valleys. The straightened dunes are the valley-marginal dunes, while the parabolic dunes are strongly curved (Kiss and Bódis 2000)

development. The evolution of periglacial forms was determined by the cumulative effects of several factors: the climate had the greatest importance, though orography (gradient) and the lithology are also significant. The Pleistocene environmental conditions (see Sect. 1.8) were quite diverse, which were the result of a 10 °C mean temperature decrease in approximately half or third of the period. This climate did not support continuously frozen subsoil (permafrost) in the Carpathian Basin, though the landform types presuppose it (e.g. existence of frost wedges). This contradiction was solved by the latest researches, which proved the existence of a so-called sporadic frost in the basin (Schweitzer 1993; Gábris 2007). The Carpathian Basin was dominated by periglacial climate and processes in the second half of the Pleistocene, as the ice sheet expanding to the Brno-Prague line approached the basin; thus, the landforms developed under periglacial climatic conditions. The evolution of every surface and near-surface landforms was connected to humidity or, more exactly, to the fact that the freezing of water involved volume growth. In Hungary, Székely (1973, 1977, 1983) was the first scientist who draw the conclusion that the periglacial landforms had got arranged into groups conforming to the different climatic, morphological and lithological conditions. This means that if a form belonging to a certain group (e.g. a cryoplanation wall) is found, the existence of the other members of the given landform group (e.g. block field) could be supposed as well. To understand the origin of these landform groups, the prevailing climatic, morphological and lithological parameters have to be evaluated and to what extent. The theoretical problem is that if these parameters are too detailed (e.g. several climatic or lithological types are considered), too high and hardly understandable number of landform groups will be distinguished. Therefore, usually one periglacial and one pre-, post- and interglacial landform types are distinguished in connection with climate, and additionally, the relief is described by steepness of the slope. Pinczés (1977, 1986, 1995) also described and analysed the landform groups on slopes. Even if the names of forms applied by different researchers are dissimilar, both approaches emphasize the importance of lithological and climatic factors (mainly the alternation of freezing and thawing), and both derive the forms that developed on slopes from the source of the material, which could be, for example, the cryoplanation wall.

The mountainous regions were under the influence of strong cryogenic effects in the Pleistocene, which resulted in the formation of local polygonal tundra and cryoplanation terraces (Pécsi 1996). The rock surfaces exposed to cryogenic effects and frost action produced a large amount of debris which covered even the lower parts of slopes by block streams and block fields. Wind, run-off and solifluction may have transported the fine-grained material further. Without considering the lithological differences (e.g. volcanic or carbonate rocks were weathered), on the southern slopes, the following form assemblages developed (see Figs. 1.40 and 1.41).

On the upper parts of slopes, block fields and debris fields can be found, just under the cryoplanation walls (Image 1.15), or towers shaped further by frost action and aeolian activity (Székely 1977; Pinczés 1986). The wind erosional forms are typical in caldera remains in the North Hungarian Mountains, where corrasion resulted in spectacular forms in the former caldera edges. However, the same forms can be found in lower orographical situations too, with a different lithological character. The frost action created large blocks and debris which may have accumulated at the bottom of the wall. The cryoplanation wall gradually receded, consuming the affected rocks, and in the meantime turned the surface into a plain (this is the process of cryoplanation). A cryoplanation terrace may have developed in front of the wall under advantageous lithological conditions (e.g. existence of a bedding plane) (Fig. 1.41; Kiss 2006), which was covered by various types of block streams and block fields (Pinczés 1977; Székely 1983).

Block streams (Image 1.16) and debris cones developed of the fragmented material of the upper part of the slope, and their movement was

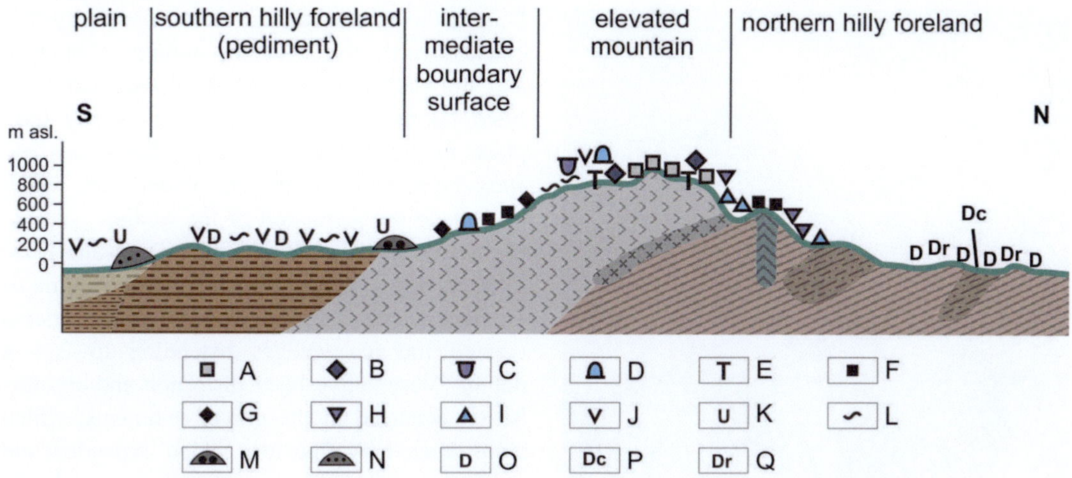


**Fig. 1.40** The typical periglacial forms of the volcanic mountains of the North Hungarian Mountains (after Székely 1977). *A*—autochthonous block field, *B*—stone polygon, *C*—large-sized stone bag, *D*—rock barrier, *E*—rock formed by frost action and wind erosion, *F*—debris cover, *G*—mixed slope cover, *H*—block stream, *I*—block field, *J*—large frost wedge, *K*—frost pockets, *L*—cryoturbation, *M*—alluvial fans of coarse material, *N*—alluvial fans of finer material, *O*—derasional valley, *P*—derasional ridge, *Q*—wide derasional ridge, *1*—Tertiary sedimentary rock, *2*—Middle Miocene rhyolite tuff, *3*—weathered Miocene subsurface volcanic formation, *4*—Miocene volcanic formation, *5*—Sarmatian and Pannonian clayey and sandy sediments, *6*—Pleistocene clayey and sandy sediments

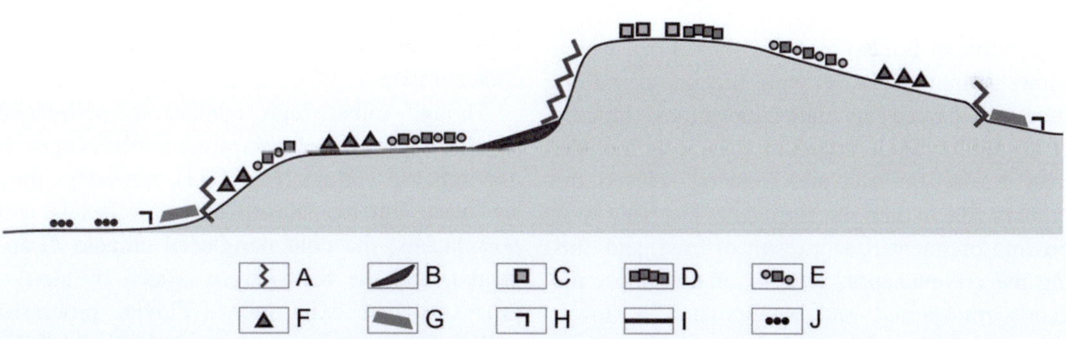


**Fig. 1.41** Cryoplanation form assemblage and its spatial distribution in the higher elevation zone of the North Hungarian Mountains (after Kiss 2006). *1*—cryoplanation wall, *2*—debris slope, *3*—autochthonous block field, *4*—debris field, *5*—mixed debris cover, *6*—debris cover, *7*—terraced slope, *8*—simple terrace on the slope, *9*—cryoplanation terrace, *10*—rock stripe

directed by gravity (and they are still active according to the analyses); thus, they got arranged depending on the gradient (e.g. larger blocks can be found on the upper slope sections). Debris cones are composed of the finer material, embedded in the lower parts of the slope. The location and form of the elluvium and the colluvium are determined by the presence of water. The debris cones can further be classified by their grain size distribution, position and morphology.

The block streams are typical in the lower parts of foothills with a smaller gradient, but the

**Image 1.15** Steep cryoplanation wall in the Zemplén Mountains (Sólyom-bérc) (own photograph)

cryopediment is also a typical landform, which shows some similarity with the cryoplanation steps and terraces in higher elevations (Pinczés et al. 1993). Their common feature is that the frost action (freezing and thawing) played the primary role in their development. The solid rock became fragmented as a result of frost, and during the cryoplanation process of the slope, the debris transported and accumulated at lower elevations. The typical forms on the gently sloping, lower slope sections are the frost wedges, the frost pockets and other cryoturbation landforms (also found elsewhere; Fig. 1.42). Frost wedges could be formed because of the low amount of precipitation (approx. 200 mm) and the low temperatures in the periglacial climate. During the winter, the ground became dry and cracked vertically, while in summer it thawed, and the melted water and the loose sediments filled up the crack. The thawing usually took place in the 2–4-m-thick active layer, and the sporadic permafrost remained underneath, which also made the formation of ice wedges possible (Schweitzer 1993; Gábris 2007). The water froze in the crack in winter, and the larger volume of the ice pushed the wedge apart. This process created the ice wedges extending to 2–4 m depths. Very active layer distortion and cryoturbation occurred in the gravel sediments, which were created by freezing, which expanded and uplifted the wet sedimentary strata (Image 1.17). Terrace II/a is characterized by smaller (ca. 1 m) frost pockets, frost wedges and pectinated soils, while in the material of terrace II/b-V, approximately 2–4-m-deep frost wedges, frost pockets, frost strips and polygons of 4–5 m diameter were identified (Pécsi 1997). Derasional (corrasional) valleys formed by areal erosion and mass movements at the edges of alluvial fans.

The landforms of slopes with northern exposure are usually not as diverse as the ones on the southern slopes. Here, fewer block stream occurrences, the consequently smaller rate of debris covers and the alluvial fans are characteristic.

Though rather high number of periglacial landform types developed ranging from pingos to assymmetric valleys (Fig. 1.43), nowadays they are often buried. During the interglacials and postglacials, the cold periglacial climate disappeared, and the forms were eroded by similar processes that exist today. Fluvial processes replaced the cryogenic processes because of the increasing amount of precipitation (however, erosional valleys developed on glacial climate too). Thus, V-shaped valleys developed, derasional valleys were reshaped by run-off erosion, intermittent gullies appeared, landslides became active, and blown sand forms evolved.

**Image 1.16** The block field at Badacsony Mt. (photograph by Kiss, T.)

## 1.7 Main Geographical Issues of the Hungarian Karst Areas

Approximately 1.5 % of the territory of Hungary is covered by karstic sedimentary rocks, but considering only the hilly and mountainous areas, their proportion is over 7 % (Fig. 1.44). If those karst areas are also considered, which are covered by sediments and soil, the proportion is even higher (see Fig. 3.25). For the general public, karst refers only to the small area of bare karst surfaces, since their ecological, aesthetic, vegetational, hydrological and geological value is easily justifiable. However, the surface and subsurface karst forms, the karst water aquifers used for drinking water are sensitive, and their contamination could lead to their drastic functional transformation. Thus, several karst forms (e.g. thousands of caves) are under *ex lege* protection by the Nature Conservation Act.

In the Carpathian Basin, karstic carbonate sediments have formed in various periods from the Devonian to the Pannonian Period, and travertine has been forming up to now. The Mesozoic (mainly Triassic–Jurassic) and the Tertiary (Eocene–Miocene) limestones and dolomites must be taken into consideration when discussing the karstic landforms. These Mesozoic surface and near-surface formations characterize the Bükk Mts., the Aggtelek Karst, the Rudabánya Mts. and West Mecsek Mts, while Tertiary rocks dominate in Transdanubia (Chaps. 10–12). These sedimentary bodies have

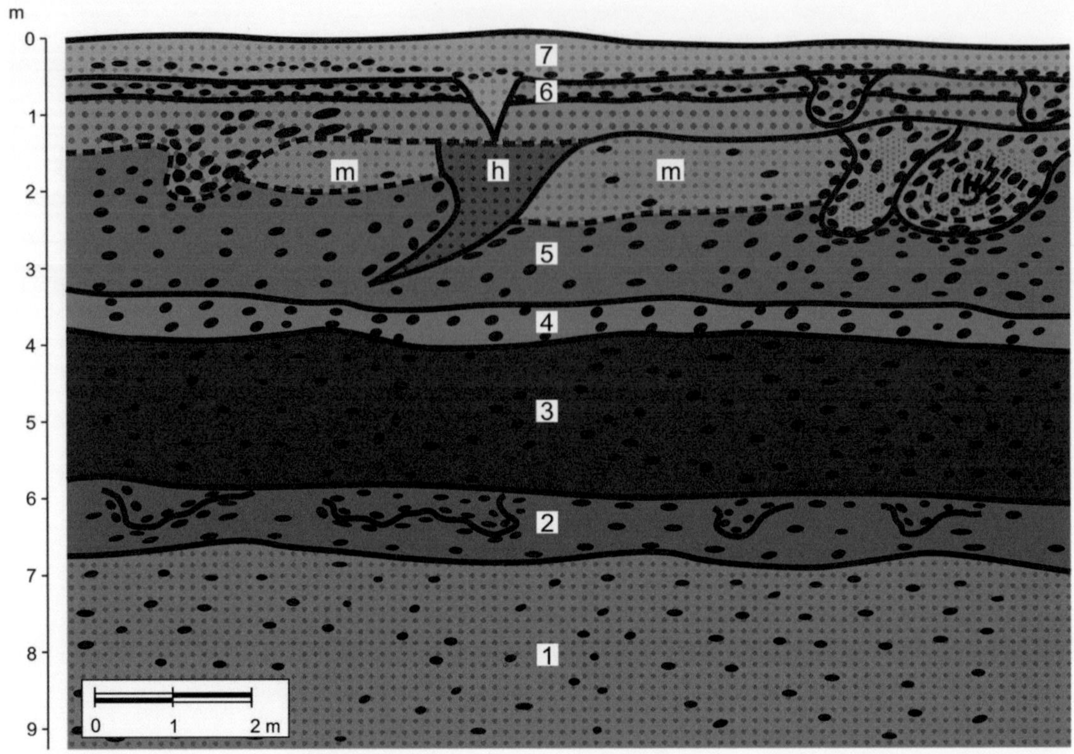


**Fig. 1.42** Periglacial forms of the oldest fluvial terrace with frost wedges and cryoturbation in Budapest (after Pécsi 1996). Günz gravel: *1*—horizontally bedded sandy gravel, *2*—slightly cryoturbated reddish brownish gravel, *3*—horizontally bedded yellow sandy gravel; Mindel gravel, *4*—discordant red brown gravel, *5*—gravels with frost wedges and frost pockets embedded in clay, younger sediments, *6*—small gravel pockets and frost wedges in sand, *7*—recent wind-blown sand, *m*—sandy dolomite powder, *h*—frost wedge filled up by carbonated sand

changed their latitude quite considerably (see in detail: Sects. 1.1 and 1.2). For example, the Triassic limestone of the Bükk was formed in the southern platform of the Tethys several thousands of kilometres away from its location today, under significantly different temperature and ecological conditions. These formations may have subducted, became covered by non-karstic sediments, then exhumated and eroded during their several-million-year-long evolution history. All together, this is the lithological framework in which the karst forms could and can develop.

It is difficult to draw a comprehensive description on the formation and evolution of the Hungarian karstic areas, because of their multidirectional movement and several hundred million years of development history. For example, although the facies of limestone and dolomite of the Mecsek and Villány Mts. were deposited ca. 1000 km away from the material of the Transdanubian Mountains, their form assemblage is similar (Hevesi 1991a), because the karstic landforms were formed in a more homogeneous environment. Wide information on the karst development of the Hungarian carbonate regions was supported by the geologists, who mapped these regions to support the extensive mining of manganese and bauxite in the 1960s and 1970s. Thus, four main periods were identified prior to the Quaternary, when palaeo-karst features developed in the Carpathian Basin (Korpás 1998): 1. Lower Cretaceous, 2. Upper

**Image 1.17** Layer distortions in the material of Danubian terrace V. in the gravel pit of Kistarcsa (own photograph)

Cretaceous–Eocene, 3. Early Tertiary and 4. Sarmatian–Lower Pannonian. The kartification during the Holocene and the Pleistocene was not considered, because in the Quaternary karst formation had a different activity, for example because the surface was bare. Other researchers distinguish seven karstification phases (Juhász et al. 1995; Fig. 1.45). It is obvious that different landform groups developed under different orographical, environmental (precipitation, temperature, vegetation), or hydrological circumstances, which landforms may get disappeared or in contrary got preserved. The groups of karst forms can slightly differ conforming to the different environments in the different locations.

### 1.7.1 Main Regional Characteristics of Karstification

Karstification takes place under clearly defined lithological, topographical, climatic and hydrological conditions both on the regional and on the

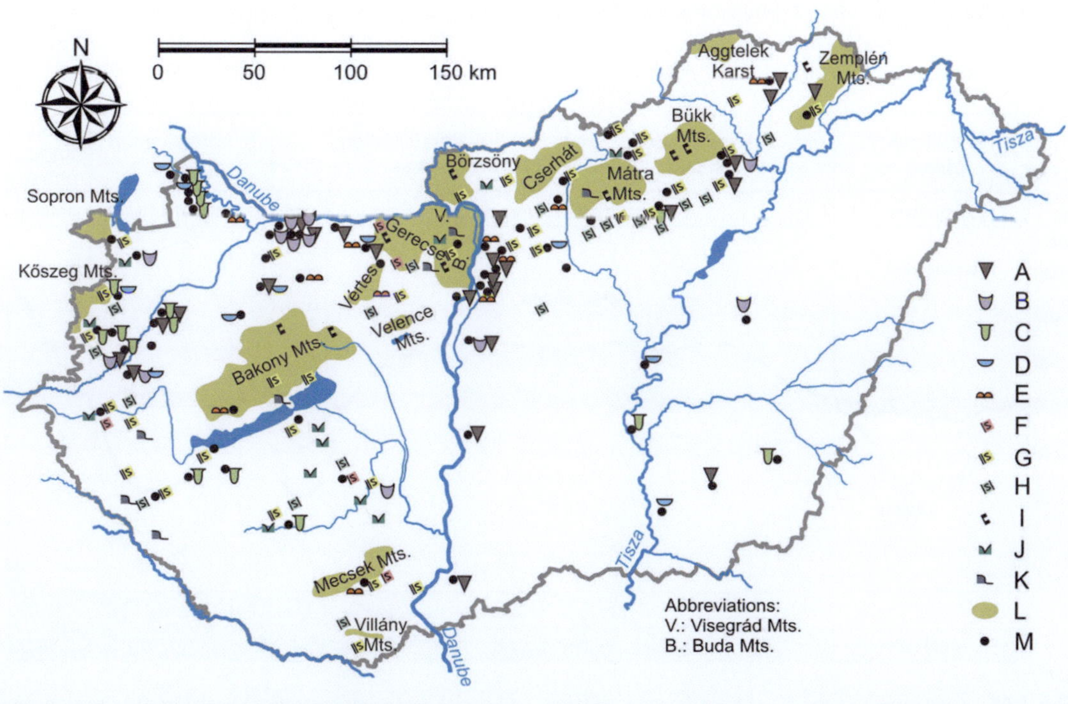


**Fig. 1.43** Pleistocene cryoturbation and other periglacial phenomena in Hungary (after Pécsi 1996). *A*—frost wedge, *B*—frost pocket, *C*—cauldron-like regular polygon, *D*—cryoturbated subsoil, *E*—cryoturbation impacts, *F*—traces of solifluction on slope, *G*—solifluction sediments on slope, *H*—solifluction sediments in general, *I*—periglacial blocks, *J*—asymmetrical valleys due to periglacial processes, *K*—cryoplanation terrace, *L*—mountains, *M*—main study sites

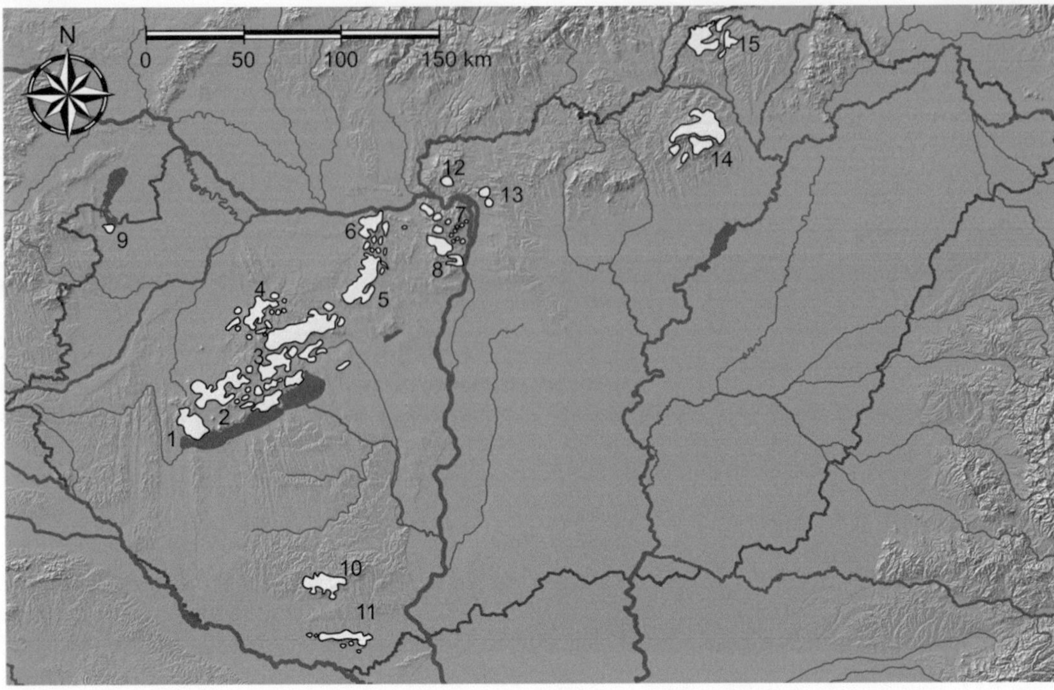


**Fig. 1.44** The surface karst areas of Hungary. *1*—Keszthely Mts., *2*—Balaton Uplands, *3*—South Bakony Mts., *4*—North Bakony Mts., *5*—Vértes Mts., *6*—Gerecse Mts., *7*—Pilis Mts., *8*—Buda Mts., *9*—Sopron Mts., *10*—Mecsek Mts., *11*—Villány Mt., *12*—Szokolya Basin, *13*—karst of the Cserhát Region, *14*—Bükk Mts., *15*—Aggtelek–Rudabánya Mts.

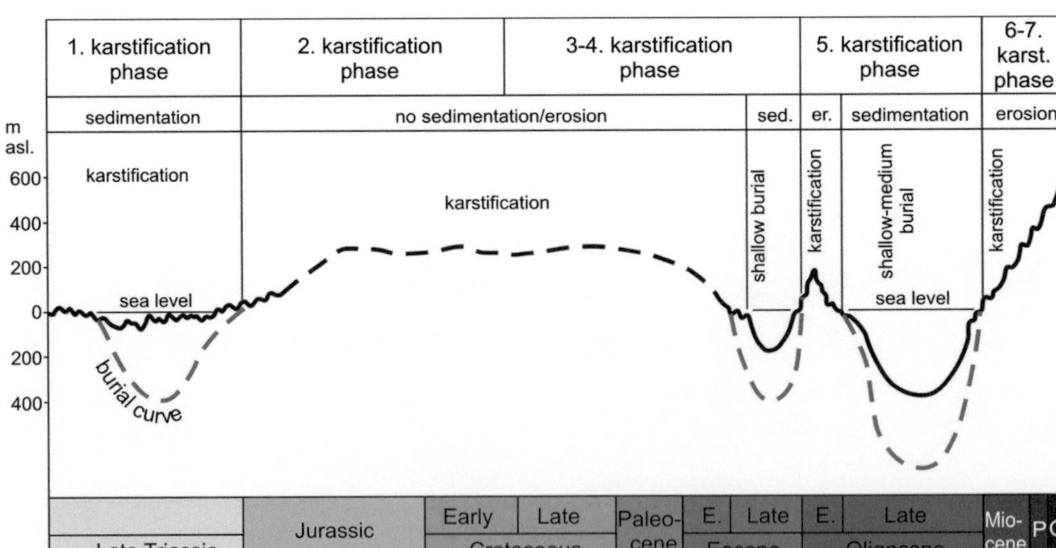


**Fig. 1.45** Karst evolution model of karstification (erosion) and burial (sediment formation) based on the development of the Naszály Mt. (after Juhász et al. 1995)

local scales (Jakucs 1971). In Hungary, the most typical karstified rocks are limestone and dolomite sediments. They have ca. 1.5 % (primary) pore volume on average, which is important from the point of view of karst water storage. Karstification is a dissolving process, which increases the porosity of the involved zone. The Paleozoic carbonates of Transdanubia were karstified when they were on the surface, resulting in a porosity of about 8–10 %. After these carbonate bodies sunk and became buried, they became effective natural oil storages; therefore, Hungarian oil mining started in Budafapuszta in 1937 from Middle Miocene calcareous carbonate sediments. The slightly different lithological composition of the karstified blocks (e.g. diverse chemical composition of limestone or dolomite) results in different rates of dissolution. The recrystallization of dolomite or the significant deviation of the $CaCO_3/MgCO_3$ ratio reduces the efficiency of dissolution. However, it does not result in different forms, just their morphometric changes according to the differences in lithological facies at micro- and meso-scales.

The influence of the relief on karstification begins when the vadose (infiltration) zone and the phreatic (permanently filled with water) zones get separated from each other (Fig. 3.25), when the area gets elevated. Accordingly, karstification could have started in the surface and near to the surface after the elevation during the past few million years. The elevation of the Hungarian mountains started in the Pliocene, and it became more active since the Pleistocene. Altogether the karstic blocks uplifted by 200–300 m; thus, newer cave levels developed, and the older ones have started to decay (Hevesi 1991a). Thus, the oldest spring caves can be found at the highest elevations; for example, one of the oldest caves is the decaying Körös-Lyuk in the Bükk Mts. (930 m asl), or in the Esztramos (see also Sect. 12.2.), where the upper caverns have already been exploited by mining. Jánossy (1979) found a Middle Pliocene vertebrate fauna in these karst caves; thus, these landforms can be the oldest forms of that karstification period. The micro-scale changes in cave levels are more difficult to identify, although several examples on new karstification levels are known from the mountains; thus, multilevel caves developed (e.g. the main and the lower cave of the Baradla Cave System), or sinkhole generations evolved in this way.

The intensity of karstification is mostly influenced by climatic (principally precipitation) and the biogenic (e.g. affecting the soil) factors. Thus, under a colder climate or in the periods of Pleistocene glaciations, though the solute content of karst waters was higher, the intensity of the process was an order of magnitude lower than under sub-Mediterranean circumstances, because lower amount of carbon dioxide is produced by the vegetation and in the soil (Jakucs 1971). Therefore, the rate of karstification decreased in glacials when the climate was drier and colder than today, and frost action and (geli)solifluction became the dominant processes. The local effect on landform development of the climate–vegetation relationship is also well known: for example, the elongation of the dolines is typically different because of the different environmental conditions of the eastern and the western slopes (Fig. 1.46). In a warmer and more humid environment with dense vegetation, the processes accelerate (Zámbó and Weidinger 2006). In fact, the erosion rate is defined by the several influencing factors to different extents. According to the measurements of Zámbó and Ford (1997), the rate of karst formation is different within the Aggtelek Mts. (though they are influenced by the above-mentioned orographical and vegetational factors); thus, the dissolution rate of plateau surfaces is 7 mm/1000 years, and it is 6 mm/1000 years in the cases of slopes, 10 mm/1000 years in the edges of the doline bottoms, and 13–16 mm/1000 years in the permeable doline bottoms.

Some researchers dated the active karstification back to the Pliocene and Miocene, and they claim these were the periods of the highest intensity, and the process declined during the Quaternary (Jámbor 2001). Very favourable climatic conditions for intensive karstification were present in the Pliocene, and karst forms remained from this period too. However, the orographical, tectonic and lithological conditions of a given

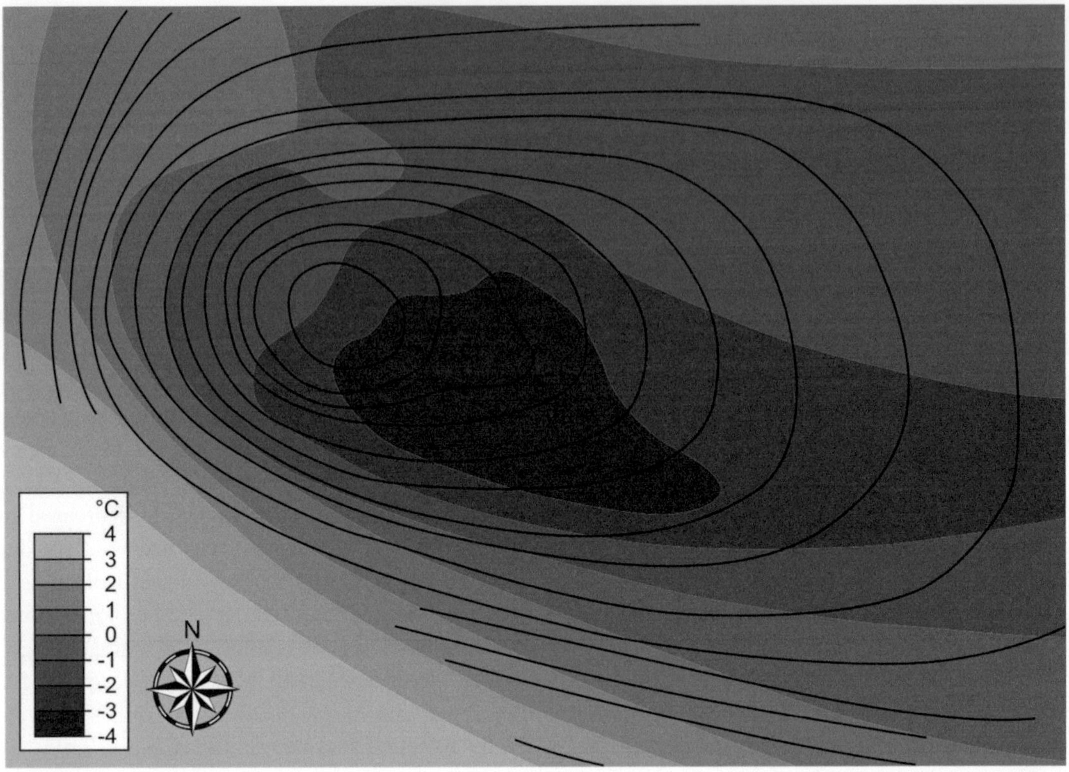


**Fig. 1.46** The different heat household of doline sides with eastern and western exposure is indicated by the night isotherms (the black lines are 1.0 m contour lines; after Keveiné 2009)

block at a given period are not well known. The water of the phreatic and vadose zones became orographically separated by Pleistocene uplift, resulting in accelerated dissolution in the boundary zone due to the intensive corrosion of the mixing waters. Simultaneously, the tectonic uplift resulted in intensive erosion of the sedimentary cover; thus, some blocks became open karst surfaces, while the evolution of the subsurface karst forms also depended on the surface erosion, as quartz gravel could be transported to the caves.

### 1.7.2 The Hungarian Karst Types

Worldwide, several different classifications of karst landscapes exist. The most obvious typology is based on climatic, lithological and pedological factors in the global and regional scales. In general, on the basis of the climatic factors, different karst landscapes are distinguishable, as tropical conical and tower karsts, karsts of a temperate zone and forms of a Mediterranean or of a cold climate. These types show conspicuous differences in mean different denudation and evolution rates and also in forms. The relationship between lithological conditions and karst denudation is more simple. The typology may be based on whether the karst is covered by soil or it is bare (open). To create a local, landscape-based typology is more difficult. In Hungary, two main karstic form groups were identified, in which both belong to the temperate zone. The general genetic classification of karst form assemblage was made by Jakucs (1977), though some similar works have been written since then (e.g. Hevesi 1991a).

The two main Hungarian karst types are the Aggtelek-type karst (its subsurface features are called allogenic by Jakucs) and the Transdanubian-type karst (or called Bakony-type karst by Hevesi and autogenic by Jakucs). Both have several unique genetic and morphological features.

The members of the **Aggtelek-type karst** group are the Aggtelek-Rudabánya Mts, the majority of the Bükk Mts. and the West Mecsek Mts. (on the regional scale). These members commonly dispose of rich surface and subsurface karst forms. On the surface, dolines are the prominent landforms. Independently of the lithology, the plateau dolines (not connected to former valleys or sinkhole series, they are in higher elevations, and developed by solution processes) cover one-third of the total surface of the Aggtelek Mts., while valley dolines developing in a line cover its quarter or one-fifth. The largest dolines emerged into uvalas characterized by a diameter of hundreds of metres, a depth of 20–25 m, and their bottom can be filled by terra rossa and terra fusca. These impermeable sediments can result in the formation of lakes (Image 1.18). In the Bükk (e.g. Nagy-mező, Létrás), polje-like karst hollows with uvalas and extensive dolines were also identified (Hevesi 1991a, b). The West Mecsek Mts. are characterized by smaller, but more abundant dolines.

Different data and approaches exist on the age of dolines in the Hungarian karst areas. Their formation may have started in Aggtelek as early as the Würm, which statement is supported by the calculations of corrosion intensity (Zámbó 1998; Telbisz 2004). In the Bükk at the Early Pleistocene, the formation of sinkholes started, as well as the development caves and the transformation of sinkholes into dolines; thus, the solution dolines evolved much later (Hevesi 1991a, b).

Dry karstic valleys are typical landforms of the Aggtelek-type karst. They used to drain water as early as the Pliocene or sometimes at the end of the Miocene, and when the karstic block uplifted,

**Image 1.18** Lake Vörös developed in a doline near Aggtelek (own photograph)

they became dry valleys and gave way to the formation of dolines at the bottom of the valley.

The sinkholes ensure the linkage between the surface (intermittent or permanent waterflows) and the subsurface systems. A sinkhole can be formed at the bottom of a doline with a larger watershed (indicated by the fact that the convex doline wall transformed into concave). In the Aggtelek-type karsts, sinkholes usually are situated at the boundary of karsts and non-karstic rocks, and they are connected to caves, whereas they follow the faults and the cracks of rocks in the Transdanubian-type karsts.

The occurrence of karren fields does not depend on the karst type; rather, it refers to intensive erosion, which exhumed the subsurface stones developing further as karr. The final karren forms are determined by the rate of different processes, as run-off and patchy vegetation (Veress 2004). Near Aggtelek, a karren field developed over a lake formed by ponor clogged with debris (Image 1.19). This area was indicated as arable land (in archives, there are notes about a vineyard) on the I. Military Survey map made in 1784. Probably, the area was ploughed—although it must have been difficult, because the slope has a high gradient—and this was exactly the cause of intensive erosion, which finally led to the abandonment of the land (Nagy 2004). According to this approach, the—often human induced—intensive erosion exhumed the subsurface karst forms in other places too.

Subsurface karst forms have also high diversity. The infiltered waters become carbonate-saturated quickly; thus, they unable to dissolve any material any more—apart from a few exceptions—but the dropping water is able to accumulate dripstones and travertine dikes. The exceptions are the result of the mixing of different karst waters, which may lead to increasing dissolving capacity in the boundary zone. The caves of the Aggtelek-type karsts are often eroded by waters and the transported bedload originating from non-karstic catchment (Image 1.20). The erosional marks of such subsurface streams are recognizable in the majority of caves.

**Image 1.19** Lake Aggtelek and the neighbouring karren field (own photograph)

**Image 1.20** The sinkhole of Zombor-Lyuk joins the Baradla Cave in its central section. During floods, the stream transports Pannonian gravel into the cave from its non-karstic watershed (own photograph)

Nowadays, the majority of the caves are in a state of decay, because they are increasingly filled in by carbonate materials (e.g. various types of dripstones). The Baradla and Béke caves belong to this category (Table 1.8). The caves of the Aggtelek-type karsts are not significantly preformed by tectonic faults; thus, thermal water activity appears only sporadically. The pollen analysis of the alluvium protected by stalagmites and accumulated on the "alluvial" terraces in the caves of Aggtelek (e.g. Baradla Cave and Béke Cave) indicates Würmian aggradation; thus, the most active erosion of the caves could be dated to the Riss-Würm Interglacial or to the Early Würm. The formation of the oldest caves can be dated back to the beginning of the Pliocene. (The Diabáz Cave is still an active cave filled by water at 870 asl; Hevesi 1984.) The intensive karstification in the caves could start when the blocks were elevated in the Pleistocene. The formation of the karst forms in a higher altitude in the Bükk Mts. was hindered by the fact that the surface had been covered by mostly Miocene sediments. Similarly, the formation of the decaying caves of Esztramos is dated to the Pliocene on the basis of fossils (Kordos 1984).

Another typical subsurface landform is the shaft (vertical) cave, which was created by infiltrating precipitation along cracks corroding the limestone. These do not function as sinkholes, but their water conducted into the main karst water bodies, and then, their water could appear in springs of the surrounding areas. In Hungary, the shaft caves are large; for example, the largest ones are the Istvánlápa Cave (253 m deep) and the Vecsembükk Shaft Cave in Alsó Hill (236 m deep; Fig. 1.47).

The **Transdanubian-type karst** areas are characterized by much more modest surface landforms and different subsurface karst forms as it was mentioned above. It could be explained by their different geological structures (predominance of dolomites, extensive tectonic impacts and non-karstic cover) and their lower elevation. The Cretaceous-Eocene palaeo-karst forms are also preserved by the cover sediments. The

**Table 1.8** The longest caves of Hungary

| Name of the cave | Mountain | Length of the cave (m) | | | |
|---|---|---|---|---|---|
| | | 1977 | 1987 | 1998 | 2011 |
| Baradla-Domica Cave system | Aggtelek Mts. | 25,000 | 23,916 | 23,916 | 23,916 |
| Mátyás-hegyi-Pál-völgyi Cave system | Buda Mts. | 4200 | 4770 | 5152 | 28,600 |
| Pál-völgy Cave (when it was known as a separate cave) | Buda Mts. | 1200 | 6415 | 12,400 | |
| Béke Cave | Aggtelek Mts. | 8743 | 8743 | 6408 | 7183 |
| István-lápa Cave | Bükk Mts. | 2940 | 4100 | 6000 | 6000 |
| József-hegyi Cave | Buda Mts. | | 4300 | 5500 | 5486 |

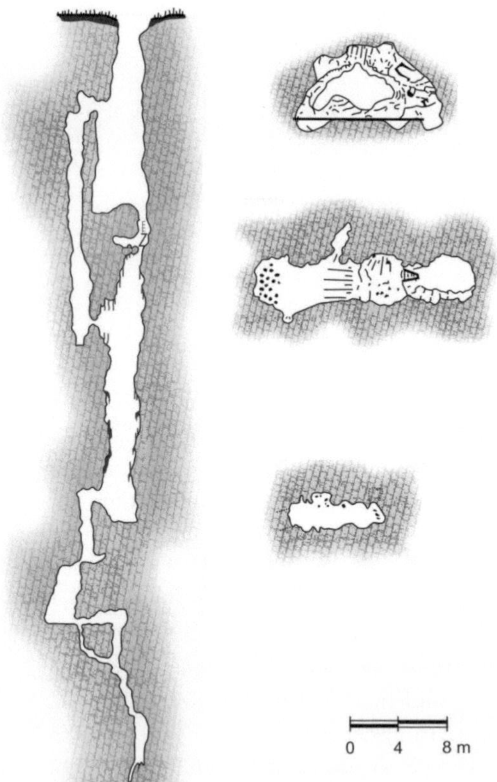


**Fig. 1.47** The vertical and cross sections of the Vecsembükk Shaft Cave (after Kósa 1992) with tenfold vertical distortion

surface forms have smaller size than those of the Aggtelek-type karst areas, partly because of the abundance of dolomites; thus, there are smaller and rarer dolines, undeveloped dolines and uvalas. At many places, the karstification is indicated by only karren forms, e.g. in Vértes Mts., in Pilis Mts. (connected to strata heads) or in the Villány Mts. The karrens are the basic forms of dissolution (Veress 2004), and they show that the karst surface is transforming, and thus, their presence does not depend on the karst type.

In case of the Transdanubian-type karst, two main subsurface processes have to be highlighted in cave evolution. One is the strong tectonic preformation of the caves: the structural lines determine the directions of cave passages; thus, they have NW-SE directions and perpendicular to it, fitting to the European main fault-line directions. This characterizes all subsurface karst forms belonging to this type; therefore, the caves have chessboard-like pattern (Fig. 1.48).

The other main subsurface processes are the direct corrosion induced by thermal waters in connection with postvolcanic activity and the dissolution through chemically aggressive solutions enlarging the cave systems. As a result, the cave systems have unique spatial structure, which sometimes consist of a series of spherical caverns (Fig. 1.49). The growth model of spherical caverns was elaborated by Szunyogh (1982). This thermal water welling and reshaping of the passages appear in all members of the Transdanubian mountains, though spherical caves have not been formed widely, but the canals of thermal waters (e.g. Ferenc-hegyi

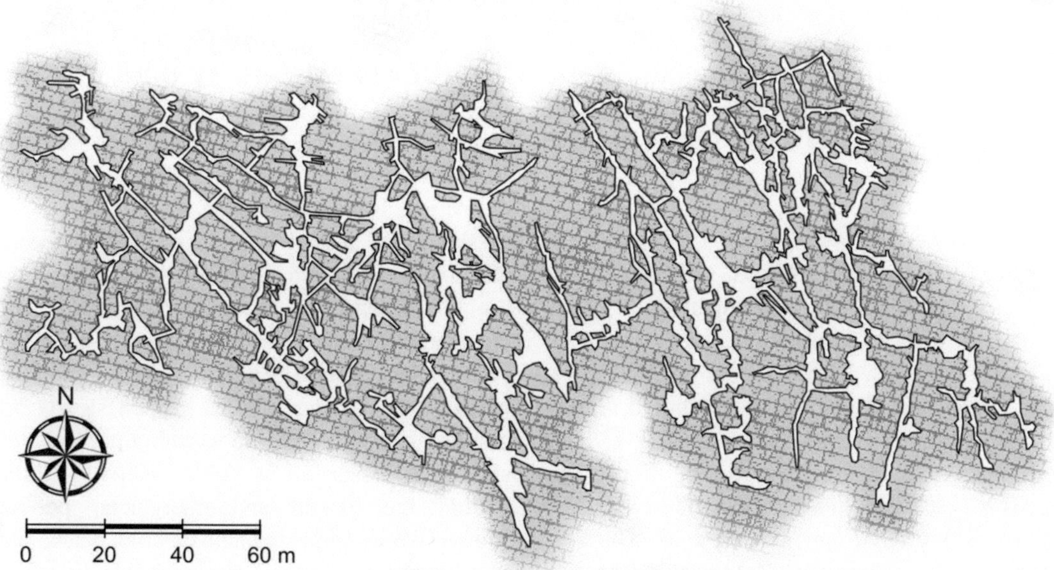


**Fig. 1.48** The map of Ferenc-hegy Cave (after Fleck and Hazslinszky 1998)

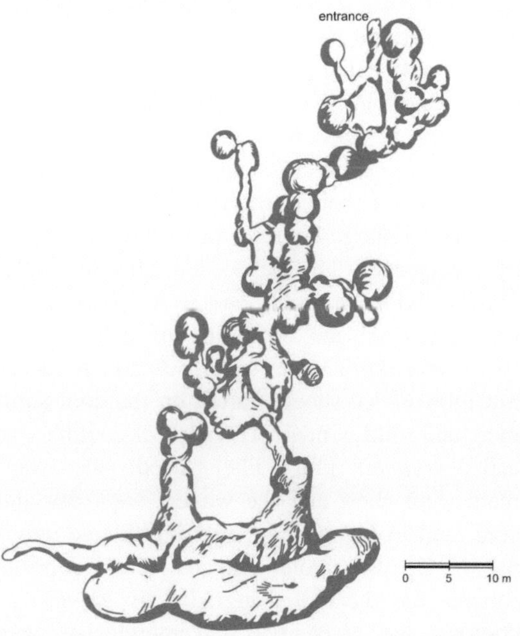


**Fig. 1.49** The model of the spherical cavern of Sátorkőpuszta Cave (after Jakucs 2000)

Cave) and crystals precipitated on the walls (e.g. Szemlő-hegyi Cave) can be found everywhere.

Transdanubian-type karst occurrences are in the Keszthely Mts., Bakony Mts., Vértes Mts., Gerecse Mts., Pilis Mts. and Buda Mts., and in certain parts of the Bükk Mts. In the Bakony Mts., the typical karst forms range from ponor dolines (17 could be found in the Tési Plateau), through karren fields referring to erosion (typically in the Northern Bakony Mts.), to caves and thermal water cavities (e.g. Cserszegtomaj Cave, the Lake Cave of Tapolca). The state of the Lake Cave of Tapolca is being restored slowly after the termination of industrial water extraction (Mezősi 2008). The Vértes Mts. mainly consists of Triassic dolomite; therefore, karst forms are rare; however, the dissolution activity of thermal waters along faults appears (e.g. the spring well at Csókakő; Hevesi 1991b). The Gerecse is rich in karst cavities, characterized by long rows of spring caves and avens in higher elevations. The karstic springs are accompanied with travertine deposits, which are important from the archaeological and palaeo-ecological point of view, e.g. Vértesszőlős. Extensive karren fields characterize the surface of the Pilis Mts., though the ruins of fissure (tectonic) caves (rock gates, passage caves) are also appear. Its characteristic caves are formed by partly or entirely thermal waters (Sátorkőpuszta Cave, Legény Cave, Leány

Cave). The Buda Mts. are the most fractured and dissected units in the Transdanubian Mountains, they are poor in surface forms, partly due to urbanization, however, they have rich subsurface karst forms. The caves formed along tectonic fault-lines and have thermal water origin.

### 1.7.3 Some Questions of the Anthropogenic Use of Karsts

The karstic landscapes are sensitive and vulnerable ecological systems (Keveiné 2009). They are susceptible to surface erosion, disadvantageous vegetation changes, resorption of dripstones or any environmental changes. The cold karst waters containing $CaMgHCO_3$ deserve special attention, because these are very good drinking water sources due to their dissolved mineral content. The rate of seepage is 1.4 million $m^3$/day, and that of water withdrawal is 0.25 million $m^3$/day. The sensitivity of karst water budget and the near-surface pollution are the greatest problems. The karst areas are especially sensitive from the point of view of the state of subsurface water where the karstic sediments are on the surface or maximum 10 m underneath (219/2004. Government Regulation).

The Hungarian caves also played an important role in the evolution of humans, since several served for dwelling, as it was proven by archaeological researches. The Pilis Mts. have been extensively researched in Transdanubia and the Bükk Mts. in the North Hungarian Mountains (Kordos 1984). We have to highlight the Suba-Lyuk Cave (Hór Valley, Bükk Mts.) where 61–71 thousand-year-old Neanderthal fossils were found, and the Szeleta Cave (Szinva Valley, at 350 m asl) inhabited throughout 100 thousand years from the beginning of the last interglacial. Traces of inhabitance have been found at two levels in the cave of Istállóskő Mt. at 600 m asl: at the lower cave level, ca. 36 thousand-year-old finds of the Neanderthaler man were discovered, and from the upper level came the 30 thousand-year-old finds of the Cro-Magnon man (representing the Bükk Culture).

## 1.8 Main Features of Climate, Vegetation and Fauna of the Quaternary (Kiss, T.)

During the Tertiary, the mean temperature of the Earth fluctuated around 20 °C, but at the end of the period, it started to decrease; thus, it became 12 and 13 °C at the beginning of the Pleistocene. Relatively great climatic fluctuations characterize the Pleistocene: the mean surface temperature changed by 8–10 °C between the glacial and the interglacial periods (within 10–20 thousand years). The cooling used to happen slowly and gradually; in contrary, the warming periods were shorter (Császár et al. 2003; Kerényi 2008). Within the glacial and interglacial periods, shorter warm (interstadials) or cold peaks (stadials) developed (Fig. 1.50 and Table 1.5).

In Hungary, different climate types appeared during the Pleistocene, as subtropical, subarctic, glacial and the Mediterranean climate types alternated. Their sequence resulted in ice-free interglacial and interstadial periods, or glaciated glacial and stadial periods. These have recurred three times in the past 130 thousand years according to the model of Milankovics and Bacsák, which based on the data of the orbital and rotational changes of the Earth (Pécsi 1993).

Several Hungarian researchers have attempted to reconstruct the Pleistocene climate. Bulla (1964) divided the cold and warm periods into two types. At the beginning of cold periods (*kryotroph*: kryo = ice, troph = feed), the accumulation of ice sheets started on the cool summers and mild winters. The peak of cooling was called *kryophil* (phil = like) period, when summers were short and the winters were dry and cold, and in Hungary, periglacial climate dominated. On this cold climate, loess deposits formed and the rivers aggraded by gravels. In contrary, the interstadials and interglacials were mild and warm periods, which were also divided: the *hyleotroph* (hyleo = forest) periods were characterized by a Mediterranean type climate with warm summers, mild winters and intensive chemical weathering, while the *kryophag* (ice consuming) periods were characterized by hot summers and dry and cold winters. Rónai (1973)

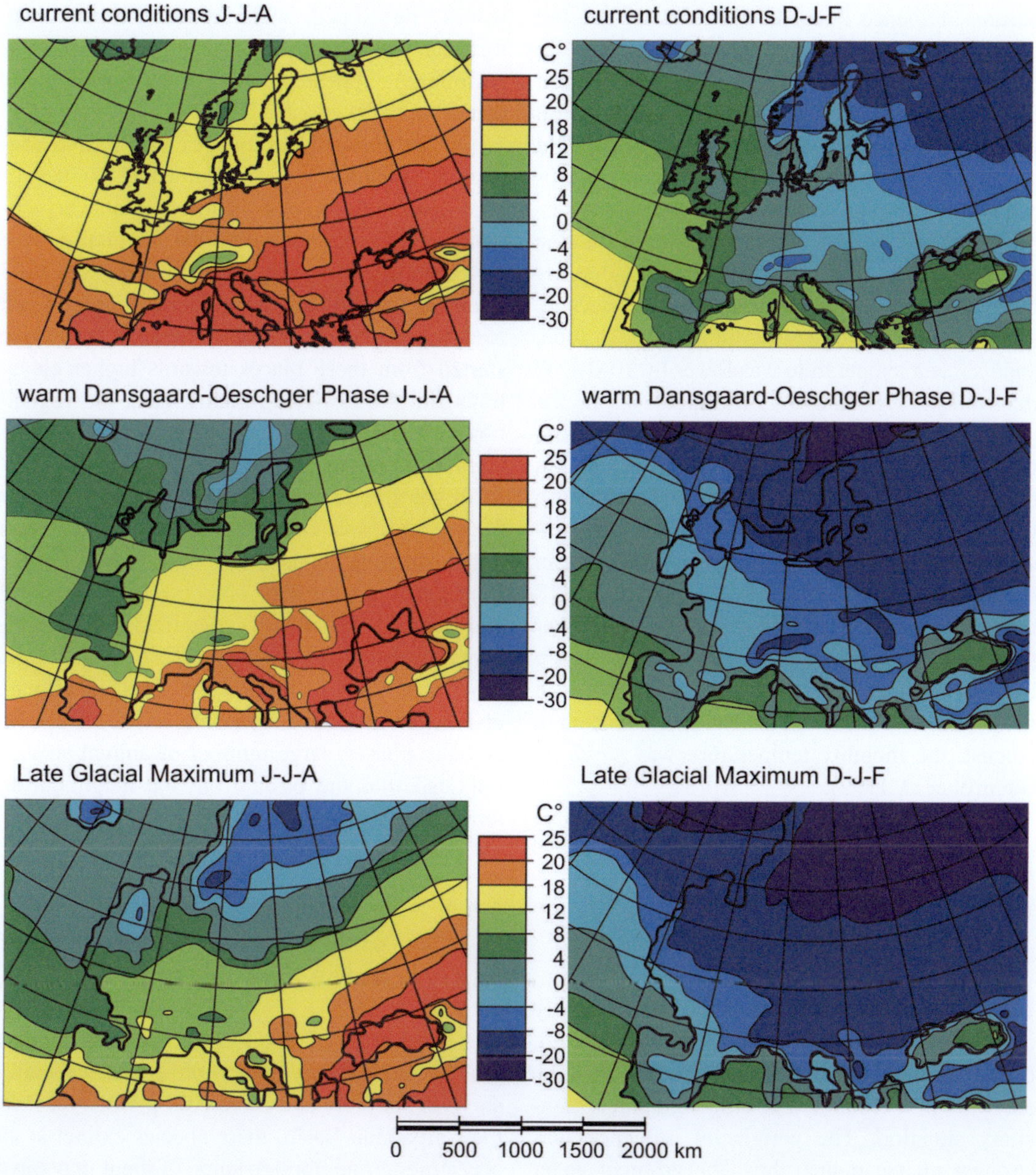


**Fig. 1.50** The temperatures in Europe in summer months (June, July, August [JJA]) and in winter months (December, January, February [DJF]), nowadays and modelled for the end of the Pleistocene. The present-day data are on the top, the warm Dansgaard-Oeschger event (ca. 40 thousand years ago) is in the middle, and the Late Glacial maximum (ca. 21 thousand years ago) is at the bottom (after Andel 2002)

highlighted the individual characteristics of each glacial and interglacial period (indicating a temperate, sometimes Mediterranean-like periods before the Günz, during the Mindel-Günz and Riss-Mindel interglacials). Székely (1977) distinguished the glacials, interglacials and periglacials based on their humidity, adapting the periods to the explanation of the periglacial processes.

However, according to Bariss (1991), the climate used to be determined by oceanic and continental influences of different intensities: the oceanic influence intensified in glacials and the continental one during the interglacials. Thus, the ice of the previous glacial period started to melt in the hot summers, and the amount of precipitation was not sufficient to replace it during the cold winters; therefore, the ice sheets gradually disappeared. According to other authors, the interglacials used to be warm and wet (Járainé Komlódi 2000). The area of Hungary had never been covered by ice sheet during the Pleistocene, since it always belonged to the periglacial zone, and its mean temperature was lower by 10–12 °C than today (Fig. 1.50). As the area was in the periglacial morphoclimatic zone, föhn winds from the northern ice sheets carried little precipitation (in the summer half-year it was ca. 200–400 mm); thus, dry and cold climate prevailed in the glacials. As the results above, in the Carpathian Basin during the Last Glacial Maximum (ca. 20–18 thousand years ago), the precipitation was lower than today by 250 mm and the mean temperature by 10–12 °C lower. Thus, in February, the mean temperature was between −12 and −14 °C, and in August, the monthly temperature was 8−12 °C (Frenzel et al. 1992).

### 1.8.1 Main Features of the Quaternary Flora and Fauna

In the Carpathian Basin, the knowledge on Early and Middle Pleistocene biogeography is limited, due to insufficient dating and the fact that the area was in a climatic and biogeographical transitory situation. The periglacial influence was stronger on the plains; thus, the arboreal vegetation receded, and the extinction was more excessive than that in the mountains, where the southern slopes provided advantageous microclimatic conditions. On these habitats, some herbaceous plants were able to survive the vicissitudes of the glacial climate, and during the warmer periods from these refugiums, plants could invade the plains ("Ancient Mátra Theory"). The fauna had also adapted to the rhythm of climatic changes; thus, "large mammal glacial waves" developed, when certain animals (e.g. reindeer, musk ox; Siberian vole or willow ptarmigan) appeared in the temperate zone (Jánossy 1979).

On the plains, loess steppes with sparse vegetation, subarctic grassy tundra and coniferous taiga forests developed on the cold and dry climates of the glacials, while the mountains were covered by alpine vegetation. During the mild interstadials, the expansion of mixed pine-birch forests and the broadleaf woodlands may have started from these places towards higher elevations and the low-lying plains; thus, the vegetation of the Basin became similar to that of today (Járainé Komlódi 1991, 2000).

In the interglacials of the Early and Middle Pleistocene, even the subtropical and warm temperate zone plants and animals returned to the Carpathian Basin. The boreal and arctic floral and faunal elements characterizing the continental steppes appeared by Late Pleistocene. Since the basin had always a transient climate and vegetation, and a mosaic relief, in certain climatic phases, large number of animal species migrated into the basin from the neighbouring territories (Varga 2006).

### 1.8.2 Historical Changes of the Flora and Fauna in the Pleistocene

The beginning of the Pleistocene was milder; thus, some Tertiary thermophile species could survive. As more and more cold periods reached the Carpathian Basin, these species extinct at an accelerating rate, though some of them may have survived even until the Riss-Würm interglacial, that is, much longer than that in the areas north of Hungary.

Coniferous forests dominated in the colder and wetter phases, which were disadvantageous for the wildlife. Some deciduous trees (alder, birch and willow) mixed into the pine forests creating taiga and swamp forests and alluvial groves. In the warmer and humid phases, mixed oak forests spread with elm and linden, and in the warm and dry periods, Mediterranean deciduous forests appeared (Járainé Komlódi 2000).

The Lower Pleistocene started with the Villányium Fauna Complex (2.5 million–760 thousand years ago), although according to some researchers it does not belong to the Quaternary. At this time, animals of steppe dominated, though some thermophile animal species from the Tertiary also survived. In North Hungary, ancient deer was common, while in the south antelope and rodent species were widespread. In the second half of the Early Pleistocene (first third of the Bihari Fauna Complex), the Tertiary animal species extinct, and the "modern" vertebrate fauna (e.g. gophers, squirrels and water voles) and the ancient forms of bison, bear and deer appeared (Jánossy 1979).

Most of the fossils from the Middle Pleistocene (760–120 thousand years ago) were found on terrace No. IV of the Danube, and they were excavated from travertine at Vértesszőlős, together with the findings of the prehistoric man (*Homo erectus seu sapiens palaeohungaricus*). The climate of the interglacials is warm and more humid; thus, some Tertiary species could survive. There are hardly any data from the Mindel Glacial; however, its interstadial flora has been studied in detail: the dominant vegetation indicates a dry and Mediterranean climate in the Carpathian Basin, and several tree species appeared that are considered to be exotic today (e.g. olive, laurel, lilac, walnut and cigar trees). The pine forests of the Mindel-Riss interglacial indicate a quite disadvantageous climate. The Great Hungarian Plain was covered by subarctic larch taiga and treeless steppe in the Riss glacial. The Bihari and the Steinheim (Oldenburg) Fauna Complex dominated in the period.

In the Riss-Würm interglacial of the Upper Pleistocene (approx. 120–11.5 thousand years ago), different pine species (e.g. Siberian spruce) invaded the Catpathian Basin, and mixed deciduous oak forests grew on the milder climate. The cold peak of the Würm was the coldest, driest and most extreme climate period of the Pleistocene. The annual precipitation (180–250 mm) was not enough for the plants despite of the cold; thus, the vegetation became sparse and poor in species. Besides, the Carpathian Basin belonged to the discontinuous permafrost zone, which was again an unfavourable factor the vegetation. Therefore, patches of open coniferous taiga forests and steppe grassland prevailed through the Würm. During the warm interstadials, forestation started with subarctic taiga (spruce, fir and larch), or with mixed deciduous taiga (birch, willow, oak, elm and linden). On the Great Hungarian Plain, oak forests with lily of the valley and Tartarian maple spread, but like today, the different regions have diverse vegetation due to the micro-climatic alterations (Járainé Komlódi 2000). The bone remains (early Pilisszántó Fauna Phase) from the Lascaux Optimum reflect that deers and roes lived in the forests, and wild horses, steppe bisons and subarctic reindeers were widespread on the open grasslands (Vörös 1987). The climate was extremely dry and cold in the stadials; thus, a treeless steppe prevailed on the Great Hungarian Plain, while pine forests with frost-tolerant, psychrophile Scotch pine and larch have been in the mountains. Other researchers claim that the landscape used to be covered by forest steppe, whose present equivalent may be the open boreal forests with the dominance of pine and birch. On the cold and strongly continental climate, cave bears and ptarmigans spread, and the reptiles disappeared. Later, during the very cold end of the Würm, cave bears disappeared, and reindeers, subarctic chamois and the arctic hares appeared (Jánossy 1979; Vörös 1987).

The Late Glacial can be divided into two mild and wet (Bølling and Allerød) and cold (Dryas) phases, which were about 5 °C colder than today. These periods lasted just for a few hundred or thousand years; thus, they were too short for significant vegetation transformations. The predominant vegetation types were the cold and dry continental steppe with diverse grass species and the mugwort steppe.

### 1.8.3 Landscape Development and Climate Changes

The above-described climatic and vegetational conditions controlled landscape development. During the humid periods of glacials, linear erosion prevailed, while during the cold peaks the humidity decreased; thus, loess formation took place, and the periglacial processes typical in discontinuous permafrost zones were active. The fluvial processes became less active due to the decrease in precipitation, and aggradation of the channels by gravels was typical. The valleys broadened because of the frost action, and alluvial fan systems developed. During the interglacials, chemical weathering, karstification and the fluvial processes became more intensive, and the alluvial fans and valley slopes became dissected. Forest and chernozem-type soils were formed, depending on the amount of precipitation (Bulla 1964; Székely 1997).

## 1.9 Late Glacial and Holocene Landforms and Processes

### 1.9.1 Palaeo-geography of the Late Glacial

The transition between the Pleistocene and the Holocene was the so-called Late Glacial, which was characterized by a climate alternately warm and cold, but with a warming tendency. The length of this transitional period depends on the indicators, as palynology, dendrochronology, sedimentary archives, or isotope data (e.g. 3-km-long ice cores from Grönland) could be considered. The most widely accepted scale was made by Meese et al. (1997), dating the Late Glacial to 11.5–14.5 thousand years (Table 1.9).

The first two cold phases of the Late Glacial are the Oldest and the Older Dryas, which were divided by the mild Bølling Interstadial. (The cold periods got their name after a small plant called *Dryas octopetala.*) These cold periods were short, as they lasted just for 100–200 years (Brauer et al. 1999), and their intensity varied from place to place; thus, their existence could be detected just by sensitive indicators. The Oldest and Older Dryases were characterized by woodless tundra and sparse trees. These periods were followed by the Allerød Interstadial, which was characterized by a cool climate. The spreading of Scotch pine was connected to this period. The last cold period of the Late Glacial was represented by the Young Dryas (11.5–12.5 thousand years ago), which is considered to be the last cold wave with an annual mean temperature of 4–5 °C. This cold millennium appeared just regionally, as it was supported by the results of the chronological analyses of the Grönland ice cores (GISP3 and GRIP; Brauer et al. 1999).

Sümegi and Hertelendi (2001) described the temperature changes during the Late Glacial in the Carpathian Basin. The vegetation had a tundra character, the vegetational period shortened to a few months, and the precipitation was just 400–500 mm. The whole basin was affected by the 1000-m decrease of the tree line (Lotter 1999).

These changes in climate, vegetation and water budget resulted in significant modifications in the surface processes. There are just some data on loess formation in Hungary, which probably continued in the Late Glacial too, though the warming and wetter climate obstructed formation of typical loess in the Carpathian Basin. The denser vegetation of the Bølling and the more intensive soil formation significantly hindered the loess formation in the longer Younger Dryas, as well.

The chronological data on blown sand movement (in the Nyírség, the Danube-Tisza Interfluve and the Bodrogköz) also indicate that the Dryases favourable conditions for active aeolian processes, and the present-day wind-blown landforms developed mostly at this time. Contrarily to earlier hypotheses, the sand movements in the younger Holocene (Boreal Phase) may have had a lesser importance in surface formation (e.g. Borsy 1991; Gábris 2003; Lóki and Félegyházi 2007; Lóki 2009).

The pattern of river changed depending on the water and sediment discharge. During the interstadials, the closed forests prevented soil erosion; thus, lower amount of sediment got into the

**Table 1.9** Chronology of the Late Glacial (after Brauer et al. 1999; Litt et al. 2007)

| Year | Chronological unit | Stratigraphical units (in Germany) |
|---|---|---|
| 11,500 | Holocene | Preboreal Phase |
| 12,500 | Late glacial | Young Dryas |
| 13,640 | | Young Allerød |
| 13,700 | | Older Allerød |
| 13,850 | | Older Dryas |
| 14,500 | | Bølling |
| | | Oldest Dryas |
| | | Transitional Phase |
| | Würm | |

rivers; thus, the huge meanders developed with high water discharge. In the Dryas interstadials, the valleys aggraded and the river pattern became braided in certain cases, though the climate change did not last long enough to create considerable changes in the relief (Gábris 1995b, 2001).

### 1.9.2 The Division of the Holocene

Two features must be highlighted in connection with the climate changes during the Holocene in the Carpathian Basin, as (1) the high rate of warming (up to 1 °C/100 years) and (2) the extent of the change. About 11.5 thousand years ago, the Late Glacial terminated, and within few thousand years, the yearly mean temperature increased by 5–7 °C. Though the beginning of the Holocene could be defined relatively accurately, its end is not. Some claims that the Holocene—the present interglacial—may continue in the next several ten thousands of years, while others believe that the next glacial will start within 200–20,000 years (Pécsi 1993). Nowadays, the anthropogenic impact overrides the previous estimations; therefore, it is difficult to formulate a grounded opinion.

The range of the temperature changes was only few °C in the whole Holocene, and the human impact just partly contributed to it, resulting in a warming which started at the end of the period. There are examples of "natural" climate changes within the Holocene; for example, the Saharan wildlife boosted approx. 4–5 thousand years ago (in the climate optimum of the Late Neolithic). Hsu (1998) reconstructed global climate cycles lasting for 600 years each, based on the written Chinese observations of the past ca. 7 thousand years. Based on these data, a cold period of the early Middle Ages happened between 60 and 600 AD, a short warm period during the Middle Ages lasted from 600 to 1280 AD, and the climate deterioration called Little Ice Age took between 1280 and 1820; finally, today's climatic optimum started in 1820.

The phases of the Holocene may provide a frame for connecting the individual environmental conditions (climate, water supply, surface formation, vegetation, etc.) to the social influences, which have had a distinguished role in the past millennia. The most accepted approach regarding the division is based on the changes of vegetation. It seems to be an applicable solution, especially in Hungary, because the chronological boundary of the Holocene is set at the appearance of deciduous forests consisting of oak and other thermophylous species. Apart from the vegetation, there are other attempts on chronological divisions, and one of the oldest is based on climatic changes. Analysing bog sediments, at the end of the 1800s, Blytt distinguished dry continental phases (Boreal) and wet oceanic phases (Atlantic). The separation of the 2.5-thousand-year-old climatic periods after the late glacial is also based on this logic. The first phase was the Preboreal Phase starting about 10 thousand years ago; it was followed by the warm dry Boreal, then by the warm wet Atlantic, then by the warm dry Subboreal, and finally by the cold wet sub-Atlantic.

The results of researches on peat and bog sediments reflect vegetational changes, and the pollen analysis also strengthens this approach. (The palynological approach is based on the fact that the environmental requirements of certain species refer to climate elements of a given period.) Some other methods also exist; however, they can be applied only indirectly in the Carpathian Basin. For example, the geomorphological method, which is based on the expansion of the glaciers (Table 1.9; Gamper and Suter 1982), provided a benchmark for the estimates of the climate change in the first half of the Holocene. Newer approach is the analysis of ice cores, which provides climate data since the Late Glacial based on the analysis of air inclusions (NOAA 2008).

Apart from these, several other division methods exist. A division based on fauna waves was less successful, as 10 thousand years are too short for their formation. The Hungarian attempt, the so-called vole thermometer (Kordos 1979), has not proved to be efficient, either. The archaeological scale has a long tradition, but the problem is that different researches include largely different divisions. As a conclusion, numerous approaches result in largely different, sometimes incompatible chronologies (Table 1.10). Therefore, it is an especially difficult task to analyse the whole, interconnected system, for example, to separate the landscape changes effected by human activity from those effected by climate change.

The division of the Holocene based on vegetation changes is based on the climatic requirements of the typical vegetation, where the problem is that none of the indicator values could

**Table 1.10** Environmental and cultural changes during the Holocene

| Age | Climate | Landforms | Vegetation | Archaeological division |
|---|---|---|---|---|
| (3) BP years | (1), (3), (6), (7), (9), (14) | (2), (4), (8), (9), (10), (11), (12), (13) | (1) | (4), (5), (9) BP based |
| Sub-Atlantic (2500–) | Wet, oceanic effects 750-mm precipitation climatic fluctuations (i.e. Roman Optimum, Little Ice Age) | Lower floodplain incised into high floodplain development of point bars during Little Ice Age: floods, landslides | **Beech II** beech disappears from plains, territory of oaks decreases, spread of domestic vegetation, forest clearances | Modern Age (300–) Middle Ages (300–1100) |
| | | | | Great migration period (Huns, Avars—1100–1600) |
| | | | | Roman Ages first sign of global pollution |
| Subboreal (2500–5700) | Mild, wet, differing data cooling, dryer "climatic optimum" (6)—900-mm precipitation over Zagyva catchment The end of the phase was dryer | Intensive fluvial activity, development of oxbow lakes, highest water stage of Lake Balaton Increasing linear erosion and deflation due to deforestation usage of plough led to the beginning of irreversible environmental changes | **Beech I** Phase of most enclosed forests, oak and hornbeam advances, beech appears on the Great Hungarian Plain, appearance of wheat and weed species | Iron Age (2850–2000) Mezőcsát Culture, Hallstatt Culture in Transdanubia |

(continued)

**Table 1.10** (continued)

| Age | Climate | Landforms | Vegetation | Archaeological division |
|---|---|---|---|---|
| | | | | Bronze Age (2800–5000 ca.) agriculture, settlements, hillforts |
| Atlantic (5700–8300) | End of the phase was warm and dryer Beginning of phase was colder and wetter (+10 % precipitation compared to present times) June +23 °C January −10 °C | Intense fluvial activity, development of river curves development of peat, rendzina Accretion of high floodplain development of alluvial fans, intense erosion, landslides During dry period: saline soils | **Oak** forestation, mainly mixed oak steppes emerge of mediterranean species (grape, ivy) | Copper Age (5000–7000 ca.) Great Plains: shepherding, Transdanubia: permanent settlements, Kurgan burial |
| | | | | Neolithic Age (7000–10,000) Körös culture (7500–8000) fishing-hunting-collecting, poor animal husbandry |
| Boreal (8300–9500) | Warm, dry Annual mean temperature 1–2 ° C warmer than nowadays January 0 °C | Slight sand movements, development of chernozem soils, fluvio-aeolian forms | **Hazel** saline steppes, sand steppes, oak forests (steppe forests) migration of Central Asian species | |
| Preboreal (9500–11,500) | Cooler, dry Annual mean temp.: 8–9 °C January −2 °C June 18 °C Precipitation: 400 mm moister than West Siberian, slightly less continental | Slight deflation increasing fluvial activity | **Pine-birch** birch forest steppes appearance of thermophilous tree species (oak, lime, ash) | Transitional Stone Age |

be defined accurately, and thus, the division cannot be precise. It is probable that the environmental changes (e.g. increase and decrease of precipitation or temperatures) resulted in surface transformations; however, to establish their extent is uncertain due to the complex, indirect calculation possibilities.

### 1.9.3 Main Environmental Changes in the Holocene

After the harsh climate of the Late Glacial, quick warming characterized the Holocene, which resulted in present-day temperatures. There were some fluctuations in temperature and water regime, as cooler periods with peaks ca. 2.8, 5.3, 8.0 and 10.5 thousand years ago altered with warmer ones at 3–4, 7 and 8–9 thousand years ago (Pécsi 1993). These meant up to 2–3 °C change in annual mean temperature. The precipitation was the most abundant in the Preboreal and Atlantic phases (Gábris 1995a), though some researchers claim that it shifted to the Subboreal Phase (Gábris 2001). The increased precipitation resulted in higher discharge and more efficient surface formation.

The tectonic features of the Pleistocene also characterized the Holocene. The central areas (the Little Hungarian Plain, the Lower Tisza Valley, the Lower Zagyva Valley and the Körös Basin) remained tectonically active; thus, nowadays, these are the most intensively subsiding areas (Fig. 1.51). It is a greater challenge to define the thickness of the Holocene sediment and the rhythm of subduction (or elevation). The

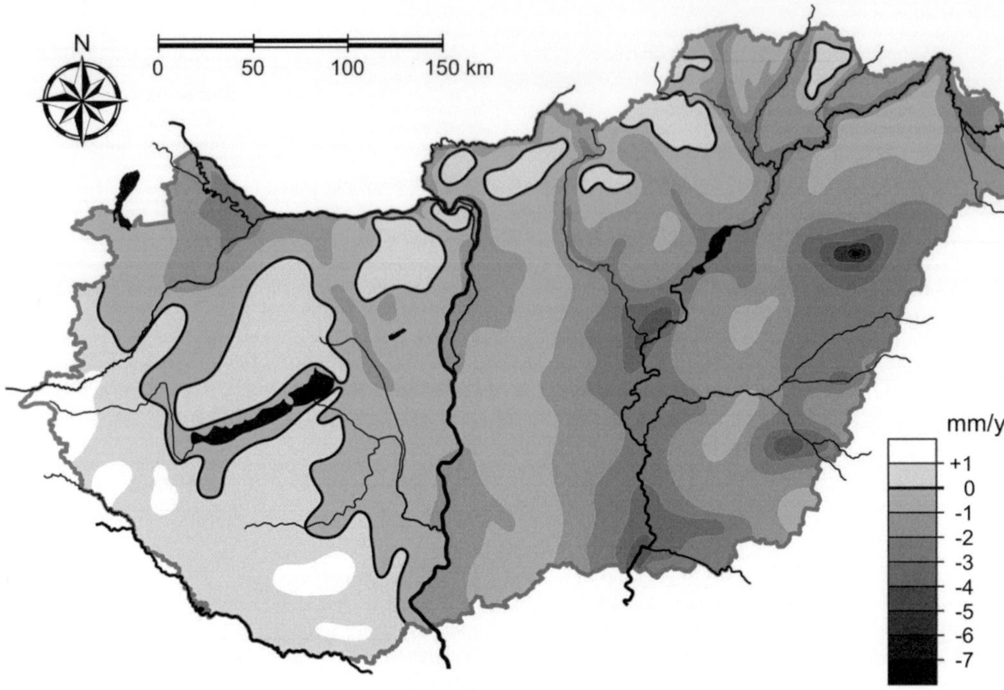


**Fig. 1.51** The rate of active subsidence and uplift in Hungary (after Joó 1998)

Pleistocene and the Holocene sediments often cannot be separated; thus, different calculations were made to estimate the thickness and the rhythm of subduction. Regarding that the Pleistocene and Holocene rates of subduction are the same (0.1–0.2 mm/year), the thickness of Holocene sediments is 1–2 m. However, based on geomorphological analyses, the elevation changed by 20 m on the Little Hungarian Plain and by 30 m on the Great Hungarian Plain (Somogyi 1973). Based on geodesic measurements, only 0.08-mm/year subduction rate and 1-m-thick Holocene sediment sequence could be identified (Szeidovitz et al. 2001). If we consider that the Pleistocene subsidence rates were similar as the present-day average rates, it could be supposed that the Lower Tisza Valley subducted by 4.4 mm/year, the Little Hungarian Plain by 2.3 mm/year and the northern part of the Körös Basin by 7 mm/year. These are 20 times higher values than those in the Quaternary; however, this magnitude cannot be supported by other tectonic or sedimentological for the entire Holocene (Joó 1998). Thus, altogether ca. 1–10-m Holocene sediment sequence developed, and the location of active movement can be identified clearly.

The information on vertical movements is important, because these subductions controlled the course and pattern changes throughout the Holocene. These were not great changes, as the Holocene is relatively short and the rivers mostly had the same directions as today. The most important changes took place along the Danube in Szigetköz, the Maros River changed its course frequently on its alluvial fan, and the Tisza joined the Bodrog River at a different location as today (Fig. 1.52).

The Tisza River (and its tributary, the Szamos River) frequently changed its channel in the subsiding Szatmár-Bereg Plain and in the Bodrogköz too. (Besides aggradation, erosion may also have played a role in the formation of these plains.) As a result of these channel changes, a dense network of palaeo-meanders remained on the surface, and due to the lateral erosion of the rivers, the majority of blown sand forms were destroyed in the west of Bodrogköz (Fig. 7.8).

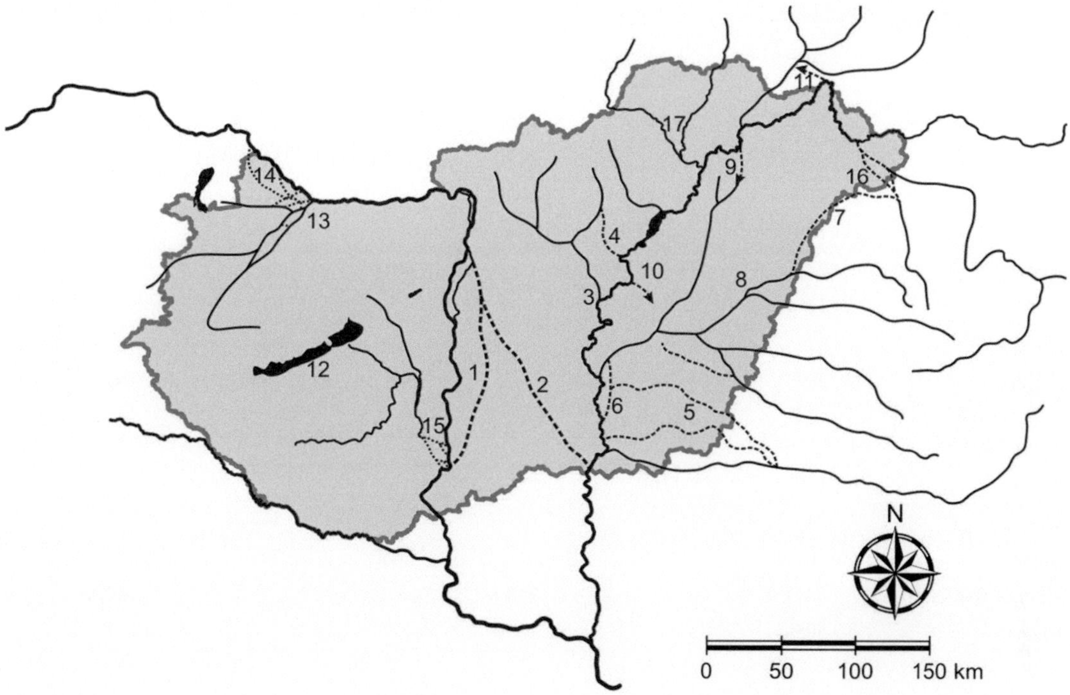


**Fig. 1.52** The most important channel changes of the Hungarian rivers in the Holocene (after Somogyi 2000a). *1*—Early Holocene side channel of the Danube at the boundary of the Danube Plain and the Danube-Tisza Interfluve, *2*—the last probable course of the Danube across the Danube-Tisza Interfluve according to Sümeghy, *3*—former conjunction of the Zagyva River, *4*—former course of the Tarna River, *5*—course of the Maros at the end of the Pleistocene, *6*—the Kurca was the conjunction of the Körös River, *7*—along the Ér Valley the Szamos and later the Kraszna rivers flow across, *8*—meandering river bed of the Berettyó River prior to channel regulations, *9–10*—flow directions of the Tisza floods drained by crevasses, *11*—conjunction of Tisza and the Bodrog Rivers, *12*—areal changes of Lake Balaton, *13*—conjunction of the Rába and Marcal rivers, *14*—channel changes of the Danube in the Szigetköz, *15*—channel changes in the Sárköz Region (Sió and Danube), *16*—channel changes in the Szatmár Plain (Szamos and Kraszna Rivers), *17*—shift of the Sajó-Hernád confluence

The Boreal channel of the Upper Tisza River was occupied by the Karcsa Brook. Similar changes took place in the Middle Tisza Region, where the River Tisza created a 20–40-km-wide floodplain. The surface is characterized by a dense network of abandoned channels in the east, extending towards the Hajdúság. It refers to a not-well-defined course direction of the Middle Tisza during the Late Pleistocene and Holocene, probably because the whole area intensively subsided; thus, probably even the Paleo-Bodrog took its course across this region (the rhythmical subsidence could have been the reason for the avulsions and channel changes).

Nováky (1991) studied the water system of the Zagyva River, concluding that even small climatic fluctuations may result in significant run-off changes. Thus, for example, a 5 % increase (or decrease) in precipitation may induce run-off change by 40–45 %, or a 0.2–0.3 °C temperature change may result in change by 10 %. The generality of this statement in the Carpathian Basin has not been proven yet, but all studies show that the Holocene climatic changes may have caused significant water discharge fluctuations, which sometimes led to considerable changes in meander size. Gábris (1986, 1995b) calculated palaeodischarge values based on the morphometry of

**Table 1.11** The annual Holocene water discharge ($\Sigma Q\ 10^6\ m^3$) of some Hungarian river sections (after Gábris 1995a)

| | Present-day | Preboreal Phase | Boreal Phase | Atlantic Phase | Subboreal Phase | Sub-Atlantic Phase |
|---|---|---|---|---|---|---|
| Upper Tisza | 6002.5 | | 1560 | | | |
| Middle Tisza | 16,742.5 | | 7230 | 174,000 | 116,000 | 90,200 |
| Szamos | 4106 | 39,600 | 1040 | | | |
| Sajó | 1087 | 10,500 | 474 | | | |
| Bódva 5 | 234 | | | 4610 | | |
| Szuha | 14 | | | | 98 | |
| Kácsi Brook | 9.5 | | | 164 | 66 | 35 |
| Zagyva | 303 | | | | 2650 | 1070 |
| Zagyva (without Tarna River) | 109 | 1040 | | | | |
| Tarna | 126 | 1160 | | | | |
| Galga | 28 | | | | 347 | |
| Tápió | 30 | | | 2050 | 250 | 177 |
| Hármas-Körös | 3317 | | 748 | | | 10700 |
| Maros | 5054 | | 1070 | | | |

palaeo-meanders and on the river bed material. The results show that in the different periods of the Holocene, the water discharge of rivers was extremely different (Table 1.11; Fig. 1.53). Thus, for example, in the Atlantic Phase, the discharge was ten times higher than today, and in the Preboreal Phase, it was quarter or fifth of today's values; consequently, the fluvial development also differed considerably. These values were slightly modified by the latest researches; thus, precaution is needed during the descriptions. The calculated water discharge of the Middle Tisza (at Polgár) was 2500 $m^3/s$ in the Preboreal Phase, 200 $m^3/s$ for the Atlantic Phase, and 700 $m^3/s$ for the Subboreal (Gábris 2001; Félegyházi and Tóth 2001). The Subboreal Phase is considered to be one of the most humid phases; thus, for example, the Száraz-ér had 240 $m^3/s$ water discharge, and it decreased to 30–40 $m^3/s$ in the sub-Atlantic Phase. Thus, the most intensive fluvial activity characterized the humid periods of the Holocene; hence, in the Preboreal, Atlantic and Subboreal phases, and the driest periods with low fluvial activity were the Boreal Phase and some periods of the Atlantic Phase. The latest palaeo-hydrological studies date the palaeo-meanders by OSL and radiocarbon dating. Based on the data, three floodplain-level development phases and two incision phases were identified along the Lower Tisza. The palaeo-channel is still visible on the surface, though they were filled up by silty and clayey sediments (see the Geological Atlas of the Great Hungarian Plain). The estimated bank-full water discharge of the Pleistocene–Holocene palaeo-meanders was 5–10 higher, than the present-day value (Fig. 1.54).

At the beginning of the Holocene, after the harsh climate of the Late Glacial, the forests were dominated by pine and birch species, which were later replaced by different oak forests. The type and areal extension of oak forests depended on the water supply. For example, in the milder Boreal Phase, oak forests appeared only in small extent on the huge steppes, though in the

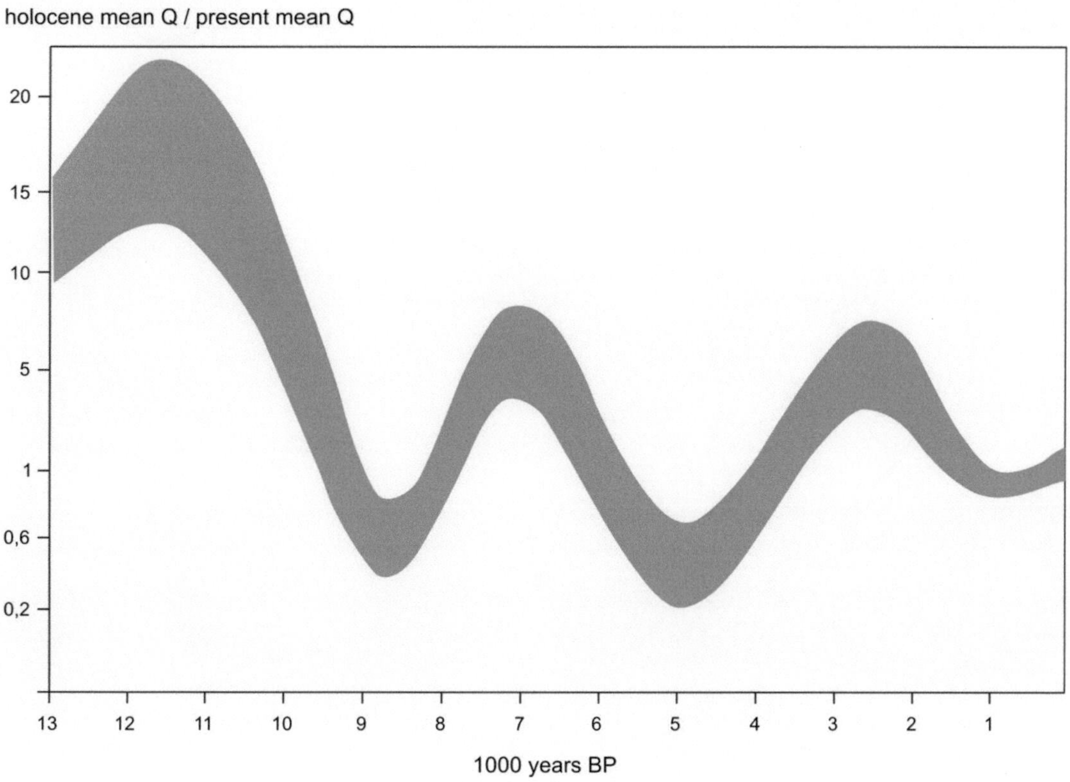


**Fig. 1.53** The changes in mean water discharge of Hungarian rivers during the Holocene (after Gábris 1995a)

following Atlantic Phase mixed oak forests invaded the area of former steppes too (Table 1.10).

Several botanists claim that the steppe of the Great Hungarian Plain does not have a climatic origin, but rather it was created by social influences, because the forestation was hindered by animal husbandry and agriculture. In the Neolithic Körös Culture, animal husbandry and shifting cultivation became well organized; thus, it was the first time when the environment became intensely exploited. Since the Copper and Bronze ages, constant settlements were established in great number at the edges of floodplains, or by natural waters (Erdősi 1987; Somogyi 1987). The first steps to protect the society against natural hazards (e.g. floods) started when more intensive agriculture and animal husbandry became widespread in the Copper Age. The serious tribal wars of the Copper and Bronze ages forced the majority of the population into defendable locations, that is why we can find traces of settlements in wetlands, on coastal dunes and natural levees, or earth fortifications were built on the slopes of hills (Rácz 2008; Visy 2003).

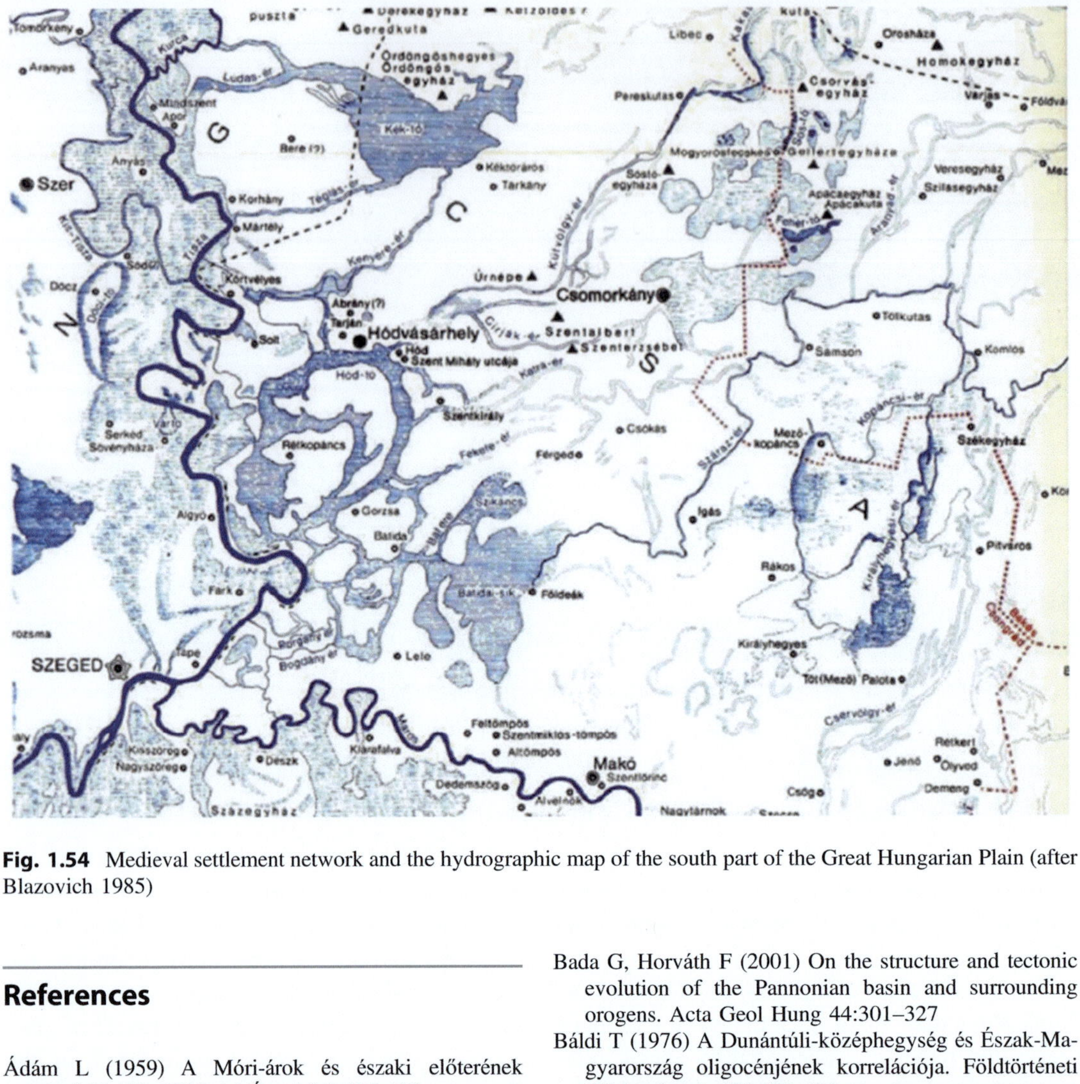


**Fig. 1.54** Medieval settlement network and the hydrographic map of the south part of the Great Hungarian Plain (after Blazovich 1985)

## References


Ádám L (1959) A Móri-árok és északi előterének morfológiája. Földrajzi Értesítő 8:417–432

Ádám L (1975) A Nyugat-magyarországi-peremvidék természetföldrajzi jellemzése. In: Ádám L, Marosi S (szerk) A Kisalföld és a Nyugat-magyarországi-peremvidék. Akadémiai Kiadó, Budapest, pp 68–81

Ádám L, Marosi S, Szilárd J szerk (1987–1988) A Dunántúli-középhegység. A–B. Akadémiai Kiadó, Budapest

Alföldi L, Kapolyi L szerk (2007) Bányászati karsztvízszint-süllyesztés a Dunántúli-középhegységben. MTAFKI, Budapest

Andel TH (2002) The climate and landscape of the middle part of the weichselian glaciation in Europe: the stage 3 project. Quat Res 57:2–8

Bada G, Horváth F (1998) APannon-medence jelenkori tektonikája. Természet Világa 1998. II. különszám, pp 18–23

Bada G, Horváth F (2001) On the structure and tectonic evolution of the Pannonian basin and surrounding orogens. Acta Geol Hung 44:301–327

Báldi T (1976) A Dunántúli-középhegység és Észak-Magyarország oligocénjének korrelációja. Földtörténeti Közlemények 106:407–424

Báldi-Beke M, Báldi T (1974) A novaji típusszelvény (kiscelli-egri) nanoplanktonja és makrofaunája. Földtani Közlöny 104:60–88

Barabás A, Barabásné Stuhl Á (2005) Geology of the lower triassic jakabhegy sandstone formation, Hungary. SE Transdanubia. Acta Geol Hung 48:1–47

Bariss N (1991) The changing paleogeographic environment during the Upper Pleistocene at Northern Mid-latitudes. In: Pécsi M, Schweitzer F (szerk) Quaternary environment in Hungary (Studies in Geography in Hungary 26.) Akadémiai Kiadó, Budapest, pp 27–35

Bérczi I, Jámbor Á szerk (1998) Magyarország geológiai képződményeinek rétegtana. MÁFI –MOL, Budapest

Blazovich L (1985) A Körös–Tisza–Marosköz középkori településrendje. (Dél-alföldi Évszázadok 1.) Csongrád megyei Tanács, Szeged

Bodnár A, Lovász Gy, Nagyváradi L (2009) A Dráva--völgy geomorfológiai kutatásának újabb eredményei. In: Szabó B, Tóth J, Wilhelm Z (szerk) Környezetünk természetitársadalmi dimenziói. PTE, Pécs, pp 29–38

Borsy Z (1961) A Nyírség természeti földrajza. Akadémiai Kiadó, Budapest

Borsy Z (1989) Az Alföld hordalékkúpjainak negyedidőszaki fejlődéstörténete. Földrajzi Értesítő 37: 211–224

Borsy Z (1991) Blown Sand Territories in Hungary. Zeitschrift für Geomorphologie 90(1–14):355

Borsy Z, Molnár B, Somogyi S (1969) Az alluviális medencesíkságok morfológiai fejlődéstörténete Magyarországon. Földrajzi Közlemények 93(17):237–254

Borsy Z, Balogh K, Kozák M, Pécskay Z (1987) Újabb adatok a Tapolcai-medence felszínfejlődéséhez. Acta Geogr Debr 23:79–104

Brauer A, Endres Ch, Negendank JFW (1999) Late-glacial calendar year chronology based on annually laminated sediments from Lake Meerfelder Maar, Germany. Quaternary International 61(1):17–25

Bulla B (1937) Der pleistozäne Löss in Karpathenbecken. I–II. Földtani Közlöny 67:196–215, 289–309

Bulla B (1964) Magyarország természeti földrajza. Tankönyvkiadó, Budapest, 424

Burján B (2002) APesti-síkság fiatal-harmadidőszaki és negyedidőszaki kavicsképződményeinek összehasonlító vizsgálata. PhD-értekezés, TFGT, Szeged

Butrym J, Maruszczak H (1984) Thermoluminescence chronology of younger and older loesses. In: Pécsi M (ed) Lithology and stratigraphy of Loess and Paleosol. Geographical Research Institute, Budapest, pp 195–199

Cholnoky J (1910) Az Alföld felszíne. Földrajzi Közlemények 38:413–436

Cholnoky J (1918) A Balaton hidrogeográfiája. In: A Balaton tudományos tanulmányozásának eredményei. I. kötet, II. rész. MFT Balaton Bizottság, Budapest

Cholnoky J (1926) A földfelszín formáinak ismerete (Morfológia). Királyi Magyar Egyetemi Nyomda, Budapest

Császár G, Haas J, Nádor A (2003) Klímaváltozások a Föld történetében. Természet Világa 134(9):389–395

Csontos L, Vörös A (2000) A Kárpát-Pannon térség lemeztektonikai rekonstrukciója. In: Karátson D (szerk) Pannon enciklopédia. Kertek, Budapest, pp 56–57, 62–64

Erdősi F (1987) A társadalom hatása a felszínre, a felszíni vizekre és az éghajlatra a Mecsek tágabb környezetében. Akadémiai Kiadó, Budapest

Félegyházi E, Tóth Cs (2001) A Tisza medergenerációinak és formakincsének űrfelvételes elemzése a Közép-Tiszavidéken. Acta Geogr Debr 35: 93–102

Félegyházi E, Lóki J, Szabó J (2003) A folyó őstörténete, a mai Tisza kialakulása az Alföldön. In: Teplán I (szerk) A Tisza és vízrendszere. MTA Társadalomkutató Központ, Budapest, pp 29–41

Fleck N, Hazslinszky T szerk (1998) Cave Capital. MKBT, Budapest

Frányó F (1992) The thickness of the quaternary sediments in Hungary. M = 1: 500.000. Geol Inst Hungary, Budapest

Frenzel B, Pécsi M, Velichko A eds (1992) Atlas of the Paleoclimates and Paleoenvironment of the Northern Hemisphere. MTAFKI–Fischer, Budapest–Stuttgart

Gábris Gy (1986) Alföldi folyóink holocén vízhozamai. Alföldi tanulmányok 10:35–52

Gábris Gy (1995a) A folyóvíz felszínalakítás-módosulása a hazai késő glaciális–holocén őskörnyezet változásainak tükrében. Földrajzi Közlemények 43(1):3–9

Gábris Gy (1995b) A paleohidrológiai kutatások újabb eredményei. Földrajzi Értesítő 1–2:101–109

Gábris Gy (2001) A folyóvíz felszínalakító tevékenysége Magyarországon. MTA, doktori értekezés, Budapest

Gábris Gy (2003) A földtörténet utolsó 30 ezer évének szakaszai és a futóhomok-mozgásának főbb periódusai Magyarországon. Földrajzi Közlemények 51(1–4):1–14

Gábris Gy (2006) A magyarországi folyóteraszok kialakulásának és korbeosztásának magyarázata az oxigénizotóp-sztratigráfia tükrében. Földrajzi Közlemények 130(3–4):123–133

Gábris Gy (2007) Földfelszín és éghajlat. Eötvös, Budapest

Gábris Gy (2008) A magyarországi folyóteraszok kialakulásának és korbeosztásának magyarázata. In: Szabó J, Demeter G (szerk) Geographia generalis et specialis. Kossuth Kiadó, Debrecen, pp 141–145

Gábris Gy, Nádor A (2007) Long-term fluvial archives in Hungary: Response of the Danube and Tisza Rivers to tectonic movements and climatic changes during the Quaternary. Quatern Sci Rev 22–24:2758–2782

Gamper M, Suter J (1982) Posttglaziale Klimageschichte der Schweitzer Alpen. Geogr Helvetica 37(2):105–114

Haas J szerk (2001) Geology of Hungary. Eötvös, Budapest

Haas J, Hámor G, Korpás E (1999) Geological setting and tectonic evolution of Hungary. Geol Hung 24. 179–196

Hámor G (1984) Paleogeographic reconstruction of Neogene plate movements in the Paratethyan realm. Acta Geologica Hungarica 27(1–2): 5–21

Hámor G (2001) Magyarázó a Kárpát-medence miocén ősföldrajzi és fáciestérképéhez. MÁFI, Budapest

Harangi R (2000) Bazaltvulkánok a Kisalföldön. In: Karátson D (szerk) Pannon enciklopédia. Kertek, Budapest, pp 306–308

Hartai É (2005) Magyarország földtana. ME, Miskolc. http://fold1.ftt.uni-miskolc.hu/~foldshe/mof04.htm

Hevesi A (1984) A karsztformák kormeghatározása. Földrajzi Értesítő 33(1–2):25–36

Hevesi A (1986) A Bükk hegység felszínfejlődése és karsztja. Kandidátusi értekezés, kézirat, Budapest

Hevesi A (1991a) Magyarország karsztvidékeinek kialakulása és formakincse I. Földrajzi Közlemények 115 (1–2):292–305

Hevesi A (1991b) Magyarország karsztvidékeinek kialakulása és formakincse II. Földrajzi Közlemények 115 (3–4):99–120

Horváth E, Gy Gábris, Juvigné E (1992) Egy pleisztocén vezérszint a Kárpát-medencében: a Bag Tefra. Földtani Közlöny 122:233–249
Horváth E, Novotny Á, Horváth E, Frachen M (2006) Bulla löszsztratigráfiája legújabb abszolútkormeghatározások tükrében. Földrajzi Közlemények 130(3–4):171–183
Hsu KJ (1998) Sun, climate, hunger and mass migration. Sci China Ser D Earth Sciences 41:449–472
Hubai J (1992) Magyarország erőforrásainak geográfiája. Tankönyvkiadó, Budapest
Jakucs L (1971) A karsztok morfogenetikája. Akadémiai Kiadó, Budapest
Jakucs L (1977) Amagyarországi karsztok fejlődéstörténeti típusai. Karszt és Barlang 18:1–16
Jakucs L (2000) Magyar eredmények a karsztelméletben. In: Karátson D (szerk) Pannon enciklopédia. Kertek. Budapest, pp 437–439
Jámbor Á (2001) Quarternary. In: Haas J (ed) Geology of Hungary. Eötvös, Budapest, pp 265–278
Jánossy D (1979) A magyarországi pleisztocén tagolása gerinces faunák alapján. Akadémiai Kiadó, Budapest
Járainé Komlódi M (1991) Late Pleistocene Vegetation History in Hungary since the Last Interglacial. In: Pécsi M, Schweitzer F (ed) Quaternary Environment in Hungary. (Studies in Geography in Hungary 26.) Akadémiai Kiadó, Budapest, pp 35–46
Járainé Komlódi M (2000) A Kárpát-medence növényzetének kialakulása. Tilia 9:5–60
Jaskó S, Krolopp E (1991) Negyedidőszaki kéregmozgások és folyóvízi üledék-felhalmozódás a Duna-völgyben Paks és Mohács között. A MÁFI évi jelentése 1989-ről, pp 65–82
Joó I (1998) Magyarország függőleges irányú mozgásai. Geodézia és kartográfia 9:1–7
Juhász Gy, Pogácsás Gy, Magyar I (2007b) Óriáskanyon-rendszer szeli át a pannóniai üledékeket? Földtani Közlöny 137(3):307–326
Juhász E, Korpás L, Balog A (1995) Two hundred million years of karst history, Dachstein Limestone, Hungary. Sedimentology 42:473–489
Juhász G, Pogácsás G, Magyar I, Vakarcs G (2006) Integrált sztratigráfiai és fejlődéstörténeti vizsgálatok az Alföld pannóniai s.l. rétegsorában. Földtani Közlöny 136(1):51–86
Karátson D, Ruszkiczay-Rüdiger Zs, Székely B (2007) Miért kanyar alakú? A Duna-kanyar kialakulása az évmilliós vulkáni formák és az évszázezredes folyóvízi erózió tükrében. Földrajzi Közlemények 131(4):289–302
Kerényi A (2008) Éghajlatváltozás: múlt, jelen, jövő. Földrajzi Közlemények 132(4):419–431
Kerényi A, Kocsisné Hodos E (1990) Löszlepusztulási formák és folyamtok kvantitatív vizsgálata. Földrajzi Értesítő 39(1):29–54
Keveiné BI (2009) A karsztok ökológiai rendszere. (Földrajzi tanulmányok 4.) JATEPress, Szeged
Kiss G (2006) A Tokaj–Zempléni-hegyvidék kifagyásos formáinak természetvédelmi szempontú értékelése. Folia Historico Nat Mus Matraensis 30:383–390
Kiss T, Bódis K (2000) Late Pleistocene drainage reconstruction of an alluvial fan covered by sand dunes. In: Dulias R, Pelka-Gosciniak P (eds) Aeolian processes in different landscape zones. University of Silesia, Sosnowiec, pp 149–163
Kiss T, Tornyánszki É (2006) Futóhomok-területek geomorfológiai vizsgálata a Duna–Tisza közén. 3. Magyar Földrajzi Konferencia, Budapest. www.geography.hu
Kiss T, Sipos Gy, Kovács F (2009) Human impact on fixed sand dunes revealed by morphometric analysis. Earth Surf Proc Land 37:700–711
Kordos L (1979) A magyarországi paleoklimatológiai kutatások módszerei és eredményei. OMSZ L. kötet, OMSZ, Budapest, p 167
Kordos L (1984) Magyarország barlangjai. Gondolat, Budapest
Kordos L, Schweitzer F (2000) Negyedidőszak-kutatás. In: Karátson D (szerk) Pannon enciklopédia. Kertek, Budapest, pp 442–446
Korpás L (1998) Paleokarst. MÁFI, Budapest
Korpás L (2000) A Visegrádi-hegység. In: Karátson D (szerk) Pannon enciklopédia. Kertek, Budapest, pp 327–328
Kósa A (1992) Alsó-hegyi zsombolyatlasz. MKBT, Budapest
KOZ (2009) http://kozpark.hu/archives/2009/01/30/Obudai_agyagbanyak_teglagyarak_tortenete
Kovács S (1989) Geology of North Hungary: Paleozoic and Mesozoic terranes. 21st Eur. Micropal. Coll. Guidebook. Hungarian Geol. Soc. Budapest, 15–36
Kretzoi M (1987) Remarks on the correlation of European, North American and Asian Late Cenozoic local biochronologies. In: Pécsi M (ed) Pleistocene environment in Hungary. MTAFKI, Budapest, pp 5–38
Litt T, Behre KE, Meyer KD, StephanJH, Wansa S (2007) Stratigraphische Begriffe für das Quartär des norddeutschen Vereisungsgebietes. Eiszeitalter und Gegenwart 56(1–2):7–66
Lóki J (1981) Belső-Somogy futóhomok-területeinek kialakulása és formái. Acta Geogr Debr 18–19:81–111
Lóki J (1994) Mezőgazdaság-központú természetföldrajzi vizsgálatok a Duna–Tisza köze északi felének példáján. Kandidátusi értekezés, kézirat, Debrecen
Lóki J (2009) Eolikus folyamatok vizsgálata Magyarországon. In: Kiss T, Mezősi G (szerk) Recens geomorfológiai folyamatok sebessége Magyarországon. (Földrajzi Tanulmányok2.) JGYF Kiadó Szeged, pp 173–183
Lóki J, Félegyházi E (2007) ABodrogköz kialakulása és mai felszíne. http://geogis.detek.unideb.hu/Kutatas/Felszinfejlodes/Bodrogkoz.pdf
Lotter AF (1999) Late-glacial and Holocene vegetation history. Veg History Archaeobotany 8(3):165–184
Lovász Gy (2006) A pleisztocén–holocén hordalékkúpok fejlődéstípusai Magyarországon. Tiszteletkötet Hahn Gy.-nek. Egyetemi Kiadó, Miskolc, 117–121
Lovász Gy, Majoros Gy (1994) Magyarország természeti földrajza. JPTE, Pécs

Majoros Gy, Péró Cs (2000) A Kárpát-medence földtörténete – újpaleozoikum. In: Karátson D (szerk) Pannon enciklopédia. Kertek, Budapest, pp 74–87
Marosi S (1970) Belső-Somogy kialakulása és felszínalaktana. Akadémiai Kiadó, Budapest
Marosi S, Szilárd J (1981) ABalaton kialakulása. Földrajzi Közlemények 29(1):1–30
Márton E, Márton P (1996) Large scale rotations in North Hungary during the Neogene as indicated by palaeomagnetic data. In: Morris A, Tarling DH (eds) Palaeomagnetism and tectonics of the mediterranean region, vol 105. Geological Society Special Publication, London, pp 153–173
Martonné Erdős K (1995) Magyarország természeti földrajza. KLTE, Debrecen
Meese DA, Gow AJ, Alley RB, Grootes PM, Ram M, Taylor KC, Zielinski GA, Bolzan JF, Mayewski PA, Waddington ED (1997) The Greenland ice sheet project 2 depth-age scale: Methods and results. Journal of Geophysical Research 102:26411–26423
Mezősi G (2008) Magyarország környezetföldrajza. JATEPress, Szeged
Mezősi G, Szatmári J (1996) Széleróziós vizsgálatok a Duna–Tisza közén. In: Szabó L. (szerk.) A termőföld védelme. GATE, Gödöllő, pp 24–33
Mike K (1991) Magyarország ősvízrajza és felszíni vizeinek története. AQUA, Budapest
MNA (1989) Magyarország Nemzeti Atlasza. Carthographia, Budapest
Molnár B (1965) Adatok a Duna-Tisza köze fiatal harmadidőszaki és negyedkori rétegeinek tagolásához és származásához nehézásvány összetétel alapján. Földtani Közlöny 95:217–225
Müller P (2000) A középső miocén első fele. In: Karátson D (szerk) Pannon enciklopédia. Kertek, Budapest, pp 124–127
MT = Glatz F szerk (2002) Magyar Tudománytár I. Föld, víz, levegő. Kossuth, Budapest
Nádor A, Thamóné BE, Magyari Á, Babinszki E, Dudko A, Tóth Z (2005) Neotektonika és klímaváltozás együttes hatása a Körös-medence késő pleisztocén vízhálózatfejlődésére. MÁFI évi jelentése, pp 131–147
Nagy D (2004) A történeti tájhasználat és felszínborítás rekonstrukciójának lehetőségei archív térképek feldolgozásával. KÉP Program, Budapest
Nagy Á, Tóth T, Sztanó O (2007) A „harmadik folyó" – Pleisztocén folyóvízi üledékek ultranagy felbontású szeizmikus szelvényeken a Tisza Tiszadob–Martfű közti szakaszán. Földtani Közlöny 137(2):239–260
Nagymarosy A (2000a) Magyaroszági eocén. In: Karátson D (szerk) Pannon enciklopédia. Kertek, Budapest, pp 110–112
Nagymarosy A (2000b) A legkorábbi miocén és a korai miocén második fele. In: Karátson D. (szerk.) Pannon enciklopédia. Kertek, Budapest, pp 116–124
NOAA (2008) http://www.ncdc.noaa.gov/paleo/icgate.html
Nováky B (1991) Climatic effects on runoff conditions in Hungary. Special Issue on the landscape-ecological impact of climatic change. Earth Surface and Landforms 7:593–600
Nyári D, Kiss T, Rosta Sz, Sipos Gy, Geiger J (2008) Emberi tevékenység következtében történt geomorfológiai változások vizsgálata egy Kiskunhalas melletti régészeti ásatás területén. 4. Magyar Földrajzi Konferencia, Debrecen. www.geography.hu
Pécsi M (1959) A magyarországi Duna-völgy kialakulása és felszínalaktana. Akadémiai Kiadó, Budapest
Pécsi M (1975) A Kisalföld természeti földrajzi jellemzése. A felszín kialakulása. In: Marosi S (szerk) A Kisalföld és a Nyugat-magyarországi-peremvidék. Budapest, Akadémiai Kiadó, pp 46–60
Pécsi M (1984) Is typical loess older than one million years? In: Pécsi M (szerk) Lithology and stratigraphy of loess and paleosols. (Elmélet–Módszer–Gyakorlat 30.) MTA FKI, Budapest, pp 213–224
Pécsi M (1988b) A Vértes–Velencei-hegyvidék. Helyzet és elhatárolás. Domborzati körzet és tájbeosztás. In: Ádám L, Marosi S, Szilárd J (szerk) A Dunántúli-középhegység. B. Akadémiai Kiadó, Budapest, pp 190–194
Pécsi M (1991) Geomorfológia és domborzatminősítés. (Elmélet–Módszer–Gyakorlat 53.) MTAFKI, Budapest
Pécsi M (1993) Negyedkor és löszkutatás. (Elmélet–Módszer–Gyakorlat 54.) MTAFKI, Budapest
Pécsi M (1996) Geomorphological regions of Hungary. MTAFKI, Budapest
Pécsi M (1997) Szerkezeti és váztalajképződés Magyarországon. (Elmélet–Módszer–Gyakorlat 57.) MTAFKI, Budapest
Pécsi M, Kretzoi M eds (1985) Problems of the neogene and quarternary in the Carphatian Basin. (Studies in geography in Hungary 19.) Akadémiai Kiadó, Budapest
Péró Cs (2000) A Kárpát-medence földtörténete – prekambrium és ópaleozoikum. In: Karátson D (szerk) Pannon enciklopédia. Kertek, Budapest, pp 74–87
Pinczés Z (1977) Hazai középhegységek periglaciális planációs felszínei és üledékei. Földrajzi Közlemények 25(1–2):29–45
Pinczés Z (1986) Periglaciális formák és üledékek térbeli rendje egy vulkanikus hegy lejtőjén. Földrajzi Értesítő 35(1):28–42
Pinczés Z (1995) A Tokaji-hegység krioplanációs hegylábfelszíneinek kialakulása és típusai. Földrajzi Értesítő 44(3–4):205–216
Pinczés Z, Martonné Erdős K, Dobos A (1993) Eltérések és hasonlóságok a hegylábfelszínek pleisztocén felszínfejlődésében. Földrajzi Közlemények 3:149–162
Prinz Gy (1936) Magyarország tájrajza. Magyar föld, magyar faj, I. kötet. Királyi Magyar Egyetemi Nyomda, Budapest
Rácz L (2008) Magyarország környezettörténete az újkorig. MTATörténettudományi Intézete, Budapest
Rónai A (1973) Proportion and character of Quaternary tectonic movements in the Hungarian Basin. Földrajzi Közlemények 97(21):153–160
Rónai A (1974) Size of Quaternary movements in Hungary. Acta Geologica Hungarica 18:39–44
Rónai A (1985) Az Alföld negyedidőszaki földtana. Geologica Hungarica Series Geologica 21, MÁFI. Budapest

Ruszkiczay-Rüdiger Zs, Dunai T, Fodor L, Bada G, Leél-Őssy Sz, Horváth E (2006) A negyedidőszaki függőleges kéregmozgások számszerűsítése a Duna völgyében a korábbi kronológiai adatok és új, kozmogén 3He kitettségi kor mérések alapján. Földtani Közlöny 135(3):373–402

Schweitzer F (1993) Domborzatformálódás a Pannóniai--medence belsejében a fiatal újkorban és a negyedidőszak határán. MTA, doktori értekezés, Budapest

Schweitzer F (2002) Pleisztocén. In: Karátson D (szerk) Pannon enciklopédia. Kertek, Budapest, pp 130–135

Schweitzer F, Szöőr Gy (1992) Adatok a Magyar-medence száraz-meleg klímájához a mogyoródi „sivatagi kéreg" alapján. Földrajzi Közlemények 106(3–4):105–123

Somogyi S (1961) Hazánk folyóhálózatának fejlődéstörténeti vázlata. Földrajzi Közlemények 9(1):25–50

Somogyi S (1973) Adatok a fiatal kéregmozgások hazai földrajzi hatásaihoz. Geonómia és bányászat 6(1–4):245–256

Somogyi S (1987) Relationship between environmental changes and human impact until the 9th century. In: Pécsi M, Kordos L (eds) Holocene Environment in Hungary. (Elmélet–Módszer–Gyakorlat 41.) MTAFKI, Budapest, pp 25–37

Somogyi S (2000a) Aterméazeti változások és a társadalmi-gazdasági folyamatok kölcsönhatása az Alföldön a Honfoglalás előtt. In: Frisnyák S (szerk) Az Alföld történeti földrajza. NYF, Nyíregyháza, pp 7–25

Somogyi S (2000b) Ősvízrajz. In: Karátson D (szerk) Pannon enciklopédia. Kertek, Budapest, pp 238–242

Stegena L, Géczy B, Horváth F (1975) A pannon-medence késő-kainozóos fejlődése. Földtani Közlöny 105(2):101–123

Sümeghy J (1944) A Tiszántúl. (Magyar tájak földtani leírása 6.) Magyar Királyi Földtani Intézet, Budapest

Sümegi P, Hertelendi E (2001) Negyedidőszak végi éghajlati változások rekonstrukciós lehetőségei őslénytani és izotópgeokémiai módszerekkel Magyarországon. Acta Geogr Debr 35:317–342

Szabó J (1995) Csuszamlásos folyamatok szerepe a magyarországi tájak fejlődésében. Kossuth, Debrecen

Szádeczky KE (1938) Geologie der rumpfungarländischen Kleinen Tiefebene. Mitteilungen der berg- und hüttenmaschinen Abteilungen

Szeidovitz G, Bus Z, Gribovszki K (2001) Research for seismogenic zones seismic hazard estimation for Budapest. Acta Geod Geoph 34:417–438

Székely A (1973) A Magyar-középhegyvidék negyedidőszaki formái és korrelatív üledékei. Földrajzi Közlemények 21(2):185–203

Székely A (1977) Periglaciális domborzatalakulás a magyar középhegységekben. Földrajzi Közlemények 101(1–3):46–59

Székely A (1983) A pleisztocén periglaciális domborzatformálás Magyarországon. Földrajzi Értesítő 32 (3–4):389–398

Székely A (1997) Vulkánmorfológia. Eötvös, Budapest

Szemán A (2009) Újabban előkerült középkori bányászszerszámok Rudabányáról. Bányászattörténeti közlemények 7:3–30

Szunyogh G (1982) A hévizes eredetű gömbfülkék kioldódásának elméleti vizsgálata. Karszt és Barlang II:83–88

Telbisz T (2004) Karsztmorfológiai elemzések különböző léptékű digitális domborzatmodellek alapján. 2. Magyar Földrajzi Konferencia, Szeged. http://geography.hu/mfk2004/mfk2004/cikkek/telbisz_tamas.pdf

Thamó-Bozsó E, Murray AS, Nádor A, Magyari Á, Babinszki E (2007a) Investigation of river network evolution using luminescence dating and heavy mineral analysis of Late-Quaternary fluvial sands from the Great Hungarian Plain. Quat Geochronol 2(1–4):168–173

Thamó-Bozsó E, Magyari Á, Nagy Z, Unger Z, Kercsmár Zs (2007b) OSL dates and heavy mineral analysis of upper quaternary sediments from the valleys of the Ér and Berettyó rivers. Geochronometria 28:17–23

Timár G, Telbisz T, Székely B (2006) Űrtechnológia a digitális domborzati modellezésben: az SRTM adatbázis. Geodézia és kartográfia 2006:11–16

Varga Z (2006) A Kárpát-medence faunatörténete és állatföldrajza. In: Fekete G, Varga Z (szerk) Magyarország tájainak növényzete és állatvilága. MTA Társadalomkutató Központ, Budapest, pp 44–76

Veress M (2004) A karszt. BDF, Szombathely

Visy Zs, szerk (2003) A magyar régészet az ezredfordulón. NKÖM, Budapest

Vörös I (1987) Large mammalian faunal changes during the Late Upper Pleistocene and Early Holocene times in the Carpathian Basin. In: Pécsi M (ed) Pleistocene Environment in Hungary. Elmélet–Módszer–Gyakorlat 42.) MTAFKI, Budapest, pp 81–103

Vörös A (2000) Triász, jura, kréta. In: Karátson D (szerk) Pannon enciklopédia. Kertek, Budapest, pp 95–107

Walker M (2005) Quaternary dating methods. Wiley, London

Wein Gy (1978) A Kárpát-medence kialakulásának vázlata. Általános Földtani Szemle 11:5–34

Zámbó L (1998) Az Aggteleki-karszt felszínalaktani jellemzése. Földrajzi Értesítő 47(3):359–379

Zámbó L, Ford D (1997) Limestone dissolution processes in Béke doline. Earth Surface Process 22(6):531–543

Zámbó L, Weidinger T (2006) A karsztkorróziós talajhatás néhány tényezőjének vizsgálata esőszimulációs kísérletek alapján. In: Táj, környezet és társadalom. Ünnepi tanulmányok Keveiné Bárány Ilona professzor asszony tiszteletére. SZTE, Szeged, pp 757–765

Zelenka T (2000) Ásványok, kőzetek, bányakincsek. In: Karátson D (szerk) Pannon enciklopédia. Kertek, Budapest, pp 178–183

Zöller L, Oches E, McCoy WD (1994) Towards a revised chronostratigraphy of loess in Austria with respect to key sections in the Czech Republic and in Hungary. Quat Geochronol (Quat Sci Rev) 13:465–472

# Climate of Hungary

# 2

**Abstract**

The climate of the Carpathian Basin reflects the combined characteristics of the neighbouring regions. Despite of the small area and the modest orographic differences of the country, the climate elements (e.g. solar radiation, temperature or precipitation conditions) have distinctive spatial and temporal characters. The result of climatic measurements and the differences in the values will be described in this chapter. Besides, the agro-economical (e.g. drought hazard) and environmental resource potential of the climate (e.g. wind energy utilization) will be introduced. The latest will be described in detail (e.g. recreational potentials and skiing, are limited in natural conditions), as well as the renewable energy sources. Based on estimations, the potential amount of solar radiation (72 PJ,), wind energy (46 PJ), geothermal energy (50 PJ) and biomass (120 PJ) is far higher than the actual usage of these energy resources, which altogether is lower than 10 PJ, if the biomass usage (ca. 52 PJ) is not accounted. In this chapter, the special climate characters of the cities will be analysed too. Slightly detailed description is provided on the potentials (mountainous medical locations) and hazards of the climate (e.g. rainstorm hazard, heat stress, increasing extremities). The past and future climate changes of the Basin are also analysed.

## 2.1 Factors Influencing the Climate of the Carpathian Basin, the Climate Type

The climatic features of the Carpathian Basin are primarily defined by its location, which influences the angle of the incoming solar radiation. The country is situated between 45°45′N and 48° 35′N latitudes, and between 16°5′E and 22°58′E longitudes. It is situated in between different climatic zones; therefore, the climate is mainly characterized by the features of the humid oceanic climate slightly varying temperatures and by the features of the less humid continental climate with more extreme temperatures. The annual mean temperature is about 2.5 °C higher than the typical value at this latitude, due to the influences (e.g. currents) of the Atlantic Ocean and the Mediterranean Sea that cause positive temperature anomaly in the Carpathian Basin.



G. Mezősi, *The Physical Geography of Hungary*, Geography of the Physical Environment,
DOI 10.1007/978-3-319-45183-1_2

The climate is transitional, whose actual character is defined by the interactions of four major pressure systems ("action centres") and the resulting strongly variable surface pressure pattern (Péczely 2002):

- Temperate maritime air masses ("Iceland low"), which transport mild air in winter, and cool but humid air in summer;
- The temperate air masses of continental origin ("Siberian high"), which cause dry and very cold weather in winter and can cause critical air pollution in spite of the northerly and easterly winds (since they can strongly hinder the ventilation of the basin in the time of their winter activity;
- Subtropical air masses of oceanic origin ("Azores high"), which are active throughout the year, and transporting humid air with southerly and westerly airflow;
- The subtropical air masses of continental origin, carrying heat in the summer with southerly airflows ("Persian trough").

In the summer half-year, 60–70 % of the air masses arriving have oceanic origin, while continental air masses are prevailing in the winter half-year. In addition to these major impacts, Hungary is also influenced by the Mediterranean effect, prevailing primarily in periods of quiet autumn rains and in the early winter snowfalls (MET2 2009).

According to the Trewartha climate classification system, which is one of the most common classifications, the climate of Hungary is a (humid) subtype of the continental climate with long-warm season and a significant temperature fluctuation. Its main feature is the four distinct seasons, which sometimes take on an extreme character. The summer is long, it lasts for at least 3 months, its mean temperature is above 18 °C, and the daily maximum is frequently above 35 °C in the middle of the summer. The winter temperature is instable, colder and milder periods alternate; however, the number of months with a mean temperature below 0 °C is not more than three. The amount of total precipitation significantly differs in Hungary, but still satisfies the water demand of the society (see Chap. 3).

Apart from the transitional character, the other key feature of the climate is its basin character. Firstly, the disadvantages of the closed basin have to be highlighted, as the significant temperature variation, or the low wind energy (and to a smaller extent, the basin character can also influence the amount of precipitation). The irradiation conditions are more advantageous, especially for the agriculture. Among the disadvantageous consequences, the negative human health effect of the increased winter air pollution has to be mentioned.

Even the slight topographic differences can cause significant local variations in the climate; e.g., in the winter half-year, the total amount of sunshine duration is 50 % higher in the mountainous areas because of the cleaner air. More than a half of the country is a lowland plain area, with an elevation less than 200 m above sea level; thus, there are no significant orographically influenced regional differences in the climate.

On the basis of the global climate classification systems (e.g. Köppen, Trewartha), regional climate differences within the country cannot be identified due to the size of the country. Therefore, a different local-scale classification system has to be applied to reveal the regional differences. Bacsó (1959) made an attempt at it, and some preliminary attempts are also described in the climate atlas of Kakas (1960). The regional climate classification was developed on the basis of Péczely's (1979) water and heat availability categories. Based on these categories, 12 regions were defined which are characterized as wet or dry from the point of view of water supply and as warm or cool from the point of heat supply.

## 2.2 Spatial and Temporal Characteristics of the Main Climatic Parameters

In this chapter, the spatial features and interrelations of the sunshine duration, temperature, precipitation and wind conditions are introduced, which define several site conditions important for the society (see Sect. 2.3.).

### 2.2.1 Regional Differences in Sunshine Duration

Although the amount of solar radiation is reduced by several factors, (e.g. half of it is absorbed or reflected by the clouds), a rather large amount of energy still reaches the surface. Bartholy (2000) calculated that the annual surface radiation balance of an area of 20 $km^2$ equals to the yearly energy consumption of capital city, Budapest (about $10^{11}$ MJ). The spatial distribution of sunshine shows that its value is higher by 10–15 % in the central part of the Great Hungarian Plain than that in the northern or western near-border areas (Fig. 2.1; MNA 1989; MT 2002; OMSZ 2003). The highest values in the central part of the Great Hungarian Plain can exceed 5000 MJ, while 500 MJ lower values are typical in the western and north-eastern border regions. The amount of energy in the summer months is 5–6 times greater than that in the winter months. The highest values can be measured in July and not in June, because of the much smaller cloud cover of the former.

The duration of sunshine varies between 1800 and 2200 h in the Carpathian Basin (Fig. 2.2). The highest values occur in the southern part of Hungary; however, they still do not reach the half of the potential latitudinal values. The relief can significantly diversify the spatial pattern; e.g., during winter periods, when the mountains emerge from the cold, foggy air that covers the basin, higher temperatures and higher sunshine duration are measured in the mountains than in the lowlands.

### 2.2.2 Spatial Distribution of Air Temperature

The air temperature is primarily influenced by the cyclones and anticyclones of the major pressure systems described earlier, but secondarily, it is

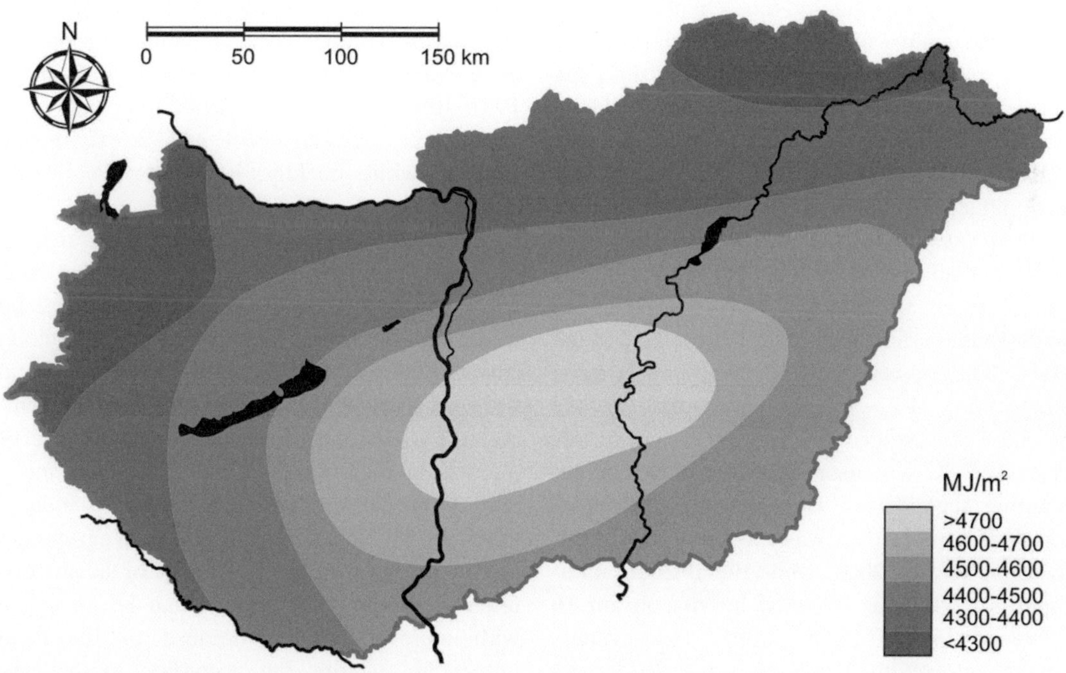


**Fig. 2.1** The spatial distribution of the yearly amount of global radiation (MT 2002)

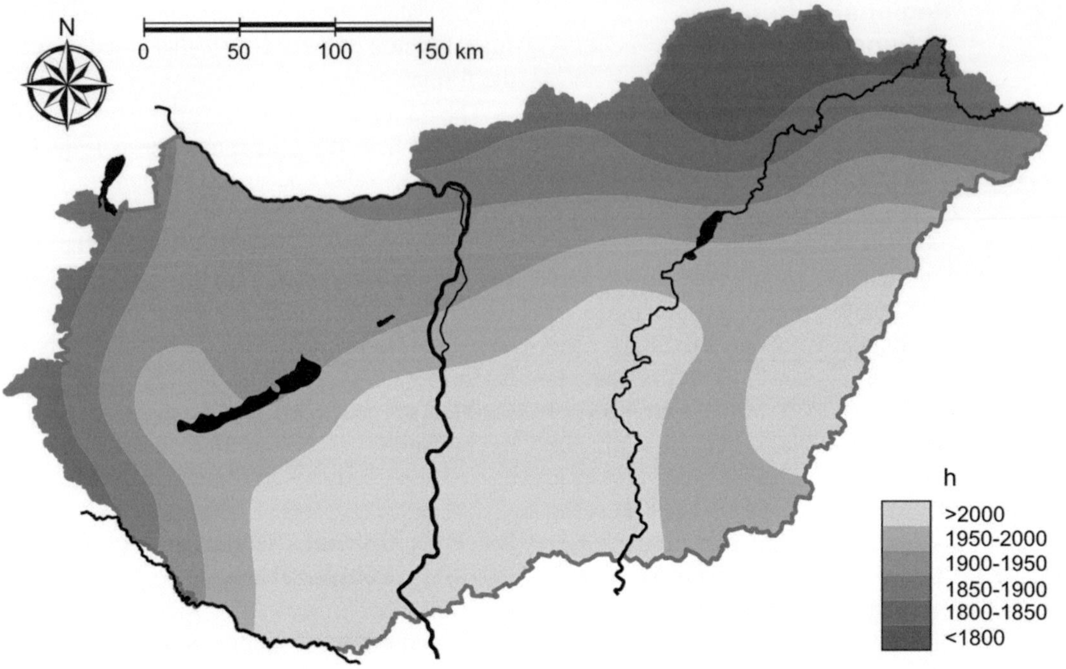


**Fig. 2.2** The total annual sunshine duration (after OMSZ 2003)

influenced by topography too. On the western part of the country and in the hilly regions, lower annual mean temperature (8–11 °C) values are characteristic, while on the majority of the Great Hungarian Plain, the annual mean temperature is higher than 10 °C (Fig. 2.3). The slope aspect also influences the temperature; generally, the southern slopes of the mountains are much warmer than the northern ones. The highest annual mean temperatures are typical in the south-south-eastern part of the Great Hungarian Plain. The observed maximum annual mean temperature (13.3 °C) was measured in the downtown of Szeged (SE Hungary), and in this year, 12.7 °C was measured also in the neighbouring settlements. The absolute maximum temperature (41.9 °C) was recorded in Kiskunhalas on 20 July 2007, while the absolute minimum (–35 °C) was recorded in Miskolc on 16 February 1940 (MET3 2009). The average annual temperature fluctuation is over 20 °C, and the absolute annual temperature fluctuation exceeds the 65 °C in the whole country (MT 2002), which shows the continental character of the climate. Based on the 30-year average of the 1961–1990 periods, the mean temperature in January is between −1.5 and 3.5°C in the country, while in July, it is between 19 and 21 °C.

The yearly temperature pattern is defined by the large-scale climatic factors (e.g. global radiation), while the daily patterns are defined by local environmental factors (e.g. relief, cloud cover; MET2 2009). Between the northern and southern parts of the country, only a slight temperature difference can be observed because of the size of the country. The 3 °C annual mean temperature difference results in the 6–9 days' advance of the phenological phases in agriculture in the southern areas. The effect of height differences of the hilly regions can be estimated with more difficulties because of the local influencing factors, but according to the estimates, it is lower than 0.65 °C/100 m.

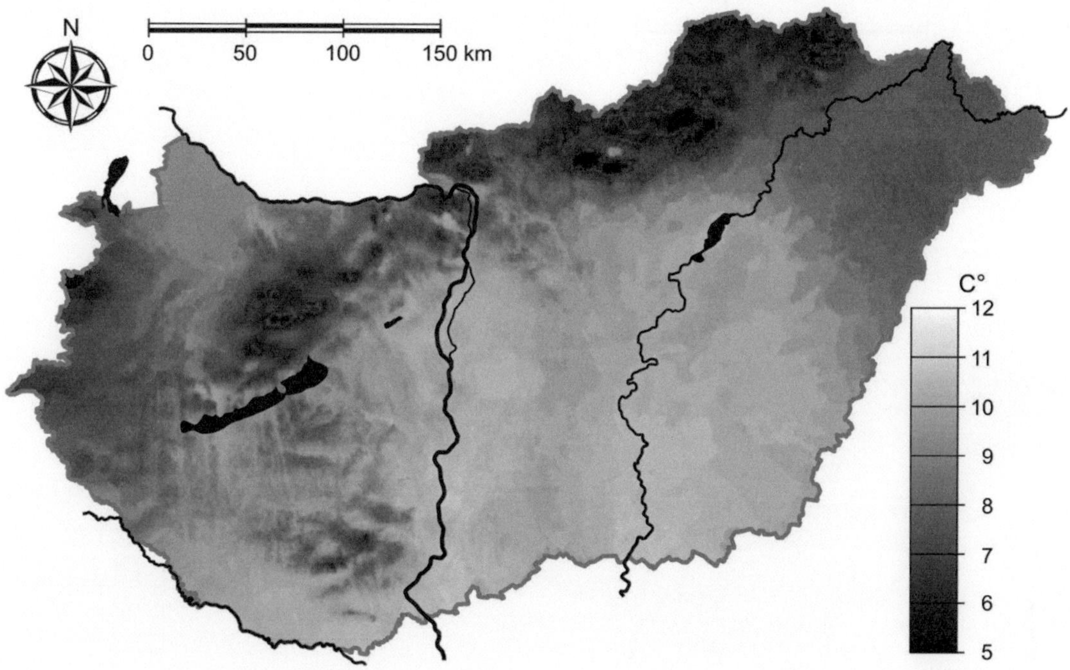


**Fig. 2.3** The average annual mean temperature in Hungary between 1901 and 2000 (after MET3 2009)

### 2.2.3 Spatial Differences in Precipitation

The 600-mm annual precipitation in the centre part of the basin generally fulfils the water demands, as in average, the evaporation rate is 89 % of the total precipitation, though the evaporation can be estimated with errors (MT 2002). In general, the amount of precipitation decreases from the west towards the east (with the increasing continental character), and it increases with the increase in the height above sea level (by 15–35 mm/100 m; OMSZ 2003). The precipitation seems to be sufficient; however, its spatial and temporal distribution and its extremes can cause problems.

Based on the 30-year average of the 1961–1990 period, the precipitation amount varies mostly between 500 and 800 mm: the centre part of the Great Hungarian Plain is the most arid, while the south-western part of the country, the foothills of the Alps, and the higher regions of the mountains are the most humid (Fig. 2.4). The precipitation exceeds 800 mm on 3 % of the area (and exceeds 700 mm on 14 %), and it is lower than 500 mm in 2 % (but it is lower than 550 mm in 27 %; Lovász and Majoros 1994). The recorded lowest annual precipitation (203 mm) occurred in 2000 in Szeged.

The precipitation amount usually exceeds the evaporation (400-600 mm) in the whole country (Weidinger and Mészáros 2000). The total amount of yearly evaporation is influenced by not just the precipitation, but also by the soil moisture (Ács et al. 2007); thus, it is the greatest in Western Transdanubia. However, the potential evaporation is much higher than the precipitation in the summer half-year; thus, it sometimes reaches the 250–300 mm in the centre part of the Great Hungarian Plain, which means 250–300 mm water shortage. In periods of lower precipitation, lower air humidity or higher temperatures, water shortage (drought) can be evolved, which often appear in the central areas of the Great Hungarian Plain. Drought has several different types, depending on the point of view of the study: thus, meteorological or atmospheric (the precipitation is lower than

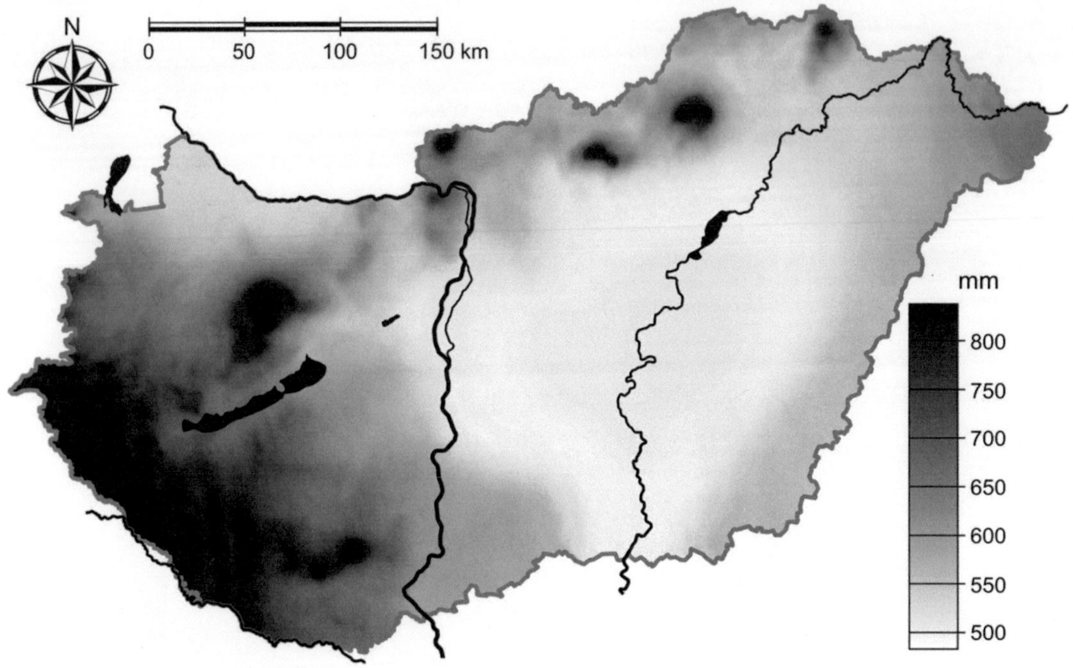


**Fig. 2.4** Spatial distribution of the mean annual precipitation between 1960 and 1990 (after OMSZ 2003)

usual, while the temperature is high), hydrological (low level of groundwater and rivers), pedological (water shortage hinders the vegetation) and agricultural drought can be distinguished (Mezősi 2008).

Drought is one of the most serious environmental challenges in Hungary, caused by several factors. The amount and distribution of precipitation (in the growing season) is a major factor, but the evaporation, which mainly depends on temperature, the fertile layer and the physical character of the soil, is also important. From the agricultural point of view, these factors can hardly be modified. The evolution of droughts also depends on land use (its water demand), the selection of culture and the moisture saving cultivation methods; thus, the intensity of drought can be artificially influenced by several ways.

In Hungary, the most commonly used index to describe drought is the Pálfai Drought Index (PAI; Pálfai 2002), since the data of the widely used Palmer Drought Severity Index are not available (Horváth et al. 2001). The PAI is suitable to describe the yearly changes of drought rate, and it integrates the April–August air temperatures and the monthly precipitation values with different weighting (summer month have the highest weights); moreover, correction factors estimate the heat periods and dry periods, and the changes of the groundwater are also used for the calculation. The typical values are between 6 and 12 (°C/100 mm) in Hungary. The areas with a value of 5–6 are mildly, while the areas with values over 10 are greatly affected by drought (Fig. 2.5).

The spatial pattern of drought occurrence shows high variability. Most part of the country is affected by mild drought that can occur in every third year on the centre part of the Great Hungarian Plain. Heavy drought occurred only in the centre part of the Great Hungarian Plain in the past 100 years, with a return period of 20–50 years, while it occurs every 50–100 years in its peripheral regions (Pálfai 2002). Serious drought events were documented as also in the previous centuries: it was well documented in 1794 and 1863, and during the period of 1779–1800, altogether 11 unusually dry years

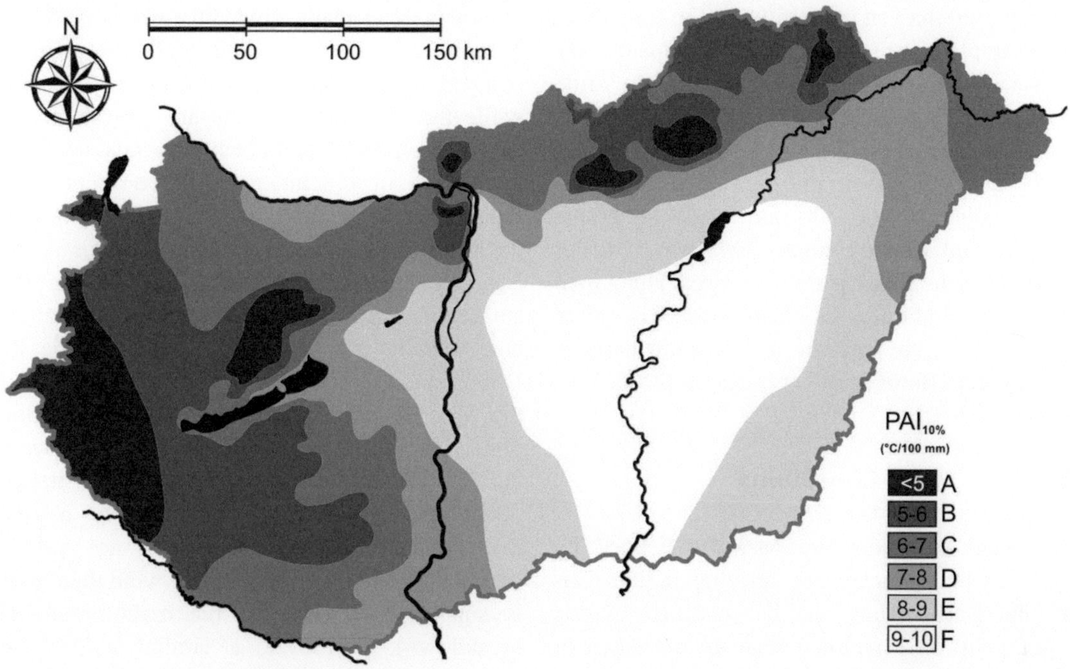


**Fig. 2.5** Drought map of Hungary (after Pálfai 2004) **a** no drought hazard, **b** low hazard, **c** moderate hazard, **d** medium hazard, **e** serious hazard, **f** extreme drought hazard

appeared. Every drought events have unique characteristics that make any future predictions related to the hazard difficult, despite the long-term data sets, besides it would require a precise, long-term weather forecast, which is not available at the moment.

The temporal changes in precipitation distribution in Hungary have well distinguishable characters: the larger proportion of the yearly precipitation occurs in the summer half-year, and the yearly precipitation has two peak periods. The first precipitation maximum occurs in May–June, which is connected to the activity of the Atlantic air masses. (This early summer precipitation maximum was formerly called as the "European monsoon", although incorrectly.) The second maximum in autumn (September–October) is caused by the influence of the Mediterranean air masses. The minimum precipitation occurs in the middle of winter (December–January), when the anticyclones cause cold and dry weather. The annual precipitation amount has decreased in the past decades (MT 2002). The significantly recorded 100-mm decrease in the Danube-Tisza Interfluve during the past 100 years is particularly alarming (Kertész and Mika 1999). The observed decrease mainly affects the spring period; however, its spatial distribution is diverse.

The extreme values of precipitation demonstrate the extremities of the climate. In some years, more than 50-day-long dry period was recorded without any precipitation in Great Hungarian Plain: it had not rained for 52 days in Gyula in 1897. On the other hand, an annual amount of over 1000 mm is not rare in the western part of Hungary. High intensity storms can also cause serious damages, as it was recorded several times in the past decade, causing floods and flash floods (e.g. in 2010), soil erosion and accumulation (e.g. at Mátrakeresztes, Bükkzsérc, Mád, Gönc, Szabó 2001), or landslides. The precipitation is much more unpredictable than the temperatures, especially in the central part of the Great Hungarian Plain, where the yearly variability is significant.

The efficiency of the agriculture depends more on the amount of precipitation (and its distribution) than on the fertility of the soil. Thus, for example, the more humid environment compensates the lower fertility of soils in the Little Hungarian Plain compared to the Great Hungarian Plain. The precipitation difference is 200 mm in the growing season and the typical precipitation amount is 400–500 mm in West Transdanubia and North Hungary, while it is under 300 mm in the centre part of the Great Hungarian Plain; in addition, its distribution is more favourable in the western and northern parts of the country (Botos and Varga-Haszonits 1974).

### 2.2.4 Wind Conditions

The wind conditions are basically defined by cyclones and anticyclones arriving in the Carpathian Basin. They can be modified by the topography on a regional scale in the cases of winds under 100 m absolute height (e.g. orography is likely to cause the variety of wind conditions in the North Hungarian Mountains). At a few kilometres height, western winds are dominant. The main wind direction is NW and N; thus, winds arrive through the Devín (Dévény) Gate reach the line of the Tisza River, while the NE and N winds are typical on the eastern part of Hungary, which winds are often the source of the intense cold convections in winter (Fig. 2.6). The typical temporal pattern of the wind direction is that east of the Tisza, the NW winter winds change to NE during the summer half-year (Tar 2007; the wind directions and velocity are based on the map made by Dobosi and Felméry in 1971). The complexity of the wind directions is caused by the circulation phases and the topographic conditions (Péczely 2002). The distribution of the near-surface directions is not constant either, and the assessment of measurement results is not simple because the relative frequency of wind directions is small (max. 27 % in the eight-directional breakdown). Moreover, a similar amount of winds proceed in the main directions, as in the opposite directions (e.g. 25 % of winds is NE and 22 % SW at Nyíregyháza).

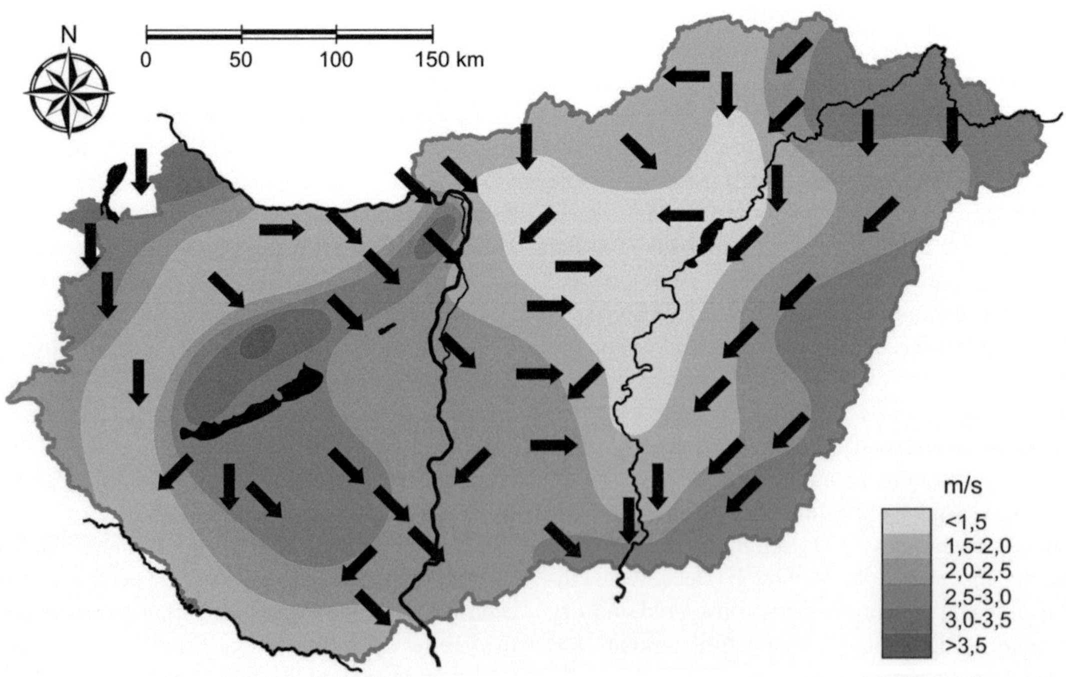


**Fig. 2.6** Prevailing wind directions and average wind speed in the summer period (after Bartholy and Weidinger 2000)

The direction and velocity of the near-surface airflow reflects the orographic influence. The annual average wind velocity is 2–4 m/s. The area of the mountains, and especially their southern parts, and the western and eastern peripheries of the country are characterized by average wind speeds over 3 m/s. Thus, the Devín (Dévény) Gate and the southern slopes of the Transdanubian Mts. and the North Hungarian Mts. can be taken into consideration from the point of view of wind energy (OMSZ 2009). The territorial differences are similar according to the values calculated by the WAsP model, indicating below 150 W/m$^2$ wind energy in average, and it exceeds the 200 W/m$^2$ only in the above-mentioned regions (Radics 2004).

## 2.3 Social Aspects of Climate

The socio-economic activities are sensitive to climate changes to different extents. Agriculture, building industry and energy production are the most affected economic sectors by climate change. Conversely, the important question is that how these sectors influence the climate (Mezősi 2008). From climate sensitivity point of view, the agriculture and its productivity are affected by several climatic parameters, e.g. the rate and distribution of precipitation and temperature, and their extremities having diverse effects in different phenological phases. The natural resources and their conditions, such as the wind or solar energy, are also important for the society. The effect of climate change, e.g. growing temperatures and increasing frequency of extremes (rainstorms or heavy rains), and climatic hazards, on these resources and their utilization, has to be considered. From this point of view, the optimal extent of adaptation to the short-term projections is a key question.

In the followings, three regional and national issues will be discussed as examples within the complex system of relationships. Afterwards, (further) features and hazards related to the climate and its changes will be introduced, which are recommended to take into consideration during geographical and environmental assessments.

### 2.3.1 Number of Snow-Covered Days: Potentials for Skiing

The number of snow-covered days is the highest in the Mátra Mts. (the observed maximum of 154 days occurred in the winter of 1943/44). Therefore, the question is whether the natural conditions are appropriate for skiing in Hungary? If we consider only the number of snow-covered days, the answer may be positive, since the average 70–100 snow-covered days of the mountains can be sufficient (Fig. 2.7). However, other natural factors make skiing possibilities questionable: for example, the average natural snow-covered depth is only 30–35 cm, which is insufficient for skiing, especially if the high variability of snow depth is also considered (there were winters, e.g. 2006/2007, when practically no snow cover occurred in the mountains). Considering the tendency of climate change, the increase in winter precipitation and temperatures (Weidinger 2000), the potentials of winter sports are more uncertain because of the decreasing snow cover. Furthermore, the precipitation increases with the increasing elevation in the winter which is only half or third of the summer values. Of course, more favourable topographic features would result in a better situation; however, the system can be sustained with appropriate maintenance and artificial snow-making equipment and with preparations for the variable winter climate. (It is also has to be considered that the artificial snow may degrade the vegetation and the soil on the slopes.)

### 2.3.2 Potentials to Utilize Renewable Climatic Energy Resources

According to the Hungarian Academy of Science Subcommittee on Renewable Energy (2006), the theoretical value of renewable energy sources is 2500 PJ/year; however, the exploitable quantity is significantly smaller, if we take into consideration the aspects of technology, economy and the expected future potentials. The realistic value

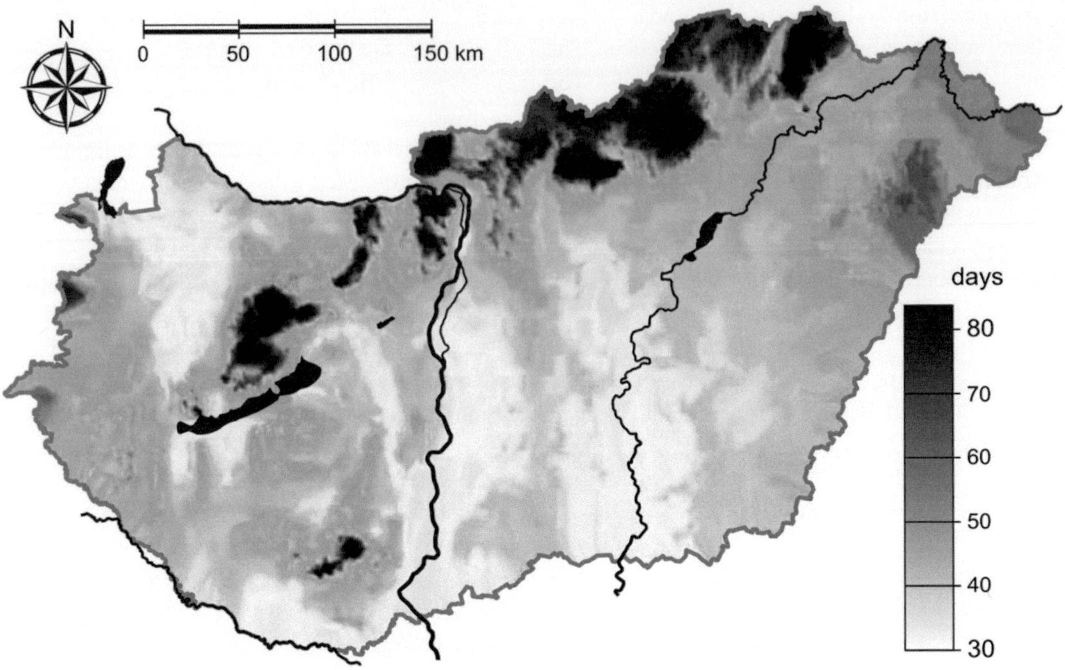


**Fig. 2.7** The number of snow-covered days (after OMSZ 2003)

**Table 2.1** The renewable energy sources in Hungary (based on http://ec.europa.eu/eurostat/data/database 2011)

| | PJ/year energy | | | | |
|---|---|---|---|---|---|
| | Biomass | Geothermal | Wind | Water | Solar |
| Potentially utilizable | 120 | 50 | 46 | 0.9 | 72 |
| Utilized at present | 52.2 | 4.8 | 2.7 | 0.8 | 0.3 |

*PJ* petajoule = $10^{15}$ J; 1 J = approx. 0.239 cal or approx. 2.7 E–7 kWh

of the useable energy potential is estimated at 100–1200 PJ/year (GKM 2007), but only part of this is used in reality (Table 2.1). This potential can be understand by comparing it to the annual energy consumption of the country, which is approx. 1000 PJ. The wind and the solar energy are the most important among the climatic energy resources. The earlier data showed a very low proportion of wind energy in energy production; however, it has grown tenfold in the past years (Table 2.1). The same tendency characterizes the (solar collector and photovoltaic) use of solar energy. Considering the present situation, even reaching of the 5 % of the potentials already would show progress into the right direction.

According to Major (in Dobi 2006), the potential of solar energy is 140 W/m$^2$, the wind energy is 5–11 W/m$^2$, the geothermic energy is 0.09 W/m$^2$, and the kinetic energy of rivers is $8 \times 10^{-5}$ W/m$^2$ in Hungary, while the estimated yearly energy consumption is about 1 million TJ (= $10^{12}$ J), which means a 0.5 W/m$^2$ energy density. In spite of the fact that this value is much lower than the surface density of solar and wind energy, the conclusion that these could easily cover the energy consumption of the country is not realistic. (Today, the EU recommendation of a 10 % share of the renewable energy out of the total energy production seems realistic).

**Table 2.2** Daily average solar radiation realized on a horizontal surface in Hungary (after Energy Centre 2006)

| Month | **Average solar radiation ($kWh/m^2/d$)** | **Total solar radiation ($kWh/m^2/month$)** |
|---|---|---|
| January | 0.776 | 24 |
| February | 1.468 | 41 |
| March | 2.733 | 85 |
| April | 4.130 | 124 |
| May | 5.171 | 160 |
| June | 5.750 | 172 |
| July | 5.807 | 180 |
| August | 4.988 | 155 |
| September | 3.820 | 115 |
| October | 2.184 | 68 |
| November | 0.826 | 25 |
| December | 0.533 | 17 |
| **annual total** | **3.200** | **1166** |

### 2.3.2.1 Solar Energy Capacity

The solar energy potential of Hungary is above the European average, since half of the potential solar radiation reaches the surface. This provides 1300 $kWh/m^2$ (Energy Centre 2006) annually (Table 2.2.), and the regional differences are only 6–8 %. The present-day solar collector technology could be replaced by solar power plants, which are more economical; however, the price of the present two technologies differs considerably. Certain surveys estimate the solar energy potential (taking into consideration also the ecological limits) to be 400 PJ/year (Vajda 2001).

### 2.3.2.2 Wind Power Capacity

Experiments on non-traditional utilization of wind power date back to more than 50 years. The measurements of wind velocity were carried out for meteorological and aviation purposes at that time, not for estimating the power of the wind. At the beginning of the 1990s, the Hungarian Electricity Company (MVM) calculated wind power at 10 points in Hungary. (The data could be used only with limitations, as the measurements were carried out near ground level.) According to these measurements, wind power reaches 120 $W/m^2$ at three places: Little Hungarian Plain, Komárom County, and on the Foothills of the Mátra Mts. (Energy Centre 2006). In the Carpathian Basin, the frequency of wind velocity over 3–4 m/s, which is enough to start a wind generator, is 56–70 %. (The strongest wind gusts could be several times stronger, since a 44 m/s wind velocity was recorded in Szarvas in 1988; however, their occurrence is low.) The operational energy of wind is proportional with the cube of wind velocity. A wind of 20 km/h can produce eight times more energy than a 10 km/h wind (Bartholy 2000). Thus, the achievable annual wind power is 1.0 TWh for Hungary, meaning an energy source of 3–4 PJ/year according to the realistic estimations, although some optimistic estimates calculate this value to be 7 PJ/year (Fig. 2.8; Vajda 2001). According to Hunyár et al. (2006), these calculations significantly underestimate the wind energy potential of Hungary, because according to them, the annual wind energy potential is 40–90 TWh, depending on the wind turbine heights. In fact, the useable wind energy is restricted by several environmental, legal, technological and economic factors. For example, it is practical to have a medium voltage cable within a 10 km distance, and it is useful if an energy storage system is available.

Wind energy utilization can trigger significant investments (in case, a state subsidy and legally guaranteed purchase of energy by the state also included). For example, companies submitted draft contracts with valid environmental permission for 1500 MW wind turbine capacity until the end of 2005, equal to the capacity of the Paks Nuclear Power Plant (Csoknyainé 2006). However, only 20 % of these proposals acquired construction permits. Although wind energy is one of the cleanest energy sources, there are several environmental problems; e.g., the operation of the wind turbines creates noise, vibration and shadow flicker; nevertheless, these can be handled.

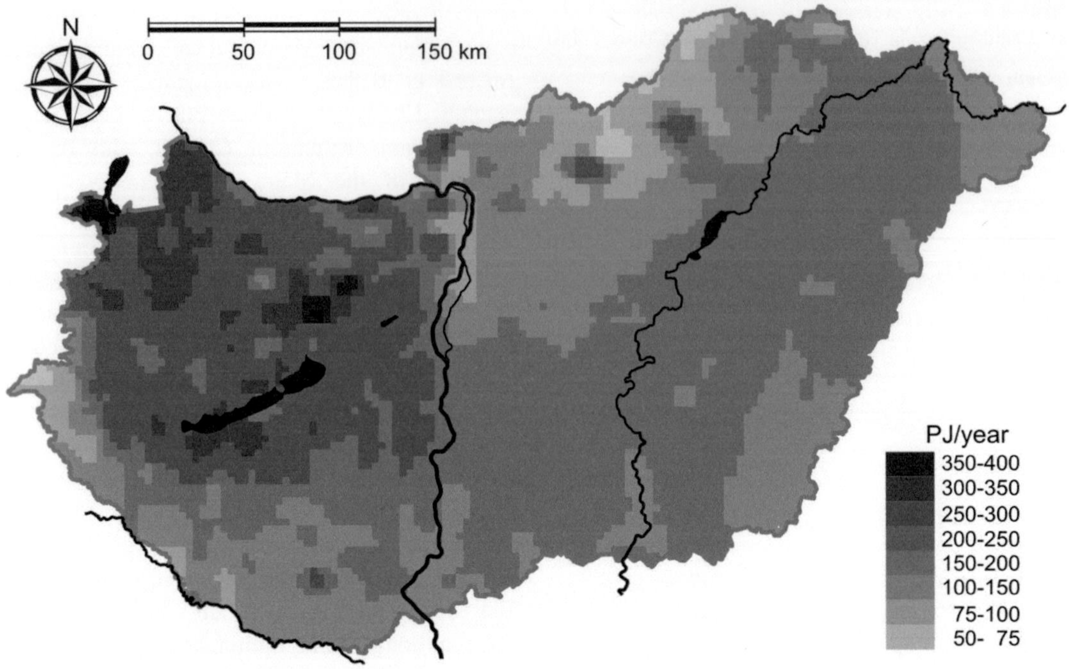


**Fig. 2.8** Wind energy potential (PJ/year) at 75 m height (after Szépszó et al. 2006)

### 2.3.3 Special Climatic Features of Cities

Larger cities can significantly modify the environmental conditions, e.g. vegetation and soil settings, or water regime (Mezősi 2007), and the climatic conditions are especially sensitive to human impacts. In general, this influence is measurable in towns with a population of more than 50 thousand, and the rate of the effects is increasing with the city size. (However, some environmental changes are already detected in the cases of smaller settlements with only a few thousand inhabitants.) A specific, so-called urban climate develops in cities, which can be characterized by well-distinguished parameters (Table 2.3), as higher temperature emerging from the different energy balance of urban and natural surfaces, which is called "urban heat island" (Unger 2007).

### 2.3.4 Further Climatic Capabilities and Hazards

(A) *Capabilities*

The conditions for climatic health resorts, which have had spectacular success in the past century abroad, are not ideal in Hungary. (The "Luftkurorts" in Austria are used and assessed similar to the places with medicinal water in Hungary.) Such climatic health resorts are situated along seasides (to cure skin and respiratory problems) or in high mountains (to cure cordial and respiratory illnesses). The main advantages of climatic health resorts are the even temperature, the clean air and the decrease in partial pressure of oxygen (Rákóczi et al. 2002). According to the Hungarian regulations, those places could be qualified as a health resort, which has a natural healing factor (e.g. climate and

**Table 2.3** Some characteristic features of the urban climate (after Sukkopp and Wittig 1993; Mezősi 2007)

| Factor | Difference between the countryside and the urban areas (%) | Explanation |
|---|---|---|
| Global radiation | −20 | Shade effect |
| Precipitation | +10 | More pollution |
| Air humidity | −60 to −30 | Lack of vegetation and natural evaporator surfaces |
| Wind velocity | −25 | Near-surface friction caused by the strong segmentation of the surface |
| Annual mean temperature | +0.5 to 1.0 | Anthropogenic impact |
| Minimum temperature | +1 to 3 °C | A maximum value of +10 °C may occur in urban heat islands |
| Aerosol | +1000 | Industry, households and traffic |
| $SO_2$ (μg/m$^3$) | +1000 | Industry and households |
| $CO_2$ | +1000 | Households and traffic |
| Dust (mg/m$^2$/d) | +600 | Traffic and industry |

water). To utilize these, an institutional background is also necessary. The climatic health resort status can be awarded for 20 years by the OGYFI (National Health Resorts and Spa Directorate General) based on the opinion of the OMSZ (Hungarian Meteorological Service) and the approval of the relevant ministries. The potentials are the most advantageous in the Kőszeg and Sopron Mts., which have a subalpine character, and in the Mátra Mts., where this type of treatment has long tradition at Kékestető. The air of several caves provides favourable conditions for medical purposes (Bálint and Bender 1995), which is beneficial not only because of the low levels of pollution and dust, but also because of the high calcium and magnesium content (e.g. Tapolca and Jósvafő caves).

(B) *Hazards*

The following sections outline natural hazards having climatic components and having continuously increasing effects because of the increasing human use of the environment. From this point of view, the question of whether the rainstorms or foggy days are becoming more frequent, or whether the air quality is worsening is important.

(a) Rainstorm hazard
Rainstorms usually occur in Hungary 20–33 times in the summer period, on average. A problem with rainstorms is the insufficient forecast system, in spite of the lightning register operated by the OMSZ since 1998, and nowadays, it is possible to follow lightning and rainstorms on the Internet. The pattern and frequency of lightning and rainstorms (Fig. 2.9) provide information about their propagation, and the released energy can also be calculated. Nowadays, the improving forecast system can reduce this type of hazard.

(b) Heat stress
Because of the increasingly warming summers, many people sense the increasing heat stress affecting their bodies. To make this perception measurable and comparable, a heat stress definition was developed. The value of heat stress is calculated on the basis of temperature and air humidity; however,

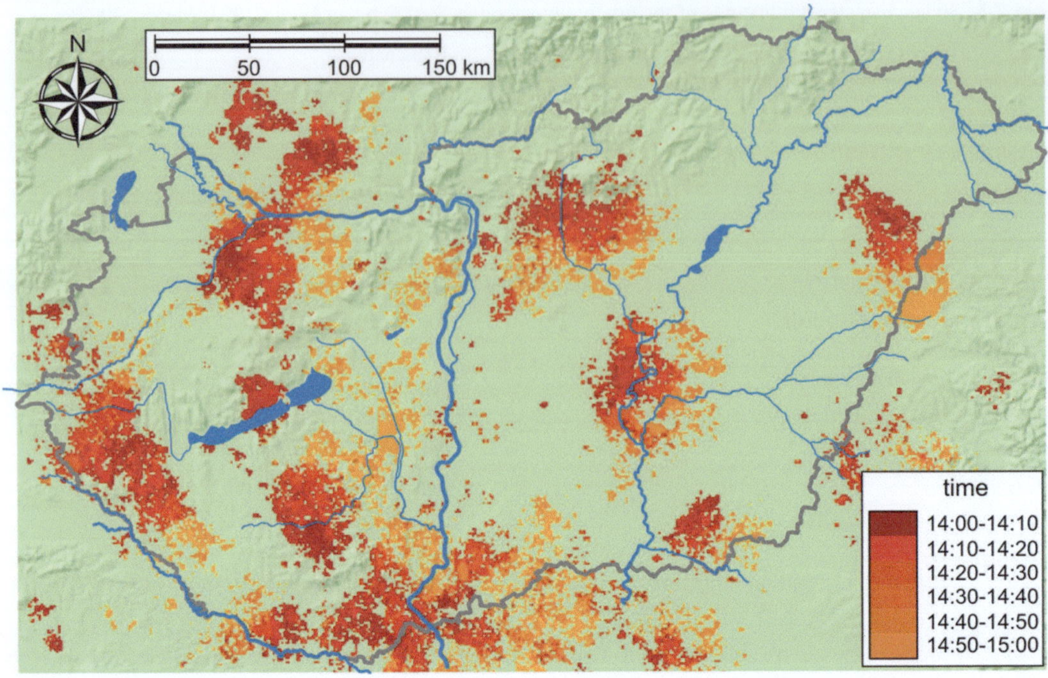


**Fig. 2.9** Hourly lightning strike map of Hungary, between 14.00 and 15.00 o'clock on 26 June 2009 (after OMSZ database)

temperature fluctuation and the extreme values are not integrated into the calculation despite of their significant effect on human comfort (Fig. 2.10; Gulyás et al. 2004).

(c) Increasing climate extremes

Everyday experience is that the frequency and rate of climate extremes, the so-called climate catastrophes, are progressively increasing. The environmental consequences of extreme climate events are becoming more frequent and affecting larger territories (Szabó 2001). The climate extreme indices related to temperature show an increasing tendency and/or a decreasing recurrence interval (Bartholy and Pongrácz 2005). The amount of annual precipitation in Hungary has been decreasing for about a hundred years; however, the frequency of extreme precipitation events has increased, e.g. intensive precipitation events occurred in 1998–2002 and in 2010, causing high and long floods on rivers. In addition, the number of days with more than 20 mm rainfall has significantly increased in the past 25 years. The growth of the total duration of the heat waves and dry periods was measured too (Pongrácz and Bartholy 2006).

The increasing extremities can be detected also in the case of the solar radiation intensity. It has different values during the day because of the changes of the angle of incidence. The very high UV radiation values (over 8.0) are dangerous, because the UV (ultra violet) radiation is the harmful component of solar radiation, and may cause skin cancer. The observed maximum value of UV-B radiation (9.4) in Hungary was measured at Kékestető (Mátra Mts.). Figure 2.11 shows a series of maps about the daily changes in the intensity of UV-B radiation. (An OMSZ recommendation on the duration of sunbathing could also be found on the original maps.)

Extreme temperature values also show high variability. The yearly maximum temperatures are generally 33–36 °C (at 200 m asl),

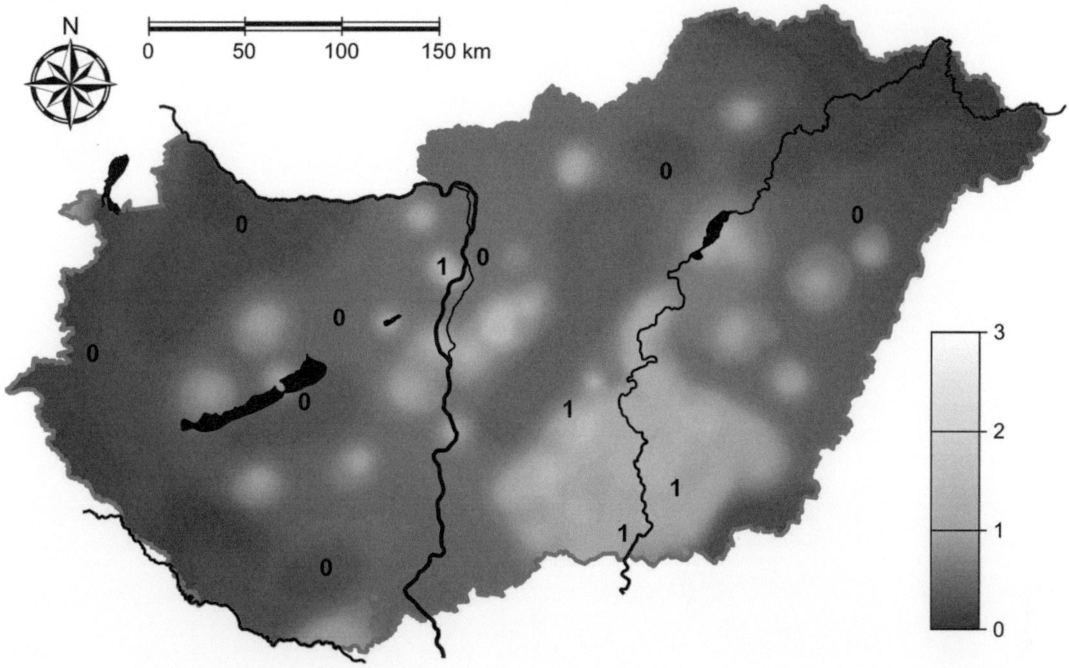


**Fig. 2.10** Heat stress map of Hungary on 1 August 2009 (after www.idokep.hu). The higher values refer to higher heat stress

while the minimum ones are usually between −16 and −19 °C. The observed absolute maximum temperature on the lowland regions is 39–41 °C, and the minimum is between −25 and −30 °C in the majority of the territory of the country. Extreme heat (above 30 °C) can occur in the whole country, even in the Kékes peak (Mátra Mts.), whereas extreme cold frequently occurs in the deeper parts of the Great Hungarian Plain and in the narrow, closed valleys in the mountains (Péczely 1979). The greatest absolute fluctuation of the temperature is 76.9 °C.

Climate projections for the twenty-first century, based on regional climate model simulations, show an increase in the frequency of extreme values (MET9 2010). These models project the changes of extreme climate indices, e.g. number of summer days ($T_{max} > 25$ °C), hot days ($T_{max} > 30$ °C), extremely hot days ($T_{max} > 35$ °C), winter days ($T_{max} < 0$ °C) and extremely cold days ($T_{max} < -10$ °C). At the end of the twentieth century, extremely hot days ($T_{max} > 35$ °C) were not common, occurred only on 10 % of the territory of the country; however, in the first half of the twenty-first century, they could occur every year at any time in the period from May to September (MET9 2010). A similar tendency can be projected in the case of precipitation: the daily precipitation intensity index will increase; thus, the amount of average precipitation on one rainy day will increase in the future. The increase is about 10 % in Hungary; however, it is only significant in the southern area of the Great Hungarian Plain.

(d) Further hazards

Among the climatic hazards, the susceptibility to fog formation and frost poses serious risks. Fog develops when the near-surface water vapour condensates (when the air temperature drops to the dew point, or when the water vapour content increases in the air and becomes saturated because of

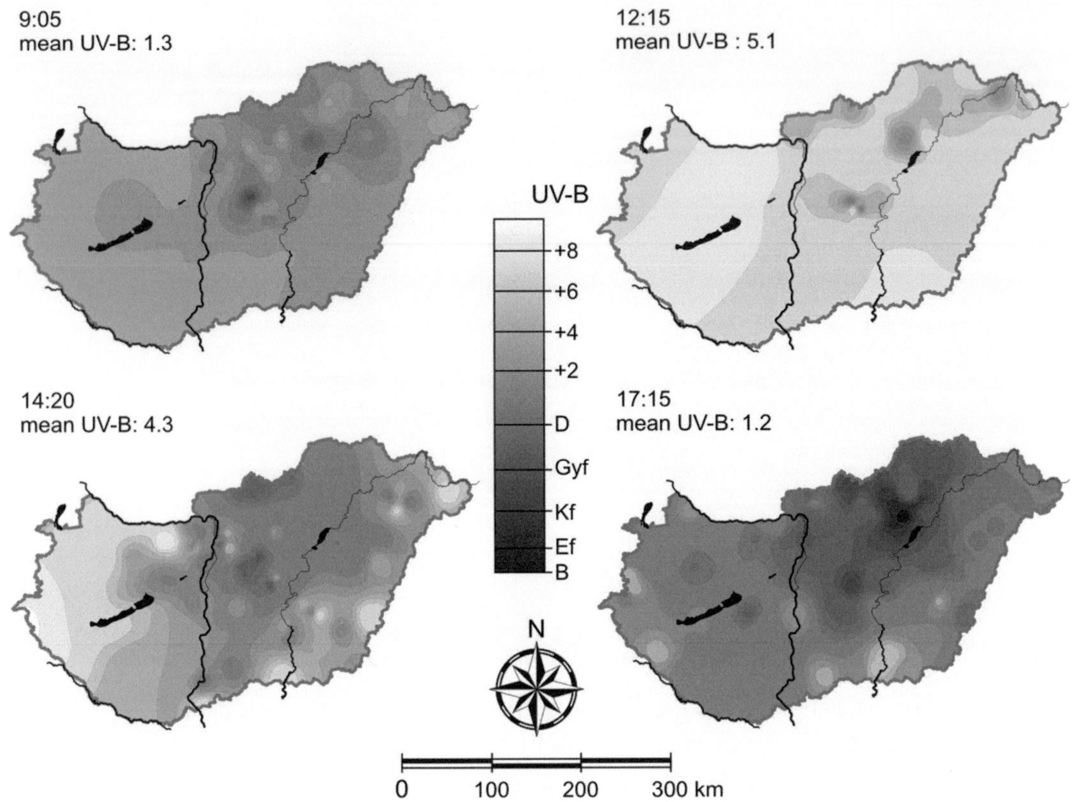


**Fig. 2.11** UV radiation values on a summer day (2 August 2009, after www. met.hu) *B* cloudy, *Ef* strongly clouded, *Kf* moderately clouded, *Gyf* lightly clouded, *D* clear sky

evaporation). The fog has several different classifications to assess and evaluate its origin and location. These classification schemes can be based on, e.g., the processes that are involved in creating fogs, the position of the fog or the horizontal visibility. Undoubtedly, one of the most critical consequences is the strong reduction in horizontal visibility for traffic; thus, sometimes, only 10–30 m is the visibility distance of the fog which is very thick (in case of radiation fogs formed near surface). The winter is the typical period of fog formation, and the periods of fog event can contribute to worsening air quality because of the decreased ventilation. Because of the deteriorating air quality and the accumulation of air pollutants caused by the fog, several larger cities had to inform (Budapest 2008, 2010; Pécs 2010) and then alarm (Miskolc 2010) the population about smog alerts.

The frost is especially harmful for the vegetation, because the water in the plants freezes; thus, the plant cells can get damaged and the leaves die (the expression “black frost” comes from this). Several other environmental hazards in connection with frost are known; for example, roads and wires can get damaged (Pinczés 1994).

## 2.4 Past and Future of the Climate

The climate change in geological times has been summarized in Sects. 1.1.3 and 1.8, while its changes in the past nearly 10 thousand years are described in Sect. 1.9. Even in historic times, significant social effects could be attributed to the ecological changes caused by the altered climate. For example, in the second–fourth centuries, a great migration started because of the cold and dry climate periods in Central Asia, because the deteriorating climate triggered limited ecological potential. Other approaches connect the migration waves to the warming and aridification of the climate in the sixth–seventh centuries, which also meant ecological limitations for the nomadic tribes practicing animal husbandry (e.g. grazing).

The short (hundred years) climate changes could be explained by the changing intensity of solar insolation. Lean and Rind (1998) reconstructed the changes of sunshine intensity until 1600, and they found that the minimum intensity had taken place between 1600 and 1800, when it decreased by 0.24 %. This corresponds to the time of the latest climate deterioration, the Little Ice Age and its continental, European maximum. It meant annual mean temperature decrease by 0.5 °C in the northern hemisphere. The Little Ice Age also corresponds to a period of low solar irradiance, which has been known for a long time: Wolf, Sporer and Maunder Minimum-like events have been recorded since the thirteenth century. During the Sporer Minimum (1402–1516) sunspots were not recorded for more than one hundred years, and during the Maunder Minimum (1645–1715), only a dozen were recorded in a 70-year period. The results of the analysis of alpine glaciers for the last cooling period also show an agreement with these observations (Jörin et al. 2006).

It is difficult to describe the climate deterioration of the Little Ice Age in the Carpathian Basin, because of the lack of data. Based on the information collected by Réthly (1962, 1970, 1998) Rácz (2008) reconstructed the climate deterioration during the Little Ice Age (between the end of the sixteenth and the end of the eighteenth centuries). Similar to other European countries, the winter and spring months became colder (March had more of a winter character), and the Junes were more humid being more similar to the spring months. These statements were also supported by various chronicles, manorial records and dendrological data (Grynaeus et al. 1994). Several analyses also draw the attention to the fact that the climatic features between 700 and 1200 AD triggered dynamic transformations in agriculture all over Europe.

The climate change of the last century was characterized by warming since the last third of the nineteenth century until the 1940s. The 1940s were especially humid, and a cooling was detected, possibly induced by volcanic activities. After this short period, the warming has been intensified due to the rapidly increasing anthropogenic influences since the 1970s. The recorded temperature fluctuation did not exceed 1 °C in the period (MT 2002).

To assess the future climate change, various climate model projections and uncertainties should be considered. Since the climate projection scenarios of the models cannot be confirmed, they are not considered as scientifically justified by several researchers. However, some main tendencies still can be outlined, owing to the more and more accurate data and information, which are appropriate to serve as a base for a "relevant" strategy of the adaptation to climate change. The climate change estimations are hindered by the fact that the natural and social factors are connected, and some factors are more easily projected than others.

The annual mean temperature will increase by 0.5 °C until the end of the twenty-first century in the northern hemisphere, if the composition of the atmosphere remains constant. However, the temperature may increase by 1.5–4.0 °C if the present economic activity maintained (ICCP 2007). It is difficult to draw definite conclusions for Hungary, because of the contradictory climate tendencies of different projections even for such a small area. Nevertheless, in longer timescale, a warming by 0.5–0.7 °C could be expected. The climate simulations project a climate alteration towards the eastern Mediterranean climate, and the anthropogenic influence is regarded as the primary cause. It is more difficult to project the changes in the amount of spatial and temporal distribution of precipitation. Some climate simulations project decreasing autumn–winter precipitation (MT 2002), whereas other models show an increase during the winters (Bartholy and Mika 2005) and autumns (Láng

2006). These differences have different effects on surface water regime and flood hazard. In general, the possible increase in evaporation due to the higher temperatures will be accompanied by the decreasing precipitation, leading to the moisture loss of soils, to limited water availability for the vegetation and to a decrease in groundwater level. There are some areas where the land use will be significantly modified in the future, because of the decreasing amount of precipitation and higher temperatures. For example, some researchers found that the Danube-Tisza Interfluve will face with water shortage periods and aridification. All climate models project an increasing frequency and intensity of extreme events causing significant difficulties in the water supply and water management. Also, the researchers highlight that a significant and extensive reduction in air pollution and the adaptation to the changing ecological conditions may considerably decrease the damages caused by climate change.

## References

Ács F, Breuer H, Szász G (2007) A tényleges párolgás és a talajvízkészlet becslése tenyészidőszakban. Agrokémia és Talajtan 56:217–236

Bacsó N (1959) Magyarország éghajlata. Akadémiai Kiadó, Budapest

Bálint G, Bender T (1995) A fizioterápia elmélete és gyakorlata. Springer, Budapest

Bartholy J (2000) Hasznosítható-e a szélenergia Magyarországon? In: Karátson D (szerk) Pannon enciklopédia. Kertek, Budapest, pp 236–237

Bartholy J, Mika J (2005) Időjárás és éghajlat – cseppben a tenger? Magyar Tudomány 7:789–793

Bartholy J, Pongrácz R (2005) Néhány extrém éghajlati paraméter globális és a Kárpát-medencére számított tendenciája a XX. században. AGRO-21 Füzetek 40:70–93

Bartholy J, Weidinger T (2000) Napsugárzás, felhőzet, szél. In: Karátson D (szerk) Pannon enciklopédia. Kertek, Budapest, pp 226–227

Botos L, Varga-Haszonits Z (szerk) (1974) Agroklimatológia és növénytermesztés. OMSz–MÉM, Budapest

Dobi I, szerk. (2006) Magyarországi szél és napenergia kutatás eredményei. OMSZ, Budapest

GKM (2007) Magyarország megújuló energiaforrás felhasználás növelésének stratégiája 2007–2020, Budapest. http://hulladeksors.hu/dokumentumok/megujulo_strategia_tars%20egyeztetes.pdf

Grynaeus A, Horváth E, Szabados I (1994) Az évgyűrű mint természetes információhordozó. Erdészeti Kutatások 7–8:203–205

Gulyás Á, Unger J, Matzarakis A (2004) A városi környezet mikroklimatikus jellemzőinek bioklimatológiai szempontú elemzése Szeged példáján. 2. Magyar Földrajzi Konferencia Szeged. http://geography.hu/mfk2004/mfk2004/cikkek/gulyas_unger_matzarakis.pdf

Hunyár M, Veszprémi K, Szépszó G (2006) Újdonságok Magyarország szélenergia potenciáljáról. In: Dobi I (szerk) Magyarországi szél és napenergia kutatás eredményei. OMSZ, Budapest, pp 94–113

ICCP (2007) http://www.ipcc.ch/publications_and_data/publications_and_data.htm

Jörin UE, Stocker ThF, Schlüchter Ch (2006) Multicentury glacier fluctuations in the Swiss Alps during the Holocene. The Holocene 16(5):697–704

Kakas J (szerk) (1960) Magyarország Éghajlati Atlasza. Akadémiai Kiadó, Budapest

Kertész Á, Mika J (1999) Aridification—climate change in South-Eastern Europe. Phys Chem Earth (A) 24 (10):913–920

Láng I (2006) A globális klímaváltozás: hazai hatások és válaszok – VAHAVA zárójelentés

Lean J, Rind D (1998) Climate forcing by changing solar radiation. J Clim 11:3069–3094

Lovász Gy, Majoros Gy (1994) Magyarország természeti földrajza. JPTE, Pécs

MET2 (2009) http://www.met.hu/omsz.php?almenu_id=climate&pid=climate_Hw&pri=5&mpx=0

MET3 (2009) http://www.met.hu/omsz.php?almenu_id=climate&pid=climate_Hw&pri=1&stt=Homerseklet

MET9 (2010) http://www.met.hu/omsz.php?almenu_id=homepages&pid=numprog&pri=9&mpx=0

Mezősi G (2007) Városökölógia. JATEPress, Szeged

Mezősi G (2008) Magyarország környezetföldrajza. JATEPress, Szeged

MNA(1989) Magyarország Nemzeti Atlasza. Carthographia, Budapest

MT = Glatz F (szerk) (2002) Magyar Tudománytár I. Föld, víz, levegő. Kossuth, Budapest

OMSZ (2009) http://www.met.hu/omsz.php?almenu_id=omsz&pid=proposal&mpx=0&kps=1&pri=3&sm0=2&dti=1

OMSZ = Mersich I (szerk) (2003) Magyarország éghajlati atlasza. OMSZ, Budapest

Pálfai I (2002) Magyarország aszályossági zónái. Vízügyi Közlemények 84(3):323–357

Pálfai I (2004) Belvizek és aszályok Magyarországon. Közl. Dok., Budapest

Péczely Gy (2002) Éghajlattan. Nemzeti Tankönyvkiadó, Budapest

Péczely Gy (1979) Éghajlattan. Nemzeti Tankönyvkiadó, Budapest

Pinczés Z (1994) A jelenkori fagy felszínformáló hatása hazánkban és ennek gyakorlati jelentősége. Kossuth Egyetemi Kiadó, Debrecen

Pongrácz R, Bartholy J (2006) A Kárpát-medence extrém hőmérsékleti paramétereinek XX. századi tendenciái. In: A III. Magyar Földrajzi Konferencia tudományos közleményei. MTAFKI, Budapest. www.geography.hu

Rácz L (2008) Magyarország környezettörténete az újkorig. MTATörténettudományi Intézete, Budapest

Radics K (2004) A szélenergia hasznosításának lehetőségei Magyarországon: hazánk szélklímája, a rendelkezésre álló szélenergia becslése és modellezése. Doktori értekezés, ELTE Meteorológiai Tanszék, Budapest

Rákóczi F, Drahos Á, Ambrózy P (2002) Magyarország gyógyhelyeinek éghajlata. Oskar, Szombathely

Réthly A (1962) Időjárási események és elemi csapások Magyarországon 1700-ig. Akadémiai Kiadó, Budapest

Réthly A (1970) Időjárási események és elemi csapások Magyarországon 1701–1800-ig. Akadémiai Kiadó, Budapest

Réthly A (1998) Időjárási események és elemi csapások Magyarországon 1801–1900. OMSZ, Budapest

Sukkopp H, Wittig R (hrsg) (1993) Stadtökologie. Fischer, Stuttgart, New York

Sz Horváth, Makra L, Mika J et al (2001) A klíma és a területhasznosítás változékonyságának kölcsönhatásai a Tisza magyarországi vízgyűjtő területén. I. MFK, Szeged

Szabó J (2001) A természeti veszélyek és katasztrófák földrajzi vonatkozásai. 1. MFK, Szeged. http://geography.hu/mfk2001/cikkek/SzaboJ.pdf

Szépszó G, Horányi A, Kertész S, Lábó E (2006) Magyarországi szélklimatológia előállítása globális mezők dinamikai leskálázásával. In: Dobi I (szerk) Magyarországi szél és napenergia kutatás eredményei. OMSZ, Budapest

Tar K (2007) A szél energiája Magyarországon. KvVM, Budapest. http://www.kvvm.hu/cimg/documents/Tar_Karoly.pdf

Unger J (2007) A város éghajlat-módosító hatása – a szegedi hősziget. In: Mezősi G (szerk) Városökológia. JATEPress, Szeged, pp 43–65

Vajda Gy (2001) Energetika és fenntartható fejlődés. Természet Világa 132(8):340–342

Weidinger T (2000) Hőmérséklet. In: Karátson D (szerk) Pannon enciklopédia. Kertek, Budapest, pp 228–229

Weidinger T, Mészáros R (2000) Csapadék, nedvesség, párpolgás. In: Karátson D (szerk.) Pannon Enciklopédia, Kertek, Budapest, 230–231

# 3 Hydrography of Hungary

**Abstract**
The potentials and hazards of surface and subsurface waters are very important elements of the Hungarian environment. The geographical location of the region (e.g. basin-like character, continental water household) determines the possibilities (e.g. great subsurface water bodies) and the hazards (e.g. extreme water regime of the rivers, flood hazard). In the first part of the chapter, the geographical background of the surface waters is introduced, describing the most important morphological, hydrological and ecological characteristics of rivers and lakes. Besides the description of the genetic lake types, a detailed analysis is given on the Lake Balaton, Lake Fertő and Lake Velence. Besides their evolutionary history, the temporal changes of their water households during the last decades are also explained, especially the responses of the lakes on climate change are emphasized. Considering the environmental problems of the subsurface waters, those of the artesian waters (e.g. recharge, water household) are discussed in detail, as in Hungary, these problems are the most important from the point of view of thermal water and drinking water utilization. The drinking water demand is fulfilled by bank-filtered wells and karst waters; thus, they are highlighted. Besides, a general picture is drawn on the human impact on waters and the way of their management in Hungary.

## 3.1 Hydrographical Characteristics of Surface Waters

### 3.1.1 Catchment-Scale Hydrology of the Carpathian Basin

The water household of Hungary is mostly determined by the precipitation supply from its closer and wider environments (Carpathian Basin). Approximately 56 km$^3$ precipitation reaches the basin directly, and about an additional 114 km$^3$ of water enters in the form of rivers. Since the basin is almost entirely composed of porous sediments, there is a significant subsurface inflow and outflow as well.

The water household is one of the most sensitive issues of the Carpathian Basin, since it determines various environmental factors apart



G. Mezősi, *The Physical Geography of Hungary*, Geography of the Physical Environment,
DOI 10.1007/978-3-319-45183-1_3

from the agriculture and water supply. The natural factors contributing to availability are precipitation, run-off and effluent drainage from subsurface water bodies. The latter factor largely depends on the rate of the run-off, which is significantly influenced by slope gradient, lithological features, vegetation, surface roughness and water-saturation of soils. Surface run-off is best characterized by specific run-off. The border between plains and mountains is set at the 2 l/s/km$^2$. This parameter has a close relationship with water availability; thus, the similar spatial pattern of specific run-off and aridity is characteristic.

Evaporation and infiltration decrease the run-off. The rate of infiltration can be 20–30 % on karst surfaces, while it is only a fraction of this in clayey areas (Fig. 3.1). (The watershed of the Tisza River is mostly covered by sediments of low permeability; thus, infiltration is insignificant.) It is obvious that the water household is a multifactorial and sensitive system. On the Great Hungarian Plain, the mean annual precipitation is 560 mm and the evaporation is about 515 mm, while the run-off is 30 mm, and there is approx. 15 mm infiltration to recharge subsurface waters. The human impacts, for example, the slight decrease of forest cover, can significantly influence several factors that determine the water household, and contributes to increasing frequency of extreme events (Szesztay 2000).

Considering that the water household has a special importance in Hungary, the changes in the future should be predicted. Various models support that the major input factor, which is precipitation, will slightly decrease, and the humid periods within a year are going to be rearranged. In the winter half-year, the precipitation will increase by ca. 10–20 %, while in the summer half-year, it will decrease by 50–70 mm, in case of a 1 °C increase in the annual mean temperature. The evapotranspiration is an output element of the system, and it is projected to grow by 10–15 % per 1 °C increase throughout the year. It should be considered too that these changes have an anthropogenic origin. On the whole, the extreme climatic conditions will also result in extremities in the water household. At the same time, several experts claim that the aridification of the Great Hungarian Plain is caused by the disadvantageous selection of the planted tree species (Szesztay 2000).

The run-off rate is low (under 5 %) on the plains, but it can reach 20 % in the hilly and mountainous areas (Fig. 3.2). Characteristically run-off supplies the rivers, but the subsurface inflow could supply half or sometimes even more of the discharge of the rivers (MT 2002). It is very difficult to monitor or quantify the amount of subsurface water entering the Carpathian Basin. Nevertheless, it has to be considered in relation with discharge, since observations have proved that, apart from precipitation, subsurface supply is also very important, especially in regions of higher orographic situation. Subsurface outflow also leaves the basin in a “hidden” way, usually through the southern borders of the country, in accordance with surface topography.

All of the Hungarian rivers belong to the catchment system of the Danube, which is 2860 km long and have a 800,000 km$^2$ drainage basin in total. (The catchment area of the Danube upstream of Mohács is more than 200,000 km$^2$.) The drainage basin, due to its large size, includes several subcatchments of highly diverse climatic, topographic and lithologic features, resulting in a water network with diverse hydrogeography. Direct tributaries of the Danube in Hungary are Rába River (18,000 km$^2$) and the Zala River (15,000 km$^2$), but its largest Hungarian tributary, the Tisza River (total area: 150,000 km$^2$), joins the Danube in Serbia (Figs. 3.3 and 3.4). The length of the Tisza in its natural state was 1419 km from its source to its mouth, but it was shortened to 962 km by water engineering works. The largest tributaries of the Tisza are the Szamos River (almost 16,000 km$^2$), the Sajó River (about 5000 km$^2$), the Körös River (27,000 km$^2$), and the Maros River (30,000 km$^2$). These catchments have very low gradient on the Great Hungarian Plain; thus, their lowermost areas were marshlands in their natural state, due to low gradient, vicinity of alluvial fans and lithological reasons; therefore, nowadays they face extensive inland excess water hazard.

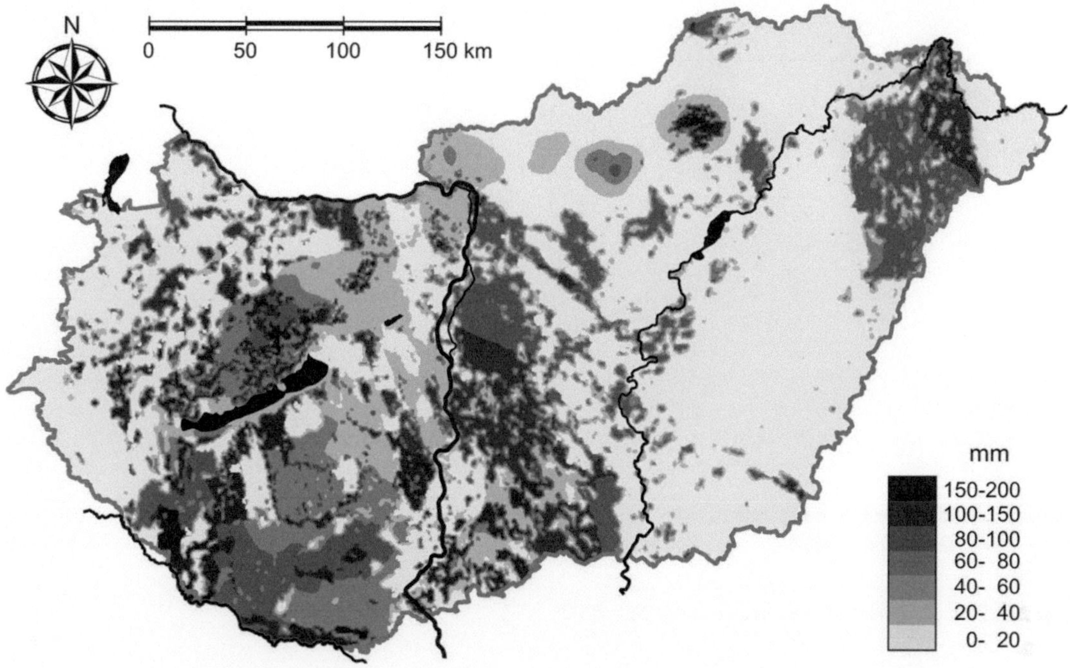


**Fig. 3.1** The spatial differences of run-off in Hungary (after Simonffy 2008)

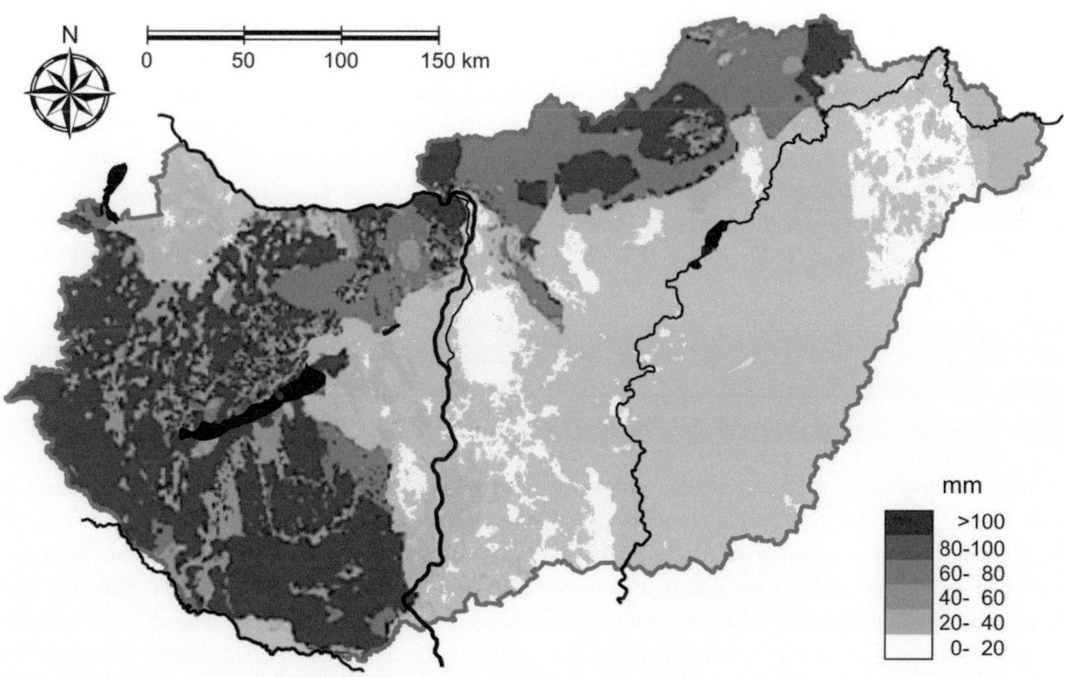


**Fig. 3.2** Spatial differences of infiltration in Hungary (after Simonffy 2008)

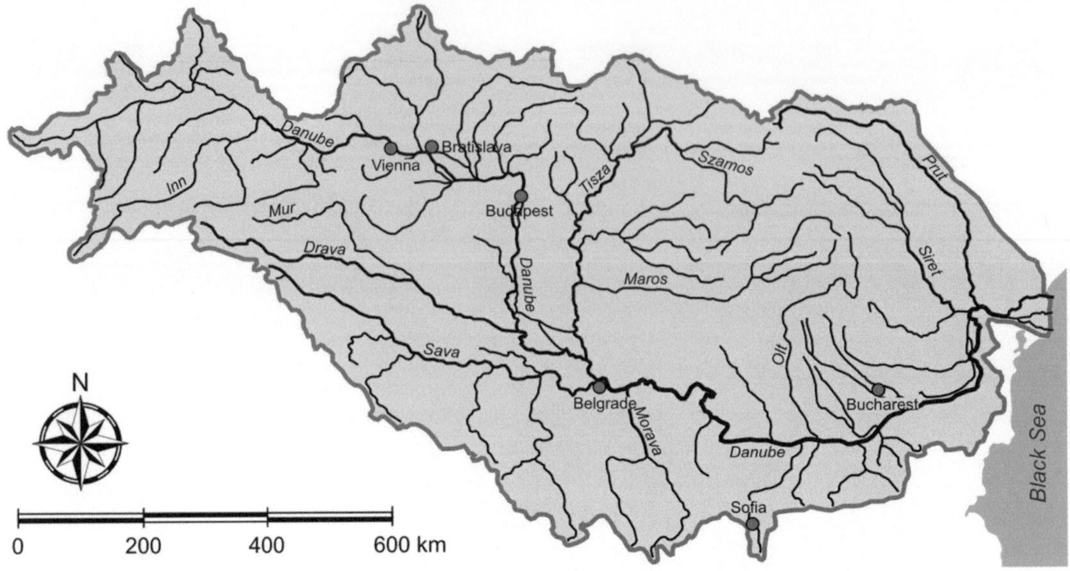


**Fig. 3.3** The catchment of the Danube (after MT 2002)

In Hungary, more than 2500 major valleys developed, which have waterflows only during short rainy periods. Therefore, just the hydrographical features of the largest rivers will be introduced here. The development of the river system has been presented in a previous chapter (Sect. 1.3). The annual discharge of all rivers is ca. 100 billion $m^3$; most of it is drained by the Danube (mean discharge at Budapest: 2200 $m^3/s$) and the Tisza River (mean discharge at Szeged: 810 $m^3/s$; Fig. 3.5 and Table 3.1). The main problem of Hungary is that about 3782 $m^3/s$ water is drained from the territory of the country, while the inflow is 3802 $m^3/s$, and additionally approx. 120 million $m^3$ is the subsurface outflow (Fig. 3.5). Only 5 % of the rivers originate in Hungarian territories; the rest flows through the country causing more or less trouble. The consequence of this is situation is that Hungary could not control the waters, neither from the quantitative, nor from the qualitative points of view. Being a downstream country, the only solution is to sign agreements with the neighbouring countries. The quantity control may pose two questions (these are analysed in details in Part II). If the amount of incoming water is too high, floods develop (e.g. very high floods were in 2006 and 2010); thus, it is important to clarify what the Hungarian authorities can do in the catchment areas of the mountains to reduce floods. As far as the upper parts of the catchments are dominantly not in Hungary, the new Vásárhelyi Plan could be the solution on the problem of floods within Hungary. It aimed to construct emergency flood reservoirs; however, even if the project has been elaborated and realized, it will not be an effective tool to reduce the increasing flood levels. The low amount of water arriving in the country in dry periods also creates a serious problem of the same extent. The basin is prone to aridification; therefore, it is important to know how much water is at disposal (e.g. for irrigation). Besides, the water quality of the incoming rivers is also determined by the neighbouring countries; thus, it is worth to follow a strategy, which includes a relevant disaster recovery plan, and restrains the activities leading to quality deterioration (taking into account the EU directives). The cyanide poisoning of the River Tisza (in 2000) or the water quality issues of Austrian origin of the Rába River have been very instructive in this respect.

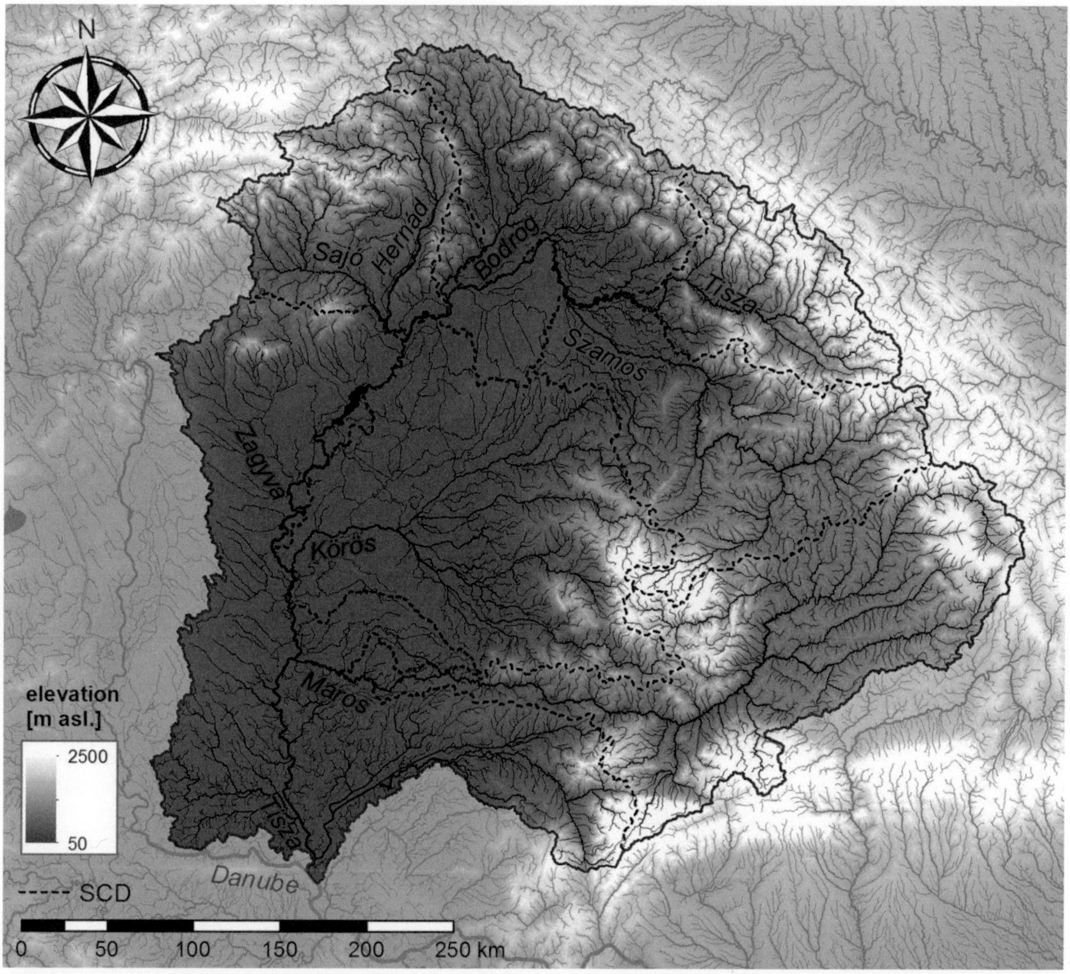


**Fig. 3.4** Relief of the catchment area of the Tisza river (after Bódis 2008)

### 3.1.2 Rivers

*Main geomorphological features of the large rivers*

From the point of view of geomorphology, the rivers could be classified in many ways. Cholnoky's classification has a significant tradition in Hungary. Cholnoky developed Heim's work at the beginning of the twentieth century, and being an engineer himself, he defined categories based on the stream power of the rivers. In this classification, the stream power that is available for the transport of sediment is compared with the amount of transportable sediment, and based on this ratio, the rivers were classified as "lower, middle and upper section-like character", respectively. The simple classification—although still applied in Hungary—has proved to be incomplete in many ways. On the one hand, the sections do not appear in the proper spatial order, for example, a river section considered homogeneous may have features of the different classes. On the other hand, the definition of these section types could not have happened along rigid rules, and in particular, the "middle section-like character" could be divided into more subclasses.

In the 1960s, Kádár developed Cholnoky's classification, applying the same logic. In his approach, the "middle section-like character" has

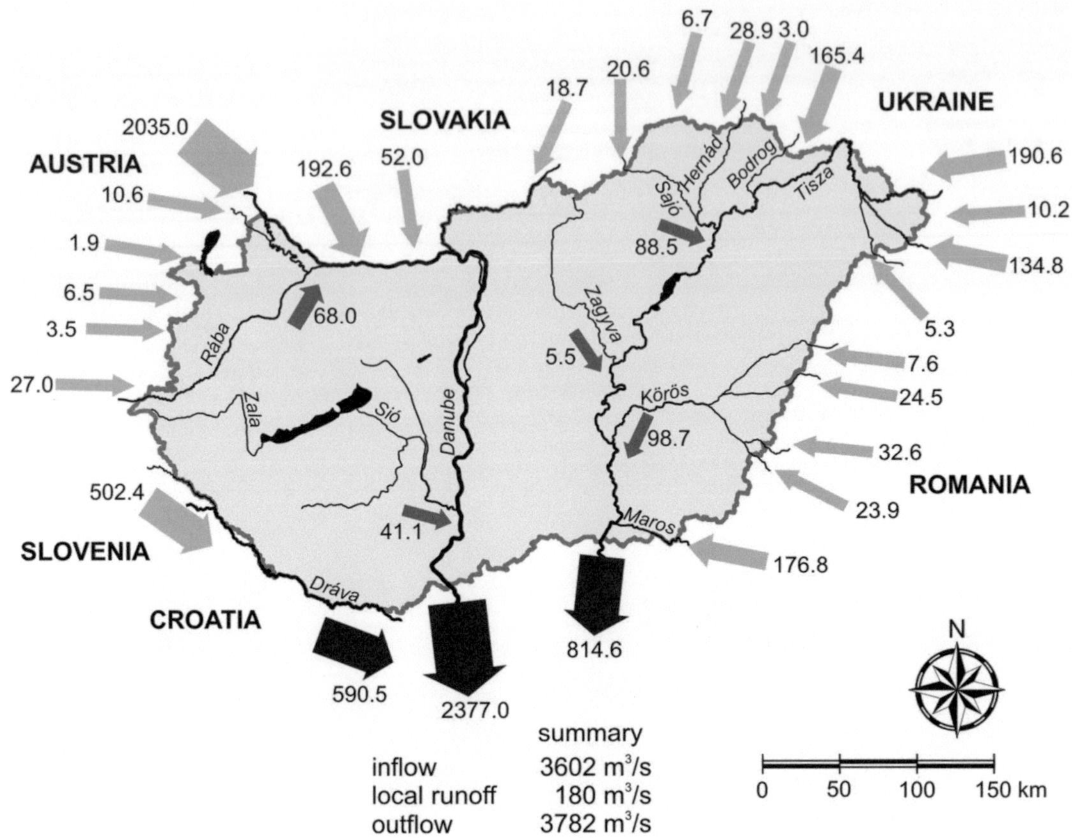


**Fig. 3.5** The average mean discharge of the largest rivers (after Konecsny 2010)

subtypes: he defined an aggrading meandering pattern which is an intermediate type towards the "lower section-like character", and an incising meandering pattern towards the "upper section-like character". The typically meandering channels were classified as "equilibrium pattern". The terms of aggrading (alluvial fan builder) and incising meander are still in use (Somogyi 1983; MNA 1989). The definitions for the section pattern provide useful information to river management in the long term (e.g. evaluation of surface forming activity of floods) in spite of the uncertainties involved.

According to this classification, the Danube has an aggrading (alluvial fan builder) pattern from the Austrian–Hungarian border to Gönyű (apart from the regulated sections), an incising meandering pattern from Gönyű to Esztergom, an incising at the Danube Bend, then an aggrading meandering pattern, afterwards at the Gellért Mt. incising meandering pattern, after that at the Csepel Island it has aggrading, the next section has mainly incising meandering pattern (with aggrading short sections), and finally at the Mohács Island, it becomes again aggrading. The Tisza River has incising meandering pattern from the Ukrainian–Hungarian border to Tiszalök, an aggrading meandering pattern from Tiszalök to Kisköre, and finally, it has incising meandering pattern again (MNA 1989). A significant pattern change can be observed in the case of both rivers due to human impact. Before the river regulations, the Tisza had an aggrading meandering pattern, but after the numerous cut-offs, it became incising meandering, since the regulation works had increased its gradient and erosional capacity. The largest morphological transformation affected the Maros

**Table 3.1** Main hydrographical parameters of the Danube and the Tisza

| | Tisza | | | Danube | | |
|---|---|---|---|---|---|---|
| | Water level (cm) | | | | | |
| | Vásárosnamény | Szolnok | Szeged | Dunaremete | Budapest | Mohács |
| Lowest stage | −203 | −272 | −224 | 188 | 62 | 113 |
| Mean stage | 21 | 150 | 215 | 373 | 287 | 397 |
| Highest stage | 912 | 909 | 960 | 688 | 845 | 984 |

| | Tisza | | | Danube | | |
|---|---|---|---|---|---|---|
| | Discharge ($m^3$/s) | | | | | |
| | Vásárosnamény | Szolnok | Szeged | Dunaremete | Budapest | Mohács |
| Lowest discharge | 30 | 54 | 57.8 | 86.5 | 580 | 656 |
| Mean discharge | 367 | 551 | 865 | 370[a] | 2265 | 2298 |
| Highest discharge | 3930 | 3030 | 3820 | 10,600[a] | 8310 | 8240 |

Water temperature (mean temperatures in 2000, expressed in °C)

| Tisza | | | Danube | | |
|---|---|---|---|---|---|
| Vásárosnamény | Szolnok | Szeged | Dunaremete | Budapest | Mohács |
| 10.7 | 13 | 13.2 | 11.3 | 11.8 | 13 |

Morphometric data (based on Bulla 1964)

| Settlement | River | Width (m) | Depth (m) | Slope (cm/km) |
|---|---|---|---|---|
| Komárom | Danube | 450 | 4 | 26.2 |
| Budapest | Danube | 290 | 5 | 6.8 |
| Vásárosnamény | Tisza | 184 | 8 | 15.1 |
| Szeged | Tisza | 250 | 9 | 2.9 |

The average sediment transport of the Tisza and the Danube between 1931 and 1940 (Juhász 1987)

| | Station | Suspended load (million $m^3$/year) | Bed load (thousand $m^3$/year) |
|---|---|---|---|
| Danube | Dunaremete | 1.387 | 103.0 |
| | Dunaalmás | 3.050 | 21.6 |
| | Duanújváros | 6.500 | 15.8 |
| | Baja | 10.162 | 24.0 |
| Tisza | Polgár | 5.400 | 3.9 |
| | Tiszabő | 3.330 | 8.8 |
| | Szeged | 12.260 | 11.0 |

*Source* www.vizadat.hu, www.vizugy.hu, www.ovf.hu
[a]The lowest water discharge data refer to the regulated river bed of the Öreg-Danube

River, which had meandering pattern before the regulations, and it became straightened almost totally, and afterwards it became aggrading (alluvial fan builder). The altered discharge affected the entire floodplain, since before the regulations the floods deposited inorganic materials, and the abundant vegetation resulted in increased organic deposition too. However, the proportion of these two accumulation types has changed due to the consequences of the regulations.

The classification of rivers is even more complex, if one considers that the patterns could not get clearly manifested, because an increased amount of sediment also accompanied the regulations. This hindered the transformation; thus, sand bars developed in several places, and the incising meandering pattern could be manifested only in few places. The meandering character of the Tisza is reflected by its length: in Hungary in the first half of the twentieth century, the length increased by 5 km due to bend development (Károlyi 1960). According to measurements on the Lower Tisza, this lengthening is still ongoing, and similar results have been obtained on the Hungarian section of the Rába River, which has become 2.5 km longer in the past 50 years.

The channel of the Danube is incising, especially between Győr and Esztergom due to the decreasing amount of bedload. A considerably proportion of transported sediment is deposited in the reservoirs of the hydroelectric power plants; however, the river has enough energy to erode material from the channel downstream of the dams. The extent of channel degradation is estimated to be several metres, which can change the habitat conditions of the aquatic life. Similar incision takes place in the area of Gemenc along the Danube, and in the length of the common Hungarian–Croatian section of the Dráva River, where 1.0 m deepening was measured in the past 30 years due to the Croatian reservoirs and in-channel gravel mining. The maintenance of the shipping route on the Danube requires serious interventions too (e.g. dredging), but simultaneously the natural values of the Danube have to be preserved too (Natura 2000 areas). The increased energy of the rivers after human intervention could trigger not only incision, but also the accelerated bank erosion; therefore, it is advisable to obtain information on the erosional state of the river bed when installing engineering facilities (e.g. pipeline passages, bridge piers or groynes).

In the past decades, the geomorphological analyses based on the pattern classification of Schumm's (1977) were applied in Hungary instead of applying the classes of Cholnoky or Kádár. This approach also considers the discharge and sediment transport capacity of rivers, even if it applies different terms and point of view. Morphology of the channel is in its focal point, but besides, it also considers the energy conditions, the water regime, the sediment conditions and the gradient of a river. As rivers adapt

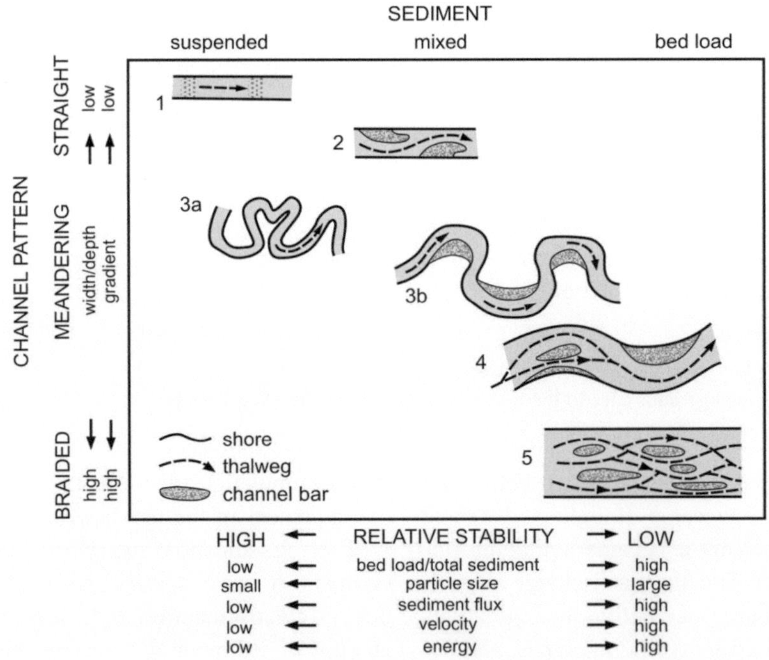


**Fig. 3.6** Channel pattern classification and its main parameters (after Schumm 1977; Sipos 2004). *1* Straight, *2* straight with alternating bars, *3a* low sinuosity with fine-grained suspended sediment load, *3b* low sinuosity with coarse-grained suspended sediment load, *4* meandering with bars, *5* braided with medial bars

to the discharge conditions, the channel pattern also fits to the hydrology in the long term. (Therefore, "energy-based" classifications developed, although their logic and the nomenclature are different.) The classification distinguishes simple and multichannel types (Fig. 3.6). The former includes the straight (e.g. the Danube between Győr and Esztergom is bar-divided, straight), the meandering (e.g. the Tisza before the regulations in the Great Hungarian Plain; Image 3.1) and the braided classes. In the latter case, sand and gravel bars are in channel dissecting the thalweg, and the channel is very wide and shallow; such pattern characterizes the Maros River between the Romanian–Hungarian border and Makó (Image 3.2). According to Nádor et al. (2005), braided channel patterns characterized some rivers in the glacial periods of the Pleistocene, while they were meandering during the interglacials. The complex channel types are represented by anabranching and anastomosing patterns, which are characterized not only by bars, but also by islands (Lóczy and Veress 2005).

The analysis of channel pattern could not be easily applied in practice. Nevertheless, it may help in constructing bank protection or to secure waterways, especially if we consider that the pattern does indicate not only the energy relations, but also the way of channel development (bend development), the stability of the channel, or the sediment transport characteristics (e.g. during the planning river engineering measures, it is important to know the rate of channel shift.)

The Danube along the Szigetköz region fits the criteria of an anastomosing channel (Image 3.3), as it has low sinuosity, several narrow and deep side-channels, a small gradient, and the appropriate type of sediment transport (twenty times more suspended load than bedload. However, there are some factors related to the sediment ratios and the tectonic conditions of the area, which makes the classification of the Danube in Szigetköz less evident (Sztanó et al.

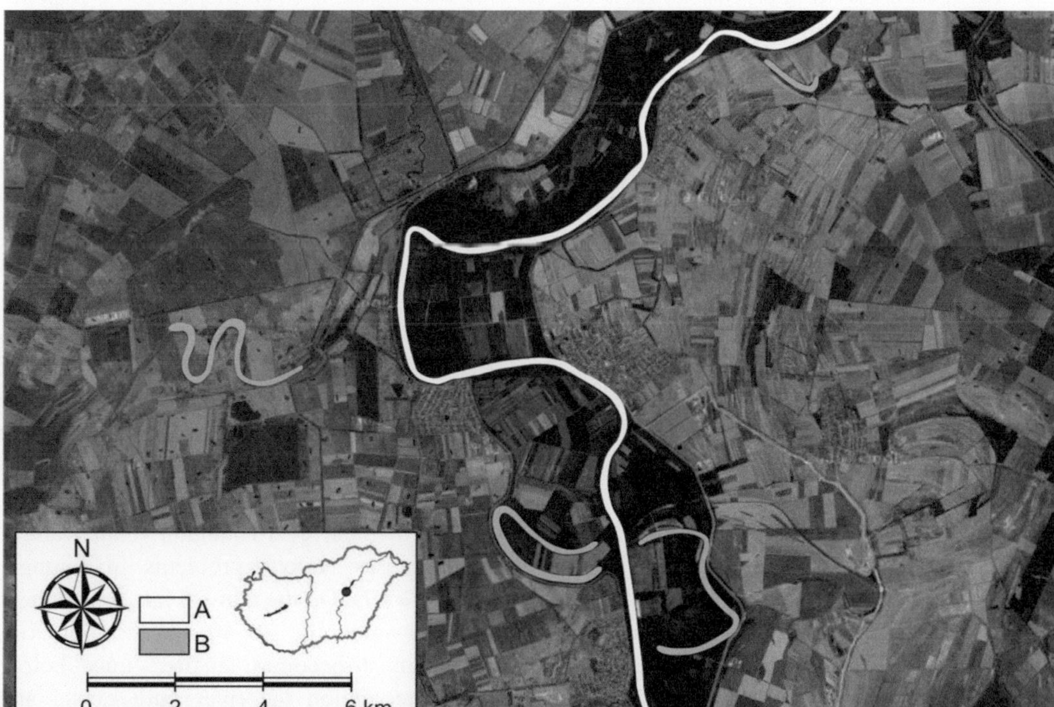


**Image 3.1** The meandering river bed of the Tisza river between Tiszabura and Tiszaroff was regulated by cut-offs in the nineteenth century (own construction based on googlemaps). *A* Main channel, *B* cut-off oxbow lake

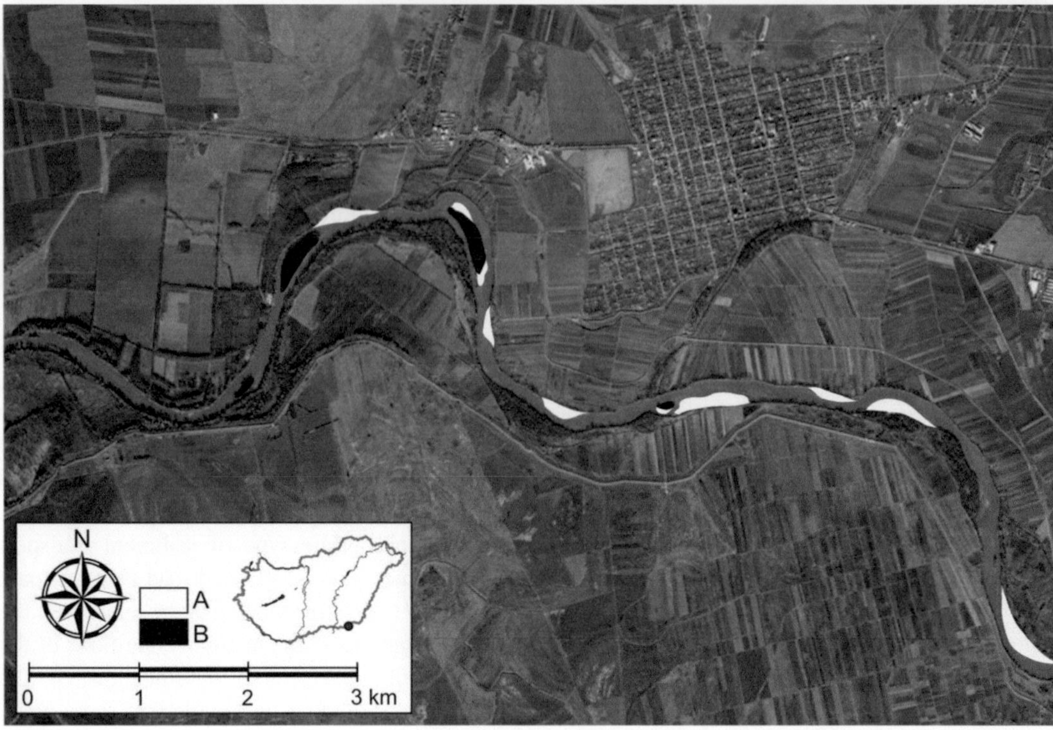


**Image 3.2** Braided channel of the Maros river at Nagylak (own construction based on googlemaps). *A* Sand bar, *B* island

2005). The aggradation and the non-lateral floodplain expansion due to the above-mentioned processes hindered the use of the river before the regulations to such an extent that the shipping stopped during the Turkish occupation, and the cargo and passenger traffic started again only in 1830. Even in 1875–1876, hundreds of barges waited in front of the gravel bars (Kaján 2005). This problem of shipping was not unusual; therefore, there had been attempts to redirect the shipping onto the Moson Danube, though they had been unsuccessful.

*Hydrographical features of the main rivers*

The Danube and the Tisza are the largest rivers of the Carpathian Basin, and their evolutionary history is described in Sect. 1.3, while their most important present-day hydrographical data are in Table 3.1. The Tisza is more deeply embedded (its average depth is 6–8 m, its mean water width is 200–300 m), which can be connected to the clayey bed material and to the river regulations. In comparison, the average depth of the Danube is 4–5 m, which is increasing due to human impacts. The width of the Danube channel is 450 m at Komárom (depth: 4 m), and 290 m at Budapest. The data of the literature show that if the value of the width-depth ratio is over 50, the river changes its channel pattern from meandering to braided. In the case of the Danube, this value prevails everywhere; however, the anastomosing pattern is likely to be connected to the alluvial fan of Szigetköz. The convex surface of the alluvial fan—which includes the side-channel system—is 6–7 m higher than the floodplain. Thus, it is understandable that the ever-changing main Danube channel causes serious problems in shipping (VITUKI 2007). On the alluvial fan, the gradient of the Danube is 30–40 cm/km; however, downstream it becomes smaller; thus, before human intervention the river built gravel and sand bars, and it had braided pattern. This pattern still exists at Nyergesújfalu; however, most of the channel transformed into meandering (MT

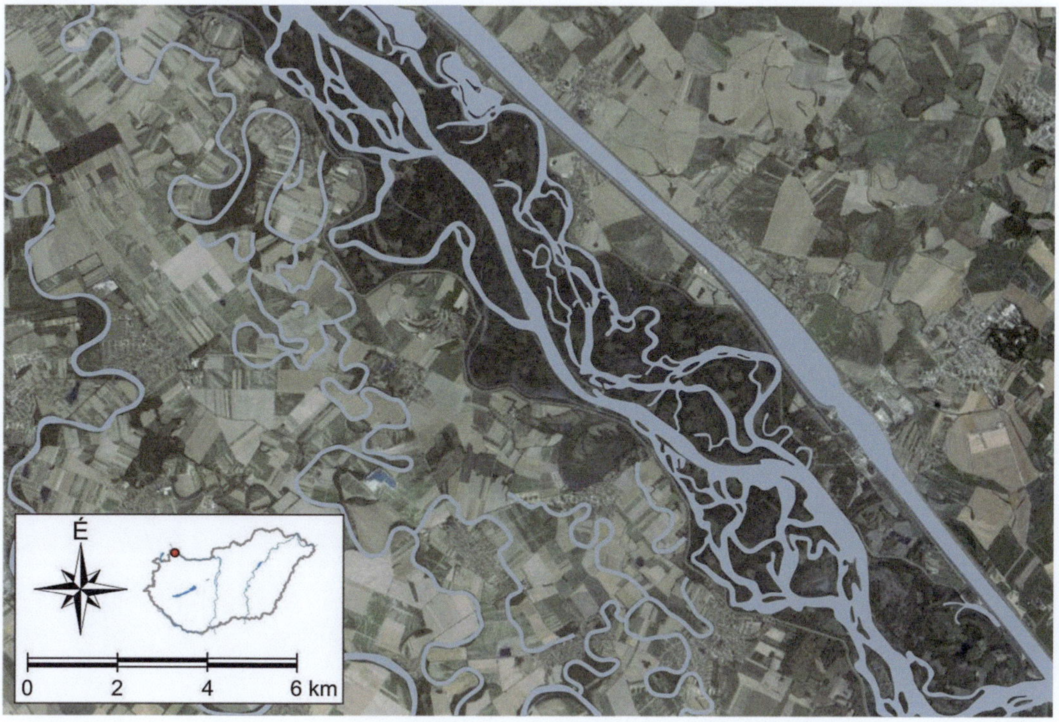


**Image 3.3** The regulated, anastomosing channel of the Danube at Bős/Gabcikovo (own construction based on googlemaps)

2002). The erosional activity of the Danube is reflected by the incision into the (volcanic) rock basement in the Danube Bend at Dömös. Downstream of Visegrád, the slope decreases considerably (approx. 7 cm/km); thus, here the meandering character is coupled with bar and island formation, which are on close relation with intensive bank erosion. The bank erosion triggers the landslides of loess walls along the high escarpments (from Ercsi through Dunaföldvár to Dunaszekcső). The prevention and mitigation of damages caused by mass movements is a serious geomorphological engineering task.

The mean discharge of the Danube at Budapest is 2265 $m^3$/s (the average of 1961–1990), and its increases by 300 $m^3$/s between the Hungarian entrance and exit points (from 2025 to 2362 $m^3$/s). The mean flow velocity is 0.6–1.4 m/s until Paks, but downstream it decreases below 1 m/s, because of the decreasing gradient. The water regime of the Danube is characterized by a discharge maximum (flood) in June, which originates from the precipitation of the marine air masses reaching the catchment area. The other, usually earlier, maximum is in connection with the melting of Alpine snow (the amount of water content of the snow, and daily stage changes could be found in the database of the Hungarian Hydro-Information service at www.hydroinfo.hu). Since some of the tributaries (e.g. Rába, Dráva) have a Mediterranean climate influence on their catchment, the precipitation and the hydrographs have peaks in the winter half-year too. Floods can occur in other seasons as results of increased precipitation, e.g. in August 2002 a flood developed on the Danube, with a peak discharge over 10,000 $m^3$/s (Table 3.1 and Image 3.4). Thus, the water regime of the Danube is characterized by three flood waves, and it has been well documented for almost one thousand years. For example, 17 devastating floods were reported in the fourteenth–seventeenth centuries (MT 2002). The most devastating flood in 1838 caused extensive damage at Budapest. To understand that flood, it

**Image 3.4** Flood on the Danube at Budapest in August 2002 (own photograph)

must be noted that apart from that the large amount of precipitation falling on the catchment area, the snow melts earlier in the upper subcatchments than in the lower ones. It could be explained by the oceanic influence on the south German subcatchments; thus, the winter ends earlier. The same happened in the early spring of 1838, when abundant precipitation reached the upper sections of the Danube, and at the same time, the river was frozen or breaking up downstream of Vienna, causing ice jams. The waterflow was blocked by the ice jams, and it was combined by the increased discharge from snow melt. The freezing of the section downstream of Budapest was not rare during the Little Ice Age, e.g. the Mongol invaders could cross the Danube in 1242. (The chronicles also described the frozen Danube in the times of King Matthias.) A similarly devastating ice flood occurred in 1956, and major floods of precipitation origin developed on the Danube in 1926, 1954, 2002, and 2006, and the flood protection was a serious professional achievement. The lowest stages occur in autumn.

Comparing the floods of the Danube and the Tisza, one could observe that none of them is able to drain high quantities of flood water without difficulties, especially if a flood develops within a short time. The important difference in the flood hazard of the two rivers is that shorter, maximum a few week-long floods evolve on the Danube, whereas the duration of floods is up to several months on the Tisza; therefore, its floods become really dangerous. Extreme amount of precipitation is always in the background of floods; however, floods also develop if the rain reaches a frozen subsoil or snow-covered terrains, and it rapidly tigers melting.

The average water discharge of the Tisza is 551 $m^3$/s at Szolnok (measured in 1961–1990), and 865 $m^3$/s at Szeged (Table 3.1). The gradient of the Tisza is small in the entire Hungarian section of the river, as in total the elevation difference is just 30 m. However, the slope has changed to different extents due to the regulations, in relation with the number of cut-offs. The gradient has increased from 1.6 to 3.2 cm/km between Tokaj and the conjunction of Sajó River, whereas it is 2.0–3.5 cm/km between Szolnok and the Körös River; finally, it decreases to 2.2–2.9 cm/km from the Körös River outlet to Szeged. The small slope results in low flow velocity (under 0.5 m/s on average), although it can be four times higher during floods (Bulla 1964).

The floods on the River Tisza follow the same temporal pattern as on the Danube. The first flood caused by spring snow melt. The majority of the catchment of the Tisza lies under the snowline, and the largest amount of snow and water accumulates at 1000–1500 m asl. The snowmelt is a key element in the water regime of the Tisza. The flood of the Tisza, which develops quickly in the mountains and then slows down on the plains, often encounter with the floods of the tributaries, as they have similar catchment conditions. The second flood wave on the Tisza occurs in June, induced by the early summer rainfall, and it may meet with the preceding spring flood. A third, autumn flood also characterize the water regime, but it is just slightly manifested in the discharge data (Lovász and Majoros 1994). The Tisza has its lowest stages in autumn, between September and November.

Vágás (1982) proved that the lower section of the Tisza (from Szolnok to the Danubian confluence) is not an independent river section hydrologically. This means that here the water level and the regime are mostly determined by the hydrological situation of the tributaries and the Danube, instead of the hydrology of the Tisza's catchment. The discharge data (even the maximum values) reflect this phenomenon, as the flood water of the Tisza often cannot drain into the flooding Danube, and the flood wave dammed upstream until Szolnok. The situation is the same in the case of the relationship of the Tisza and its tributaries. Fortunately, the flood maximums of the Danube and the Tisza rarely coincide. The regime of the Tisza and the Danube could be also characterized by the ratio of highest and lowest discharge values. If this ratio is above 1:20, the regime of a river could be characterized as "extreme". The ratio on the Danube is 1:18 at Bratislava, 1:14 at Budapest, and 1:12 at Mohács. However, on the Tisza it is 1:120 at Tiszabecs, and 1:66 at Szeged, referring to the extremities of low stages and floods.

### 3.1.3 Lakes

The present-day extension of lakes is considerably different from their former, historical extension. On the one hand, their territorial extension used to change to a great extent depending on the precipitation supply; thus, they specifically occupied a larger territory than today (e.g. Lake Balaton). Nowadays, the lakes cover 1 % of the surface, whereas this value must have been an order greater earlier. The chronological analyses revealed that there have been years when the water household was significantly positive, and the territory of lakes increased two or threefold. On the other hand, the constancy of the large extension of lakes was related to the unregulated water system of the Great Hungarian Plain, and the consequent long-time inundation. The genetic classification distinguishes various lake types:

(a) The most spectacular, largest lakes have a tectonic origin, and their evolution can be understood on a geological timescale (e.g. Lake Balaton and Lake Velence were thought to definitely belong to this type earlier).
(b) The meandering rivers on the floodplains have created an extremely high number of oxbow lakes since the Late Glacial, which gradually filled up mainly by organic material. However, large number of abandoned meanders and oxbow lakes has remained referring channel shifts (e.g. Lake Szelidi along the Danube). High number of abandoned meanders was created by artificial cut-offs of large meanders during the course of the river regulations. After the construction of artificial levees, the cut-off

oxbow lakes on the active floodplain rapidly started to aggrade (e.g. all oxbow lakes has been filled along the Maros River within decades), but those on the protected side have better retained their natural state (with open water surface) constituting a significant natural value. The richness of oxbow lakes is reflected by the data that 237 such oxbow lakes are which have larger territory than 4 ha (Pálfai 2001). They are not so important from the hydrological point of view, but their conservation and recreation value are significant. The preservation of the natural state of their environment and the sustenance of a good water quality sometimes prove to be difficult, mainly because their water supply is problematic due to the incision of the main channel.

(c) Few spring lakes exist in Hungary; however, Lake Hévíz is well known for its natural value and environmental conflicts. The source of the 38-m-deep thermal lake is a spring of 17.2 °C, and another one of 40 °C (Fig. 3.7). The water of the colder spring originates from the Keszthely Mountains, while the South Bakony Mts. supplies the warmer spring. Fault lines dissect the Triassic carbonate basement of the lake, as well as the area of the former underground mine at Nyírád; thus, the conduction of subsurface karst waters has one main direction. The karst water flow from the South Bakony has a SW direction, and it turns to SE due to the existence of an impermeable layer at Nagylengyel. Along this long circulation route the karst water warms up, and finally swell up as a spring on the bottom of the lake (MÁFI 2005). The cold and the hot spring enter the lake at the boundary of the karstic rock and the covering clay-sandstone layers. The 38.8 °C thermal water recharging the spring cave is formed through the mixing of the two waters of different temperatures

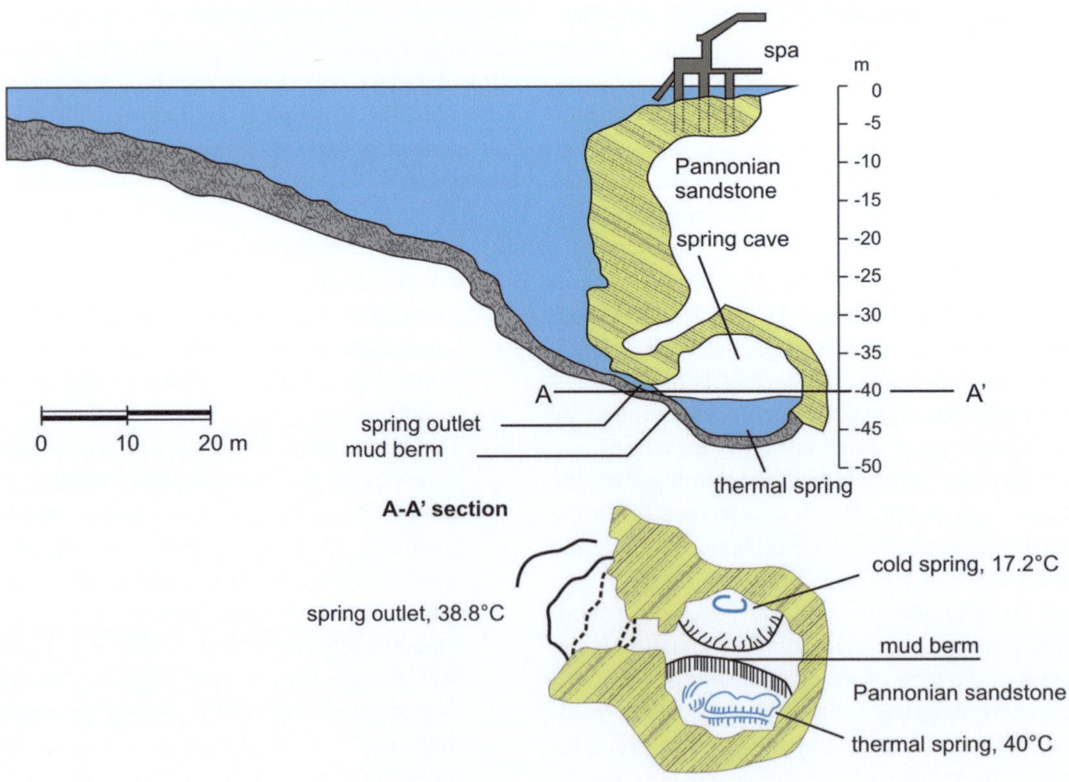


**Fig. 3.7** The structure of Lake Hévíz (after MÁFI 2005)

(Kordos 1984). The age of the waters was dated back to ca. 10 thousand years (where the colder spring has proved to be younger). The environmental conflict has been caused by the decreasing discharge of the warm spring, which was connected to the nearby water extractions (Mezősi 2008). Nowadays, the water system is in a regenerational state.

(d) Mass movements are common processes in the Hungarian hills and mountains due to their lithologic, topographic and climatologic features. Common feature of landslides—like rotational, translational or block slides—that over the glide plane water could drain towards the toe of the slope, where it accumulates behind the slipped sediment or toe (Fig. 3.8). This humid environment is manifested usually in the vegetation, but sometimes (too) lakes are formed too. In this way developed was the Lake Arló at Ózd or Lake Csikóvár on the volcanic tuff slope of the Visegrád Mountains (Szabó 1996).

(e) Another genetic lake type is created by the formation of barriers. Some of the above-mentioned types fit into this class, at least partially, e.g. the oxbow lakes are blocked by sediment plugs, or the lakes behind the too are also blocked by the debris of the mass movements. In addition, lakes are formed in enclosed deflational depressions, between alluvial fans and natural levees, or in areas blocked by aeolian accumulational forms. The largest barrier lake in Hungary is Lake Fertő, which is mostly surrounded by alluvial fans. The Ecsedi-láp (bog) is blocked by levees, while the basin of Lake Fehér at Szeged or Lake Sós at Nyíregyháza was created by deflation, and the outflow of the lakes is blocked by aeolian sand ridges.

(f) The saline lakes constitute a special group. They cannot be classified on the basis of their genetics, since the basin of the lakes developed in several different ways, their common feature is the similar chemistry of the water. For example, saline lakes developed in the Danube-Tisza Interfluve in deflational landforms (e.g. Szappanos-szék), the Lake Balaton has a tectonic origin, Lake Fertő is blocked, but all of them have a saline character to different extent. Among them the lakes in the Danube-Tisza Interfluve dispose the most unique features that originate from the saline character, and in their case, it determined their evolution history as well.

The high salinity of the water is the result of the very strong summer evaporation, accompanied by the increasing concentration of the water,

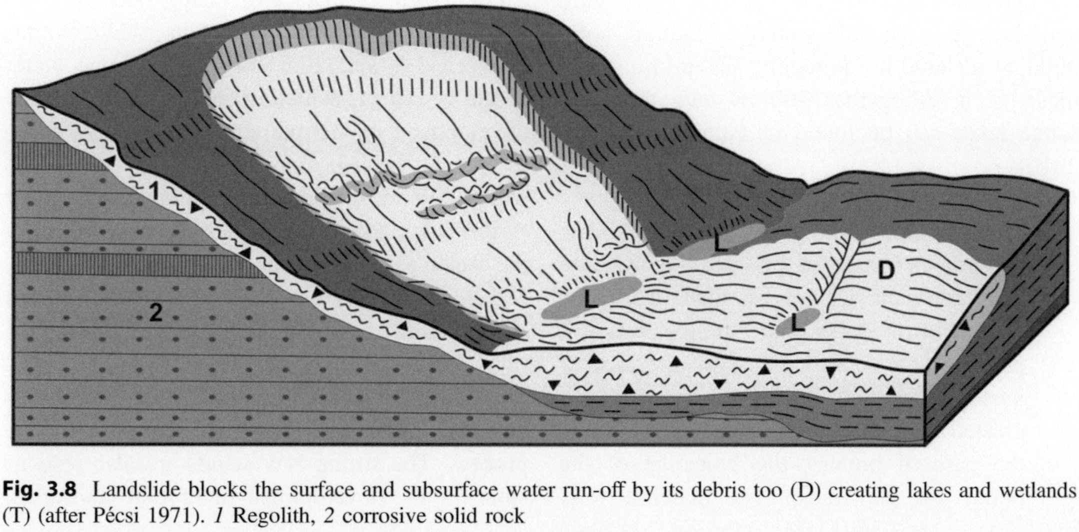


**Fig. 3.8** Landslide blocks the surface and subsurface water run-off by its debris too (D) creating lakes and wetlands (T) (after Pécsi 1971). *1* Regolith, *2* corrosive solid rock

**Table 3.2** Water household of Lake Balaton (after KDTVIZIG 2009)

| Lake mm[a] | 1987 | 1993 | 1996 | 2002 | 2003 | 2004 | 2008 |
|---|---|---|---|---|---|---|---|
| Precipitation | 778 | 554 | 688 | 541 | 509 | 660 | 526 |
| Inflow | 1225 | 435 | 1212 | 293 | 552 | 688 | 360 |
| Evaporation | 827 | 866 | 875 | 922 | 982 | 778 | 808 |
| Outflow | 891 | 1 | 1014 | 1 | 0 | 0 | 39 |
| Water withdrawal | 35 | 32 | 31 | 31 | 30 | 24 | 19 |
| Change of the natural water household | 1176 | 123 | 1025 | −88 | −180 | 434 | −12 |

[a]10 lake mm ~ 5.9 million $m^3$

which induces the genesis and diagenesis of dolomitic lime mud with a 90–95 % $CaCO_3$ content (since the parent material disposes of abundant carbonate). According to Molnár (1980), the $CO_2$ abstracting activity of the vegetation further increases the high salt concentration. The autumn precipitation reduces it; however, the proportion of $Mg^{2+}$:$Ca^{2+}$ increases to 7–12, and magnesium abundant calcite precipitates.

Lake Fertő is the westernmost saline lake in Europe, and its touristic promotion is based on this fact. The Mg:Ca proportion of the mud in the lake is around 7 at present, that is, the primary carbonated sediments mainly consist of calcite with a high magnesium content (Balogh 1991). The material created through diagenesis is unique and is called "meadow limestone" (though it is dolomite). This material is 20–30 cm thick, and as it can be found in several places in the Danube-Tisza Interfluve, it is easy to work with, and since it is the only source of natural, solid building material in the region, several buildings made of it. Altogether 460 *ex lege* protected saline lakes can be found in Hungary (TERM 2010).

(g) Lakes can form in karst areas too, in dolines and ponors (see Sect. 1.7). Doline lakes are created when clay is eroded from the neighbouring slopes into a doline, where it prevents the infiltration into the deeper karstic rocks; thus, a lake or a marshland is created (e.g. Lake Vörös at Aggtelek). In the case of ponors, the blockage of the ponor by the transported material (e.g. wood, eroded sediment) creates the conditions of lake formation. Therefore, the existence of these lakes is temporal; it lasts until the clearance of the sinkhole and the restoration of its original function (e.g. Lake Aggtelek). In the connected caves, the periods of drainage appear as floods, which are indicated by series of notches.

(h) The artificial lakes constitute a particular group, which were created, e.g. for recreational, economic or hydrological purposes. Road construction requires more and more gravel and sand; thus, the pits created in this way get filled by subsurface water and could deteriorate the groundwater conditions of their environment through evaporation.

*The hydrographical features of Lake Balaton*

Lake Balaton is the largest lake in Central Europe with its surface of nearly 600 $km^2$, and its length is 75 km, while its width varies between 1.5 and 12.5 km. The average water depth is 3.5 m, which is not enough for the development of thermal stratification. The deepest point of the lake is 11.5 m at Tihany. The strong NW winds played role in the development of this pool, since they induce currents and push the water towards the eastern basin of the lake, and when the current passes the strait between Tihany and Szántód, it erodes the lakebed (therefore, there is no sedimentation here). Standing waves could form between the western and the eastern basins, which may enhance the process. The strong NW winds are also responsible for the development of several landforms, e.g. the sand spit system developed along the southern shore (the spit triangle at Szántód got an international recognition by being included as an example in Davis's geomorphological reference book), or the large number of submerged offshore bars. The meridional valleys running towards the lake have been transformed into closed bays by NE-SW barrier sand spits. The bays represent a high value from the point of view of nature conservation; thus, they belong to the Balaton Uplands National Park.

The Lake Balaton is supplied by precipitation, groundwater and stream inflow; their total discharge is 10–12 $m^3/s$. A half of it is provided by the Zala River (during floods it transports ten times more discharge). The input side of the water household consists of the precipitation falling onto its catchment area (684 mm/year in the average of 1921–2003), and onto its surface (617 mm/year), while the run-off coefficient of the lake is 15 %. The output side consists mainly of the evaporation (902 mm/year in average; Table 3.2). The decreasing amount of precipitation in the past decades and the consequent run-off decrease lead to a negative water balance, which was first recorded in 2000 (Varga 2004).

The alternations of the lake level clearly indicate the sensitivity of the water household. In a historical timescale, quite high stages were characteristic approx. 2400 years ago, during the Tatar Invasion (it is also possible that the water level was artificially elevated for defence purposes; MT 2002), and in the humid period of 1732–33 (Figs. 3.9 and 3.10). Low water levels occurred in the first half of the 1860s due to the small amount of precipitation, which characterized the entire Carpathian Basin, and then, the situation was even worse in the beginning of the 1920s (Varga 2004). It could be concluded based on the available data that since the formation of the lake the water level changed between +6 and −1 m compared to the present level in accordance with environmental changes (Cserny 2002).

The storm waves erode the southern shoreline; thus, it retreated by several hundred metres to south. However, a significant amount of sediment was transported into the lake from the catchment area of the Zala River before the usage of the Kis-Balaton system, causing intensive aggradation. The sedimentation is a natural process of all lakes, its rate in case of the Balaton is 0.4–0.5 mm/year on the average of hundreds of years. This rate is in accordance with the sedimentation rate (0.6 mm/year) calculated by Lóczy in 1913 (he estimated the age of the lake to be 5000 years).

*The evolution of Lake Balaton and its main hydrographical questions*

According to Lóczy (1913), the lake developed in the Pliocene (which could be considered as Lower Pleistocene today), and it is situated in the depression series developed along fault lines (Image 3.5). On the basis of the analysis of the erosional terraces of the streams flowing into the lake from the north, which can be dated just with difficulties, Bulla (1943) claimed that the Balaton was created in the Riss-Würm (100–200 thousand years ago). They recognized that the second capture of the Zala River can be connected to the formation of the basin of Lake Balaton. Zólyomi (1992) dated the lake to the end of the Würm or the beginning of the Holocene, on the basis of pollen analysis.

The evolution of the basin and that of the lake was distinguished (Marosi and Szilárd 1981; Juhász and Marosi 1998). Before the formation of Lake Balaton, from the end of the Pliocene until the Middle Pleistocene the area was covered by the alluvial fans of streams whose catchment was in the elevating Transdanubian Mountains. The rivers run in the meridional valleys towards SE, and their erosional base was the sedimentary basin of the Felső-Kapos-Kalocsa Depression. Both the terraces of the meridional valleys and the sediment accumulated in the depressions refer to this flow direction. Marosi called this period "Pre-Balatonian". He supposed that a significant rearrangement of the river network started at the end of the Middle Pleistocene, as probably the basin of Lake Balaton started to subside, and the Felső-Kapos-Kalocsa Depression filled up. The subsidence of the basin is an essential issue, because it could determine the age of Lake Balaton. The difficulty lies in the fact that the subsidence may have happened in

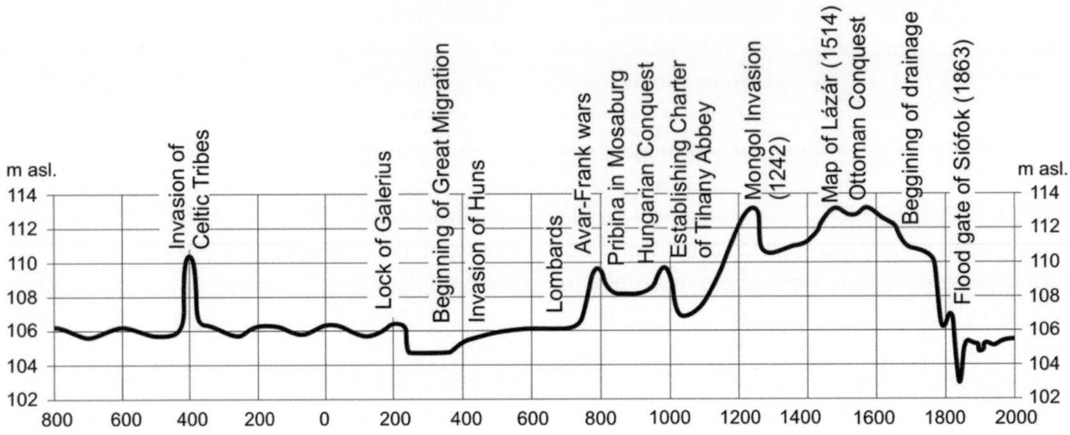


**Fig. 3.9** Water level changes of Lake Balaton (after Bendefy and Nagy 1969)

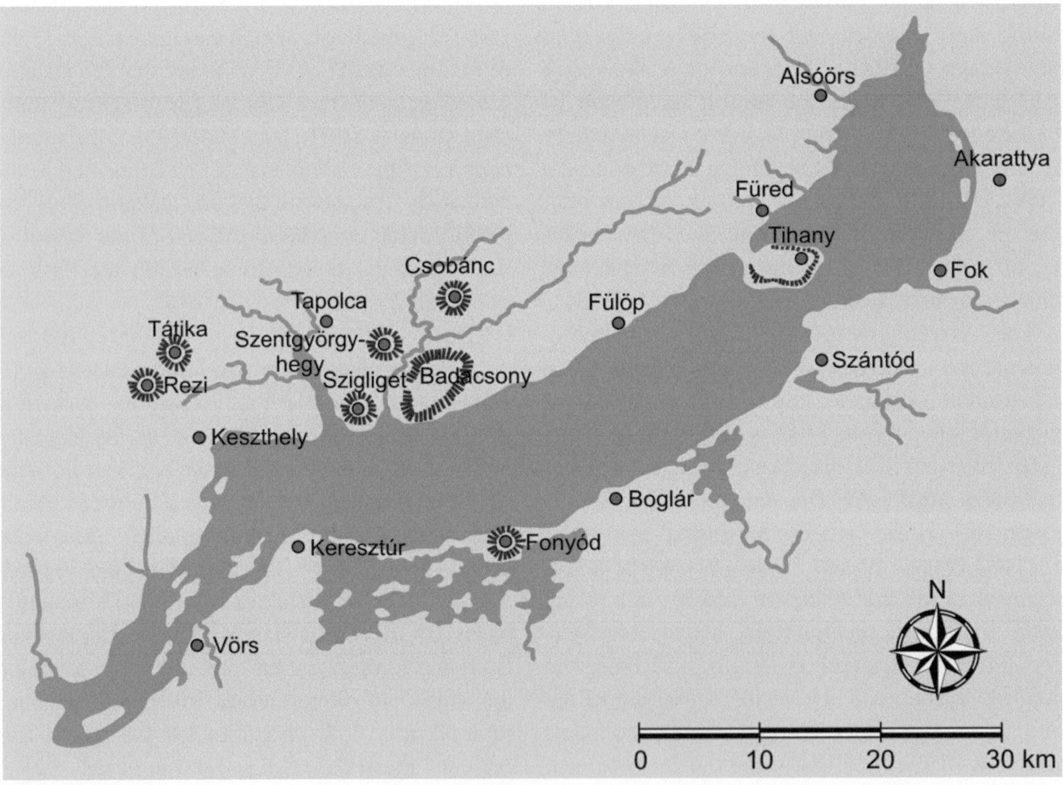


**Fig. 3.10** The Mikoviny's map was made in the extremely wet year of 1732 about the Lake Balaton

several separate parts of the basin (Lóczy has recognized four of them). Marosi and Szilárd (1981a) and Juhász and Marosi (1998) identified three subsidence phases; however, the latest studies refer to a continuous subsidence since the Riss-Würm (Cserny et al. 2002, 2008). Furthermore, the existence of subsidence became questioned by several researchers (it will be described later). The basin of the Balaton gradually became the local erosional base of the Transdanubian

**Image 3.5** The Lake Balaton and the basalt volcanoes in the Tapolca Basin, a view from Szépkilátó (own photograph)

streams, replacing the sedimentary basin of the Felső-Kapos-Kalocsa Depression (though in Fig. 3.11, it is not indicated) from the Riss-Würm. In the SE sloping meridional valleys, divides were formed; thus, the northern section of the valleys turned towards NW, and their streams run towards the Balaton. Their obsequent character is usually indicated by the terrace directions, but it is demonstrated only by the young terraces No. II/a and II/b towards the Balaton. The Felső-Kapos-Kalocsa Depression gradually filled up until the Middle Pleistocene; thus, only fine-grained sediments could be transported in from the mountains. The latest seismic measurements did not detect fault lines around the lake. Thus, the basin of the Balaton may have been a continuously elevating and consequently, eroding surface, even if the extent of elevation was less intensive than of the Transdanubian Mts. It means that the basin was primarily created by erosion, apart from tectonic preformation (Magyari 2000; MT 2002).

The existence of sub-basins in the area of Balaton was confirmed by Góczán (1962), in connection to the capture of the Zala River. According to this model during the intensive subsidence period, the first third of the Würm may have resulted in the capture of the River Zala, which flew into the lake basin through a valley described to be erosional. Marosi and Szilárd (1981a) explained the evolution of the uniform water body by a regression between Würm I and III (and by the blockage of the Sió River and the growing amount of precipitation). According to them, largest extent of the lake (thus the uniform lake) was probably 20 thousand years ago. On the basis of the present knowledge, three or four smaller depressions can be identified stretching from W to E, where smaller lakes developed 15–17 years ago, adjusting to the environment, and later, they became peat bogs in the Late Glacial (Fig. 3.12). Since that time, the Lake Balaton has been such a system, whose hydrography, water quality and water levels are determined by the climatic conditions (temperatures, precipitation, evaporation). Until the beginning of the Pleistocene, this multiple lake system existed, characterized by peat bog and clastic sedimentary deposition. Since then, the chemically precipitated clayey

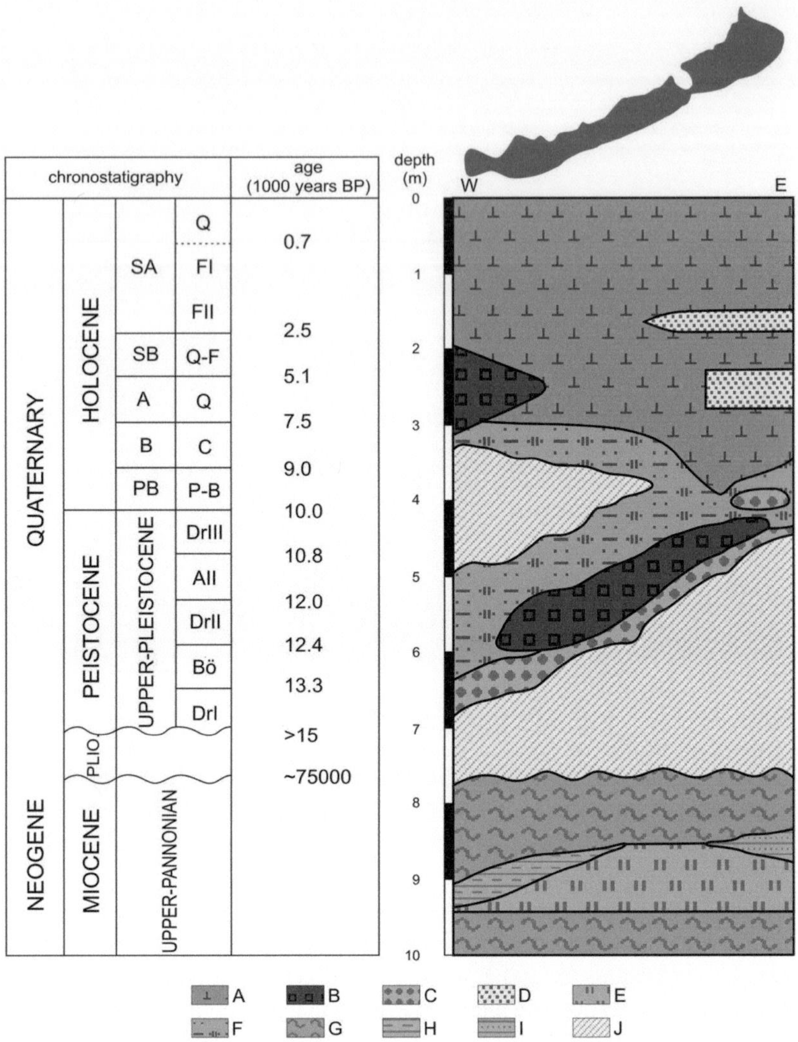


**Fig. 3.11** The theoretical sedimentary sequence of the Quaternary lacustrine sediments of Lake Balaton (after Cserny 2002). *A* Clayey lime mud, *B* peat, *C* gravel, *D* sand, *E* rock flour, *F* clay, *G* sandy clay, *H* clay marl, *I* marl, *J* sandstone

lime mud has become predominant in the sediments younger than the Pliocene, having an average thickness of 7 m (therefore, Lake Balaton is classified as a saline lake). Cserny et al. (2008) believe that the unified lake surface may have been created at the beginning of the Subboreal Phase (about 5000 years ago), when the lacustrine abrasion activated under more humid climatic conditions, it eroded the barriers separating the sub-basins. On the basis of the analyses mentioned above, the lake must have had its largest volume (and its greatest absolute height) at Late Glacial—Holocene boundary. Thus, the supposed Late Würm origin of the typical relief forms of Lake Balaton, such as abrasional terraces, sand spit systems and bays, also becomes uncertain, as they may be formed in the Holocene.

Some alternative ideas have been also formed on the genetics of the lake. Cholnoky's nearly one-hundred-year-old opinion is that deflation deepened the tectonic basin of the Lake Balaton. It may sound strange today, though at the beginning of the twentieth century the monoglacial theory prevailed, and the wind played a special role in the explanations of the evolution of landforms. The formation of the meridional valleys was also explained by wind erosion, supposing arid conditions. Several scientists believe and consider the desert conditions to be

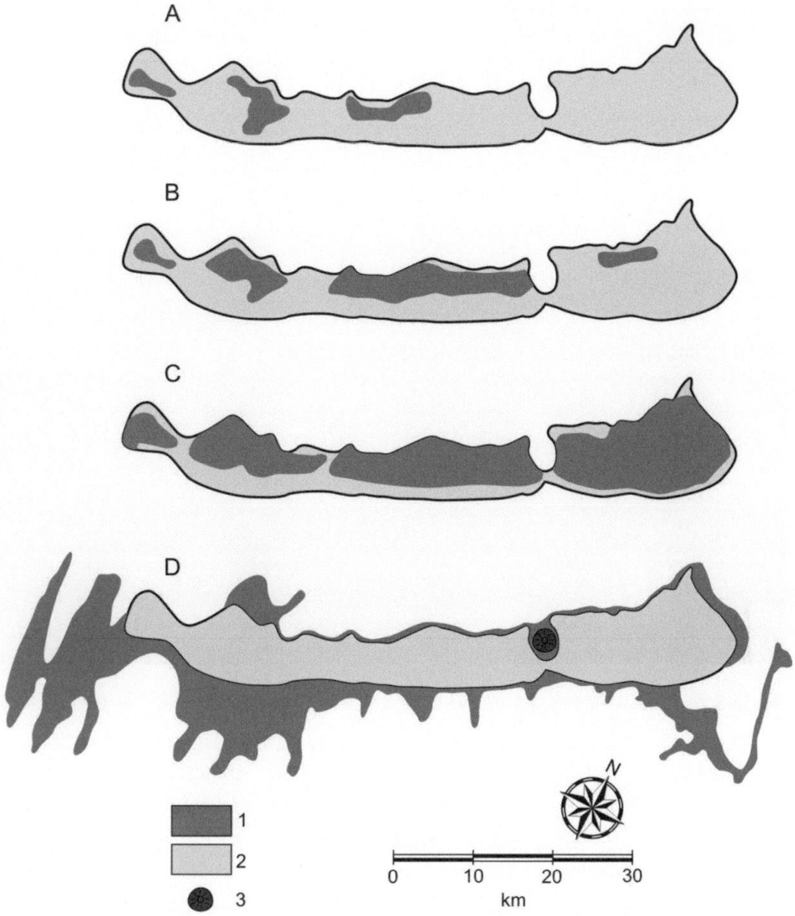


**Fig. 3.12** The evolution of Lake Balaton (after Cserny 2002). *1* The largest extension of the lake and its sub-basins, *2* the present-day extension of the lake, *3* island; *A* 15,000 years ago, *B* 12,000 years ago, *C* 9000 years ago, *D* 5000 years ago

proven, but in the Pliocene–Miocene boundary (about 5.5 million years ago), and not in connection to the formation of the much younger Lake Balaton. Certainly, Cholnoky dated the development of the lake before the Ice Age. The uniform southern shoreline also indicates a tectonic evolution. The Pliocene development of the Danube was previously described (see Sect. 1.3), and in connection with it there is a hypothesis on that Lake Balaton is a palaeo-channel of the Danube (Mike 1980).

Today, more scientists support the idea of the formation of the basin at the end of the Pleistocene, the existence of separate lakes in the sub-basins, and the development of the uniform water table in the middle of the Holocene. At that time, as a result of climate change, the coniferous forests were replaced by deciduous ones, and instead of the debris accumulation limy mud sedimentation started, and the volume of the mud have been 1.5–2 times greater than the amount of water in Lake Balaton (Cserny 2002).

*Correlation between the water quality and quantity of Lake Balaton*

The connection between the quality and quantity of the water in a lake has been known for a long time. Thus, it has always been an important issue in the water management to achieve higher (optimal) water level, which presupposes better water quality. In the case of Lake Balaton, an equilibrium water level can be calculated, which is about 107 m asl, if the evaporation and the water input are approximately equal (Varga 2004). The 1990s and the beginning of the 2000s (2003) were especially dry; thus, the water level dropped to the 1920s minimum level (−38 cm; Varga and Pappné Urbán 2007); therefore, the necessity of extra water

**Image 3.6** Lake Velence and the wide reed stands along its shoreline at Pákozd (own photograph)

supply emerged (see in details in Sect. 11.3). The next section of the present work describes the same problem through the example of Lake Velence, where the water shortage and the water level decrease in the extremely dry year of 1993 were solved by supplying karst water into the lake.

*Evolution and a hydrographical characteristics of Lake Velence*

Lake Velence was created by tectonic preformation and erosion, similarly to the Sárrét in Fejér County or the area of Tikacs located in the same mountain foreground depression; however, nowadays these do not have open water surface. The supposed subsidence probably happened at the end of the Pleistocene or the beginning of the Holocene (Martonné 1995), though some researchers dated the sinking to the Holocene (Lovász and Majoros 1994), or to the end of the Pleistocene (15–17 thousand years ago), but it is quite sure that it was simultaneous with the formation of Lake Balaton. The relative sinking was probably quite slow process in the area, similarly to the other depressions (e.g. Bodrogköz). Thus, it is impossible to date the birth of the lake basin exactly, though it is proved that the depressions were occupied by water from the Late Glacial to the Holocene, in periods when the area had a positive water household.

The area of the lake itself is 24 km$^2$, but its extension greatly varies depending on the water level. The average depth of the 10 km long and 2.5 km wide lake is less than 1 m. The Lake Nádas (or Reed Lake) belongs to the basin of Lake Velence (Image 3.6); half of its territory is covered by reeds; therefore, it is a valuable nature conservation area. The water supply of Lake Velence is mainly originates of precipitation and by the small brooks, like Császár-víz and Pázmándi-víz (Fig. 3.13). The Dinnyés-Kajtor Canal is connected to the lake near to the main inflowing brook, ensuring run-off. This is not an ideal spatial setting from the point of water quality, since the inflowing water cannot refresh the water body of the lake, and the modest water exchange does not improve the water quality. It varies considerably, and the salt content of the lake may reach 2000 mg/l in some places, due to the intensive evaporation and increasing salt concentration (MT 2002).

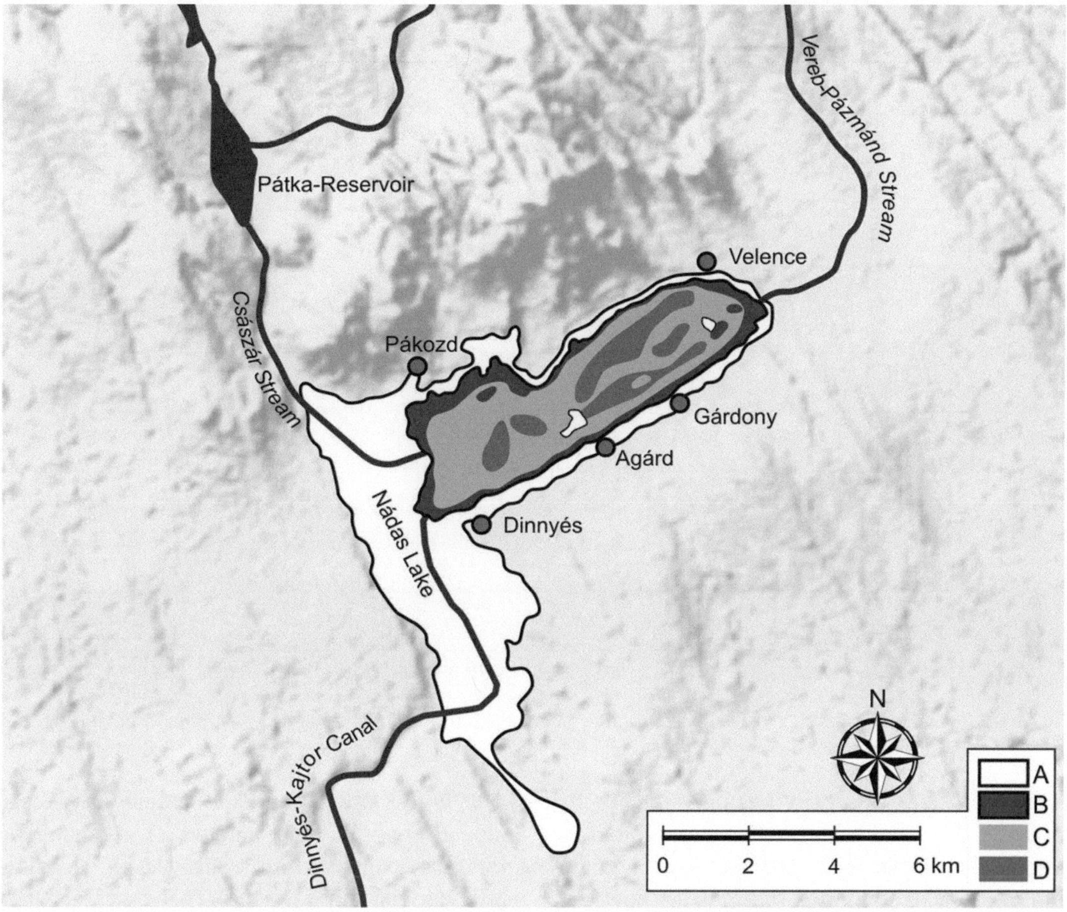


**Fig. 3.13** Geographical position and depth conditions of Lake Velence. *A* Reeds surrounded by dry land, *B* <1 m water depth, *C* 1–2 m water depth, *D* 2–3 m water depth

**Table 3.3** Water household of Lake Velence (after Simonics 2005)

| Lake mm[a] | 1931–70 | 1971–96 | 2001 | 2002 | 2003 | 2004 |
|---|---|---|---|---|---|---|
| Precipitation | 610 | 514 | 528 | 509 | 366 | 635 |
| Inflow | 735 | 339 | 266 | 264 | 227 | 232 |
| Evaporation | 925 | 879 | 918 | 955 | 960 | 765 |
| Outflow | 415 | 77 | 0 | 0 | 0 | 0 |
| Water withdrawal | – | 42 | 39 | 23 | 22 | 22 |
| Change of the natural water household | 430 | 126 | 99 | −147 | −367 | 102 |

[a]10 lake mm—240 thousand $m^3$

The water household of Lake Velence is highly sensitive. In the earlier centuries, it was not a recreational centre, and in dry years, plans were made to drain the area and use it for agriculture. The idea was supported by the fact that the lake sometimes had a mire character, e.g. between 1792 and 1794, then between 1835 and 1837, and finally in the second half of the 1860s,

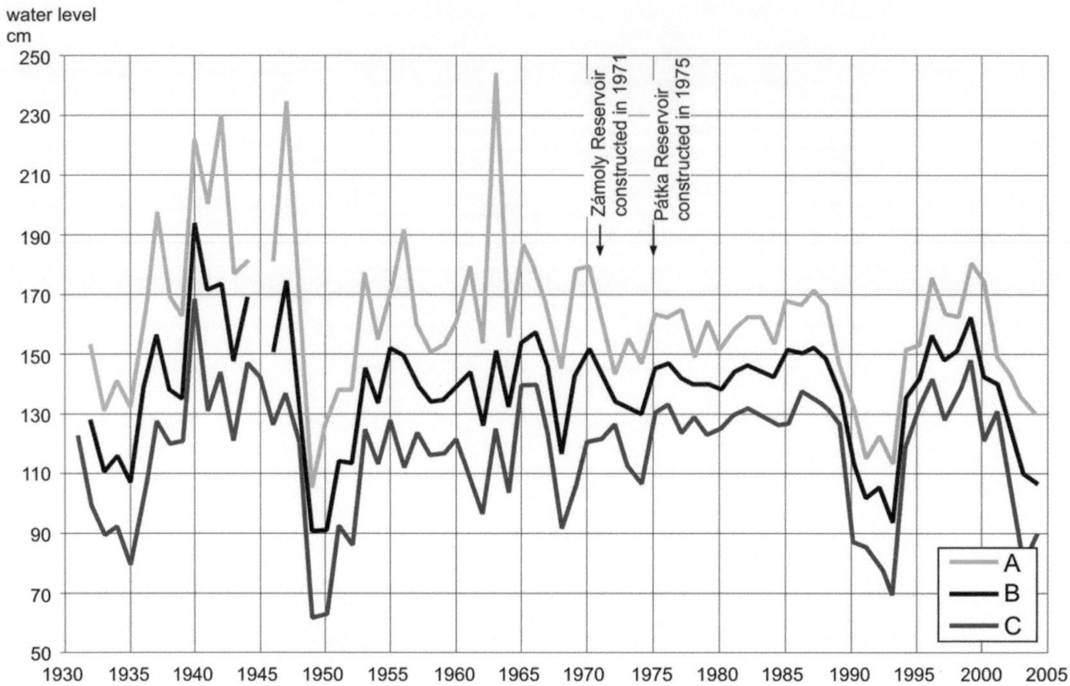


**Fig. 3.14** Typical water levels of Lake Velence (after Simonics 2005). Annual water levels: *A* high, *B* medium, *C* low

when there was no water resupply due to the droughts. The lake basin was already plotted for later use as arable land. The intensive aggradation of the lake supported these plans; however, the drainage did not start in spite of technical difficulties. The water household became critical again in the last recent years (e.g. in 1993 and 2003), when there was a negative balance due to decreased precipitation and increased evaporation (Table 3.3 and Fig. 3.14). In such years, even the artificial reservoirs at Zámoly and Pátka were not able to support enough water supply, although they are able to store 8–9 million $km^3$ water. The water shortage of 1993 (Fig. 3.14) was managed by water supply from outside sources, i.e. from the Vértes Mts. drinking-water-quality karst water was withdrawn from a mine (abandoned since then at Rókahegy).

High water stages also occur: they were recorded eight times since the fifteenth century, the last one in 1963.

The water quality of the Lake Velence has deteriorated due to the decreasing amount of water in the 1990s, and its remediation was also hindered by the high amount of organic and inorganic sediments in the lake basin, deposited partly due to natural reasons. The water exchange of the lake along its longitudinal axis, through Császár-víz Brook, and the hydrological connection between the rivers and their floodplains are regarded the most important tasks of the Central Transdanubian Water Management Directorate. The problem of organic material and nutrient input (from animal husbandry, municipal waste water) transported by rivers into the lake is also very important. Though the chemical composition of rivers is not extremely bad, the nutrient content of Lake Velence is still high, and the annually algae growth can be observed (KÖDUVIZIG 2008). The action plan to preserve the touristic and natural values, to maintain the water quality of the lake is in the regional development programme (1117/2003. -XI. 28) of the Lake Velence–Vértes Recreational Area, which may provide a reliable framework for the preservation of a good environmental state (Zseni 2002).

*Evolution and a hydrography of Lake Fertő*

Lake Fertő is the second largest (309 km$^2$) saline lake of the Carpathian Basin, its surface extension greatly varies, depending on the water supply: according to the calculations, 1 cm water level fluctuation means $\pm 3$ km$^2$ territorial change (MT 2002). It is shallow (1.0–1.5 m deep); thus, third of lake (Fig. 3.15), that is, 80 % of its Hungarian territory is covered by reed (Image 3.7). The water input of the lake originates from precipitation and groundwater, its main inflowing brook is the Vulka, with a modest water discharge. Hunfalvy remarked in the 1870s that "Lake Fertő consists of open water, and the so-called Hanság". In fact, a large wetland with a W–E extension is connected to the lake having a N–S extension. But the wetland was drained in the 1950s, and its rehabilitation started about fifteen years ago.

The shallow, sedimented lake was formed in a closed, natural depression enclosed by three natural objects (some researchers describe it as a postglacial, tectonic depression; Pécsi 1975). The lake is enclosed by the Leitha Mountains from W–NW, the Parndorf Plain (the old alluvial fan remnant of the Danube) from NE, and other alluvial fan remains of the Danube (being reshaped by the Ikva Stream) from south. (The old gravels of the Danube can be seen even on the surface at Fertőd.) Thus, Lake Fertő is originally a lake without an outlet from the hydrographical point, which is destined to sedimentation and evaporation. In the basin, large amount of sediment was accumulated, which fills a half

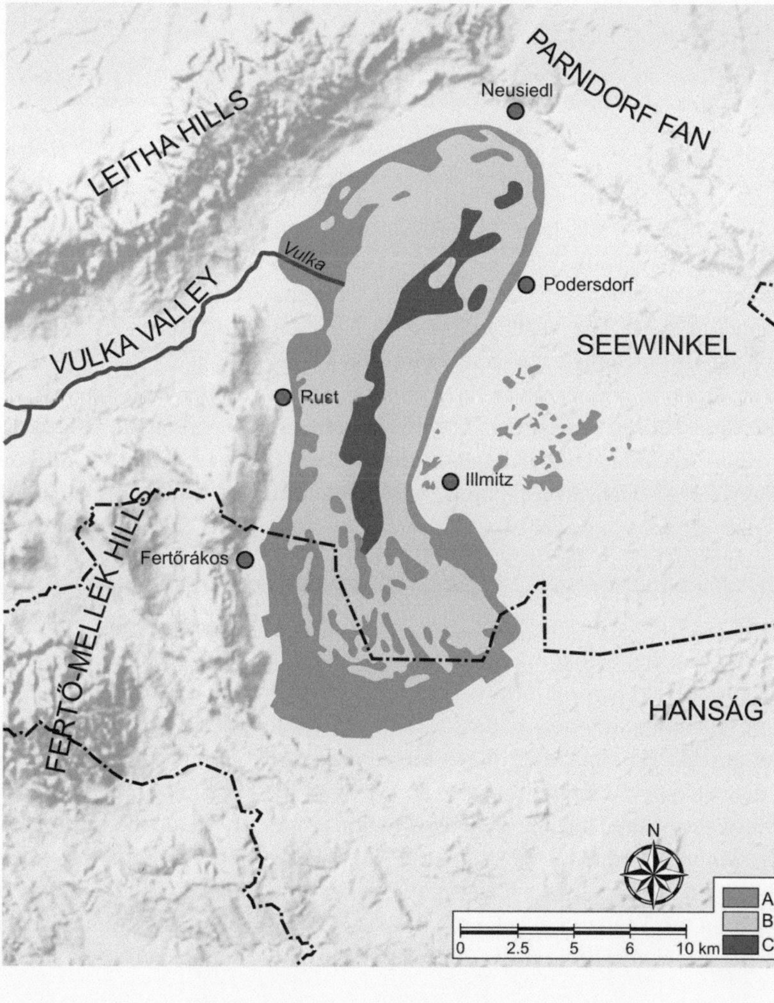


**Fig. 3.15** Location and depth conditions of Lake Fertő (after VITUKI) depth: *A* <1 m, *B* 1–2 m, *C* 2–3 m

**Image 3.7** The extensive reeds along the southern shore of Lake Fertő (own photograph)

of the volume of the lake (MT 2002); however, it is still a low specific value compared to the Lake Balaton. The name Fertő also originates from the aggrading state. The water of the lake is saline with a hydro-carbonate character, which makes its water slippery, and results in the formation of the extensive saline vegetation. The high salt and organic material content causes its turbidity (Nagy 1996). The salt content used to be tenfold of the present-day value (1500 mg/l), before the construction of the lock and the provision of the possibility of water exchange (Somogyi 2000).

The Hanság, belonging to the water system of Lake Fertő, has low run-off, its surface is covered with Holocene peat accumulated on fluvial sand and gravel. The flat relief of the former wetland is dissected by 50 mounds, which are elevated by 2–3 m above the lake and they used to constitute the southern boundary of the basin (Fig. 3.16). The relief slightly differs towards east, in the direction of Tóköz. Lake Fertő used to have a close, mainly hydrological, connection to the wetland of the Hanság Basin (length: 50 km; width: 7 km). The Hanság used to be the local erosional base of the region, as the Ikva, Répce and Rábca rivers joined it, and the peat absorbed all the water like a sponge. Based on this, it is understandable that at the beginning of the 1800s the drainage of the wetland was an important issue (mainly local farmers urged, as, e.g., the wheat from Osli used to be a famous product). The drainage works were made in 1825, and the area of wetlands was reduced by the half (about 20,000 ha), and later, in the middle of the 1950s the remaining wetlands were also drained. The Fertő/Neusiedl–Hanság National Park (1994) aimed to preserve the wetland habitats. The low elevation differences caused (the eastern part is 109 m asl, the western is 112 asl, while the Győr Basin is 109 m asl) that the 4–5 m high floods of the Danube could reach Lake Fertő through the Hanság, and vice versa, high water of the Lake Fertő could outflow from the basin through the Hanság (Kogutowicz 1936). The “strait” between Pamhagen and Fertőd was one of the few

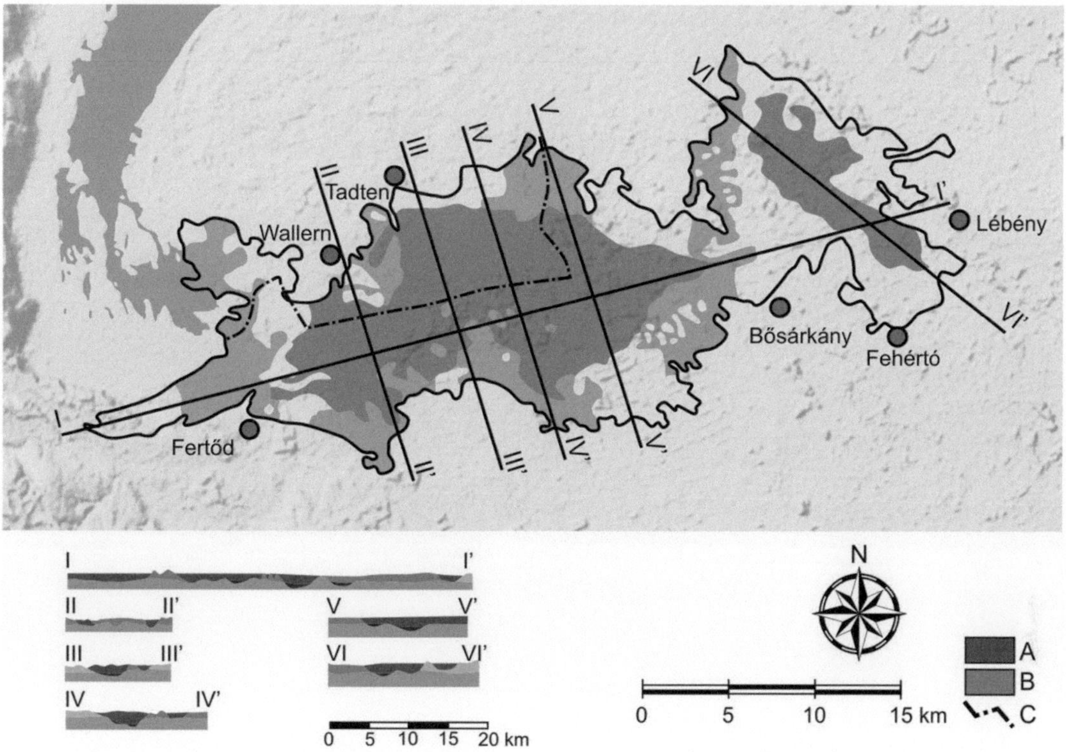


**Fig. 3.16** Simplified morpho-lithologic map of the Hanság (after László and Emszt 1915; Pécsi 1975). *A* Peat, peat mud, moor, sand dunes, *B* wetland with sandy, clayey islands, *C* national border

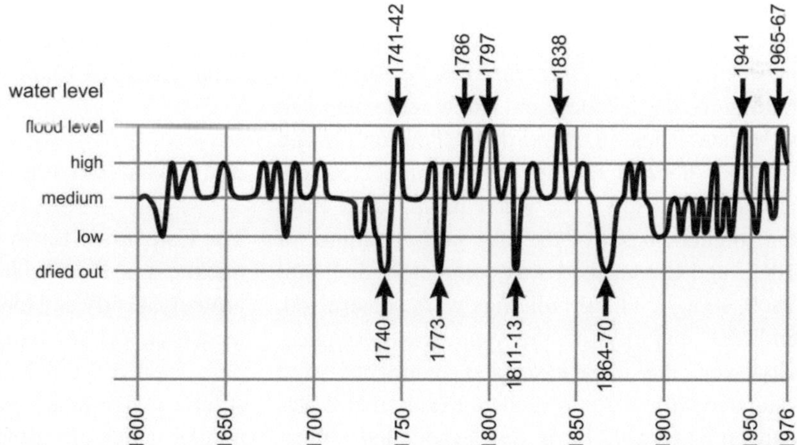


**Fig. 3.17** Water level changes of Lake Fertő (after Kiss 2004)

waterways that joined the Hanság and toll could be collected on it. The lock of the Hanság was built here, ensuring a modest water exchange.

When considering the environmental status of Lake Fertő, the water-level changes are the most important. The strongly precipitation-dependent water household is reflected by the water level of the lake, which indicates repeated dry periods or the hazard of drying. The reconstructed water-level changes correlate with the regional processes of the Carpathian Basin, and with the similar tendencies of other lakes (Fig. 3.17). The

**Image 3.8** Recreation facilities along the shoreline of Lake Fertő in the neighbourhood of Fertőrákos (own photograph)

last, completely dry lakebed was reported between 1868 and 1872, when even a village was built in the dry lakebed, and a new settlement had to be established in the northern shore only when the lake flooded the area. The name Neusiedl also refers to this. The water disappeared from the southern basin of the lake in the summer of 1867, and the lakebed was completely dry until 1869, so agricultural buildings were constructed, and the majority of the area was ploughed. However, the clayey surface sediments were saturated with water to such an extent, that it was dangerous to walk on it. Salt evaporated ("blossomed") along the eastern shore, which was used in soap production by the inhabitants (Nagy 1996). It can clearly be mapped that in times of drying outs (the recession of the lake) the settlements and the agricultural buildings followed the water, and temporary accommodations were created in the lakeshore at the given time (Kiss 2004). Nowadays, the lake has high touristic value (Image 3.8).

## 3.2 Subsurface Waters

The Carpathian Basin was mainly filled by large thickness of Pannonian, Pliocene and Pleistocene gravelly, sandy and clayey sediments of fluvial or lacustrine origin, which ensured water storage pore capacity; thus, the subsurface waters constitute a significant asset. The lowermost Pannonian layers are usually fine grained of marine origin, consisting of sand, sandstone or clayey marl layers, which are distinguishable from the upper, up to 700-m-thick Pleistocene–Holocene layers. The latter typically intruded into the basin in the form of alluvial fans, referring to overall fluvial sedimentation in the basin. Due to the spatially and temporally gradual sedimentation towards the centre of the basin, the sediments are getting finer towards the centre as the coarser sediments had already been deposited along the edges.

The subsurface waters can be classified on the basis of the following parameters (Tóth 2006):

- pore type of the medium (crystal water, pore water, karst water), and its equilibrium water level (groundwater level, potentiometric surface);
- temperature (warmer waters than 30 °C are considered as thermal water in Hungary);
- chemical composition (freshwater, if the soluble material content is below 1000 mg/l; mineral water, if the soluble material content is above 1000 mg/l, or above 500 mg/l with a significant $CO_2$ content or a significant concentration of biologically active elements; while medicinal water is a mineral water type whose healing effect can be proven).

The traditional nomenclature applies groundwater, artesian water, crack water and bank-filtered waters in the basin. This didactic classification has been broadly criticized, partly, because the subsurface waterflow is continuous inside the basin; thus, it cannot be divided into such categories. The division primarily has a practical point of view, and it does not consider the origin of the water. The second criticism reckons that the impermeable layers, necessary to define for the basic categories, do not really exist according to the latest studies (Mádlné Szőnyi and Tóth 2003). Therefore, water yielding aquifer layers (e.g. sand, gravel, limestone) are distinguished, where the porosity ensures the infiltration and leakage, and where wells with high outflow discharge can be constructed in the water conducting layers. In contrary, impermeable layers (e.g. mud, clay) can also be identified, which are able to store and conduct water too, but their water yield is lower by orders of magnitude. Finally, watertight layers (or storage layers) are also distinguished, but they exist only in theory, because each rock has some water conducting capacity.

### 3.2.1 Position and Spatial Changes of the Groundwater

The groundwater is situated the closest to the surface in the basin, and it is the water above the first impermeable (or water retaining) subsurface sediment layer. It is connected to the atmosphere, and the climate directly influences its state (Fig. 3.18). The groundwater level is situated between the unsaturated (where the pore pressure is lower than the atmospheric pressure) and the saturated (pore pressure is higher, respectively) zones, where the two pressure values are nearly equal. There are two main uncertain points of this interpretation. The first is originated from its name: Why is the first subsurface water level called groundwater? What shall we call the water in the fertile soil? The latter is often called soil moisture in order to distinguish it from the former. Secondly, the first impermeable layer in the Great Hungarian Plain is on the surface on at least a half of the territory, even so the term of groundwater is also used in these places too (although the concept should not be applied here). Several scientists have emphasized that the groundwater is a quite general concept, involving various, highly differing, near-surface waters, which fact is also shown by the extremely different yields of the wells found here (MT 2002).

Some sediment layers (e.g. gravel) are able to store water in large quantity. Their great pore volume results in a large water storage capacity; thus, the groundwater table is located close to the surface (2–3 m) because of the small capillary action/cohesion rate. The groundwater is typically situated at 4–5 m below the surface in sandy sediments, and at 25–30 m in loess. The cohesion is small between loose materials, and much greater between fine-grained materials. The capillary action of loam and clay is much smaller (>40.5 h/mm) than that of sand or gravel (<300.5 h/mm; Stefanovits 1981); therefore, the location of the groundwater below the surface could be computed, without considering any water movement with a different direction. The depth of the groundwater clearly reflected the lithological conditions in the Danube-Tisza

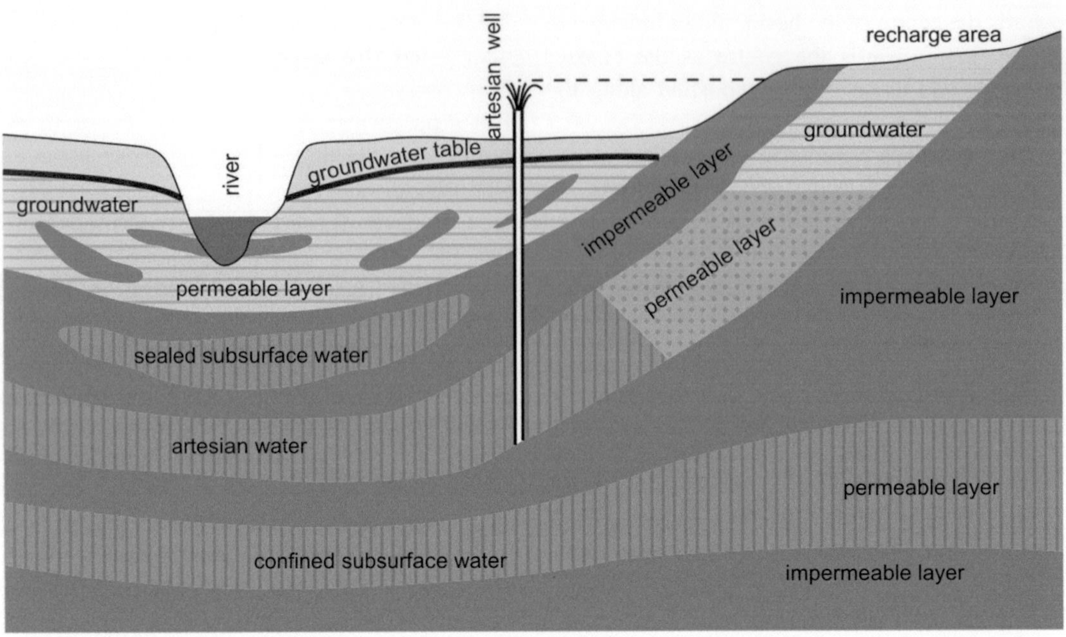


**Fig. 3.18** Main types of subsurface waters (after Juhász 1987)

**Fig. 3.19** Groundwater table depth in the Danube-Tisza Interfluve between 1950 and 1955 (after Rónai 1956). *A* Below the surface by less than 3 m, *B* 3–6 m, *C* 6–10 m, *D* deeper than 10 m

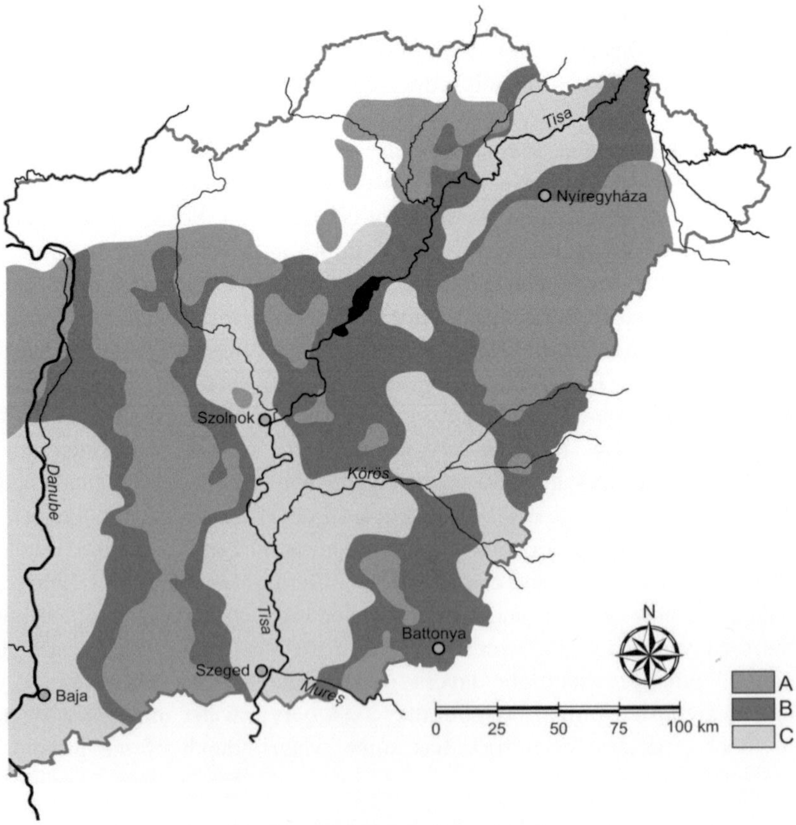

Interfluve (and in other places, too) before the large-scale anthropogenic interventions (levelling the surface, drainage, etc.) in conformity with the above values (Fig. 3.19).

The depth of the groundwater level follows the variability of the relief too. For example, the central part of the Danube-Tisza Interfluve is 30 m higher than the neighbouring Danube or Tisza Valleys; therefore, the groundwater table has an hourglass-like surface. According to Rónai (1972), it could be explained by the pressure of the surrounding sediments on the water body, and where the smaller rock cohesion (e.g. gravel) allows, the water rises from under the pressure until the outside atmospheric pressure balances it. According to another explanation, the higher groundwater levels of the central part of the Danube-Tisza Interfluve correspond to the different pressure originating from the elevation difference of the orographically higher infiltration area (MT 2002).

It is not easy task to determine the amount of the disposable groundwater supply, because groundwater does not only exist in the "classical" gravel aquifers, where lateral recharge also apparent, but also in clay and sand, though these are not real aquifers. The existence of the inflow from below is confirmed by observations, though it cannot be really measured. (Though the physical parameters of aquifers are known, the water household could not be calculated.)

An important issue in connection with groundwaters is their supply, which is even less clarified than in the case of artesian waters, in spite of the numerous researches. The groundwater is supplied by precipitation in the uncovered territories, and the rate of vertical infiltration changes depending on the environmental conditions. However, its rate is slower than of the horizontal infiltration/flow. The groundwater may dissolve different salts from the surrounding sediments during its flow; thus, valuable and special medicinal waters can be created with the growth of the concentration. (The average salt content of groundwater in the Great Hungarian Plain is 3–4 g/l.) For example, the water of Nagyigmánd was extracted from dugwells, but its withdrawal had to be terminated in the 1980s due to the extensive use of fertilizers and pesticides. Similar occurrence is the bitter water with glauber salt at Tiszajenő, which is marketed under the name "Mira".

When considering the resupply of the groundwaters, it must be noted, that the vertical infiltration rate is slow; thus, it takes 3–7 months until the infiltrated precipitation reaches the groundwater level, depending on the depth of the latter. According to Rónai's (1972) estimate, it is valid in one-third of the Great Hungarian Plain because of the lithologic conditions. At the same time, groundwater exists in the dry years too, even if it is situated deeper. The explanation for this is that some resupply originates of run-off and infiltration (one-third or on sixth of the total volume of the groundwater) of the precipitation fallen on the edges of the mountains. It could be explained by the existence of aquifer layers inclining towards the centre of the basin, which are able to conduct water due to their structure. Such sedimentary structures are the alluvial fans and alluviums. Recently, the researchers have tried to measure and verify the connection between water bodies without an open water table, and the subsurface flow towards the basin. The groundwater level fluctuation (vertical changes of the groundwater level) has yearly and monthly patterns, which are determined by the amount of precipitation, but they reflect the petrographic conditions too; as the groundwater can be found in different depths in gravel, sandy and loess areas (see Rónai's classical Fig. 3.19).

Along the rivers, if the evaporation or water withdrawal involves the groundwater body, the groundwater can be resupplied from the rivers (especially during floods), while groundwater may flow into the river in times of low water, contributing to the water discharge of the river. Depending on the lithologic and orographic conditions, this connection becomes looser with the distance from the river, and it exists just in a zone of 1–2 km usually (Kaszab 1987).

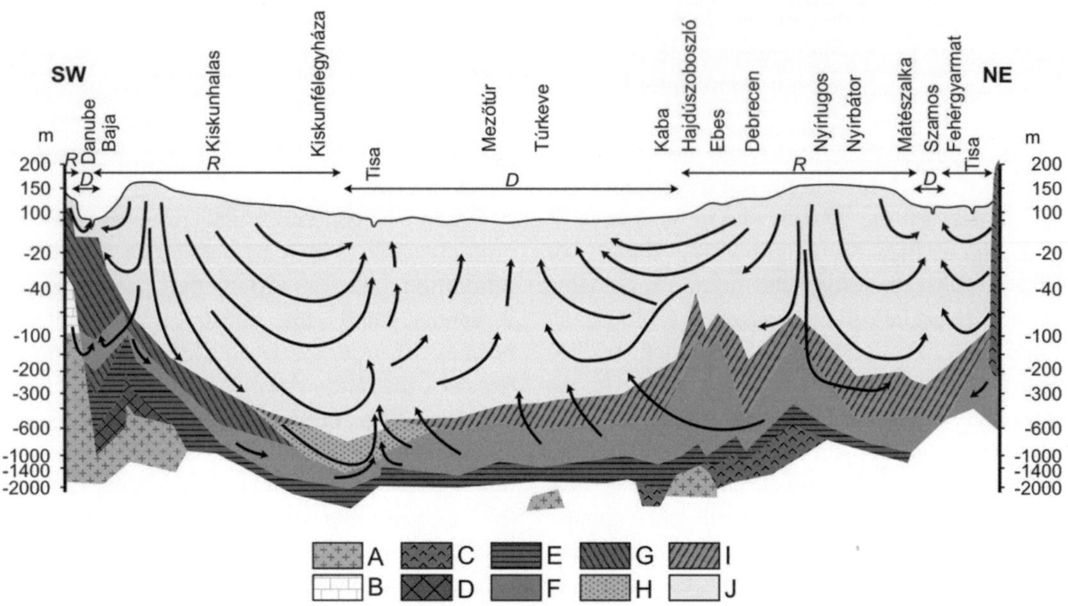


**Fig. 3.20** Theoretical profile of the subsurface flow system of the Great Hungarian Plain (after Erdélyi 1979). *A* Crystalline basement rock, *B* Mesozoic limestone, *C* Miocene volcanic rock, *D* Miocene sedimentary rock, *E* impermeable or slightly permeable Lower Pannonian layer, *F* Upper Pannonian aquifer sediments, *G* slightly permeable Lower Pannonian sediments, *H* highly permeable Upper Pannonian sediments, *I* Pleistocene slightly permeable sediments, *J* Quaternary sediments

## 3.2.2 Flow Conditions and Water Household of Artesian Waters

The artesian water is located under the groundwater; thus, in a lower orographic situation, and it is separated from the groundwater by an impermeable sediment layer. The artesian waters could get to the surface through drilled wells due to the pressure within the aquifer sedimentary body. This is called a positive (artesian) well, while in contrary negative (subartesian) conditions also exist.

A circulation of the subsurface waters in the Carpathian Basin has been analysed for a long time (Fig. 3.20). The waters create different subsurface circulation systems (e.g. Erdélyi 1979; Tóth 2006; Mádlné Szőnyi 2006, 2008); thus, areas could be identified where the positive, and areas where the negative wells are more prevailing, referring to inflow (recharge) and outflow areas (Fig. 3.21).

The experience shows that the main inflow areas of the basin are the Danube-Tisza Interfluve and the Nyírség, where the subsurface waters can flow to larger depths and can infiltrate towards the lower layers of the basin, and move from there even to deeper or towards the peripheral regions. The model of the negative and positive wells also has its origin in this phenomenon, and thus, the artesian water pressure is lower in the areas of downward infiltration, and higher in the areas of upward flow.

The existence of the flow system is not debated by scientists, since the evolution of such a circulation system is understandable in water bodies of a several kilometres of thickness under different pressure conditions. However, the question is that on the surface or near to the surface to what extent the differences of the vegetation, the chemistry of soils, or the water supply, etc., are connected to the large-scale flow systems (Tóth 2006; Mádlné Szőnyi and Tóth 2007). Recently, it is supposed that the system

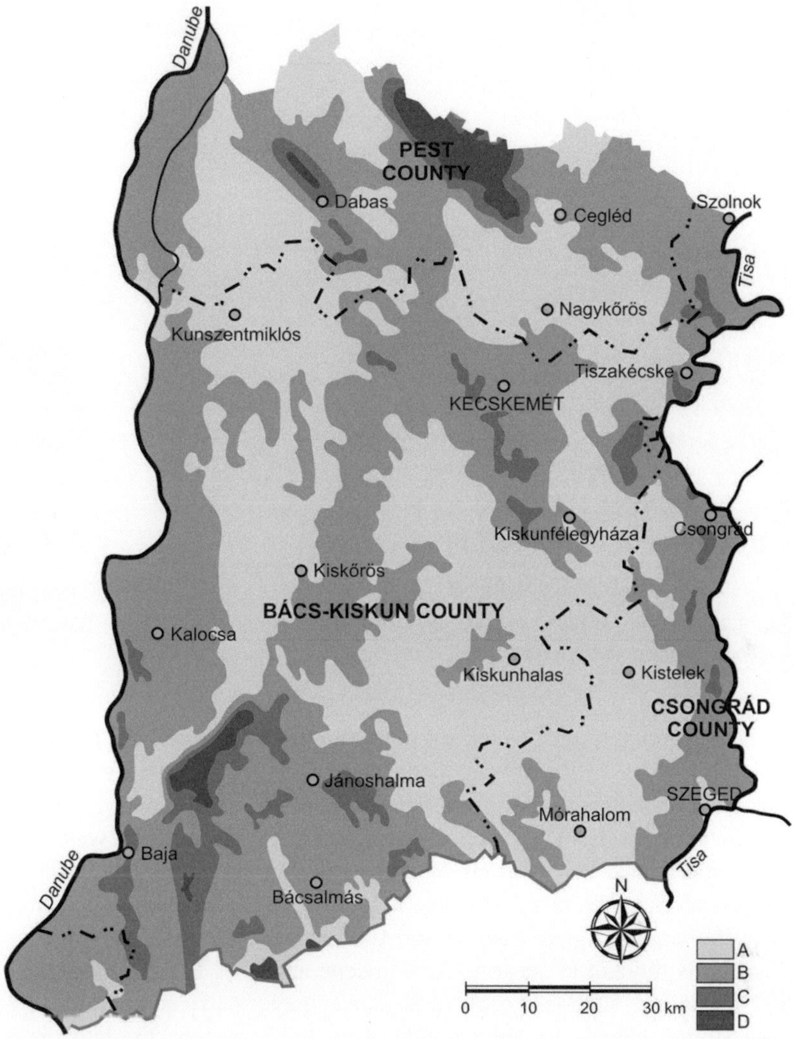


**Fig. 3.21** Spatial distribution of hydraulic systems in the Great Hungarian Plain (after Tóth and Almási 2001). *1* Recharge area, *2* area of through-flow, *3* area of outflow

works in small-scale too, and it may have geomorphological, pedological, chemical and botanical consequences on the surface (where the upward flow reaches the surface; Mádlné Szőnyi and Tóth 2007). Some researchers connect the subsurface circulation systems to the patterns of vegetation and water supply in the Danube-Tisza Interfluve (Kiss and Szalma 2006), to the well-defined spatial distribution of different habitats, through their typical geological, geophysical and hydrogeological processes (Fig. 3.22).

The upward and downward flows of the subsurface circulation system in the basin are well-described, but the supposed pedological, hydrographical or geomorphological connections of the large-scale model have to be confirmed by measurements and further data from drillings. High number of measurements, i.e. the ages of groundwater and artesian waters, the location of saline areas and the explanation of their evolution have not clearly supported the application of the model. Theoretically, it is true that an impact affecting the vegetation may be expressed, for example, by the spread of drought-tolerant species in the areas of infiltration and downward flow, whereas, the hygrophilous, salt-tolerant species would prevail in the areas of upwelling flow, where the environment is rich in nutrients. Similarly, from the point of view of soil science:

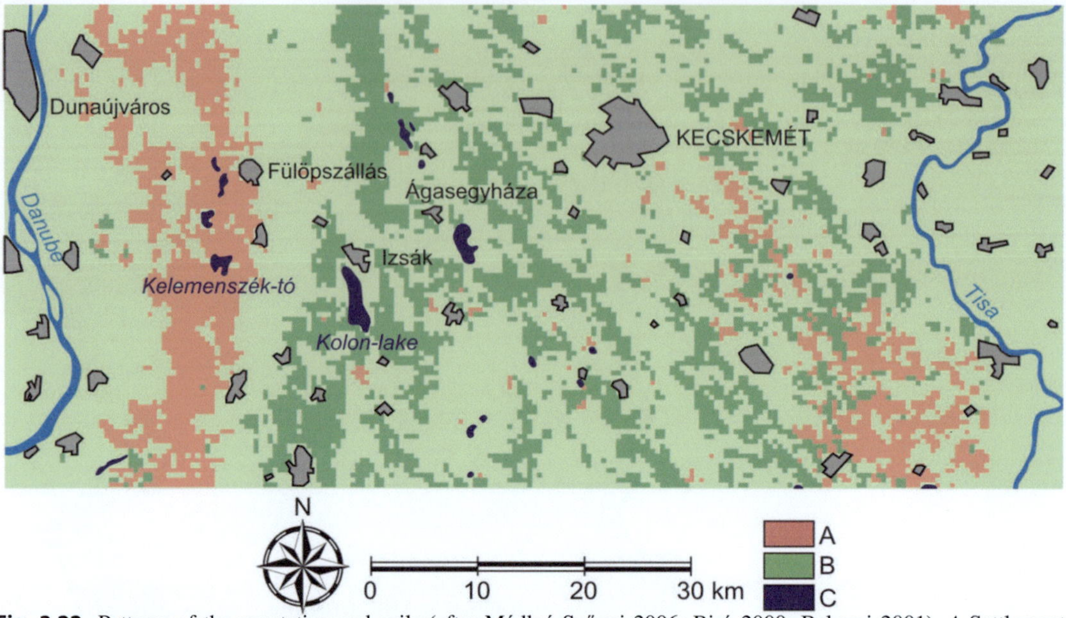


**Fig. 3.22** Patterns of the vegetation and soils (after Mádlné Szőnyi 2006; Biró 2000; Bakacsi 2001). *A* Settlement, *B* lake, *C* saline area, *D* wetland

the Na, Cl and $SO_4$ ions, which are transported in a dissolved form in the upward flow, precipitate and create saline soils, apart from the typical bog and meadow soils. The connection of single, vegetation-related observations on larger than a micro-scale requires more evidence too.

In the large sediment mass of the Carpathian Basin, it is difficult to separate the Pliocene and the Pleistocene sediment sequences, since they have similar habits, but their thickness is 1.5–2 km, where the Pleistocene series can only be up to 700 m (see Sect. 1.3). These highly porous, thick sediment strata can absorb a huge quantity of water. Porosity is highly dependent on the grain size, though its average is approx. 30–35 % between the surface and 1 km depth, and then, this value decreases to 10–20 % until 2 km because of compaction (Liebe 2000a). The water of the basin may exchange continuously and of course, slowly, by influent seepage and infiltration. The numerous water conducting layers facilitate this movement, even if with different rate. The infiltration rate can reach some 100 mm/year in the water conducting, mainly horizontal, layers; while the rate of downward infiltration is magnitudes smaller in the clayey-marly layers, though, it exists due to their discontinuity. (This can explain that the age of the artesian waters in a few thousand metres of depth is 10–30 thousand years.) Thus, it is not expedient to count with Pliocene or Miocene waters in the basin.

The supply of the artesian layers (the dynamic part) is provided by precipitation and surface waters; the latter is definitive in the case of the bank-filtered water containers. The bank-filtered water bodies may have a continuous hydraulic connection to the river bed (influent or effluent waterflows could alternate similarly to the groundwater bodies). The infiltration has a key importance in the recharge, and its maximum rate is 300 mm/year in the Pannonian layers and through the peripheries of the basin, while it is 20 mm/year in the Pleistocene layers through the qualifier layers (Kovács 2007). Thus, the infiltration much smaller than presupposed, especially if one considers the 100-mm real infiltration and the 80 mm evaporation from the groundwater (Szanyi 2004). The value is much lower in dry years. The artesian waters can drain the groundwater and the karst water containers locally, and the alluvial fans extending the country borders may also constitute serious resources.

Static and dynamic water resources are distinguished among the artesian waters, for practical purposes. The dynamic water takes part in the hydrological circulation, while the static water does not. This also means that the former is continuously resupplied and renewed; while the static waters have circulation just on the geological scale; therefore, its exploitation must be strongly limited.

The Pleistocene and Pannonian fluvial and lacustrine layers can be found in several hundred metres of thickness on the two-third of the Hungary and represent a 6000 km$^3$ water yielding aquifer volume. On the basis of the data of the Hungarian Geological Service and the National Environmental Protection and Water Management Inspectorate, the subsurface water extraction is about 900 million m$^3$/year, its 90 % is "cold water", and 10 % is thermal water (which means the total yearly extraction is under 1 km$^3$/year), which means a 10 mm/year withdrawal is projected onto the whole area of the country. The water conduction of the aquifer areas is 10–50 m$^3$/day, which sets the limits of the withdrawal; thus, the extractable quantity and the actual extraction are much lower than the limit-value (Mádlné Szőnyi and Tóth 2003). When evaluating the water consumption, the 20 mm/year natural supply must be compared to the 10 mm withdrawal. A mere 5 % of the disposable water supply is from the subsurface, but the majority of water used belongs to surface waters.

Artesian wells were drilled in high numbers in the nineteenth–twentieth centuries, the first in 1830 at Ugod. The Zsigmondy deep well in the Margaret Island (Budapest) should be highlighted here, which was the first artificial thermal spring in Hungary. The first artesian drinking fountain started to operate at Hódmezővásárhely in 1880, since at that time the majority of the groundwater was not potable. The approx. 60 thousand drilled wells produce drinking water; some of them also qualified as mineral

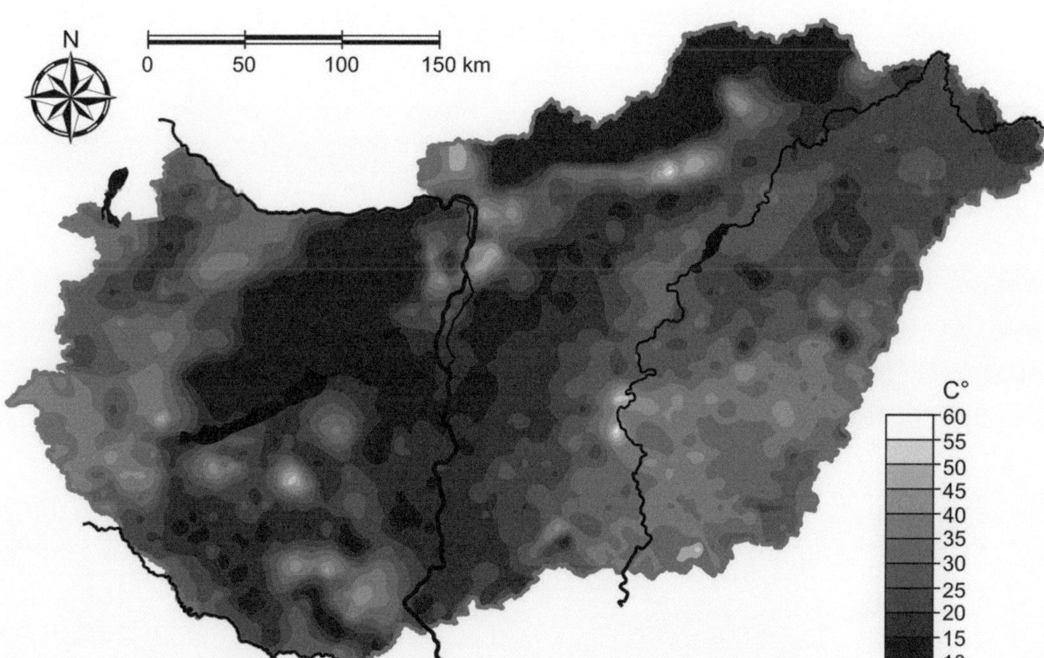


**Fig. 3.23** Temperature of the artesian waters in the bottom of the Quaternary layers (after KvVM 2003; Szanyi and Kovács 2007)

water wells, and their water is still a popular and easily exported product since the beginning of the twentieth century.

The geothermic gradient in the Carpathian basin is 4–6 °C/100 m, which means that the water becomes thermal in the depth of 500 m, and the temperature of the thermal water is 60 °C at 1 km, and 100 °C at 2 km—it is half of the global average of 3.5–4 km (Hurters and Haenel 2002). The water of the wells cools down as it flows through the pipeline towards the surface; thus, its temperature rarely exceeds 100 °C. The thermal water is a really valuable asset in the basin (similarly to the Paris Basin and the Australian Basin), which can be utilized in various fields, from the heating of greenhouses to balneologic applications. Several wells produce thermal water for these purposes, and they often benefit from the barren natural gas and oil wells deepened in the 1970s (MT 2002). The production of thermal water warmer than 30 °C is feasible on 70 % of the territory of the country (Fig. 3.23; KvVM 2003). The thermal energy is available, but its exploitation requires expensive technology (since recycling is possible only if the cooled water used in a closed system).

### 3.2.3 Quantitative Characteristics of Karst and Crack Waters

Limestone disposes of 1.5–2 % porosity due to its genetics, which value can increase by five times through karstification. Only 2 % of the surface of Hungary is covered by carbonated sediments (Fig. 3.24), but it ensures serious subsurface water storage capacity, especially in the case of the Upper Triassic limestone and dolomites. After the events of water breaking into mines in the 1950–1970s, and later their disappearance may indicate that the pores are not completely filled with water. Other volcanic and fragmented rock types (from the Palaeozoic through the Mesozoic to the Quaternary) have porosity and cracked water yielding layers too, but they are not classified as karst water bodies.

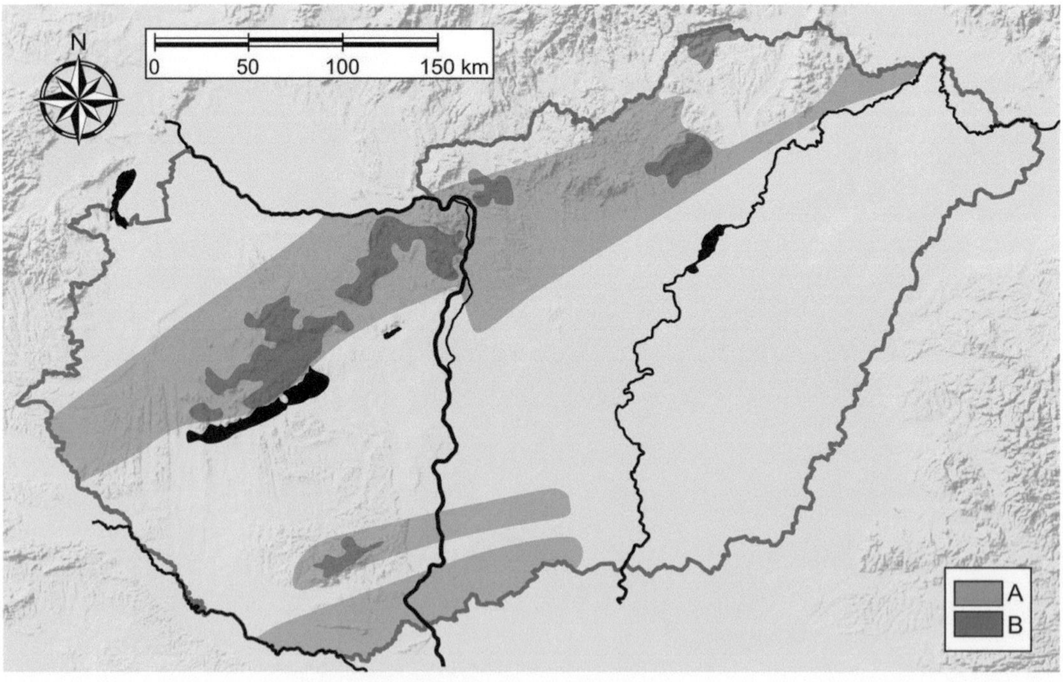


**Fig. 3.24** Spatial distribution of open and covered karst surfaces (after Kovács 2007). *A* Covered karst, *B* karst on the surface

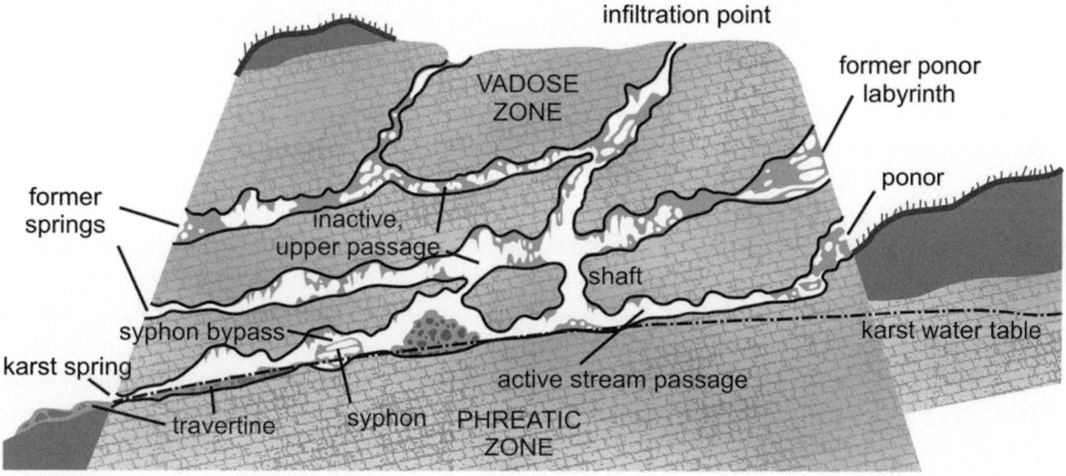


**Fig. 3.25** Theoretical speleologic structure of a cave system interwoven by passages, divided into various levels above the karst water level (after GEOL 2006)

A hydrodynamic connection is possible between the formations deposited on each another, but due to water withdrawal their level is often decreased (Kovács 2007).

Karst waters are decisively resupplied by infiltration of the precipitation; therefore, in the vadose (infiltration) zone the water infilling of the pores is partial, and it becomes complete just in the phreatic zone (Fig. 3.25). In the Aggtelek area, the infiltration rate of the precipitation is 25–30 % according to the measurements of the 1970s and 1980s (150–200 mm on the Hungarian average). From the place of supply, the karst water may reach the spring in a few days (classical method to follow the route of the water is its dyeing, which helped to explore several flow systems and caves), but the flow between the input and output points of the system may last for some thousands of years. The karst water level bulges like an hourglass between the points of supply and outflow, its height varies in connection with the rate of water supply. For example, in the rainy 1940s the typical karst water level was at 150–180 m asl in the Transdanubian Mountains, while in the 1970s when the water withdrawal was intensive it decreased to 130–140 m asl, later in the 1990s it decreased further to 110–130 m asl. Nowadays, the karst water level is rising due to the abandonment of mines.

On Hungarian karst areas, the total volume of infiltration is 1.4 million $m^3$/day (counting with 197 mm annual precipitation infiltrating into the system on ca. 2500 $km^2$ open karst areas, and the water absorption of covered karsts). The mean infiltration rate is 1026 thousand $m^3$/day in the Transdanubian Mts.; it is 31 thousand $m^3$/day in the Mecsek and Villány Mts., 55 thousand $km^3$/day in the Aggtelek Karst, and 288 thousand $m^3$/day in the Bükk Mts. (Kovács 2007).

The output side of the karst water household is ca. 200 thousand $m^3$/day outflow, which is manifested in springs along the mountains. For example, the volume of outflow is 57 thousand $m^3$/day in the Bükk Mts., 10 thousand $m^3$/day the Buda Mts., and 29 thousand $m^3$/day in the Mecsek Mts., respectively. On the other hand, the water penetrating towards the artesian and groundwater bodies must be considered too, which is, for example, 4 thousand $m^3$/day in the Roman Bath region (Budapest), 9 thousand $m^3$/day around Bicske, 43 thousand $m^3$/day in the western edge of the Bakony Mts., and 3 thousand $m^3$/day in the region of Várpalota. Finally, the flow into the surface waters must be considered too, which is about 34 thousand $m^3$/day in the Esztergom–Dunaalmás region. The "reserved" water supply is about 300 thousand $m^3$/day (e.g. for the operation of karst springs and thermal wells; the reserved water supply of Lake Hévíz is 30 thousand m$^3$/day, that of Budapest is 60 thousand m$^3$/day, and that of Lake Malom is 17 thousand m$^3$/day), and the rest is the available water amount, which is about 1 million m$^3$/day (VKKI 2009). The most significant exploitation area can be found in the Transdanubian Mountains and the Bükk Mts. Very similar values were obtained by Kessler and Bulla (1964), thus about 900 thousand m$^3$/day, which equals to 8 l/s/km$^2$.

The proportion of cold and warm karst waters in the Carpathian Basin is indicated by the distinction of 13 cold karst, and 15 thermal karst bodies (MÁFI 2005). The warm waters usually move along tectonic lines and spend much longer time flowing underground than the cold waters. It is undoubtedly proven by tritium and radiocarbon datings. Based on these measurements the water of the thermal springs of Eger started to infiltrate 7300 years ago, from precipitation fall onto the surface of the Bükk Mts., and it flowed in the crack system of the rocks (Deák 1989). At Egerszalók (developed into a wellness centre) travertine deposited from the water of an artificial well, here the water is about 13 thousand years old, while at the nearby Bogács the thermal karst water is about 15 thousand years old.

*Main hydrographical features of the springs*

Springs usually appear at the edges of mountains or along the tectonic lines, and they tend to have cold water in the former case, whereas they are often thermal karst water bodies in the latter (above 30 °C). The temperature of the springs (Fig. 3.26) is usually inversely proportional to the altitude of their source; thus, the coldest ones can be found in the plateau regions (6–7 °C). The cold ones have larger discharge, and they dispose of an appropriate buffer zone due to their mountain edge location or their cave water connections, to provide relatively steady discharge and chemically filtered water. Therefore, these springs are typically used as drinking water suppliers. The Garadna and the Szinva Springs in the Bükk Mts. are among the few but abounding springs connected to karst, the Jósva and the Komlós Springs are in the Aggtelek Mts., the springs at Tata are in the Vértes Mts., and the Tettye Spring is located in the Mecsek Mts. (Their strong anthropogenic impact is analysed in the following section.) The steadiness of the discharge of the larger springs is relative, since their water quantity depends on the varying precipitation amount and the vadose water, but the storage capacity of the larger springs is higher, which buffers the quick but small discharge changes.

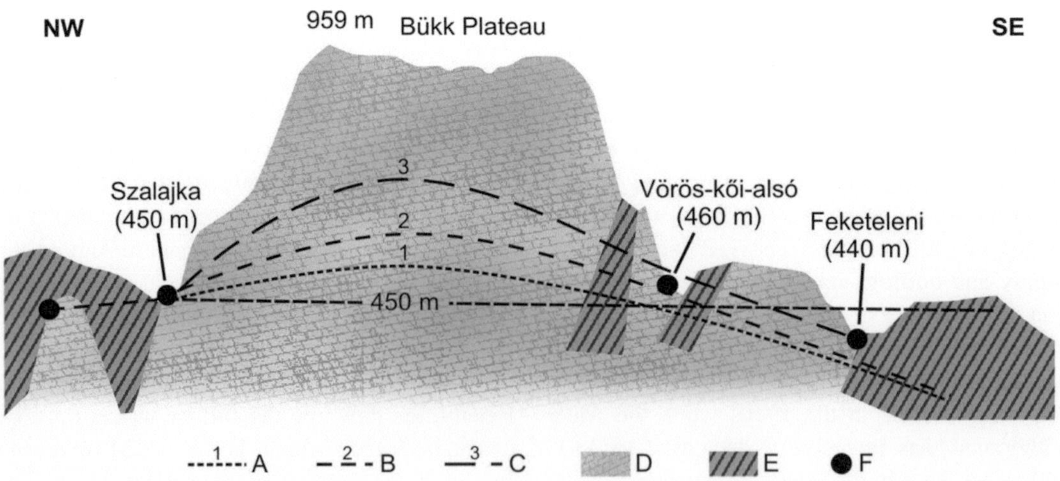


**Fig. 3.26** Structural outline of the temporary karst springs of the Bükk Mts. (after Tóth 2001). *A* Low karst water level, *B* elevated karst water level, *C* high karst water level, *D* limestone, *E* clay shale, *F* karst spring

The karst springs rising along the tectonic lines are typically used for balneologic purposes, a lot of them have medicinal water (e.g. the thermal baths of Buda, Lake Hévíz, the springs at Eger and Harkány). The most problematic issue is their sensitivity, which may be caused by the decreasing input due to the anthropogenic impact, the changes in the surface cover, or the pollution. The mixing of cold and warm karst waters is well demonstrated by the spring of the Lukács Thermal Bath at Budapest, which is composed of warm water (6–7 thousand years old) upwelling along the thermal line of Buda, and some cold springs. The upwelling warm karst water has created an active thermal water cave in József-hegy, with one of the largest active and submerged cave in the world: the Molnár János Cave (www.cave.hu). This highly sensitive water asset supplies the Lukács Thermal Bath.

Apart from the loose sediments and the porous, carbonated rocks, the Carpathian basin older rocks that have undergone several phases of burial and exhumation could be found, which have become fragmented because of the pressure changes. The consequent crack systems have a porosity of 1 % both in the basement of the Great Hungarian Plain and the Zala Hills (MT 2002). Similar crack systems have been formed in the mountains too, but for different reasons, and they have a smaller water storage capacity; thus, the discharge of the springs is modest.

### 3.2.4 Bank-Filtered Waters

The bank-filtered waters represent a particular type of subsurface waters, especially if we consider their utilization. They get the water supply mostly from surface waters (typically from a river, sometimes from a lake) and from the precipitation to a smaller extent. Morphologically, they are connected to river beds or alluvial fans, where large subsurface water bodies can be formed in the gravelly sandy Pleistocene material. These water bodies have a hydrological connection with rivers (see Sect. 3.2.2), providing effluent or influent waterflow directions depending on the water level of the rivers.

The amount and the utilizable proportion of bank-filtered water resources are difficult to assess. The calculations involve a 25-m-wide sediment zone along the riverbanks above the impermeable subsurface layer, at mean water level of the river. However, the amount of bank-filtered water is also determined by the discharge of the river and the constancy of the water level. On average, a 50 $m^3/s$ value is calculated (Liebe 2000b), which means more than 4.3 million $m^3$/day exploitable water. Its importance is indicated by the fact, that Budapest could not ensure its 1 million $m^3$/day water need otherwise. The water resource is connected to the gravel terraces of the Danube and the gravel cover of the floodplains; however, it is significant in the alluvial fans too, for example, in the Szigetköz or in the alluvial fan of the Sajó and the Hernád Rivers. The territories along the Danube and the sandbank islands to the north of Budapest dispose of significant water resources that are extracted with 10–15-m-deep collector wells. The bank-filtered waters are of high quality, and their self-purification takes place in one or two months in the gravel, or through the microbiologic filtering layer of the wall of the river bed, if the recharge is not speeded up artificially (the bacteria die within 10–60 days; Kovács 2007). The bank-filtered waters are significant, good-quality and cost-efficient forms of satisfying the need for drinking water.

## 3.3 Some Anthropogenic Impacts on the Water System

To give a summary on the hydrographic changes caused by human impact is not a simple task for several reasons. On the one hand, it is difficult to separate the natural changes from the changes induced by human activity. For example, several opinions exist on the groundwater level decrease; however, everyone agrees that the social factors play a role in it, apart from the natural ones. On the other hand, the assessment of the extent of the consequences of human intervention and those of the natural processes is also difficult, since we deal with a unified system; thus, a small-scale anthropogenic change may induce significant transformations because of the systemic character. Thus, for example, a minor pollution of the infiltrating water in karst areas result in chemical changes in the karst water, which may make the dripping water more aggressive, resulting in the dissolution of the dripstones.

The social use of surface and subsurface waters has an effect on water quality and water household too. The anthropogenic use is often connected directly to the surface, thus to the water system. Moreover, the human impact may change continuously (e.g. changing land-use), and the system may rehabilitate after the cessation of the intervention. The human impact may trigger undesirable changes, for example, the rise of the groundwater level due to irrigation, the transformation of the water system due to the drainage of wetlands, urbanization or the construction of dams. At the same time, the change may be beneficial, e.g. some measures increase the water resources or improve their quality. Single interventions may trigger an effect becoming more intensive or weaker with time. The anthropogenic impact may involve the whole catchment area: the soil (extensive deep ploughing, irrigation), the vegetation (forestation, deforestation), the relief (landscaping, sewer construction, run-off control), and the climate (evaporation), where the consequences may become long-lasting (Szanyi and Kovács 2003).

The surface and subsurface waters are serious assets and may become even more valuable in the long term, because of the need for drinking water and alternative energy sources. The situation of the Carpathian Basin is exceptional in Europe, if we look at the possibilities of water use, owing to the quantity of surface and subsurface waters. However, these important assets are limited by various factors (e.g. flood, inland water, drought, sensitive water resources). In spite of the difficulties, drinking water can be ensured practically in unlimited quantities in the long run with the right water management.

Water quantity is related to water quality, that is, larger quantities ensure a better quality (Sect. 3.1.3). However, if we take into consideration that the complete water exchange of Lake Balaton takes two years, we must see the sensitivity of the system. Another hazard in connection to surface and subsurface waters is their chemical and biological deterioration, which is partly caused by the industry, but also, to a significant extent, by the character of the water use.

The preservation of the valuable surface and subsurface waters and the rationalization of their use are of particular importance in the life of Hungary; therefore, a detailed description of the applied hydrological issues seems necessary. We will deal with them in part II, within the regional analyses. The anthropogenic impacts can be measured quite accurately through the changes of water quality, and their detailed summary is the theme of another book (Mezősi 2008).

*Anthropogenic impact on surface waters*

The human impact may modify the water system and the water household through the transformation of the surface and landforms. In the past centuries, there have been large-scale activities that have significantly affected the entire water system and its usability. Some major human impacts on the surface water system of the Tisza River, the Danube and Lake Balaton will be described in the regional part of the present volume (see Chaps. 7, 8, and 10). The common factor of the interventions may be that all one of them were destined to extend and ensure the usability of rivers and lakes. Such engineering works include flood control, river regulation mainly on the Tisza, and some concepts aimed to utilize the energy of the rivers, but these are often combined with the former.

As it was mentioned earlier, two main problems of opposite directions must be solved in Hungary in connection to the surface waters: the abundance or the lack of water, which sometimes occur simultaneously affecting different areas. They mainly have natural origins, and the anthropogenic activity aims to reduce the hazard. In the case of abundant water household, the intervention is connected to flood prevention. This is analysed in detail in Sect. 7.3.

(Nevertheless, flood hazard extends beyond the Great Hungarian Plain, e.g. in 2010 floods seriously damaged the hilly and mountainous areas too.) The problem of inland excess waters is also included in this problem-group. The different types of inland excess water are created by natural factors, but the prevention of the hazard is feasible mostly by human impact (Rakonczai et al. 2003; Kozák 2006). The situation is similar in the case of water scarcity, as the different types of drought hazards (e.g. meteorological, soil scientific, agricultural) are in connection with the lack of precipitation and water shortage.

*Anthropogenic impact on the subsurface waters*

The quality of the subsurface waters is determined by the geographical environment (e.g. circulation, flow, temperature) and by the near-surface pollution. The pollution hazard is great in the case of groundwater, because its water body is highly reactive to the pollution from the surface; thus, the utilization of groundwater is strongly limited (especially as drinking water), and it is just possible only in the case of continuous supervision (80–85 % of the groundwater bodies do not comply with the health requirements of drinking water). The limited hydraulic connection between the groundwater and the artesian waters, combined with the changing flow conditions due to water withdrawal, also hinders its use. The problem is that the—potentially polluted—groundwater can mixed with the artesian water due to the existing hydraulic connection; moreover, the flow between the groundwater lenses can transport the pollution to faraway places. Since the number of dug wells, which was 2 million in the eighteenth century, has dramatically decreased, the possibility of the quantitative and qualitative monitoring based on them has become strongly limited (Kuti et al. 2002).

The groundwater level decrease has been detected for 50 years in the plains of the entire basin. Only a part of its causes (a half of them; Pálfai 1994) have anthropogenic origin (e.g. changes in land-use, accelerating water utilization, water regulation), the other part is natural, mainly meteorological. The best known and documented case of groundwater level decrease takes place in the Danube-Tisza Interfluve (see Chap. 7).

The social use of karst water has been discussed in Sect. 3.2.3. The most significant environmental conflict in connection to karst water has been caused by the water extraction to support mining. A typical example of this is the karst water level decrease in the Transdanubian Mountains (in details see Sect. 10.3).

The basic problem in the utilization of artesian water is that, although its household is positive (if projected onto the entire territory of the country), but at some locations the exploitation is more extensive than the natural resupply. The infiltration towards artesian waters could be limited, due to, e.g., the lack of precipitation, and the increased evaporation accompanied by climate change. The artesian waters are used for drinking and industrial purposes. A third of the use of thermal waters (Table 3.4) serves balneological purposes, about a quarter accounts for drinking water supply, and a half of the use is devoted to geothermic energy production (e.g. agricultural heating and drying). The water treatment has a special importance in the cases of medical and geothermic uses (the salty and warm water is qualified as hazardous waste from the environmental point of view). According to the applicable legislation, the water has to be reinjected into the thermal water yielding layer (which is very expensive) to maintain the layer pressure (KvVM 2003).

The artesian water level decrease has been detected since the 1970s, due to natural effects and the water withdrawal. In the layers yielding drinking-water-quality water, the level decrease has been 5–10 m, accompanied by the decrease in the pressure and the discharge (e.g. the discharge of the Ann Well at Szeged has decreased from 600 to 450 l/s, its temperature from 52 to 48 °C). The water extraction showed a decreasing tendency at the beginning of the 1990s, resulting in a slowdown in the sinking, but even so, the regional artesian water pressure decrease is still 0.1–0.2 m/year (KvVm 2003). The water household of the artesian system is shifting towards a non-equilibrium state, because the

**Table 3.4** Temperature of the thermal waters at the outflow and their utilization in 2002 (Lorberer 2002)

| Temperature (°C) | Number of wells | | Utilization (number of wells) | | | | | | | | | |
|---|---|---|---|---|---|---|---|---|---|---|---|---|
| | Number | % | B | D | A | I | C | V | R | L | O | W |
| 30–39.9 | 584 | 44.8 | 60 | 183 | 73 | 29 | 1 | 9 | 0 | 87 | 40 | 102 |
| 40–49.9 | 289 | 22.2 | 93 | 23 | 16 | 18 | 2 | 20 | 0 | 43 | 45 | 29 |
| 50–59.9 | 137 | 10.5 | 46 | 9 | 17 | 10 | 2 | 14 | 4 | 16 | 12 | 8 |
| 60–69.9 | 121 | 9.4 | 34 | 0 | 17 | 6 | 1 | 25 | 7 | 18 | 3 | 10 |
| 70–79.9 | 70 | 5.4 | 8 | 0 | 23 | 4 | 6 | 16 | 2 | 8 | 2 | 1 |
| 80–89.9 | 50 | 3.8 | 4 | 0 | 33 | 3 | 2 | 1 | 0 | 6 | 1 | 0 |
| 90–99.9 | 48 | 3.7 | 4 | 0 | 31 | 1 | 5 | 0 | 0 | 5 | 0 | 2 |
| >100 | 3 | 0.2 | 0 | 0 | 1 | 0 | 1 | 0 | 0 | 1 | 0 | 0 |
| Total | 1303 | 100 | 249 | 215 | 211 | 71 | 20 | 85 | 13 | 187 | 103 | 153 |

*B* balneology, *D* drinking water, *A* agricultural, *I* industrial, *C* communal, *V* other purposes, *R* recycled, *L* closed well, *O* observation well, *W* not working

infiltration is reduced; thus, the supply of artesian waters has halved (Liebe 1998; Török 2000). This may influence the decreasing tendency of groundwater levels too. The main task is to find a solution, e.g. for the storage of the thermal water used for heating, which usually has much higher salt content and temperature than its environs. Its drainage into surface reservoirs is disadvantageous for the water quality and the wildlife (Mádlné Szőnyi and Tóth 2003).

## References


Bakacsi Zs (2001) A kiskunsági alluviális síkság talajtani-domborzati sajátságainak térinformatikai alapú vizsgálata, különös tekintettel a feltételezett talajvízszint-változásokra. Agrokémia és Talajtan 50 (3–4):371–382

Balogh K (1991) Szedimentológia. I. Akadémiai Kiadó, Budapest

Bendefy L, Nagy I (1969) A Balaton évszázados partvonalváltozásai. Műszaki Könyvkiadó, Budapest

Biró M (2000) A Duna–Tisza köze aktuális élőhely-térképe. 1:25000. Kézirat, MTA-ÖBKI, Vácrátót

Bódis K (2008) Digitális domborzatmodellek és alkalmazási lehetőségeik az árvízi kockázatkezelésben. PhD-értekezés, Szeged

Bulla B (1943) Geomophologic observations in the Balaton Highland. Földrajzi Közlemények 71:18–45

Bulla B (1964) Magyarország természeti földrajza. Tankönyvkiadó, Budapest 424

Cserny T (2002) A balatoni negyedidőszaki üledékek kutatási eredményei. Földtani Közlöny 132 (különszám), pp 193–214

Cserny T, Nagyné Bodor E, Tullner T (2008) A Balaton földtani kutatásának eredményei. Hidrológiai Közlöny 6:39–42

Deák J (1989) Dating the thermal waters in the Eger area. In: 10th international congress of speleology. Budapest, pp 570–572

Erdélyi M (1979) A magyar medence hidrodinamikája. (VITUKI Közlemények 18.) VITUKI, Budapest

GEOL (2006) http://oktatas.barlang.hu/alapfoku-jegyzet/geol4.html

Góczán L (1962) A Marcal-medence. Földrajzi Értesítő 11(1):58–61

Hurters S, Haenel R (eds.), (2002) Atlas of Geothermal Resources in Europe-EU Commission. Leibnitz Institute, Hannover, p. 261

Juhász J (1987) Hidrogeológia. Akadémiai Kiadó, Budapest

Juhász Á, Marosi S (1998) Geomorphic evolution of Lake Balaton. In: Kertész Á, Bassa L (eds) Windows on Hungarian geography. MTAFKI, Budapest, pp 37–52

Kaján I (2005) A "Rajkai-zsilip" építése. http://www.dunamuzeum.hu/modules.php?name=News&file=article&sid=25

Károlyi Z (1960) A Tisza mederváltozásai – különös tekintettel az árvízvédelemre. VITUKI, Budapest

Kaszab I (1987) Építésföldtani összefüggések Szeged és környéke felszínközeli üledékeiben. MÁFI, Budapest

Kiss A (2004) Utilization of the inundation area at Lake Fertő before regulation works. Acta Geogr Szeged 38:39–49

Kiss J, Szalma E (2006) Szikes élőhely sorozatok makromintázatának botanikai és geofizikai vizsgálata. Előadás, MTATAKI, Budapest. http://www.taki.iif.hu/3kija.pdf

KÖDUVIZIG (2008) http://www.kdtvizig.hu/

Kogutowicz K (1936) Hanság. In: Kogutowicz K. Dunántúl és Kisalföld. II. köt. M. Kir. Ferenc József Tudományegyetem, Szeged, pp 257–267

Konecsny K (2010) http://cholnoky.ro/termvil/konecsny.pdf
Kordos L (1984) Magyarország barlangjai. Gondolat, Budapest
Kovács B (2007) Hidrogeológia – Felszín alatti vízkészletek. Kézirat, Szeged
Kozák P (2006) A belvízjárás összefüggéseinek vizsgálata az Alföld délkeleti részén, a vízgazdálkodás európai elvárásainak tükrében. PhD-értekezés, Szeged
Kuti L, Vatai J, Müller T, Kerék B (2002) A talajvíztükör mélységének változása a Duna-Tisza közi hátságon. Földtani Közlöny 132:317–325
KvVM (2003) Felszín alatti vizek. http://www.kvvm.hu/szakmai/karmentes/kiadvanyok/fav/favm/favm02.htm
László G, Emszt K (1915) A tőzeglápok és előfordulásuk Magyarországon. A Magy. Kir. Földtani Intézet alk. kiadványai, pp 1–155
Liebe P (1998) A hévízhasznosítás helyzete Magyarországon. Vízügyi Közlemények 2:205–225
Liebe P (2000a) Az Alföld felszín alatti vízkészlete. In: Rakonczai J (szerk) A víz szerepe és jelentősége az Alföldön. Nagyalföld Alapítvány, Békéscsaba, pp 105–117
Liebe P (2000b) Felszínalatti vizek. In: Karátson D (szerk) Pannon enciklopédia. Kertek, Budapest, pp 253–257
Lóczy L (1913) A Balaton környékének geológiai képződményei és ezeknek vidékek szerinti telepedése. MFT Balaton Bizottsága, Budapest
Lóczy D, Veress M (2005) Geomorfológia. I. Dialóg Campus, Budapest
Lorberer Á (2002) Hévízkészleteink és idegenforgalmi-balneológiai hasznosításuk. VITUKI, Budapest. http://www.mgte.hu/04szakmai_anyagok/balneo.htm (2009–03)
Lovász Gy, Majoros Gy (1994) Magyarország természeti földrajza. JPTE, Pécs
Mádlné Szőnyi J (2006) A Duna–Tisza közének vízföldtani Tisza közének vízföldtani típusszelvénye és a szikesedés. Kézirat, ELTE, Budapest. www.taki.iif.hu/2szju.pdf
Mádlné Szőnyi J (2008) http://applied.geology.elte.hu/oktatas/alkhidro/2008_alkhidro%204ea.pdf
Mádlné Szőnyi J, Tóth J (2003) Bevezetés a hidrogeológiába. Egyetemi jegyzet. ELTE, Budapest
Mádlné Szőnyi J, Tóth J (2007) A Duna-Tisza köze típusszelvény és a szikesedés összefüggései. Földrajzi Közlemények 81(4):343–360
MÁFI (2005) http://www.mafi.hu/mafi/hu/node/11 (2009–04)
Magyari O (2000) Vizeink szeizmikus "röntgenképe". In: Karátson D (szerk) Pannon enciklopédia. Kertek, Budapest, pp 242–244
Marosi S, Szilárd J (1981a) A Balaton kialakulása. Földrajzi Közlemények 29(1):1–30
Marosi S, Szilárd J (1981b) A Dunántúli-dombság domborzatának kialakulása és általános jellemzése. In: Ádám L, Marosi S, Szilárd J (szerk) Dunántúli--dombság – Dél- Dunántúl. Akadémiai Kiadó, Budapest, pp 93–103
Martonné Erdős K (1995) Magyarország természeti földrajza. KLTE, Debrecen
Mezősi G (2008) Magyarország környezetföldrajza. JATE Press, Szeged
Mike K (1980) Ősmedernyomok a Balaton környékén. Földrajzi Értesítő 29:313–334
MNA (1989) Magyarország Nemzeti Atlasza. Carthographia, Budapest
Molnár B (1980) Hiperszalin dolomitképződés a Duna–Tisza közén. Földtani Közlöny 111(1):45–64
MT = Glatz F (szerk) (2002) Magyar Tudománytár I. Föld, víz, levegő. Kossuth, Budapest
Nádor A, Thamóné BE, Magyari Á, Babinszki E, Dudko A, Tóth Z (2005) Neotektonika és klímaváltozás együttes hatása a Körös-medence késő pleisztocén vízhálózat- fejlődésére. MÁFI évi jelentése, pp 131–147
Nagy Zs (1996) Természetvédelem Magyarországon. Ökológiai Intézet a Fenntartható Fejlődésért Alapítvány, Miskolc. http://www.ecolinst.hu/letoltok/kiadvanyok/termeszetvedelem_magyarorszagon.pdf
Pécsi M (1971) Az 1971 évi földcsuszamlás. Földrajzi Értesítő 20(3):233–238
Pécsi M (1975) A Fertő-Hanság medence. In: Marosi S (szerk) Kisalföld és a Nyugatmagyarországi- peremvidék. Akadémiai Kiadó, Budapest, pp 78–80
Rakonczai J, Mucsi L, Csató Sz, Kovács F, Szatmári J (2003) Az 1999. és 2000. évi alföldi belvízelöntések kiértékelésének gyakorlati tapasztalatai. Vízügyi Közlemények Különszám 4:317–336
Rónai A (1956) A magyar medencék talajvize. MÁFI Évkönyv 46/1. Műszaki Könyvkiadó, Budapest
Rónai A (1972) Az alföldi sekély vizek ingadozásai. In: Szádeczky-Kardoss E, Pécsiné Donáth É (szerk) 2. Agyag és Energiaáramlási Ankét. Akadémiai Kiadó, Budapest, pp 95–105
Schumm SA (1977) The fluvial system. Wiley, New York
Simonffy Z (2008) Afelszín alatti vizek védelme
Simonics L (2005) A Velencei-tó vízgazdálkodása. Előadás, Agárd, 2005. augusztus 22
Sipos Gy (2004) Medermintázat és zátonyképződés homokos medrű síksági folyószakaszon (Maros 31–50 fkm). http://geography.hu/mfk2004/mfk2004/phd_cikkek/sipos_gyorgy.pdf
Somogyi S (1983) A magyar folyóhálózat szakaszjelleg típusai. Földrajzi Közlemények 31(3–4):220–229
Somogyi S (2000) Tavaink. In: Karátson D (szerk) Pannon enciklopédia. Kertek, Budapest, pp 250–252
Stefanovits P (1981) Talajtan. Mezőgazdasági Kiadó, Budapest
Szabó J (1996) A csuszamlásos folyamatok szerepe a magyarországi tájak geomorfológiai fejlődésében. Kossuth, Debrecen
Szanyi J (2004) Magyarország hidrogeológiája (összefoglalás). MGSZ Szakhatósági Főosztály, Szeged. www.gama-geo.hu/kb/okt/hg/HidrogeolBSC12_mohidrogeol.pdf
Szanyi J, Kovács B (2003) Vízkutatás. III. Kézirat, Szeged

Pálfai I (szerk) (1994) A Duna–Tisza közi hátság vízgazdálkodási problémái. Nagyalföld Alapítvány, Békéscsaba

Pálfai I (szerk) (2001) Magyarország holtágai. KVM, Budapest

Szanyi J, Kovács B (2007) http://www.gama-geo.hu/kb/okt/hg/HidrogeolBSC8_fa_vizkeszletek3a_2007.pdf

Szesztay K (2000) Az Alföld vízháztartása. In: Pálfai I (szerk) A víz szerepe és jelentősége az Alföldön. Nagyalföld Alapítvány, Békéscsaba, pp 7–16

Sztanó O, Haas J, Mindszenty A (2005) Szedimetológia. http://geo.elte.hu/szedi/1_Szedi_bevezet_ny.pdf

TERM (2010) http://www.termeszetvedelem.hu/index.php?pg=menu_2771#szikesTIR http://geo.kvvm.hu/tir

Török K (2000) Hévízgazdálkodás az Alföldön. In: Rakonczai J (szerk) A víz szerepe és jelentősége az Alföldön. Nagyalföld Alapítvány, Békéscsaba, pp 180–188

Tóth G (2001) ABükk felszínfejlődése és karsztosodásának összefüggései. 1. Magyar Földrajzi

Tóth J (2006) Bevezetés a hidrogeológiába. Előadássorozat az ELTE-n. 20061_03 Hiderkorfoly.pdf és 20061_05FAVelof.pdf (2008–11)

Tóth J, Almási I (2001) Interpretation of observed fluid potential patterns in a deep sedimentary basin under tectonic compression: Hungarian Great Plain, Pannonian Basin. Geofluids 1:1–27

Vágás I (1982) A Tisza árvizei. Vízdok, Budapest

Varga Gy (2004) A Balaton vízháztartásának a közelmúlt években tapasztalt szélsőségei. Földrajzi Közlemények 128(1–4):1–10

Varga Gy, Pappné Urbán J (2007) Szélsőségek a Balaton 2000–2003. évi vízháztartásában. VITUKI, Hidrológiai Intézet, Budapest. http://www.aquadocinter.hu/themes/Vandorgyules/pages/3szekcio/varga_pappne.htm

VITUKI (2007) A Duna hajózhatóságának javítása tárgyú projektet megalapozó tanulmány. Kézirat, Budapest. http://www.vituki.hu/files/dunahajozhatosag/new/Project%20Final%20Report%20in%20Hungarian/Duna_hajozhatosag_vegleges.pdf

VKKI (2009) www.vkki.hu/index.php?mid=329

Zólyomi B (1992) A Balaton múltja a pollensztratigráfiai vizsgálatok tükrében. In: Biró P (szerk) 100 éves a Balaton-kutatás. MTA Balatoni Limnológiai Kutatóintézete, Tihany, pp 25–32

Zseni A (2002) Vízminőségvédelem. SZE, Győr

# Soils of Hungary

# 4

**Abstract**

In this chapter, the soils of Hungary are grouped following the widely accepted Hungarian genetic and pedological classifications. The application of the FAO system has not been realised yet (though such map is presented in the chapter), and it will be probably published in 2016/2017. However, probably the change from the genetic to FAO system will be a very slow process, as in the data bases this traditional classification is applied. The orographical conditions of the country greatly reflected in the distribution of zonal genetic soil types (e.g. chernozem and forest soils), though they are also influenced by climatic parameters. The intrazonal soils developed under different hydrological and geological conditions. In the chapter, the soils are evaluated, and the most important human influences on soils are also described, which mostly decrease the fertility of soils, though soils as natural resource and their protection are very important in Hungary. The most important human impacts are the physical degradation (soil erosion is the greatest hazard) and the chemical degradation (e.g. salinization especially in connection with flood protection).

Independently from the scale, the soil map of Hungary has a mosaic-like pattern. The main reason for this is that the surface—despite of the low relief—has diverse geological and geographical features and a varied physical and chemical character.

To analyse the soils from the point of view of their development and geographical conditions, the genetic and soil geographical classification system of Stefanovits (1981) can be applied. This classification distinguishes 9 main soil types and 38 subtypes. The soils, which were formed under similar environmental conditions, were characterized by the same processes, reached similar stages of soil formation and belong to the same soil type. The main soil type integrates the types which were formed in similar geographical environments, thus the geographical approach plays a significant role in the classification.

From a zonal point of view, the Carpathian Basin is characterized by meadow soils in the plains, and by brown forest soils in the hills and mountains (Fig. 4.1), which are influenced by the climate. The intrazonal soils occupy a significant area of Hungary, and they can be divided into two groups considering their parent material and water household as the dominant soil forming factors: lithomorphic soils (e.g. redzinas, erubase





G. Mezősi, *The Physical Geography of Hungary*, Geography of the Physical Environment,
DOI 10.1007/978-3-319-45183-1_4

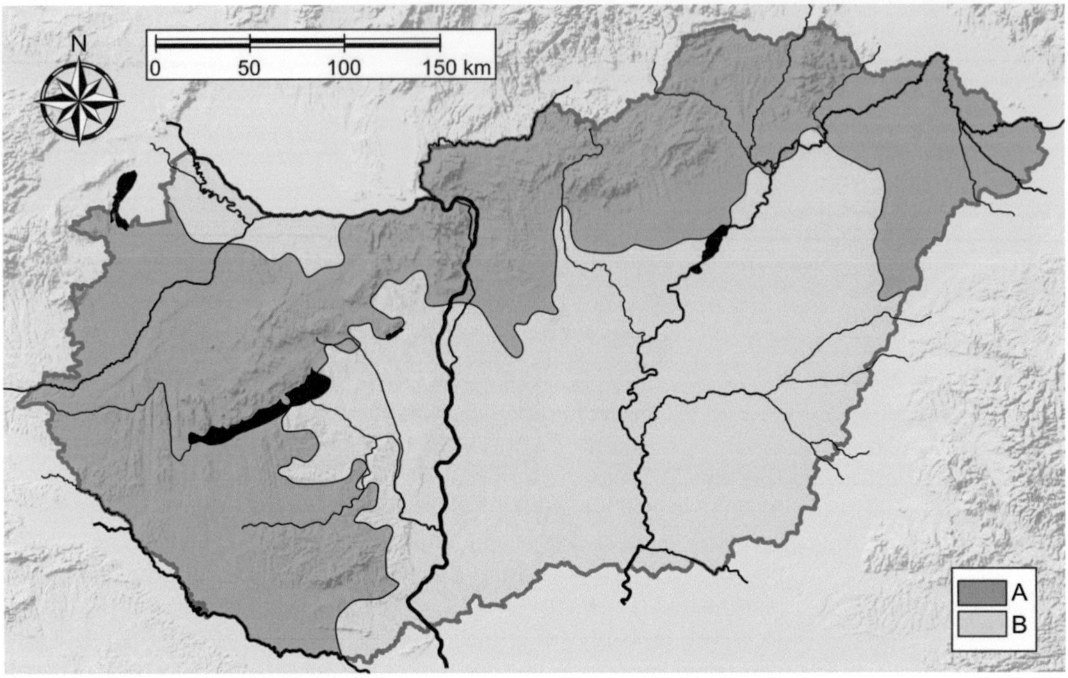


**Fig. 4.1** Distribution of meadow and forests soils in Hungary (after Stefanovits 1981) *A* forest soil, *B* meadow soil

soils) and hydromorphic soils (meadow, salt-affected and peat soils). Some small-extent azonal soils can also be found, where the zonal influence has not manifested yet, thus, they are characterized by the initial stages of soil formation: these are the skeletal soils, the alluvial, and slope sediment soils.

## 4.1 Zonal Soils Characterised by the Dominant Influence of Climate

The chernozems are characterized by humus accumulation due to the decomposition of vegetation, crumb structure, and two-directional movement of the calcium-rich soil solution. The latter may result in secondary carbonate precipitation (coating/caliche) in the B horizon. In the Carpathian Basin, the pseudomyceliar chernozem typically occurs in areas covered by typical loess (e.g. Mezőföld, Bácska Plains and Tolna Hills), while that of the lowland calcic chernozem appears in areas covered by infusion loess (e.g. Körös-Maros Interfluve, Hajdúság, Nagykunság; Fig. 4.2). They were named after the above mentioned carbonate coating observed in 30–70 cm depth, which is the result of the leaching process of different rates and the precipitation of the calcium carbonate. From the agricultural point of view, those soils are the most valuable, which have a high humus content (3–4 %) in the A horizon are neutral or slightly acidic, dispose of an excellent crumbly texture and have a good water and nutrient management. Their biological activity is high, which is also demonstrated by the frequent animal burrows (krotovinas) in the B horizon. The meadow chernozems are characterized by less humus accumulation and the effects of the near-surface groundwater. The water effect can be the result of accumulating inland excess water in depressions or due to the plough pan, which frequently causes yield loss in the lower-lying areas in the Great Hungarian Plain and the Little Hungarian Plain.

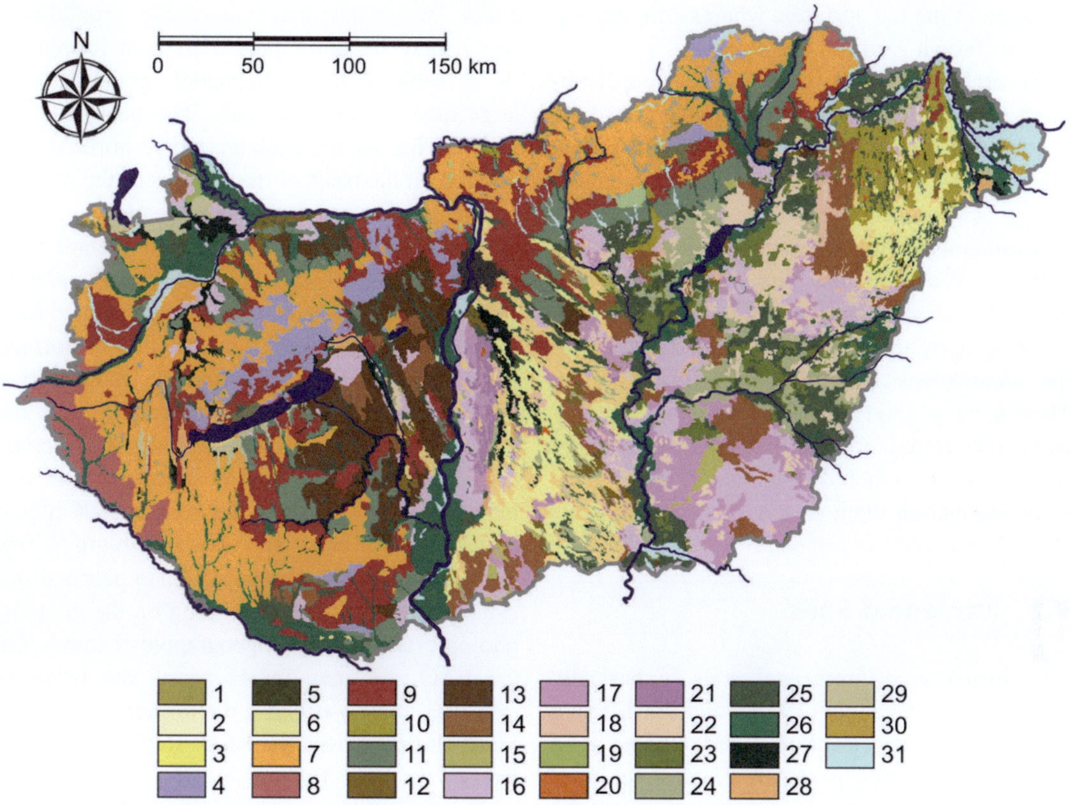


**Fig. 4.2** Genetic soil map of Hungary (after Magyarország atlasza 1999) *1* skeletal soils, *2* blown sand, *3* humic sandy soils, *4* rendzinas, *5* erubase soils, *6* acidic, non-podzolic brown forest soil, *7* brown forest soils with clay illuviation, *8* pseudogleyic brown forest soil, *9* brown earth (Ramann's brown earth), *10* banded brown forest soil, *11* chernozem brown forest soil, *12* chernozem type sandy soils, *13* pseudomyceliar chernozem, *14* lowland calcareous chernozem, *15* lowland calcareous chernozem with salt accumulation in deeper layers, *16* meadow chernozem, *17* meadow chernozem with salt accumulation in deeper layers, *18* meadow chernozem with salt accumulation in deeper layers, *19* terrace chernozem, *20* solonchak, *21* solonchak-solonetz, *22* meadow solonetz, *23* steppe meadow solonetz, *24* solonetz meadow soils, *25* meadow soils, *26* alluvial meadow soils, *27* peat-meadow soils, *28* peat bog soils, *29* meliorated and cultivated peat bog soils, *30* peat swamp forest soils and *31* raw (recent) alluvial soils

Forest soils are formed by the joint effects of micro- and soil climate in forests and forest steppes, together with the decomposing organic plant material on the surface. The forest soils are characterised by good water management, and their humus content exceeds 3 % in the A horizon. Their genetic levels are clearly distinguishable, and their nutrient content is medium. Forest soils occur in the North Hungarian Mts., the Transdanubian Mts., the Transdanubian Hills and the Foothills of the Alps (Fig. 4.2). The forest soil type that has a relatively high extension in Hungary is the Ramann's brown earth, where leaching is accompanied by a slight acidity and an intense clay formation. The difference between the clay content of A and B horizons is slight, and the type disposes of a good water management (or rather, a good water retaining capacity). Their environmental conditions are similar to brown forest soils, but they can be more frequently found in the drier areas, basins and southern slopes.

Brown forest soil with clay illuviation is created if the processes of humification, leaching and clay formation are accompanied by the translocation of clay and by a slight acidification. It can be observed in the hills that the brown forest soils with clay illuviation are more common in the flat surfaces, whereas the same is valid to brown earth on the slopes.

Several further soil types can also be distinguished on the basis of the specific features of the pedological processes. For example, if a brown forest soil with clay illuviation shows the signs of strong acidification, the decomposition of the clay minerals, or podzolisation, it is classified as podzolic brown forest soil. The chernozem brown forest soils show a relationship with the forest soils by the process of leaching and with the chernozems by the intense humification. They have a good nutrient supplying capacity and can be found in larger extensions in the hilly borders of the Great Hungarian Plain and the Little Hungarian Plain (Fig. 4.2.).

## 4.2 Intrazonal Soils

The formation of intrazonal soils is basically influenced by the effect of water and the different rock type, and they can be classified into four main groups: the meadow, peat, salt-affected and lithomorphic soils. Due to the high humidity and the organic matter in anaerobic environment the A horizon of the meadow soils is black. The meadow soils have a medium nutrient content, become highly cracked in dry periods, but their water management can be considered to be good, apart from the humid spring periods. Depending on the parent material, there can be significant differences between the A and B horizons in their carbonate content, thickness of humus layer and water management. The typical occurrences of meadow soils are strongly connected to the alluvial sediments, for example, to the calcareous sediments of the Danube and Hernád River, and the acidic sediment of the Tisza, Rába and Körös Rivers.

Peat soils are temporarily or permanently covered by water, and they were formed and developed under the vegetation of former peats and bogs. The peats were covered by grass and moss, while the bogs by reed and sedge. Accordingly, the organic material is dominant in the upper and the inorganic material (clay and sand) in the lower soil horizons. The remains of the vegetation (e.g. moss, reed, sedge) decompose under water-saturated, anaerobic conditions, which may be accompanied by peat formation. As a result, the organic matter accumulates in large quantity in these soils. The peat soils are soils having an AC horizon, where horizon A is divided on the basis of the features of the organic matter, the rate of the humification and the oxidative degradation. Soil formation starts with peat formation, which is followed by an oxidative stage. The anaerobic condition is interrupted for a while, and the organic matter is transformed under aerobic conditions, that is, humification takes place. These soils can easily dry out if the groundwater level decreases, and they become erodible. The surface and near-surface peat formation can be dated to the second half of the Holocene, the bogs are also maximum a few thousand years old. The formation of peat and soil usually get separated, depending on the geomorphologic and the hydrogeographic features. On the FAO soil map for Hungary, the histosols (Fig. 4.3) mostly occur in the former bogs, in the Fertő-Hanság Basin, the Körös Region and in the bays of the Lake Balaton.

The third main group of intrazonal soils is the salt-affected soils. Sodium salt occurs both in dissolved and solid forms, and its quantity, quality and distribution along the soil profile defines the rate and character of salinity. The fertility, physical and chemical features deteriorate in parallel with the increasing salinity. The salt accumulation is the key to the formation of salt-affected soils, which is caused by the dry climate and the near-surface groundwater that contains dissolved salt. Due to the intensive evapotranspiration, the soil water content progressing upwards gets into the atmosphere, and the salt remains in the soils. The salts resulting from the weathering process cannot infiltrate because of the precipitation shortage, thus, the less soluble calcium and magnesium salts precipitate, while the sodium becomes concentrated in the soil solution.

Based on the location of maximum salt accumulation, two salt-affected soil types can be distinguished as follows: solonchak soils are characterized by near-surface maximum salt concentration (their occurrence is frequent in the Danube-Tisza Interfluve), while solonetz soils are

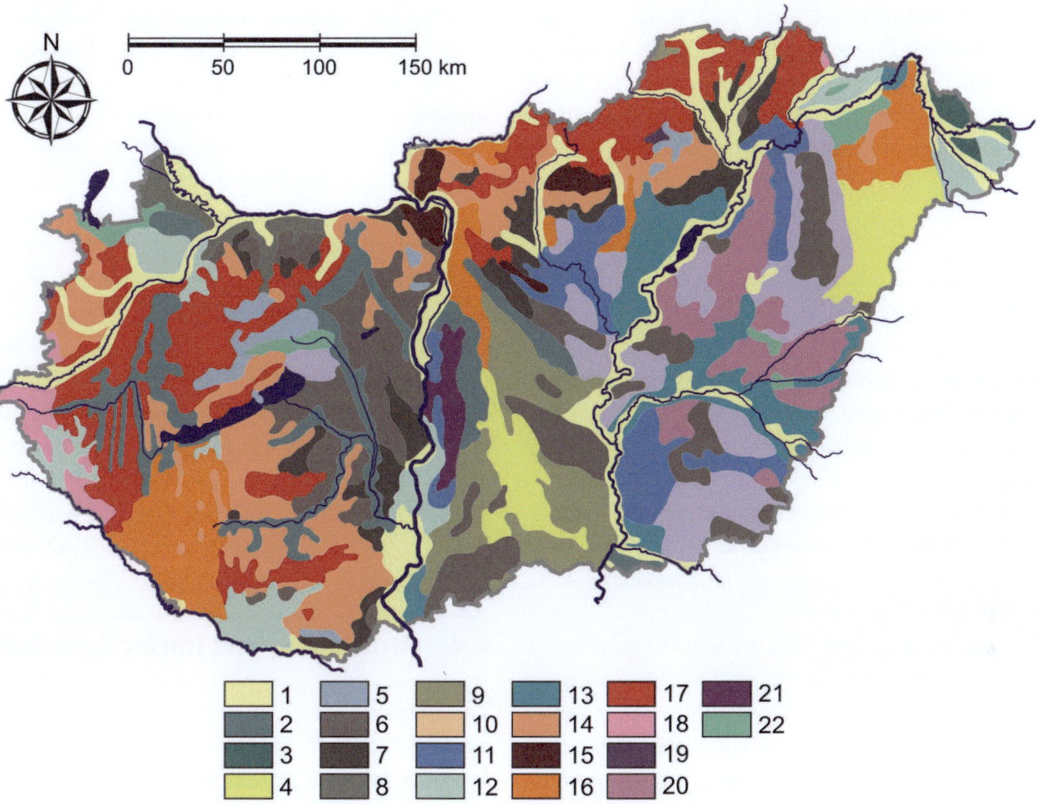


**Fig. 4.3** Soil map of Hungary based on the FAO classification (MNA 1989) *1* alluvial soils (fluviosols), *2* peat meadow soils (gleysols), *3* soils of swampy forest, *4* blown sand skeletal soil (regosols), *5* rendzina (rendzinas), *6* pseudomyceliar chernozems (chernozems), *7* brown forest chernozems (chernozems), *8* terrace chernozems (chernozems), *9* sandy chernozems (chernozems), *10* meadow chernozems (phaeozems), *11* chernozem with salt accumulation in deeper layers (phaeozems), *12* alluvial meadow soils (phaeozems), *13* meadow soils (vertisols), *14* brown earth (cambiosols), *15* brown earth formed on tuff (cambiosols), *16* banded brown forest soils (arenosols), *17* brown forest soils with clay illuviation (luvisols), *18* pseudogleyic brown forest soils (pseudogleys), *19* acidic brwon forest soils (acrisols), *20* meadow solonetz soil (solonetzes), *21* solonchaks (solonchaks) and *22* peat bog soils (histosols)

characterised by salt accumulation in the B horizon. This intensive leaching occurs more frequently in the Hortobágy, the Körös Region and the Tisza Valley (Fig. 4.2). The solonetz soils have A, B and C horizons; horizon B is alkaline characterized by high sodium content and has a well-recognizable columnar structure. On its surface, solid can form, which is the result of the separation of the skeletal part and the colloids.

Salt-affected soils used to have a large extension before the regulations, but the regular floods inundated them and diluted the concentrated solution that caused the salinization. The salt-affected areas could be used as scant pastures. Today, the area of secondary salinization is growing due to human impact. In these areas besides the morphologic features of the original soil type, the salt concentration and the replaceable sodium content increase. The reasons for the secondary salinization can be the badly designed irrigation system or the inadequate irrigation water quality.

The lithomorphic soils are a specific type of intrazonal soils, which are characterized by intense humification and mineral accumulation, depending on the soil-forming rock. A particular lithomorphic soil type is the redzina. It is formed on carbonated rocks, has a thin fertile layer, and shows intense humification. It disposes of good nutrient management, and bad water management features, since it freezes quickly in winter, and dries out in summer, due to the thin fertile layer. Lithomorphic soils can also be formed on loose sediments (humus-carbonate soils), or on the debris of volcanic rocks (erubase soils), which show strong humification.

## 4.3 Azonal Soils

The group of azonal soils includes skeletal soils, alluvial soils and soils of slope sediments. A characteristic feature of the skeletal soils is that biological processes has limited role in their formation due to the parent material or other factors.

The skeletal soils are classified into two large groups. The first includes the stony-skeletal soils, which typically occur in the mountains, where the solid rock has not weathered or fragmented to such an extent that the vegetation could colonise it, and thus, humification could start. Due to orographic reasons, soil erosion can play a significant role in this case, since the thin fertile layer is more vulnerable. On the surfaces where the coarse skeletal part is more than 50–60 %, the low amount of the fine fraction inhibits the soil development, therefore, only a skeletal soil is formed (e.g. Kemeneshát, or the floodplains and river terraces). The skeletal soils on sandy sediments, showing only the initial signs of soil development (e.g. slight humification), constitute a particular group. The vegetation is sparse, the water quickly infiltrates, thus, the soil dries out easily. The typical occurrences of this soil type are the Nyírség, the Danube-Tisza Interfluve and the young blown sand areas of Inner Somogy.

The other group of azonal soils is the alluvial soils formed on recent sediments of rivers and lakes. The surfaces are colonised by rich vegetation after becoming dry, however, due to the possible future inundations, the soil forming processes are not effective, or only for a short time.

The soils formed on slope sediments are those where the soil layers are not genetically connected, because they are not the results of local soil formation, but they were eroded and transported from higher elevations. The eroded material accumulates in valleys and at the bottom of slopes, often in a thickness of several metres.

The soil map illustrated in Fig. 4.2 is not satisfactory from several points of view and does not match the soil classification of other countries. The scientists have initiated the establishment of a new classification (Stefanovits and Michéli 2005). The basis for a more uniform classification could be the FAO map (Fig. 4.3, its original scale is 1:1,000,000). There were attempts in the 1989 edition of the National Atlas of Hungary to harmonize the Hungarian and foreign soil categories. A few examples were highlighted to demonstrate that the correspondences are far from being evident, that is, the similarities become questionable if we magnify the map, and the parallels are often constrained.

## 4.4 Anthropogenic Impacts on Soils

The human impacts on soils (Mezősi 2008) have been well known since decades. The formation of recent zonal soils started in the Carpathian Basin when the climate became similar to nowadays and not in the Ice Age or in the following warmer periods. Soils were also formed in the Quaternary and afterwards under the prevailing environmental conditions at a given time (Image 4.1), that is confirmed by series of loess and palaesols. The appropriate climate for the formation of chernozem was the Boreal Phase of the Holocene (about 8–9 thousand years ago). Therefore, the upper part of the kurgans (Neolithic tells) could transform into chernozem (Joó et al. 2003). The groundwater table sank due to the warmer climate of the Copper Age (5–7 thousand years ago), which effected the formation of the calcareous accumulation level of meadow soils. The anthropogenic activity on the surface, such as the burning of forests (its pedological evidences can be traced back to 3–4 thousand years), soil cultivation, trench digging or tell building strongly influenced the soil formation. The soils also transformed genetically. For example, the salinization can progress in a few hundred years or the skeletal soils can transform in decades to such an extent that a new soil type can be identified. The anthropogenic impact influences

**Image. 4.1** Brown forest soil in the Mecsek Mts. (**a**) and pseudmyceliar chenozem and (**b**) in the Körös-Maros Interfluve (photo by Barta K)

also soil formation processes (e.g. the surface of kurgans) and can change the direction of the development (e.g. the inappropriate fertilizers can lead to soil degradation).

The soil quality becomes a key question in the past two centuries, because the soil is the most valuable natural resource of Hungary. Earlier, the yield potential and the price expressed in monetary terms ("Gold Crown"—referring the monetary value of soils) used to be important. The out-of-date concept of the Gold Crown is still used in the soil qualification. Today, the determination of the value of the soil (involving its usage and management) is diverse and multi-purpose. Several approach exist for the valuation, for example economic value (e.g. yield potential), environmental conditions (e.g. ecological) or social aspects, and it is important to find compromise between the different approaches.

The most significant limiting factors of soil usage are divided into two groups. One group includes the soil degradation processes, which can be regarded of natural origin, though human activity can greatly modify the influencing factors. The other group involves soil contaminations.

Degradation occurs if the potential soil functions (e.g. yield potential) are hindered by certain factors (Várallyay 2001), or the normal operation of the functions is endangered. Therefore, soil protection strategies have two objectives: they attempt to prevent, eliminate or reduce the extent of the disadvantageous or harmful effects, and on the other hand, to minimize the effect of the factors hindering the expected functions. Apart from the erosion, which deteriorates the yield amount, soil quality and yield potential can also be limited by various physical, chemical and biological processes of degradation (Németh et al. 1997).

Some typical examples of physical degradation are structural deterioration or compaction; extreme water conditions can be caused by extreme precipitation distribution, inland excess water cover or over-moistening. In fact, the problem in this case is the extremely light or heavy

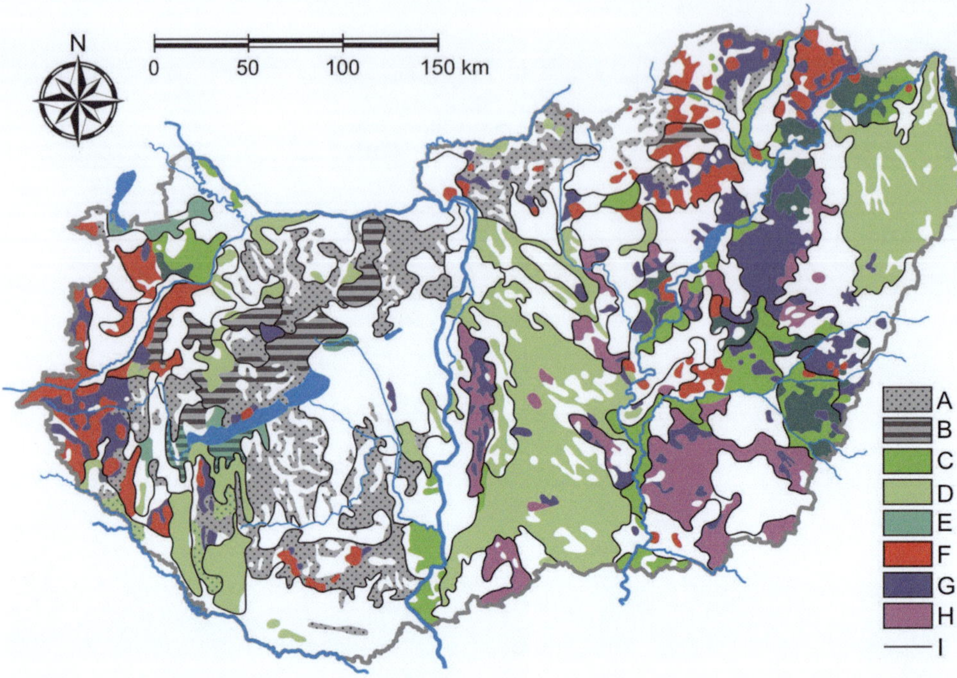


**Fig. 4.4** Factors limiting soil fertility (after TAKI 1997) *A* erosion, *B* thin fertile layer, *C* heavy mechanical composition, *D* light mechanical composition, *E* water cover, *F* acidification, *G* salinization, *H* salinization in deep soil horizons and *I* borders of land degradation

mechanical composition of affected soils and the extreme water supply (Fig. 4.4). The protection against physical degradation can be the improvement of the soil structure applying appropriate agro-technology, or a more effective soil moisture control (e.g. amelioration, soil loosening).

Soil erosion is the most extensive among the recent processes of soil degradation in Hungary (Fig. 4.4.), which affects approx. 2 million ha arable lands (Stefanovits 1999). Erosion (deflation) and accumulation caused by the wind occurs not only in sand covered areas, the degradation of the fertile topsoil is also common.

The chemical degradation of soils is basically caused by acidification (as a result of intensive leaching), salt accumulation and salinization. Among these, acidification can be reduced by adding carbonate, while the soil amendment related to salt accumulation and salinization requires significant time and financial investments: groundwater level control, addition of soil improvers (carbonate or gypsum) and proper irrigation. In the lack of irrigation secondary salt accumulation (due to the elevated, stagnant, salty groundwater) can result in salinization. Improper irrigation can also cause secondary salinization in large areas or even peat formation (Fig. 4.5).

The main contributing factors to the biological degradation processes are the erosion, the drought sensitivity or the intense use of chemicals. The combination of physical, chemical and biological degradation leads to the reduction of the filtering and buffer capacity of the soils. The rate of the phosphorus load in surface waters and the nitrate load in subsurface waters clearly indicates this. An appropriate nutrient supply system (e.g. environmental friendly nutrient management) limits the possibility of biological degradation.

After 1990, the focus of the research on soils shifted from the yield potential to the water and nutrient storage capacity of soils. The changes in soil water management result in alterations of nutrient and salt transport, which can result in more intense salt accumulation and salinization processes in the Great Hungarian Plain. Nearly a

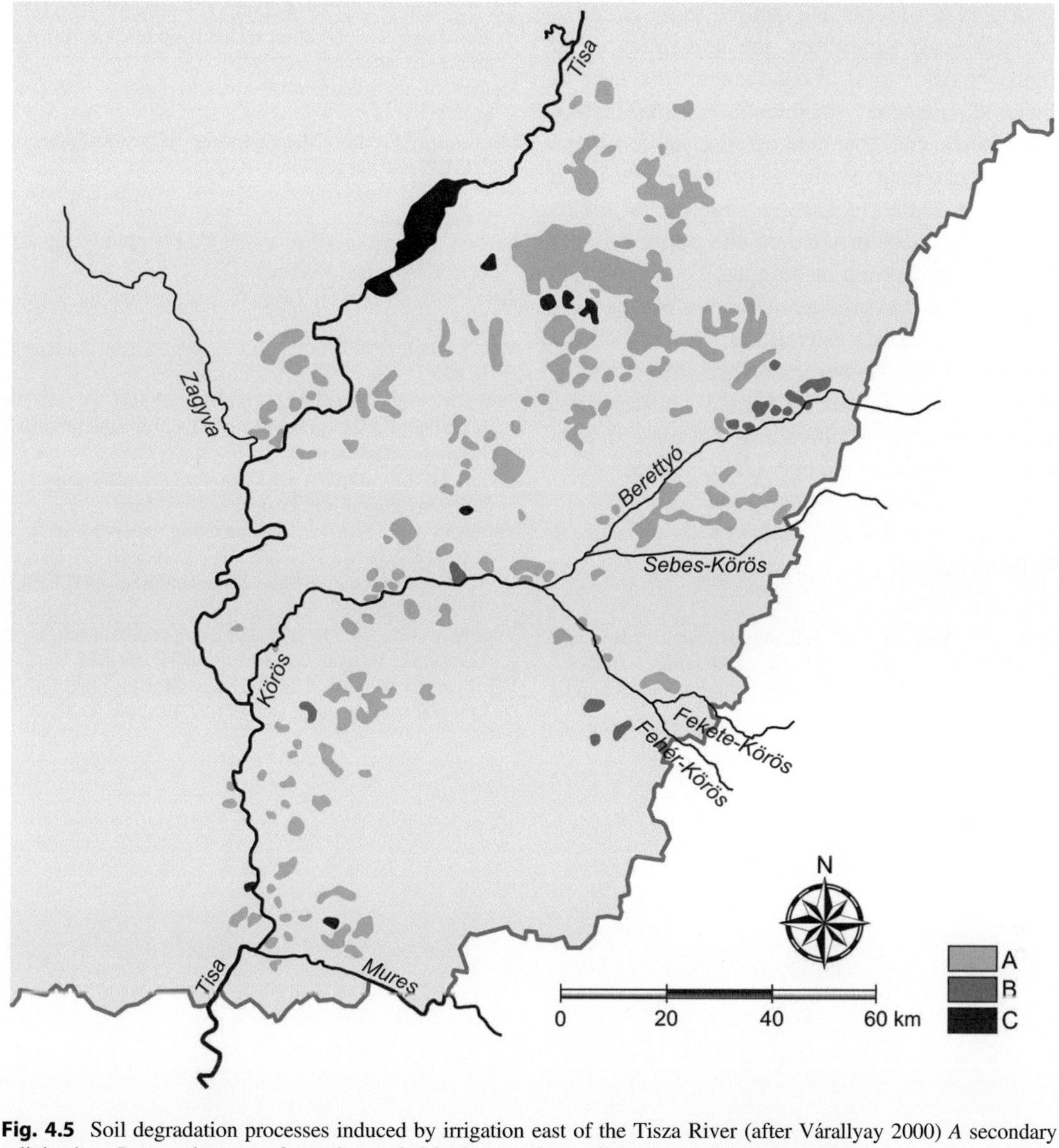


**Fig. 4.5** Soil degradation processes induced by irrigation east of the Tisza River (after Várallyay 2000) *A* secondary salinization, *B* secondary peat formation and *C* both secondary salinization and peat formation

third of the Hungarian soils can be characterised by a good, another third by bad water management (the reasons are the high sand and clay content), while ca. 10 % is characterized by salinization (Várallyay 2004).

The human impact (which is the most common impact) on the water regime triggers two-directional changes in the soils. One direction is the steppe formation due to the decreasing water supply, for example, flood control, which results a shift from meadow soils towards chernozems; the other direction is in the opposite, when the higher water supply results in shifts towards meadow soils (Várallyay 2000).

The soils can be polluted directly by non-hazardous agricultural, food industrial and communal waste (e.g. sewage sludge). The largest part of the annual waste production (tens of millions of tons) is transported to waste disposals, where the appropriate landfill construction can significantly reduce the risk of direct contact with soils.

Instead of soil loss and quality decrease due to the large-scale agriculture, the consequences of improper cultivation also appeared: the lack of drainage network maintenance, melioration deficiencies and the reduced use of fertilizers (which decreased by 50 % between 1990 and 2000, then started to increase slightly, but seems to be insufficient in spite of the application of precision agricultural techniques).

Altogether, environmental loads affecting soils stagnates or show a decreasing tendency (Bulla and Vári 2002). However, the growing extent of greenfield investments and road constructions reduces the area of cultivation, and these human activities accelerate the degradation processes.

## References

Bulla M, Vári A (2002) Magyarország környezeti jövőképe – egy évtized távlatából. Műhelytanulmányok OKT, Budapest. http://www.socio.mta.hu/konyvek/bulla-vari.pdf

Joó K, Barczi A, Zs Szántó, Horváth M (2003) A hortobágyi Csipő-halom talajtani vizsgálata. Agrokémia és Talajtan 52(1–2):5–20

Magyarország atlasza (1999) Cartographia, Budapest, p.132

Mezősi G (2008) Magyarország környezetföldrajza. JATEPress, Szeged

MNA (1989) Magyarország Nemzeti Atlasza. Carthographia, Budapest

Németh T, Stefanovits P, Várallyay Gy (1997) Talajvédelem. Spácium, Budapest

Stefanovits P (1981) Talajtan. Mezőgazdasági Kiadó, Budapest

Stefanovits P (1999) Soil science. Mezőgazda, Budapest, p. 320

Stefanovits P, Michéli E (szerk) (2005) A talajok jelentősége a 21. században. MTA Társadalomkutató Központ, Budapest

TAKI (1997) http://www.kvvm.hu/szakmai/karmentes/kiadvanyok/talaj_tajekoztato/kozep/1.htm

Várallyay Gy (2000) Magyarország talajviszonyai a folyószabályozások előtt. In: Somogyi S (szerk) Folyószabályozások és ármentesítések. MTAFKI, Budapest, pp 109–125

Várallyay Gy (2001) A talaj vízgazdálkodása és a környezet. Magyar Tudomány 46(7):799–815

Várallyay Gy (2004) A talaj vízgazdálkodásának agroökológiai vonatkozásai. AGRO-21 Füzetek 37:50–70

# 5 Natural Vegetation of Hungary

**Abstract**

The species composition of the vegetation in the Carpathian Basin is determined by the climate, relief, lithology/soil and hydrological conditions. Two main zonal vegetation types appear in Hungary: the zone of deciduous forests, which is dominated by oak and beech (especially in hilly and mountainous areas), and its most widespread forest type is the turkey oak–oak forest. The other zonal vegetation is the forest steppe on lowland areas, which includes different sand and loess forest steppes, and herbaceous associations (e.g. lossy and sandy steppe grasslands). Based on the reconstruction of potential vegetation, approximately 30–50 % of the Carpathian Basin was covered by forest before the intensive human impact, while just approx. 20 % of Hungary was covered by forest. The Carpathian Basin is very rich in species (ca. 3000 plant and 43,000 animal species), and it has colourful and various ecological systems; thus, it is one of the areas with the richest vegetation in Europe. Therefore, the questions of nature protection and conservation are highlighted in the chapter. Highly protected areas are the peatbogs, marshland forests, mixed oak forest and lowland oak forest, which are endangered by natural and human factors.

## 5.1 Geographical Features of the Zonal Vegetation in the Carpathian Basin

The species composition and the flora of a given area are determined by the environmental conditions. This also means that vegetation responses to the changing conditions at different speeds. For example, the vegetation of a meadow can adapt to the changing environment (e.g. wetter or drier and colder or warmer conditions) in one or two decades, while the adaptation of a forest takes one or two centuries. The process is not so simple, since the existing vegetation cover also reflects the features of the past, for example, that is why there is no natural spruce forest on top of the Mátra, even if the present conditions would make it possible (Pócs 1981). It is difficult to describe the former (potential) natural vegetation cover, because different associations have occurred depending on how far we go back into the past. The Tertiary vegetation of the Carpathian Basin was mostly displaced by the drastically altered Pleistocene climate conditions



G. Mezősi, *The Physical Geography of Hungary*, Geography of the Physical Environment,
DOI 10.1007/978-3-319-45183-1_5

(in the glacial periods the mean annual temperature decreased by 10 °C, and the precipitation was one-third of its former amount) (Járainé Komlódi 2009). Only a very small part of the Tertiary vegetation could adapt to the environmental changes, only in certain areas with an advantageous microclimate (see Sect. 1.8). According to the Ancient Mátra Theory of Borbás from the 1930s, during the Pleistocene some herbaceous plants migrated from refugium areas and from the southern slopes of the mountains to the plains, which lost their rich vegetation during the coldest periods.

In the periods of glaciation, the climate zones shifted to the south considerably; thus, some elements of the Pannonian vegetation cover appeared in the recent Mediterranean environment, and the former Mediterranean elements appeared on the edges of the recent Sahara. Only few plant species survived the glaciations and the interglacial warm periods. The glacial vegetation (e.g. *Comarum palustre*) remained in the cooler gorges of the mountains and in the bogs of the Great Hungarian Plain, while the interglacial/stadial vegetation (e.g. *Ferula sadleriana*, *Paeonia banatica*—Mecsek Mts; *Carpinus orientalis*—Vértes Mts) survived on the southern slopes of hills and mountains with a warmer microclimate. The endemic flora survived in a small extent after the Pleistocene (their proportion is about 2 % in the recent Hungarian flora); thus, after the glaciation, the vegetation of the neighbouring areas of the basin could spread. Consequently, apart from the Eurasian species (approx. 23 %) other flora elements can also be found in the Carpathian Basin: Pontic-Mediterranean (approx. 20 % in total), Central European (about 9 %), Alpine-Carpathian-Dacian (11 %), cosmopolitan and continental (14 %), European and circumpolar (about 9–8 %), Atlantic (3 %), and American, African or Asian adventive (3 %) species (Soó 1964).

A hierarchic spatial order can be established in the vegetation of a region on the basis of flora development and the extent of vegetation similarities. The essence of this logic is that the vegetation is characterized by a similar plant association, and often by a similar dominant plant species in a similar environment. The similarity becomes smaller as we proceed towards the higher levels of the hierarchy. The following floristic levels can be distinguished in the vegetation of Hungary, proceeding from the largest to the smallest: floristic kingdom (*Holarctic*)—floristic region (*Central European*)—floristic province (*Pannonicum*)—floristic district (it practically corresponds to the macro-regions, e.g. the Great Hungarian Plain to the *Eupannonicum*)—and floristic subdistrict (e.g. the majority of areas east of the Tisza River, *Crisicum*).

The species composition of the Carpathian Basin is formed by the climate, the relief, the lithology/soil and the water supply. Two zonal forest types characterize landscapes influenced primarily by the climatic conditions (Fig. 5.1); however, these also correspond to other landscape factors:

1. zonal deciduous forests: beech and oak forests dominate in hills and mountains;
2. zonal forest steppe: various forests (for example, sand forest steppe), and extensive treeless plant communities (for example, loess steppe grassland) on plains.

Transitional communities may form between the zones: extrazonal (e.g. peatbog) and intrazonal communities (e.g. reeds) under special environmental conditions.

The floristic district of the Great Hungarian Plain (including the Little Hungarian Plain) belongs to the forest steppe zone; it is characterized by small forest cover in natural conditions. Along the rivers, however, large gallery forests occur: hardwood gallery forest is characterized along the Danube, whereas softwood, riverine willow and poplar forests characterize the Tisza and the Körös Rivers. The largest areas used to be occupied by sand, loess and salt-affected steppes. The pioneer species on blown sand, fixing its loose sediment, are *Bromus species* and *Secale sylvestre*, followed by *Festuca vaginata* and *Stipa borysthenica*. Open sand steppes characterize the driest sand region, the Kiskunság. Willows, junipers and white poplars can occur in the interdunal depressions, and along with the progress in succession, oak forest can be formed on sand-covered areas. The

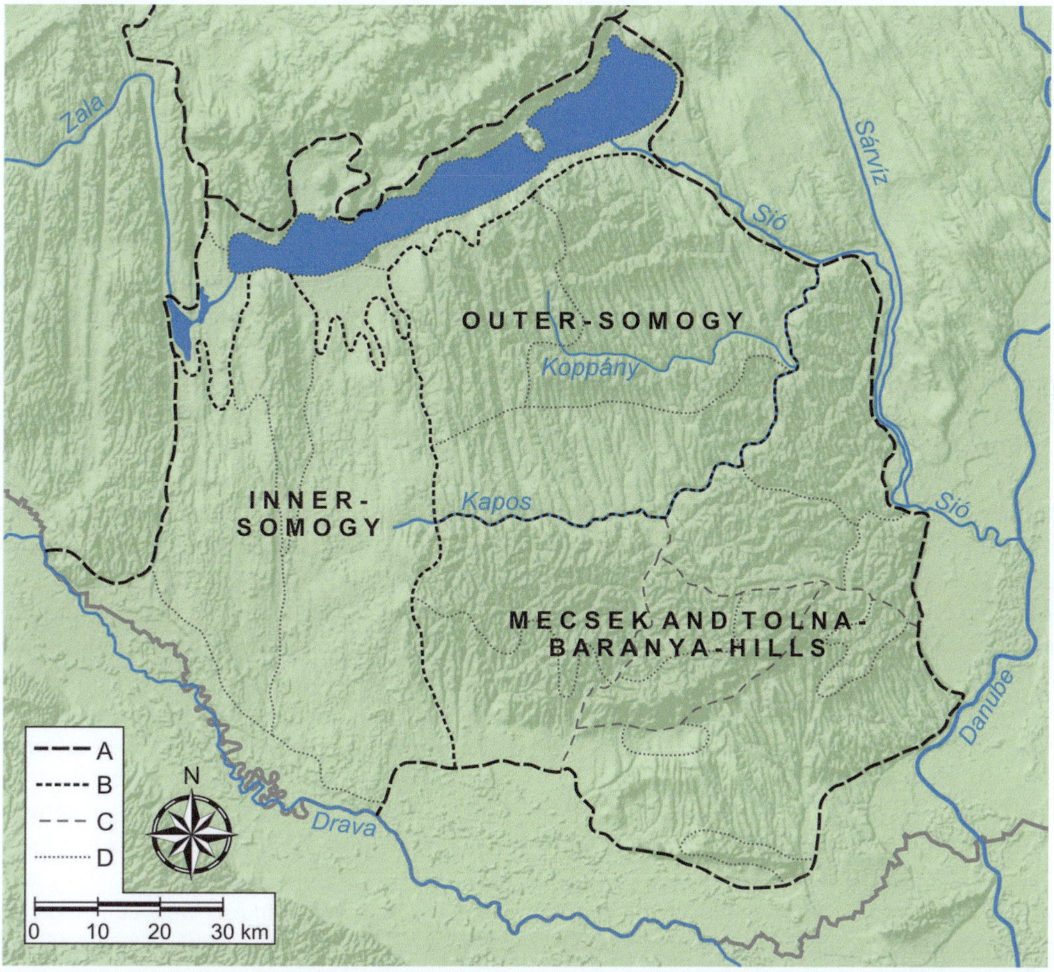


**Fig. 5.1** A comprehensive map of the Hungarian vegetation (after Zólyomi 1989 MNA). *A* Sand steppe, sand forest, *B* loess steppe, oak forest, *C* saline wetlands, *D* riparian gallery forest, *E* peat forest, peat moss, *F* downy oak, *G* mountain hornbeam–oak forest, *H* beech forest, *I* acido frequent coniferous forests, mixed oak forest, *J* turkey oak–oak forest

Nyírség is characterized by open sand steppes dominated by *Festuca vaginata*, and conforming to the slightly different ecological conditions *Corynephorus canescens* and *Festuca pseudovina* also occur. The oak forests were formed on soils with favourable water regime and used to be diversified by *Tilia tomentosa*, most of which has disappeared by now; open steppe oak forests formed in dryer habitats. The typical vegetation of salt-affected areas is *Puccinellia limosa* at *Puccinellia* swards and *Plantago* species.

The lower elevated regions of the floristic district of the North Hungarian Mountains are covered by turkey oak and oak forests (up to nearly 600 m asl on the southern, and up to 400 asl on the northern side). In the higher areas, hornbeam and beech forests, only beech forests prevail. In karst regions, such as Bükk and Aggtelek Mts., karstic shrub forests and species-rich rocky grasslands have been formed on the shallow topsoil of the rocky slopes. Sub-Mediterranean vegetation can also be found in landscapes formed on volcanic rocks. The carbonate rocks of the Transdanubian Mts. are characterized by turkey oak and hornbeam forests. Extensive beech forests can only be found in the Bakony Mts. The sub-Mediterranean influence prevails especially in the western part of the macro-region, and its signs are *Ruscus aculeatus*, or the lavender, grown in the neighbourhood of the ash cones of Tihany Peninsula (at Lake Balaton); however, the extrazonal karst shrub forests also show this influence, for example, by *Cotinus coggygria*. In South Transdanubia, the zonal association is the turkey oak–oak forest, the hornbeam and oak forest in more humid areas, but the subalpine influence—decreasing northwards—can be observed, as well. The Villány Mt. is especially rich in subalpine species (e.g. *Festuca dalmatica*, *Colchicum hungaricum*), and the kartstic shrub forests on the southern slopes of the Mecsek Mts. are also complemented by Balcan (Illyrian) vegetation. The West Transdanubian Floristic District has both alpine and boreal elements; however, these are typical only in some territories of the Lower Alps meso-region. Besides the spruce, Scotch pine also occurs in cooler valleys, and peat bogs are common. The beech and hornbeam forests become more typical towards east.

## 5.2 Typical Plant Communities in Hungary

The vegetation is characterized by smaller units adapting to the environmental conditions, which show similarities in their structural features and species composition, and have a relatively typical appearance. These are the plant communities or associations, which are usually characterized by a dominant species.

Since the environmental conditions of different associations can be very similar, the species composition of two different communities may be similar too. The analysis of the characteristic features of communities (such as the number of species and individuals, dominance, associative ability or frequency) is the main task of Phytosociology.

According to the reconstruction of the potential vegetation based on the physical geographical background, 70–80 % of the basin used to be covered by forests before the extensive anthropogenic impact (Jakucs 1981). Today the proportion is the opposite: 20 % of the country is covered by forests. The present book does not undertake to characterize the more than 300 identified plant associations and their physical geographical or environmental background. Only the most widespread communities and the conditions of their development are described.

The most widespread forest on hills and mountains up to 300–400 m is the turkey oak–oak forest (*Quercetum petreae-cerris*), which appears as a closed forest on brown forest soil with carbonate illuviation and on Ramman's brown earth (Fig. 5.2). These areas are valuable from the agricultural point of view as well; therefore, most of the forests were cut. A smaller territory is covered by calciphilous oak forests (*Orno-Quercetum pubescenti-cerris*), whose occurence is primarily defined by the relief, where the forest floor has also sub-Mediterranean

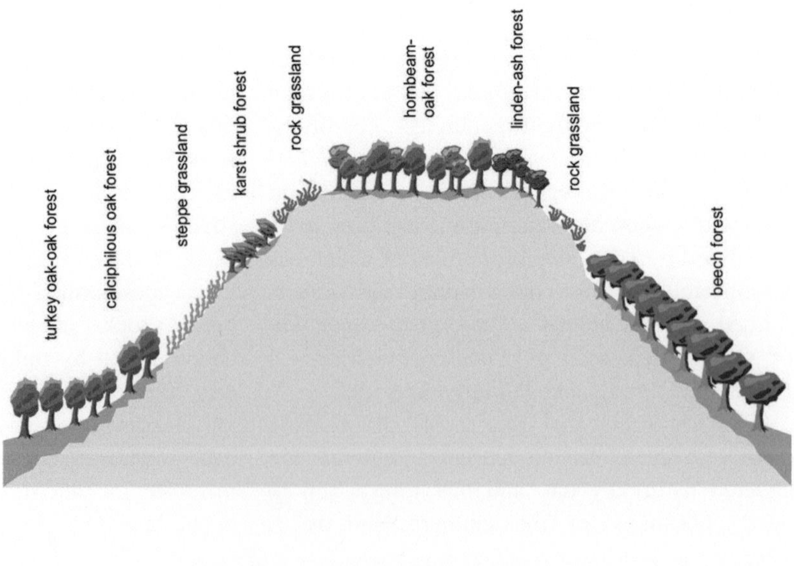


**Fig. 5.2** An idealized profile of mountain plant communities (after Jakucs 1960)

herbaceous species in the West, and Carpathian herbaceous species in the East. The northern carbonated slopes of the Transdanubian Mts. are characterized by mixed karst forests (*Fago-Ornetum*), and the main tree species are the psychrophile beech and ash, which also indicate the postglacial relic character of the community. On redzina soils downy oak shrub forests can be found. The surfaces of the gentle slopes of brown forest soils in the mountains are occupied by hornbeam–oak forests in 400–600 height. The rocky surfaces (e.g. block fields) of extreme ecological features are dominated by linden-ash (*Tilio-Fraxinetum*) communities, which are considered to be the remnants of the postglacial mixed forests. Beech forests cover the areas above 600 asl, and the northern, cool slopes of the mountains (their proportion is 15 %). Their floor is poor in species and consists of mainly shade-tolerant species. The vegetation of the orographically high, volcanic and carbonated (limestone, dolomite) surfaces, with thin fertile layers, is especially dependent on the lithologic conditions and the water regime. The vegetation adapted to the disadvantageous ecological conditions of skeletal soils by forming unique species compositions.

The sand and loess grasslands represent the largest territorial proportion among the treeless communities. They can have various species compositions depending on water supply, gradient and microclimate. The species of the loess steppe reached the Carpathian Basin from East in the drier Boreal Phase of the Holocene. The closed grasslands were used in the agriculture early, and their remains can be found on kurgans, mottes and steep loess walls. The closing community of the succession is the Tatarian maple-loess oak forests in the plains. The grasslands and woodlands could survive due to the dry, warm microclimate in the sand areas of the plains. The communities representing different succession stages are often mixed in the same area creating mosaic-like pictures (Fig. 5.3). The primary community on sand is the open sand steppe, which often appears on bare blown sand surfaces, followed by several communities, and oak and white poplar forest is the closing community of the succession. As it was already mentioned, the salt-affected areas formed in the Boreal Phase, but their current extension is the consequence of water regulations and deforestation. Because of the variable micro-relief and the groundwater table changes, various plant communities can develop on salty soils (Fig. 5.4). On surfaces characterized by disadvantageous physical and chemical characters (e.g. alkaline depressions and annual salt pioneer swards), only strongly salt- and drought-tolerant species can

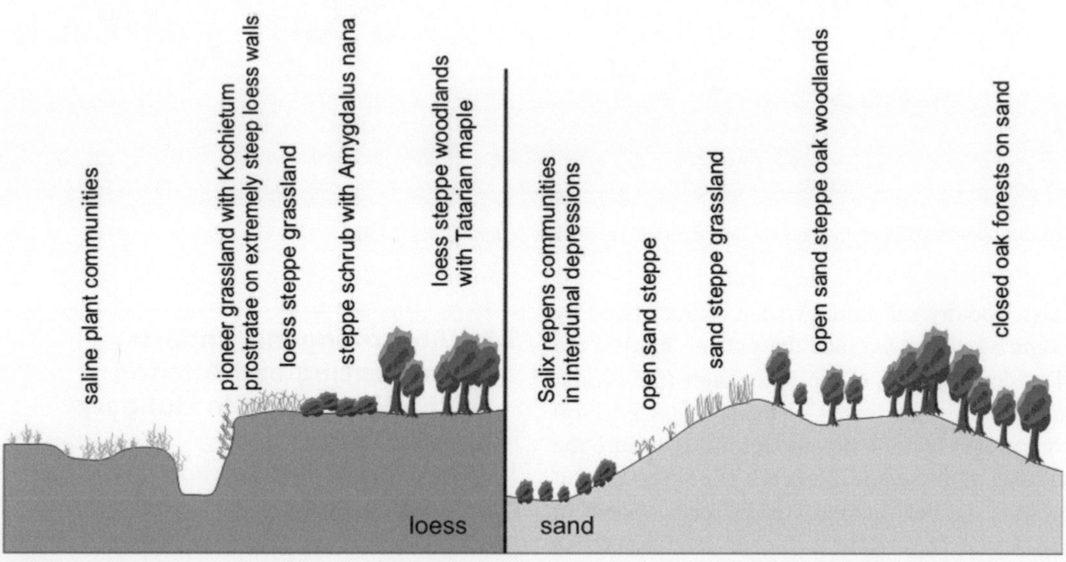


**Fig. 5.3** Steppe plant communities also adapted to the small-scale relief changes (after Jakucs 1960)

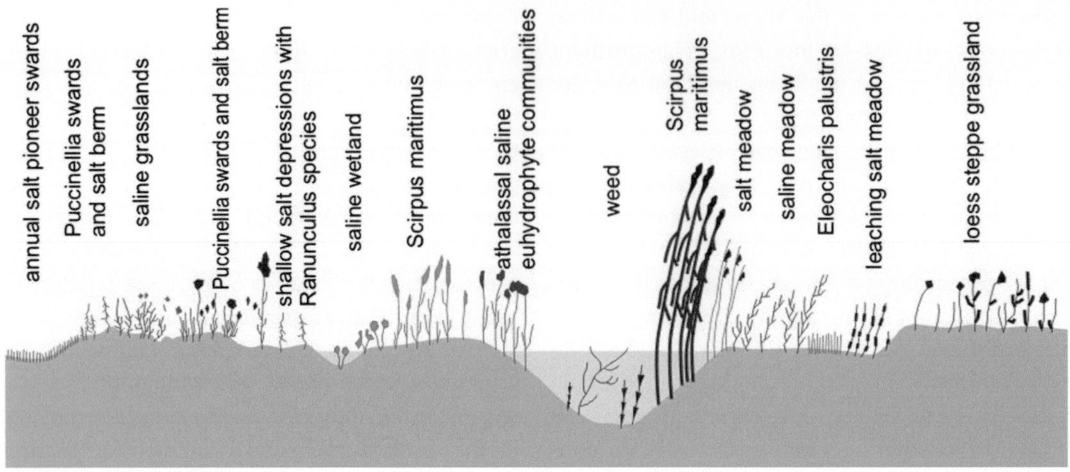


**Fig. 5.4** Saline vegetation (after Soó 1973)

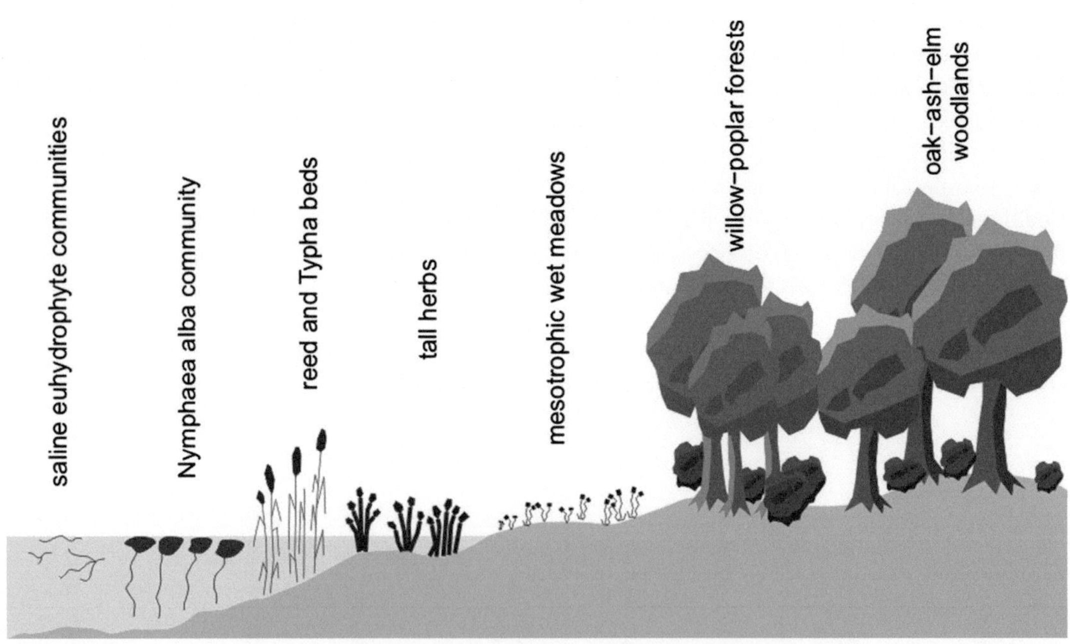


**Fig. 5.5** Succession of the aquatic and riparian vegetation (after Jakucs 1960)

exist. The dry salt meadows are characterized by *Festuca pseudovina*, and the grazing hinders the slow development of the succession (the formation of saline oak forests). Treading, grazing, soil compaction and the organic fertilizers ensure the survival of the steppe character. The succession of the aquatic and riparian vegetation is shown in Fig. 5.5.

## 5.3 Anthropogenic Impacts Influencing and Altering the Vegetation in Hungary

The Carpathian Basin has the richest vegetation in Europe due to its varied and diverse ecological system and biodiversity (3000 plant and 43,000 animal species). The society in the basin utilized this environment, influencing the state of the vegetation. The first traces of more intense influence can be dated to the Bronze Age, 3–4 thousand years ago. The expansion of the more and more efficient agricultural activities resulted in decreasing natural character of the vegetation, its transformation to be semi-natural and anthropogenic. The diversity of the vegetation has also decreased due to the human impact (for example, the monocultural cultivation is frequent today), and the balanced cycle of the elements is often disturbed. The vegetation adapts to the changing environment through natural selection, whereas the human intervention has also transformed the species diversity (by breeding). According to the scientific studies, the plant communities are able to function for a few hundred, maybe, a thousand years, while this time period is reduced to its one-hundredth due to human interventions.

The anthropogenic influence is well-demonstrated by the number of observed communities: 350 out of the described 470 Hungarian plant communities are natural or semi-natural, 120 have an anthropogenic origin, and 114 have a weedy character (Borhidi and Sánta 1999; Járainé Komlódi 2009). Due to the intensifying anthropogenic impact, some valuable plant communities become endangered. Due to the Hungarian nature conservation, the number of protected communities and plants has been doubled in the past decade (it is 600 today, a tenth of which is highly protected). Red Books were published about the extinct and endangered plant and animal species in Hungary (Rakonczay 1990; Borhidi and Sánta 1999).

The majority of natural plant communities are forests. The health of trees has deteriorated in the past years: on the basis of the lost foliage only 35.6 % of the broadleaf and coniferous trees were healthy in 2003, while 41.9 % were slightly, 17.1 % moderately, and 2.8 % were intensively damaged; at the same time, 2.6 % died (HKA 2005). As a result of the (mistaken) Forestry Act in 1968, almost 90 forest communities are endangered, and 18 are highly protected. There was a significant forestation in the past decades; thus, the proportion of forests today exceeds 20 %. Although, the vast majority cannot be considered as forests according to several

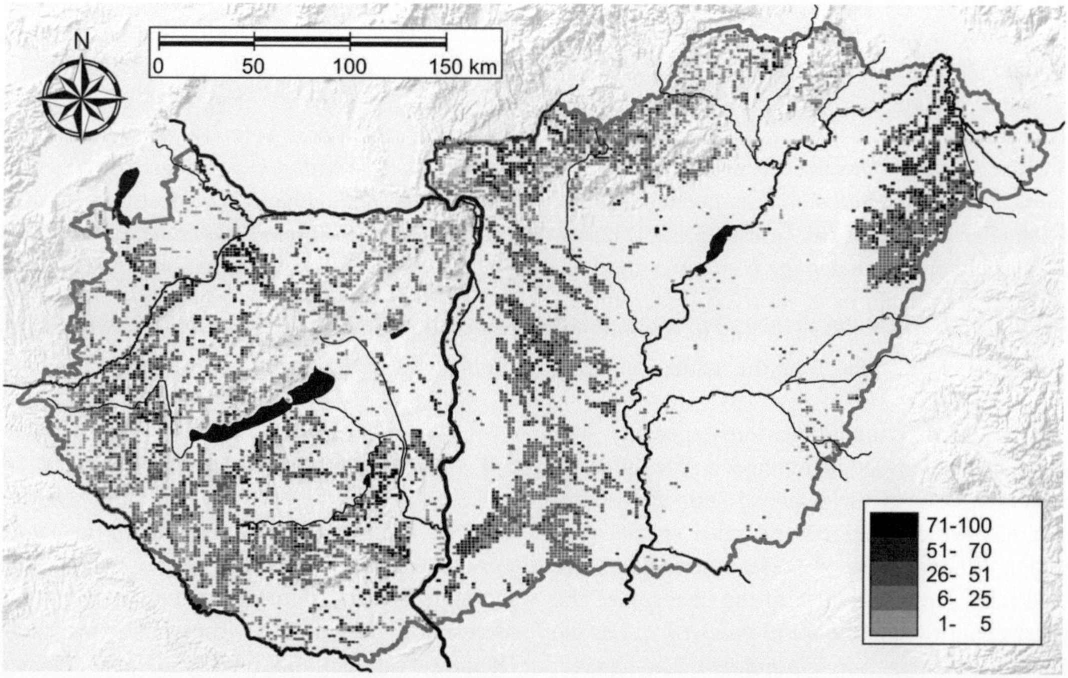


**Fig. 5.6** *Proportion of Robinia pseudoacacia in Hungarian forests* (after State Forest Service 2006)

opinions, because it does not comply with all the ecological criteria of a forest (these areas are rather cultivated areas where trees are grown). [The history of *Robinia pseudoacacia* (locust tree) is interesting. It was introduced in Hungary in the 1750s, first as an ornamental plant. Since it is a good melliferous plant and fixes the sandy soil from deflation excellently well, it was planted extensively; thus, its population is the largest in Europe in Hungary. Unfortunately, the understory and floor of *Robinia pseudoacacia* is poor in species, and the forest quickly depletes the soil (Fig. 5.6)]. Two-thirds of the Hungarian forests serve economic purposes, and the non-endemic species have been favoured; thus, only a half or one-third of the forest territories are considered as forests from the ecological point of view. That is why the forest reserves are important, and even if they only cover a small territory (about 13,000 ha), they are valuable. The Nature Conservation Act and the Forest Act of 1996 include also species-level protection (Standovér and Primack 2001).

Plant communities worthy of or proposed to be highly protected (Borhidi and Sánta 1999):

- bogs and peat forests (e.g. alder woodland, peat-moss alder marshes, willow carrs, birch mire forests), which may be damaged by water shortage, amelioration and water regulations;
- mixed karst forests that occur on steep slopes of dolimite, and erosion may endanger them;
- smaller patches of oak forests in the Danube-Tisza Interfluve and in the Northern part of the Hungarian Plain have survived. Their endangered by the aridification due to the groundwater level decrease, and the habitat decrease.

The aquatic communities are especially sensitive to the anthropogenic impact. Altogether 6–8 communities of the seaweed and sedge communities are endangered and need special protection. The species of these communities are especially sensitive either to the changes of the water quantity (e.g. *Hottonia palustris* L.) or the water quality (e.g. *Salvinia natans*). The aquatic, marsh and meadow communities are threatened by the limited water availability. (The hydrophile communities occupied 45 % of the Great Hungarian Plain in 1870, before the water regulations, while their territory shrank to 3–6 % by 1960.)

The recent development of vegetation is seriously influenced by the spread of invasive species. Plant invasion previously occurred in the Carpathian Basin as well, for example, at the end of the Pleistocene. The difference compared to nowadays is that invasive plants are currently spreading due to artificial, and not natural reasons. These date back to the Age of Discovery, when several species were imported to Europe, on purpose or accidentally. They degrade mainly the natural communities, though the typical natural habitats (dry steppe or dry forest) have higher resistance to invasion (Botta-Dukát and Mihály 2004, 2006). Furthermore, the efficient cultivation of crops—as a renewable energy source—is also hindered by certain invasive species (e.g. *Ambrosia artemisifolia*). Only 3 % of the nearly 3000 species of the Hungarian flora are invasive, but they hamper the utilization of hundreds of hectares, and the protection against them is difficult. Many people know *Amorpha fruticosa*, appeared in 1724 and was used to stabilize the riverbanks, but nowadays it spreads without any control, damaging the structure of artificial flood-protection levees, decreasing the transport capacity of irrigation canals, and decreasing the flood velocity on the artificial floodplains. *Solidago canadensis*, which appeared in 1865, enhances the destruction of the original vegetation cover; *Ailanthus altissima* has already appeared in five national parks, and *Ambrosia artemisiifolia* reached Hungary in the 1960s, and millions are suffering from its health implications (i.e. pollen allergy).

## 5.4 Nature Protection in Hungary

Hungary has a high-quality regulation in nature conservation on the European scale, and undoubtedly, nature conservation is the most successful field within environmental protection. The preservation of biodiversity, the maintenance of ecological diversity and structural character of habitats can be highlighted among the targets of protection and management. The professional background is provided by the monitoring systems established in the past decades. One of these was the Hungarian Biodiversity Monitoring System, which was prepared by 1997, and started its operation in 1998. The most important achievements (mapping and classification of habitats) are summarized in a 10-volume series (Fekete et al. 1997). The MÉTA (GIS Database of the Hungarian Habitats) mapping (MÉTA 2009) includes the important elements of vegetation monitoring. The result of detailed field surveys is a reliable digital data set of low resolution (36 ha), which may serve as a basis of several further analyses.

The protection of species and associations is not feasible without the protection of their physical habitat. Four levels are distinguished within the 850,000 ha protected area on the basis of the level of protection. According to the latest legislation, the area of the 10 national parks is about 480,000 ha, and that of the 36 landscape protection areas is 324,000 ha (Fig. 5.7). The lower levels are nature reserves and unique landscape values. The core areas of the national parks have international importance with their 6000 ha biosphere reserves. (The landscape management of the natural parks is based on 3 main units: the core areas, which theoretically cannot be used for economic purposes, the buffer zone, and the outer zone under less strict regulations.)

In the past ten years, high number of habitat protection systems were compiled (sometimes they are overlapping). Due to the Act on Nature Conservation in 1996, several objects became legally protected (*ex lege*), and required an accurate register. Thus, the following cadastres have been compiled:

- caves (more than 3500, out of which 130 are highly protected);
- marshlands (at present, 1251 *ex lege* protected marshlands are recorded);
- saline lakes (336, Lake Fertő and Lake Velence are the largest);
- kurgans (1700 kurgans with nearly 400 motte remnants);
- springs and sinkholes (thousands of items are included in the cadastre).

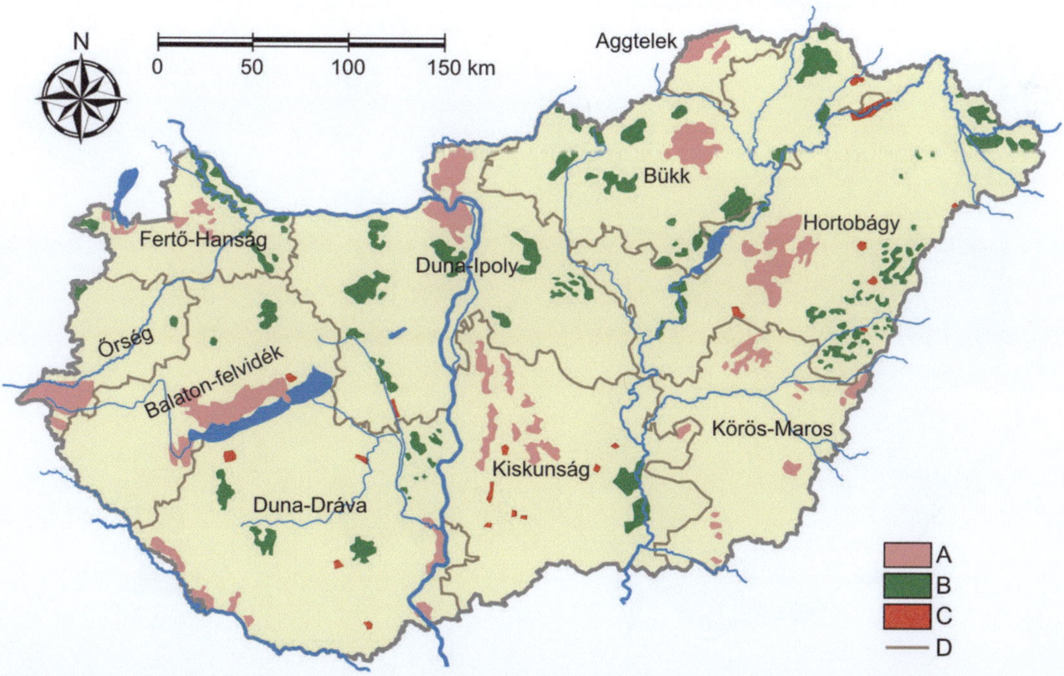


**Fig. 5.7** Protected areas of national importance 2004 (after HKA 2005). *A* National park, *B* protected area, *C* nature reserve, *D* national park border

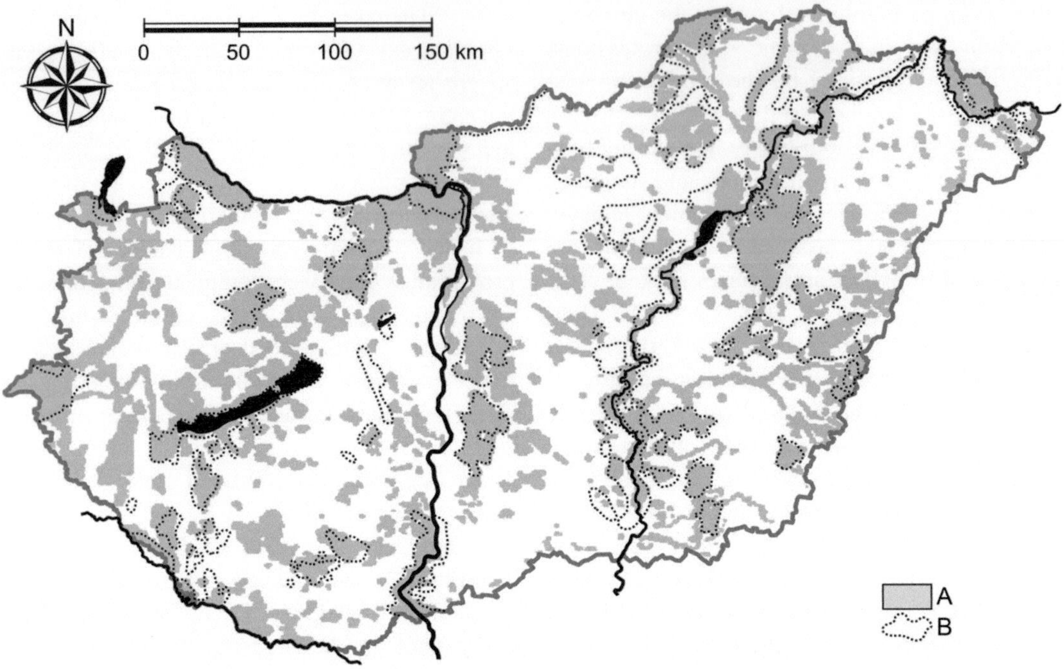


**Fig. 5.8** Natura 2000 Areas of Conservation (after KvVM 2006). *A* Special areas of conservation (SACs), *B* special protection areas (SPAs). *Source* http://www.natura.2000.hu/indexphp?p=termegorze&nyelv=hun

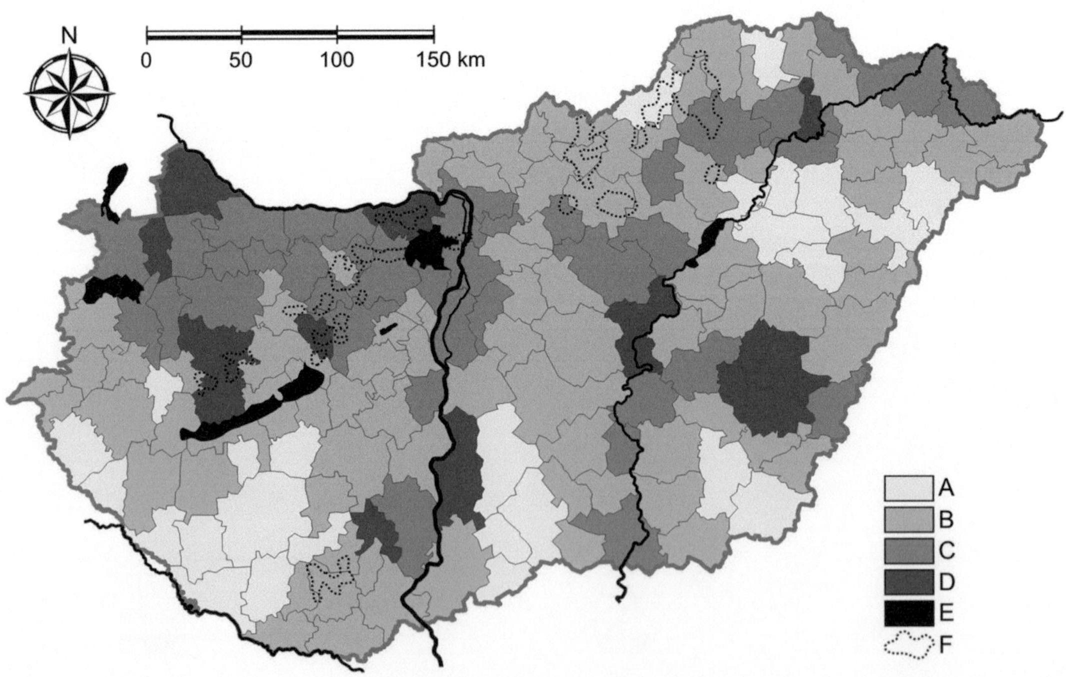


**Fig. 5.9** Number of anthropogenically transformed landscapes in the micro-regions (pcs/km$^2$) (after HKA 2005). *A* 3–8 pcs/km$^2$, *B* 9–17 pcs/km$^2$, *C* 18–26 pcs/km$^2$, *D* 26–35 pcs/km$^2$, *E* 36–41 pcs/km$^2$

Apart from the cadastres, a programme for the protection of wetland habitats (Ramsar Convention areas) was also created (180,000 ha in 2003), and for the identification of sensitive natural areas (*National Programme* of *Environmental Protection* in *Agriculture*), where the preservation of the environmentally friendly cultivation also facilitates the preservation of habitats. Further programmes targeting nature conservations are, for example, the UNESCO MAB biosphere-reserve programme, the oxbow lake rehabilitation programme, and the habitat classification system (Fekete et al. 1997).

In 2006, the regulation on the Natura 2000 network was finalized, in accordance with EU targets. In this way, approximately one-fifth of the territory of the country is under special protection (it is the EU average) assigning areas of valuable natural habitats, or bird sanctuaries (Fig. 5.8).

Approximately 15 thousand abandoned but un-recultivated mines, mine yards and pits destroy the Hungarian landscape. They are registered in a national database (e.g. Nature *Conservation Information System* (NCIS)—http://geo.kvvm.hu/tir), and their rehabilitation is state responsibility. The data of the detailed surveys are also available in the case of 3000 items in the database (Fig. 5.9).

## References

Borhidi A, Sánta A (szerk) (1999) Vörös Könyv: Magyarország növénytársulásairól. 1–2. Természetbúvár, Budapest

Botta-Dukát Z, Mihály B (szerk) (2004) Özönnövények: biológiai inváziók Magyarországon. I–II. Természet-Búvár Alapítvány Kiadó, Budapest

Botta-Dukát Z, Mihály B (szerk) (2006) Özönnövények: biológiai inváziók Magyarországon. I–II. Természet-Búvár Alapítvány Kiadó, Budapest

Fekete G, Molnár Z, Horváth F (szerk) (1997) A magyarországi élőhelyek leírása, határozója, a Nemzeti Élőhely-osztályozási Rendszer. Magyar Természettudományi Múzeum, Budapest

HKA (2005) Hazánk környezeti állapota. KvVM, Budapest

Jakucs P (1960) Növényföldrajz. In: Pécsi M, Sárfalvi J (szerk) Magyarország földrajza. Akadémiai Kiadó, Budapest, pp 74–85

Jakucs P (1981) Növénytársulástan. In: Hortobágyi Simon T (szerk) Növényföldrajz, társulástan és ökológia. Tankönyvkiadó, Budapest, pp 225–265

Komlódi Járainé M (2009) Hazánk növényvilágának kialakulása. Természet Világa 140(5):1–6

KvVM (2006) http://www.natura.2000.hu/index.php?p=termegorze&nyelv=hun)

MÉTA (2009) http://www.novenyzetiterkep.hu/meta/altalanos.shtml

Pócs T (1981) Növényföldrajz. In: Hortobágyi–Simon T (szerk) Növényföldrajz, társulástan és ökológia. Tankönyvkiadó, Budapest, pp 27–168

Rakonczay Z (szerk) (1990) Vörös Könyv: a Magyarországon kipusztult és veszélyeztetett növény- és állatfajok. Akadémiai Kiadó, Budapest

Soó R (1964) A magyar flóra és vegetáció rendszertani-növényföldrajzi kézikönyve. I–VI. Akadémiai Kiadó, Budapest

Soó R (1973) A magyar flóra és vegetáció rendszertani-növényföldrajzi kézikönyve. I–VI. Akadémiai Kiadó, Budapest

Standovér T, Primack RB (2001) A természetvédelmi biológia alapjai. Nemzeti Tankönyvkiadó, Budapest

Zólyomi B (1989) Természetes növénytakaró. In: Pécsi M (szerk.) Magyar Nemzeti Atlasz. Cartographia, Budapest, pp 89–89

# Part II

# Regional Physical Geography of Hungary: An Overview

# 6 Division of the Natural Landscape Regions

**Abstract**

This chapter describes the hierarchic system of the Hungarian landscape units and its structure. The smallest unit is the ecotope, and the group of ecotopes forms the micro-region, which constitute meso- and macro-regions. This system is the basis of the regional description of Hungary, introduced in Part II of the book. Based on the accepted Hungarian classification of regions, the regions are determined by their physical geographical factors, but in connection with them, social (and economical) processes also modify their structure, function and characteristics. Therefore, in the regional descriptions, some environmental values and conflicts of each macro-region are highlighted. Besides the structural analysis of the regions, some landscape historical data are described.

The geographical landscape is a separate, individual, complex and regional unit of the surface of the Earth, determined by its specific structure and functions (Pécsi et al. 1972; Mezősi 2003). Mainly, physical geographical approaches have become common when defining the landscapes, as it is often argued that the landscape is an environmental–historical category (Pécsi and Somogyi 1967), where human activity also plays a role. This point of view definitely excluded the non-natural elements; thus, for example, it does not use ethnographic and historic categories (such as Matyóföld, Torontál, Vendvidék and Jászság). Therefore, such units (and their names) do not appear in the cadastre of micro-regions (Marosi and Somogyi 1990).

The concept of landscape is applied to several fields in science—from landscape architecture through ethnography and to ecology, in accordance with their points of view, and it is also commonly used in everyday life; therefore, its conceptual system became quite confused. There is an approach that deals with a landscape as an epistemological category, where a plain, a floodplain or a volcanic mountainous landscape can clearly be separated in the first approximation, but the anthropogenic impact may override these categories. Among the various approaches, some opine that the landscape is fictitious and it does not exist as an independent object, as it is just an aesthetic category. From the scientific point of view, the concept of landscape is clearly described. The requirement is that the different assessors identify the same units, through the accurate application of the techniques to describe the landscape units (and types).

In geography, we claim that their framework is defined by natural factors; thus, the natural



G. Mezősi, *The Physical Geography of Hungary*, Geography of the Physical Environment,
DOI 10.1007/978-3-319-45183-1_6

landscape-forming factors jointly create a well-describable (specific) piece of the surface that possesses typical values, together with the cultural attributes, in interaction with the artificial elements. In this approach, the priority of the natural factors is expressed together with the modifying effect of other landscape-forming elements. The social activities are integral parts to landscape, since various landscapes are "cultural", which are the results of human activity (e.g. the artificial steppe of the Hortobágy).

The interpretation of the natural and social factors, and in particular, the assessment of the landscape-forming role of the anthropogenic impacts, leads to various appropriations. The landscapes can be interpreted as some integrated units of the landscape-forming factors; still, **natural landscapes** (composed of natural factors definitely) must be distinguished, and its analysis (could) be made by typical scientific methods, due to the similar parameter constituents. They are non-existent today, owing to the extensive human impact. On the other hand, the landscapes can be considered as units mainly created by anthropogenic activity, and in this case, the notion of **cultural landscape** could be used, which denotes landscapes whose processes are mostly regulated by human activities.

## 6.1 History of the Hungarian Landscape Division

The concept of landscape can be traced back to Hunfalvy; however, it became popular at the beginning of the twentieth century, when the ethnographic points of view prevailed. Teleki was the first scientist who wrote about the importance of interpreting what a landscape was and proposed a scientific analysis of the units on the basis of landscape-forming factors (e.g. Mezőség, Nyírség). The first Hungarian landscape divisions—since the scientifically grounded knowledge about surface evolution and ecology did not exist at the time—had popular and political motivations, and served cartographic and spatial separations, and the identification of places (Lukács 2003). Cholnoky (1926) attempted to introduce some artificial denominations, apart from the traditional ones. He distinguished 9 marco-regions in the historical territory of Hungary: Great Hungarian Plain, Transdanubia (Dunántúl—the name had been used at least since 1621; Kiss 1980), Transylvanian Basin, East Hungarian Mountain Range, Northwestern Uplands, Northeastern Uplands, Eastern Carpathians, South Transylavian Uplands and Dinarian Alps, although some of these names were not widely used. Prinz (1936) understood the concept of landscapes to be the same as today's macro-regions, that is larger territories that have significantly different natural conditions from those of their neighbours. In his opinion, there was no need and experience to define large landscape units, especially, because some had no historical sense at all (e.g. Great Hungarian Plain). Elaborating on Cholnoky's units, he distinguished 13 large landscapes and smaller territories within them (today, these are the meso-regions) in the territory of the historical Hungary. Bulla and Mendöl (1947) applied the development of the landscape approach and its scientific view (through the characterization of the landscape-forming factors, the genetics, the climate and the hydrography) when they created a landscape map of the entire Carpathian Basin, which also included a hierarchy levels of units.

Bulla (1964) formulated the basis of the division still in use by following Pécsi and Sárfalvi's (1960) division of four macro-regions, and he described 6 macro-regions in the territory of the country: Great Hungarian Plain, Little Hungarian Plain, North Hungarian Mountains, Foothills of the Alps, Transdanubian Hills and Transdanubian Mountains. Apart from the traditional units (such as the Great Hungarian Plain, the Little Hungarian Plain and Transdanubia), the designation of other units became necessary due to the border changes after the Trianon Treaty, when the territory of Hungary decreased by two-third. Therefore, for example, the North Hungarian Mountains was used, instead of the name of a larger unit (Mátra Forest). The division used today is essentially the same (Fig. 6.1, Pécsi and Somogyi 1967; Marosi and Somogyi 1990), although it included some differences. For example, (1) the Mezőföld and the Dráva Plain which are in Transdanubia were joined to the Great Hungarian Plain; (2) the border between the North Hungarian Mountains and the Transdanubian Mountains was not drawn along the Danube Bend, but in the Pilisvörösvár Valley, and thus, according to the new division, the Visegrád Mts. belong to the North Hungarian Mountains; (3) the foothills of the Alps were classified as a micro-region, whereas its macro-region was called Marginal Region of Western Hungary; and (4) some valley regions were also identified (e.g. Kapos Valley and Danube Bend) due to the consideration that it is not providential to draw the borders along rivers.

This landscape structure in the entire Carpathian Basin (Appendix A) provided more complete and understandable units, and the system created in this way fits into the region, which comprises several countries (Hajdú et al. 2000).

## 6.2 The Structure of Landscapes and Their Hungarian System

From the point of view of natural sciences, the ecotopes are the constructive units of a landscape, thus the integrated basic units of the landscape-forming factors (mostly of soil, lithology, climate, vegetation, relief, surface and subsurface waters). The ecotopes are homogeneous units from the point of view of each landscape-forming factor; thus, for example, the windward side of a sand dune possesses the same relief, (micro)climatic, hydrographic, etc., conditions (Mezősi 2003). The landscapes are built of these homogeneous units and can be organized into well-defined hierarchical systems. The micro-regions (and micro-region fragments) are composed of ecotopes, where the factors are similar, but not the same. On higher hierarchy levels, the similarity of the units becomes less. Independently of the hierarchy, a landscape is regarded as an integrated category, which is not applied for the separate landscape-forming

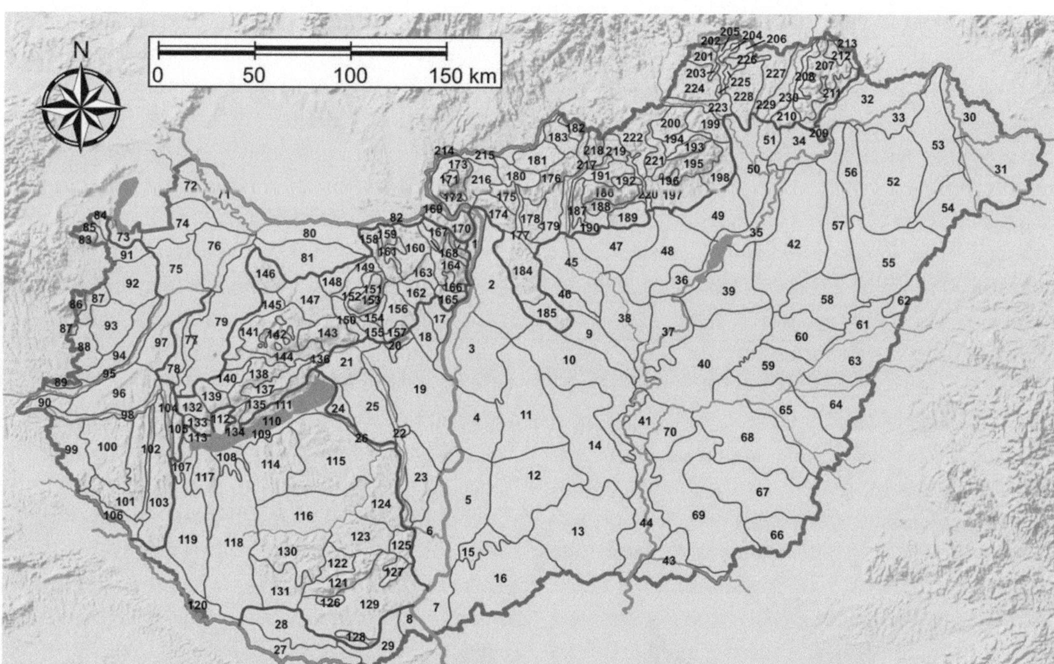


**Fig. 6.1** The natural landscape units of Hungary on the basis of the division made in 1990. The list of the numbered micro-regions is included in Appendix A, and the *bold lines* indicate the border of the macro- and meso-regions

factors (e.g. soil and vegetation). (Thus, a vineyard or an urban landscape can be understood only functionally, that is whether the surface is used for grape production or as residential area.)

The micro-region cadastre made in 1990 includes 230 items, and their territory varies between 10 and 100 km$^2$ (Marosi and Somogyi 1990). The micro-regions constitute meso-regions (35 items, area: 100–1000 km$^2$), and they can create macro-regions (6 items, area: 1000–10,000 km$^2$). The range of the sizes of landscapes indicates that their area is smaller in the hilly and mountainous regions and much larger on the plains, which is in accordance with the fact that the factors defining the structure (e.g. relief and lithology) are much more homogeneous and extensive on the plains than the factors (e.g. soil and vegetation) defining the character of the more segmented surfaces. In the rest of the book, this hierarchical structure of natural landscapes will be applied. The discussion concentrates on the description of macro-regions, and occasionally, this can be followed by a meso- or micro-regional analysis if the subsequent environmental impact analysis requires this.

An important issue of the scientific analysis of landscapes is the establishment of their borders. The importance of the issue also depends on the understanding of a certain landscape. If the categories will be used in regional development and planning, the errors have to be defined, while it is not as important if the aim is only to identify a landscape. The issue of landscape borders was discussed in detail by Prinz (1936), who declared that a landscape border is not a "divider", but instead, it follows the administrative lines which are the products of several natural and social effects. The landscape definition has changed a lot in the past 60–80 years and even more the professional environment of its interpretation (e.g. ethnographical by Kogutowitz, or morphological by Bulla). The landscape divisions (landscape structures) based on an older concept coincide with the landscape divisions of the past 20 years at the macro- and meso-regional levels. Our ancestors in the Early Medieval Times (Hungarian occupation of the basin) regarded the landscape (also) as a macro-region, that is, a basin limited by the forests surrounding it. Later, with the decreasing mobility of the population, the approach concentrated on county-sized or, maximum, meso-sized territories.

Each of the landscape-forming factors is changing dynamically; thus, no simple borders could be drawn, especially if one considers that the borders of the different factors several times do not coincide. Nevertheless, the landscapes show stability at the lower and middle levels of the hierarchy. It is demonstrated, for example, by the fact that the change in the vegetation in a meso-region (which can be the quickest change, apart from the especially mobile anthropogenic factors) means only a few kilometres' shift along the borders of the core area. Instead of borderlines, some researchers proposed the use of border zones (ecotones), while others do not wish to treat the borders strictly (Csorba 2008).

Landscapes are specific units which are also expressed by their different names (which is usually unique in the hierarchical levels higher than the micro-region). Their uniqueness is especially conspicuous when one compares them to another category, to the landscape type (Pécsi et al. 1972). It is built up by homogeneous ecotopes, and its detailed definition includes the current state of each landscape-forming factors; thus, the definition of a type at the middle hierarchic level could be, e.g., "alluvial fan plain covered by blown sand, cultural steppe with vineyard-orchard and forest mosaic with medium and low groundwater levels". Thus, surface units of the same type fall into one category, independently of their location. The surface described above as an example may occur in Inner Somogy, in the Danube-Tisza Interfluve, or in the Nyírség, and that is why the landscape types show a mosaic-like distribution. On the middle levels of the hierarchy, 14 types were distinguished (Pécsi 1989) (Fig. 6.2 and Appendix B). Recently, the classification was based on landscape characterization, following mainly in the English terminology (Konkolyné 2003).

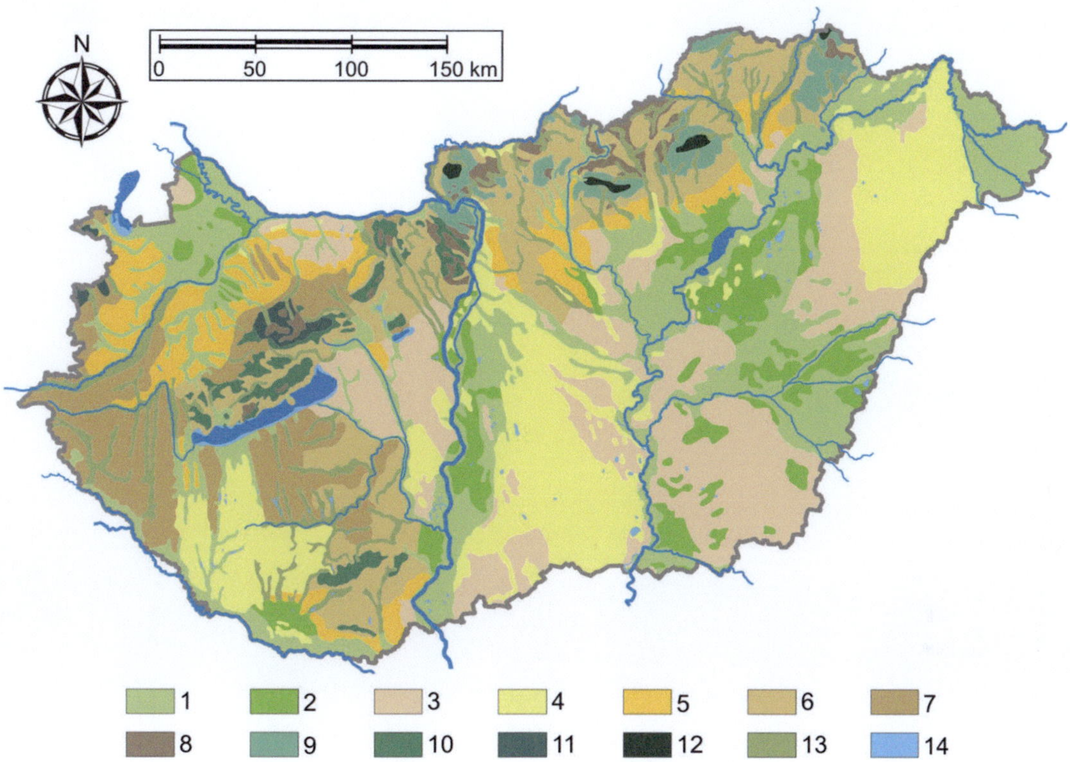


**Fig. 6.2** Landscape types of Hungary (based on MNA 1989). The detailed description of categories is included in Appendix B

## References


Bulla B (1964) Magyarország természeti földrajza. Tankönyvkiadó, Budapest 424

Bulla B, Mendöl T (1947) A Kárpát-medence földrajza. Egyetemi Kiadó, Budapest

Cholnoky J (1926) A földfelszín formáinak ismerete (Morfológia). Királyi Magyar Egyetemi yomda, Budapest

Csorba P (2008) A tájhatárok kijelölése és változása. Földrajzi Közlemények 132(2):220–226

Hajdú-Moharos J, Hevesi A (2000) A Kárpát-Pannon térség természeti tájbeosztása. In: Karátson D (szerk.) Pannon enciklopédia. Kertek, Budapest, pp 274–285

Kiss L (1980) A földrajzi nevek etimológiai szótára. Akadémiai Kiadó, Budapest

Konkolyné Gyuró É (2003) Környezettervezés. Mezőgazda Kiadó, Budapest

Lukács L (2003) Prinz Gyula tájszemléletének és tájrajzi neveinek kartográfiai vonatkozásai. Kézirat, ELTE, Budapest. http://lazarus.elte.hu/hun/digkonyv/lukacs/ll.pdf

Marosi S, Somogyi S (1990) Magyarország kistájainak katasztere. I–II. Magyar Tudományos Akadémia Földrajztudományi Kutatóintézet, Budapest

Mezősi G (2003) A földrajzi táj. In: Borsy Z (szerk.) Általános természetföldrajz. Nemzeti Tankönyvkiadó, Budapest, pp 807–818

MNA(1989) Magyarország Nemzeti Atlasza. Carthographia, Budapest

Pécsi M (1989) Magyarország geomorfológiai térképe. Magyar Nemzeti Atlasz. Kartográfia, Budapest, pp 30–31

Pécsi M, Sárfalvi B (1960) Magyarország földrajza. Akadémiai Kiadó, Budapest

Pécsi M, Somogyi S (1967) Magyarország természeti földrajzi tájai és geomorfológiai körzetei. Földrajzi Értesítő 16:285–302

Pécsi M, Somogyi S, Jakucs P (1972) Magyarország tájtípusai. Földrajzi Értesítő 21(1):1–12

Prinz G (1936) Magyarország tájrajza. Magyar föld, magyar faj, I. kötet. Királyi Magyar Egyetemi Nyomda, Budapest

# Physical Geography of the Great Hungarian Plain

# 7


**Abstract**

The Great Hungarian Plain is the largest macro-region in the Carpathian Basin. In the chapter, the evolutionary models of the region will be described, and also a general picture will be drawn on the evolution of its lithology, morphology, hydrology, climate and vegetation. In a meso-scale, the general descriptions of landscape parameters will be given, typifying the landscapes into two groups: meso-regions formed on alluvial fans and meso-regions formed on floodplains and floodplain levels. Quite detailed analysis is on the most important environmental values and risks of the Great Hungarian Plain. In a separate subchapter, the changes in human impact on the macro-region during the last millennia will be introduced. Detailed studies discuss the descending level of groundwaters, the different ways of mitigating flood hazard and risk, and in connection with these topics, the ways of possible solutions, their weak points and probable consequences will be presented. The river channel regulation and flood protection works in the Tisza River's system are analysed in detail, and also the environmental consequences of these works are presented, as these works caused the greatest environmental change in the Great Hungarian Plain.


## 7.1 Location and the Main Evolutionary Periods of the Macro-Region

It is easy to identify the Great Hungarian Plain as a relief unit than a landscape geographical entity: this plain occupies almost the half of the territory of Hungary, its elevation is under 200 m asl, and it represents a region united by genetics (subsidence, accumulation of alluvial sediments and loess), basin situation, and the resulted climatic, hydrographical, vegetational and pedological characteristics and features. Therefore, it is not always separable from its surroundings. The surface is characterized by two relief types (Marosi 2000): alluvial fans and floodplains (both could be covered by loess). This structure is the consequence of its evolution, which was described in detail earlier (see Sect. 1.3). The development of the present landforms of the Great Hungarian Plain is in connection with tectonic activity (mainly subsidence) and with



G. Mezősi, *The Physical Geography of Hungary*, Geography of the Physical Environment,
DOI 10.1007/978-3-319-45183-1_7

the accumulation of Tertiary and Quaternary sediments, which initially had a marine, then a lacustrine, and since the beginning of the Pliocene, a fluvial origin. Mainly during the Quaternary, these sediments built up alluvial fans and floodplains. The plain may have undergone further, secondary changes caused by the changing intensity of subsidence, aeolian activity and accumulation of loess.

The regions of the Great Hungarian Plain match these main relief forms, but the alluvial fans had slightly different development history, depending on the variations in tectonics, lithology and climate. Among the largest alluvial fans, the Körös-Maros Interfluve is partially covered by loess and has a fluvial character (see Sect. 1.3.3), whereas the foundation of the Bácska Plain is built of loess and sand, which can be separated territorially, the Mezőföld is mainly built of loess, the Nyírség is covered by aeolian sand, while all three sediment types deposited onto the Danube-Tisza Interfluve. Of course, smaller fans also developed, for example Hajdúhát or the Nagykunság, which are also covered partly by loess. The identification of the northern alluvial fans is neither an easy task from a landscape geographical aspect. The other typical relief form in the Great Hungarian Plain is the floodplain, which appears at different levels (e.g. high or low) along the rivers. These forms were usually created by sinking or incision between the alluvial fans; thus, they are relatively young, typically Holocene formations. In the literature, the higher floodplain sections are usually called "terrace No. I.," but the application of the notions of floodplain levels is more appropriate. The elevation difference between the levels is a few metres, and it has serious ecological and economical consequences. The high floodplain is typically a flood-free surface (inundated only exceptional floods), while the low floodplain is inundated by the regular floods; thus, settlements were found often at the edge of the high floodplain. Since the two floodplain areas have different water supplies (water household, groundwater level, which is the main influencing factor of the ecology of floodplain levels), their ecological conditions are different; thus, diverse soil subtypes develop on them (though the former hydromorphic soil formation processes are shifted towards chernozem-like; Marosi 2000). The majority of the lower floodplains are younger: they were formed by incision into the high floodplain, or by the obstruction of fluvio–aeolian forms, which created wetlands (e.g. Ecsedi-láp, Rétköz). The marshes provided protection for the villagers in the times of wars (e.g. Ottoman invasion), for example in the Körös Region or in the Ecsedi-láp.

The borders of the natural factors determining the landscape structure are fuzzy; thus, uniform integrated landscapes cannot clearly be identified. Therefore, the landscape units of the Great Hungarian Plain could not be outlined with a "mathematical precision," even if the Great Hungarian Plain has a distinguished role on the macro-regional scale. In a foreigner's mind, the Hungarian landscape corresponds to the clichéd world of the Great Hungarian Plain: grass steppes, a plain with a mirage in the background, and some associated social-ethnographic specialities. All these describe well the character of the Great Hungarian Plain as a macro-region. However, to determine and assess the units of the Great Hungarian Plain is not easy, because though the geographical landscapes are separated on a natural basis, they cannot be precisely defined and the distinction between some landscape units of different levels become uncertain. The basis of division is the distribution of natural factors; therefore, the Mezőföld or the Dráva Plain belongs to the Great Hungarian Plain instead of Transdanubia (Fig. 7.1). Their relief, genetic, climatic and vegetational features match the typical values of the landscape factors of the Great Hungarian Plain. A different approach to landscapes would have different results in the division.

In the present book, a summary at the macro-regional level will be provided; however, in some cases, meso-regional levels will be discussed to, because of the mosaic character of the Great Hungarian Plain.

The basement of the Great Hungarian Plain is composed of the mosaic-like fragments of strongly eroded mountain ranges and grabens (see Sects. 1.1–1.3). During the Palaeogene (Eocene,

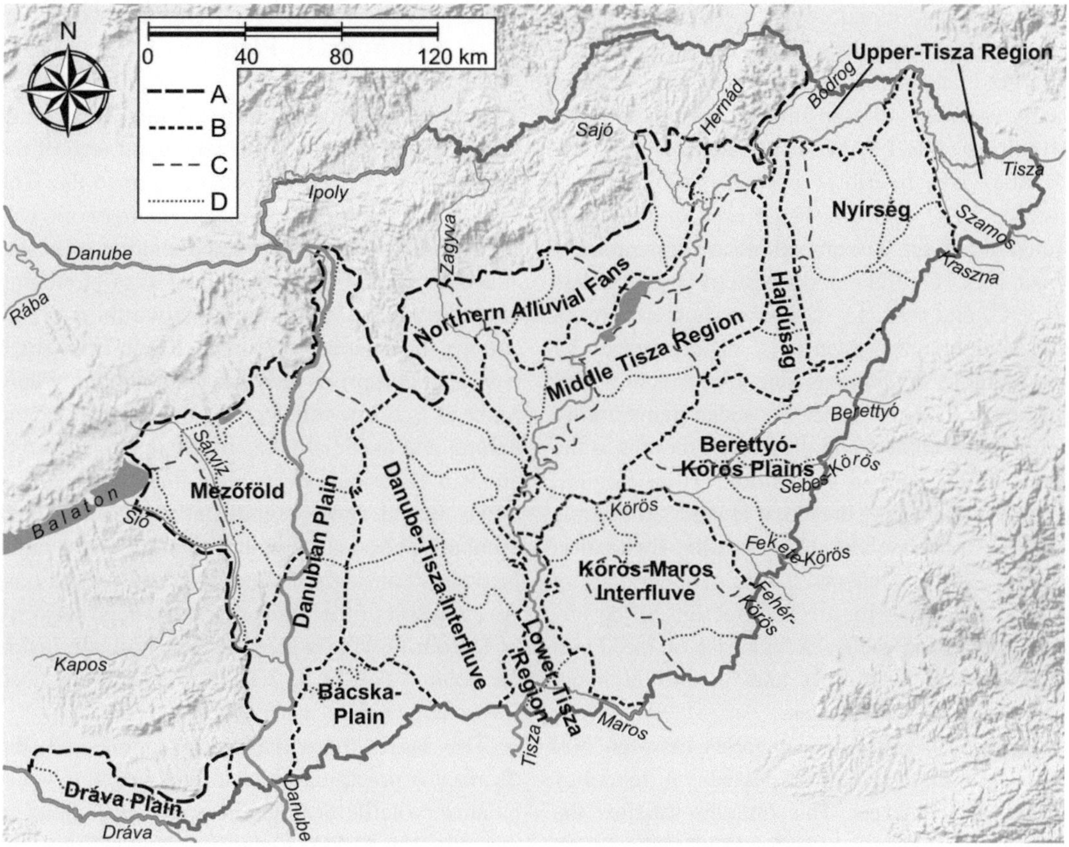


**Fig. 7.1** Division of the Great Hungarian Plain macro-region (after MNA 1989). *A* Macro-regional border, *B* meso-regional border, *C* micro-region group border, *D* micro-region border

Oligocene) and the Miocene, this basement relief used to be terrestrial surface, and later it had an archipelagic character. The subsidence of the entire Great Hungarian Plain started in the Middle Miocene, being the most intensive in the Pannonian Period. The sea in the basin gradually receded due to crust movements and cyclic aridification, and the rivers arriving from the surrounding mountains filled up the basin with their sediment at an accelerating rate. The ongoing subsidence has an uneven rate, consequently, several hundred, or in certain places, a thousand metres' thick alluvial and terrestrial sediment became deposited onto the marine layers.

The micro-regional units of the macro-region can be classified into the groups considering their climatic, hydrographic, vegetational and soil properties, besides the meso-regions (like the Mezőföld or Nyírség) are in harmony with the archaeologic, ethnographic and other social practice.

### 7.1.1 The Climatic Conditions of the Great Hungarian Plain

The general character of the climate of the Great Hungarian Plain is determined by its situation on the globe, its distance from the seas and action centres, while its special characters originate from its basin situation. The foehn character and the basin location result in smaller cloud cover, higher number of sunny hours and dryness (less precipitation) accompanied by extreme temperature fluctuations.

The average cloud cover is 50–55 %, lower values characterize the Middle Tisza Region, the Mezőföld and some parts of the Nyírség. The yearly amount of sunshine duration is over 2100 h in the Lower Tisza Region and the Danube-Tisza Interfluve (Péczely 2002), but it exceeds 1950 h in the whole area of the Great Hungarian Plain. The annual mean temperature is about 10.5 °C (in the north-eastern part, it is 10–10.5 °C, and over 11 °C in the southern parts). The temperature pattern is characterized by considerable fluctuations due to its continental character. The difference in mean temperature between the coldest and the hottest months is the greatest (24.5 °C) in the Middle Tisza Region, while the average of the Great Hungarian Plain is 23 °C; the absolute temperature fluctuation exceeds 70 °C (Justyák 2002). The prevailing wind is modest, with a NW direction in the largest part of the plain, except east of the Tisza River (called Tiszántúl) and in the Nyírség, where NE wind dominates.

The yearly precipitation varies between 500 and 650 mm, showing a decreasing tendency during the centuries. This amount satisfies the water requirements at a moderate level; however, in the central and southern areas, a water shortage occurs in summer. In accordance with the basin situation, the central part of the Great Hungarian Plain is the driest: here, the amount of precipitation sometimes does not even reach 500 mm. The most humid month is June (60–80 mm) and the driest ones are January and February (25–40 mm), and there may be a secondary maximum in late autumn. The precipitation conditions have changed in the past 100 years in various ways:

- The amount of precipitation has decreased by about 60–100 mm (although the significance of this change has not been proven yet; Kertész and Mika 1999);
- The temporal pattern of the precipitation shifted, as the winter period is becoming wetter;
- The spatial distribution and the forms have become more extreme (OMSZ 2003).

### 7.1.2 Hydrography of the Great Hungarian Plain

The overwhelming majority of the Great Hungarian Plain belongs to the catchment area of the Tisza River, and a much smaller part to the. The evolution of the present-day river network was influenced by the differentiated sinking of basins and the development of alluvial fans stretching into the basin. Approximately, two-third of the country lies between 75 and 200 m asl, which basically comprises the Great Hungarian Plain. Since it is a low-relief plain, the number of the natural and artificial lakes of the macro-region is about a thousand, even after the water regulations, which have significantly reduced their number. The area of natural waters and marshlands has considerably decreased by regulations. As the result of the basin characteristics, majority of the out-flowing waters do not originate inside the country (see Sect. 3.1), highlighting the need for an appropriate water management.

The large (potential) evaporation and the decreasing precipitation result in a negative water balance, volatile in time and space, especially in the central part of the Great Hungarian Plain. This state may have anthropogenic causes too, for example the altered surface cover, or during land-use planning water management was not considered. Due to the above-mentioned factors, this disadvantageous state does not only concern the surface waters, but also the subsurface waters and the drinking water stored in the alluvial fans. The rivers have a significant impact on these water bodies, both from qualitative and quantitative points of view. The richness of the Great Hungarian Plain in thermal waters is also the consequence of the basin character (see Sect. 3.2).

### 7.1.3 Soils and Vegetation of the Great Hungarian Plain

The chernozem soils on the Great Hungarian Plain evolved under a climatic influence, and they usually develop on loess or loess-like sediments (e.g. in Mezőföld, Bácska Plain, Hajdúhát, Körös–Maros Interfluve), but this soil type also appear on sandy loess. The different types of brown forest soils are also common in sand-covered areas. The other main group of soils is not zonal, and their character is determined by the water household of a soil (e.g. peat and meadow soils) or by the lithology (e.g. sandy soils of the Danube-Tisza Interfluve or the Nyírség). The different orographic situations also influence the development of different soil types. Since the deposition of loess and blown sand terminated at the end of the Pleistocene, the meadow and forest soils of the Great Hungarian Plain could be dated to the Holocene (for example chernozems develop on warm and dry climate, which appeared just in the Boreal Phase). The formation of meadow and alluvial soils requires shorter time, just a few thousand years. Different salt-affected soils are also common: the solonchak soils with a near-surface salt concentration occur mostly in Kiskunság, and the solonetz soils characterized by sodium accumulation in the deeper layers are typical in Hortobágy. The anthropogenic impact on waters (see Chap. 4) triggers steppe formation on one hand (if the area has become drier) and meadow formation on the other hand (Várallyay 2001). The mosaic-like picture may be further diversified by active deflation and erosion (Stefanovits 1981).

The natural wildlife of the Great Hungarian Plain belongs to the forest steppe zone characterized by low forest cover. It is diverse, since the factors influencing the vegetation type change quickly within small areas: just consider the patch-like appearance of different soil types (e.g. salt-affected, sand and alluvial soils), or the differences in water supply, sometimes caused by a few decimetres' elevation difference (see Chap. 5). However, nowadays only some patches exist of the natural herbaceous vegetation (e.g. steppe formed on loess and sand), which are used to cover the largest area and of the forests (oak forests, or the sand oak forests as climax communities), since human activity created settlements, plough lands and hayfields, etc. during the last several thousand years. The remnants of the natural vegetation have also transformed, as they became poorer in species, and new (invasive) species invaded it. Along the rivers, the most diverse associations are the oxbow lakes and marshes, while the sand and gravel bars are at the initial levels of the succession. The riparian forests beside the rivers usually consist of willow and poplar species, and dense liana climbs these softwood trees. The bogs represent a special value, thus the peat moss of Csaroda or the floating bog of Kállósemjén are all protected areas. The patches of grasslands developed on sand are the most typical in Kiskunság, while the largest saline grassland developed in Hortobágy. The oak forests steppe—the westernmost occurrences of forest steppe in Europe—have survived just in tiny patches (e.g. Pusztavacs, Csévharaszt), no anything of similar exists outside of the Carpathian Basin; thus, all of them are strictly protected. Unfortunately, the majority of the forests of the Great Hungarian Plain are planted; thus, these forests are poor in species and rarely represent a real botanical or zoological value.

## 7.2 Overview on the Evolutionary Types of the Regions of the Great Hungarian Plain and Their Physical Geographical Characteristics

The landscape geographical division of the Great Hungarian Plain should be made on evolutionary basics (alluvial fan or floodplain), as far as the development and the structure of landscapes influence their characteristics. The other landscape forming factors (e.g. pedology, climate and vegetation) complement the division only to slight extent, while human factors are able to transform a landscape essentially (particularly on a micro-regional scale).

(A) *Meso-regions formed on alluvial fans*

The alluvial fan of the **Danube-Tisza Interfluve** stretches from the Gödöllő Hills to the Bácska Plain, being the largest of the kind (7400 km$^2$; Lóki 2000) and constituting a meso-region (frequently called Kiskunság). The slightly undulating surface is mostly covered

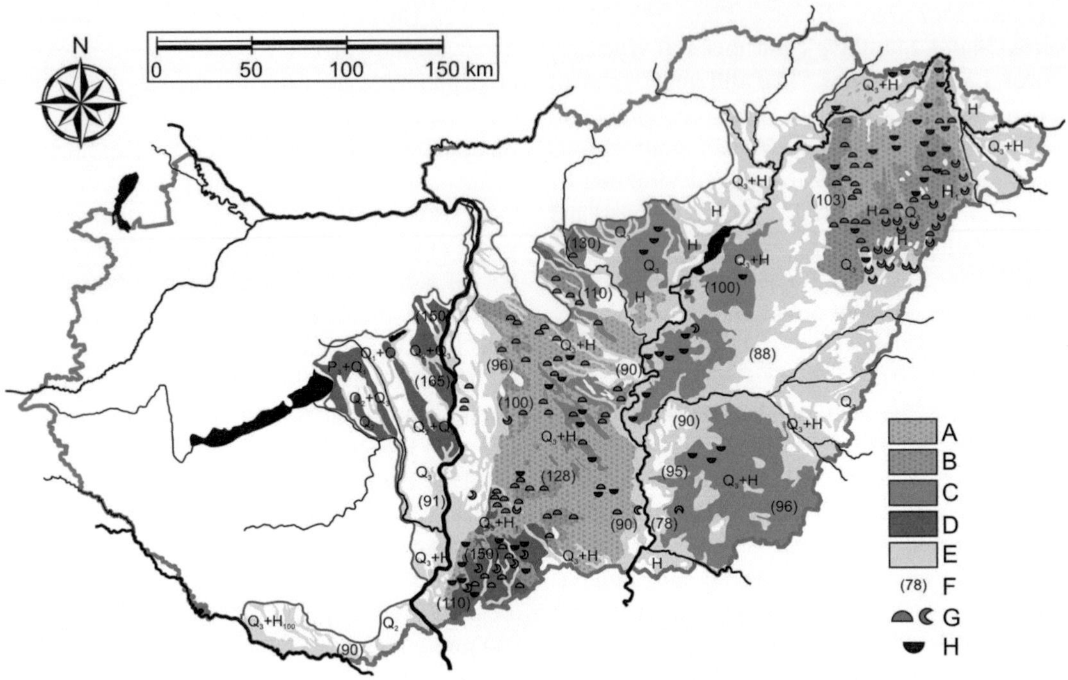


**Fig. 7.2** Main landscape elements of the Great Hungarian Plain (after MNA 1989). *A* Blown sand plain, *B* plain with blown sand dunes, *C* loessy, alluvial fan plain and loess-covered high floodplains, *D* dissected loess plain, *E* low floodplain and plain with impeded drainage, *F* typical height, *G* parabolic and elongated dunes, *H* blowouts and hummocks

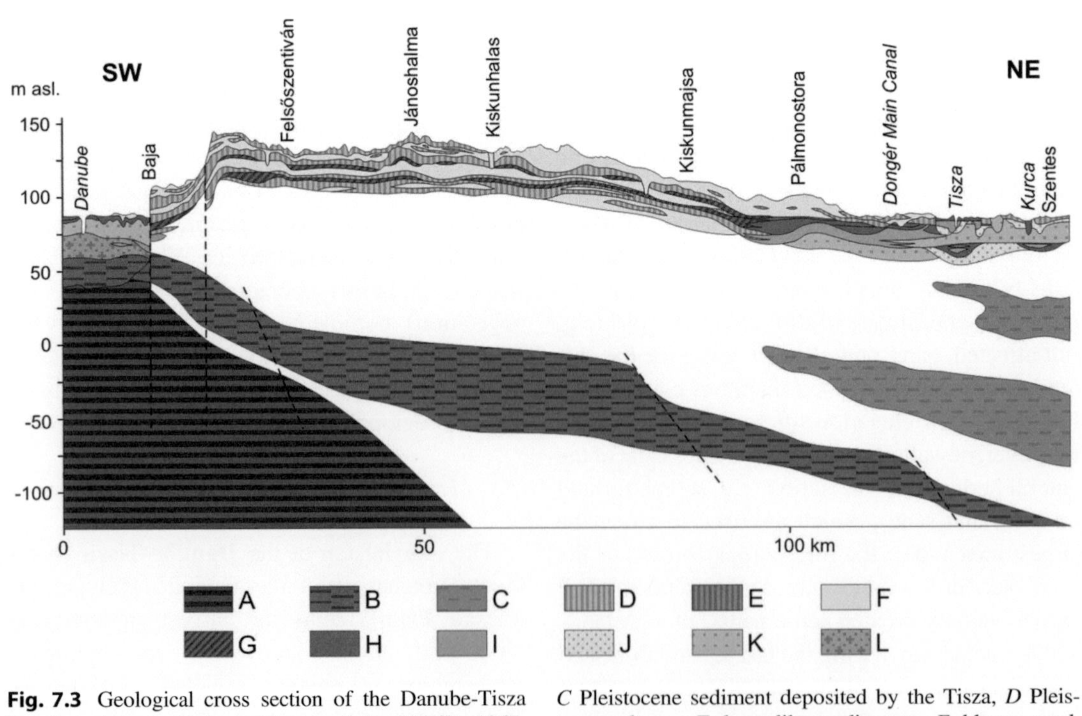


**Fig. 7.3** Geological cross section of the Danube-Tisza Interfluve between Baja and Szentes (after Mihâltz 1967; Molnár 1977). *A* Upper Pannonian marine sediment, *B* Pliocene sediment transported by the Danube, *C* Pleistocene sediment deposited by the Tisza, *D* Pleistocene loess, *E* loess-like sediments, *F* blown sand, *G* palaeosols, *H* alluvial sediments, *I* aleurite, *J* fine sand, *K* medium-sized grains, *L* coarse sand

by sand sheets and sand dunes, and partly with loessy sediments. This meso-region has sharp relief border, as its hourglass-shaped NW-SE central axis elevates 30–40 m above the floodplains of the Tisza and the Danube (Fig. 7.2). Even though this is typical for alluvial fans, there have been geomorphological debates on the genetics of Kiskunság for a century. The first problematic issue was the age of the Kiskunság. According to Cholnoky's (1910) monoglacial theory, the Danube used to have a steady N-S flow direction, similar to that of today, and the Danube-Tisza Interfluve (and the whole Great Hungarian Plain) used to be covered by loess in the Ice Age, which was later covered by blown sand deflated from the Danube valley by NW winds. Several scientists shared this evolutionary model based on cross-sectional sediment profiles (Mihá ltz 1967; Molnár 1977), but they dated the drying of the alluvial fan to the Riss Glacial. In addition, they hypothesized the existence of a 100- to 120-m-thick sand formation in the central part of the unit (Fig. 7.3).

Sümeghy (1944) was the first to claim that the region morphologically is an alluvial fan, making an assumption on the NW–SE flow direction of the Danube. Later Bulla (1964) proved the pertinence of this theory by some sediment profiles. Loess and blown sand layers deposited onto the alluvial sediment of several hundred metres thickness. The flow direction changes of the Danube in Danube-Tisza Interfluve caused that the alluvial sediment were found in greatly differing depths (typically 3–70 m; Borsy et al. 1981). The surface of the whole Danube-Tisza Interfluve had become dry approximately 30–40 thousand years ago, when the Danube shifted from it, since that time loess and blown sand layers cover it, being intercalated in certain places.

The mostly stabilized dune series are typical of the Danube-Tisza Interfluve, formed by NW winds (Image 7.1). According to Pécsi (1996), the

**Image 7.1** The sand dune area at Bugac (*source* http://knp.nemzetipark.gov.hu/index.php?pg=menu_1426)

material of the blown sand dunes and sand sheets originated from the sediments of the Danube, and the sand was transported by north-westerly winds during the Quaternary. However, this does not provide an explanation for the deeper stratification, and sand sorting (refinement) along NW-SE transects. Nowadays, still some active blown sand patches could be found. Some NW–SE-oriented wide strips of loess cover are still present on the surface of the Danube-Tisza Interfluve, like in the Kiskunság Loessy Plain micro-region.

In summary, the Danubian fluvial material, the blown sand and the loess sediments can be found in highly different thicknesses in the near-surface layers; thus, the blown sand and loess cover alternate. However, the main character of the Danube-Tisza Interfluve is determined by fixed blown sand forms.

On the sand surfaces hummocks (and scattered hairpin-like parabolic dunes), deflational flats and blowouts are predominant forms, which were formed by NW wind. The fixed blown sand areas can be found in patches and are elevated from their surroundings (e.g. Bugac or Fülöpháza are both protected areas). The shallow, basin-like depressions (called "semlyék") among the sand dunes represent special geomorphological units. Water could accumulate in them, which may become oversaturated in summer, when the evaporation is very intensive; thus, carbonate crust or, more exactly, dolomite can form in this way at the bottom of the salty lakes (Image 7.2). The carbonate layer hinders cultivation and increases the hardness of the groundwater, which is disadvantageous, and it may even cause inland excess water damage in humid periods (Molnár 1977). On the other hand, this rock was the only natural, solid building material in the region (its protected exploration is at Csólyospálos); thus, it constituted a significant value in Kiskunság (which was densely populated by the Cumanians in Medieval Times).

The **Bácska Plain** is located south-west of the Kiskunság, and just its minor part is located in Hungary. Its geological evolution is the same as that of Kiskunság. The alluvial fan was built by streams arriving from the Mecsek Mountains

**Image 7.2** Saline lake in the Kiskunság (Böddi-szék) (photograph by Kovács, F.)

(Pécsi 1996) or the Paleo-Sárvíz (Lóki 2000). Its northern part is covered by active blown sand forms (the highest is at Ólom-hegy, 174 m asl). The territory of Illancs is a patch-like sand area, which is similar to the ones described in the Kiskunság. The fluvial sediments were covered by blown sand, which was arranged into NW-SE-oriented hummock series, and the present-day river network of the region also follows this direction (e.g. Kígyós-ér). The southern part of the meso-region is covered by a few metres' thick loess sheet, which can be regarded as the continuation of South Mezőföld. The connection between the Bácska Plain and the Transdanubian areas to the west terminated when the Danube appeared in the region with the present-day N-E flow direction.

The **Nyírség** is the most similar meso-region to the Danube-Tisza Interfluve. It is also alluvial fan-based, covered by blown sand and characterized by erosional and accumulational aeolian landforms. It does not have an active large river, on its surface blown sand skeletal soils and brown forest soils (they are carbonated in the Danube-Tisza Interfluve and acidic with iron pans in the Nyírség) were formed, and it was covered by forest-steppe communities. Both meso-regions have dry character, but the water shortage is much smaller in the Nyírség, since the greatest amount of precipitation due to the orographic influence of the mountains at the edge of the basin.

The sandy material of the alluvial fan (in 150- to 300-m thickness) accumulated in the Quaternary by the rivers belonging to the water system of the Tisza–Szamos and Bodrog (it composed of Tapoly, Ondava, Laborc, Ung and Latorca rivers). The former one built the southern part of the Nyírség, while the latter mainly constructed its central part, and their common erosion base was the Körös Basin. Their former, N-S flow directions are indicated by present-day wet valleys (so-called nyírvízlapos). The independent relief evolved 16–20 thousand years ago, when the region elevated slightly, while its environment sank (e.g. Upper Tisza Region). As a result, due to an avulsion, the Tisza turned north and flowed around the Nyírség, attaching all the rivers previously flowed through the alluvial fan. This explains the flow direction changes of the Tisza and Szamos rivers described in Sect. 1.3 (Borsy 1989; Lóki 2000; Gábris and Nádor 2007).

In the western part of the Nyírség, the sand dunes are covered by a thin loess sheet, forming a transitional zone towards the Hajdúhát, which is covered by an extensive loess cover. Towards east, the surface sediments gradually become blown sand. On the north-south divide is located

**Fig. 7.4** Parabolic dunes with an undeveloped western wing, and valley-marginal dunes reaching 1–2-km length and 5–12 m height at Nyírkáta (after Marosi and Szilárd 1969)

the highest point of the Nyírség and of the Great Hungarian Plain (Hoportyó, 183 asl). The eastern and north-eastern parts of the alluvial fan are the oldest, and since they were the first to loose their fluvial environment, and the surface is also the most diverse here, as this region is characterized by blowouts, deflational flats, hummocks as high as 20 m. Special aeolian forms of the south-east Nyírség (Fig. 7.4) are the parabolic dunes and the valley-marginal sand dunes (along the wet valleys; Borsy 1961, 1977).

The central part of the Nyírség also covered by blown sand just in the Danube-Tisza Interfluve, and peat bogs were formed in the interdune areas and in the valleys characterized by impeded drainage among the NE-SW-oriented dune ranges. These could preserve the natural vegetation of the Great Hungarian Plain (Lóki 2000). The swamp of Bátorliget has become a nature reserve. It was formed in a wet blowout protected by birch, and later by oak forests. The area is characterized by special micro-climate, which made it possible that one of the richest and most diverse Late Glacial and Holocene plant and animal communities of the Carpathian Basin could survive (Sümegi 2005).

**The Körös-Maros Interfluve** is an alluvial fan formed in the south-east part of the Great Hungarian Plain, in a low orographic situation. Its surface was covered mainly by Quaternary and Early Holocene fluvial sediments. It is only slightly higher than the present-day floodplain levels of the rivers. On the fluvial sand and gravely sand sediments, a thin layer of floodplain loess (infusion loess) was deposited referring to humid surfaces, hiding some previous fluvial forms of the alluvial fan. The surface of the flat alluvial fan is diversified by the remnants of palaeo-channels of former meandering or braided rivers, and along the former channels the fragments of other fluvial or fluvio-aeolian forms could be identified (Fig. 7.5). Before the river regulations, a rich water network facilitated the water traffic of shallops between the settlements (e.g. between Hódmezővásárhely and Békéscsaba).

The Maros mostly drained its water into the Körös Basin, though sometimes it run other directions, as it is indicated by Würmian terraces along the Temes and Béga rivers. (This drainage direction could function nowadays too, and it was used during the devastating floods of the 2000s). The sinking of the South Tisza Valley and the avulsions on the alluvial fan resulted in frequent channels shift of the Maros across the alluvial fan, and ca. 2000 years ago into its present flow direction. The almost totally straightened channel is the result of river regulations. The Száraz-ér (e.g. Andó 1971, 2002) is a remnant of an Early and Middle Holocene palaeo-channel. The sandy (gravelly in the SE) alluvial deposits are located near to the surface, the groundwater is high, and the inland excess waters are abundant in connection to the alluvial fan morphology; thus, the uppermost clayey and sandy loess layer has salinized in smaller or larger patches. Fertile chernozem and meadow chernozem soils developed on the loess cover.

The relief of the large **Mezőföld** fits into the Great Hungarian Plain. This meso-region is a series of mostly NW-SE-oriented, Pleistocene sandy alluvial fans arranged along strips, and a series of lowland plains (evolved on alluvial fans) covered by thick loess, which were deeply dissected by tectonically preformed streams and derasional valleys. The alluvium deposited onto the sandy–clayey sediments of the Pannonian Lake; this sediment sequence of the lake regression is well observable in the north-east shore of Lake Balaton (in the vicinity of Akarattya and Kenese) in the strata of a high escarpment wall. The main structural lineaments referring to intensive tectonic movements have been known for a long time, and they are well marked by the flow directions of Szent László-víz, Váli-víz and Sárvíz (Ádám et al. 1959). Since the Upper Pliocene, the fluvial layers (100–200 m) of the alluvial fan began to deposit onto the lacustrine sediments. Later, the surface was covered by loess in different thicknesses: on the western part, it is thinner (intercalated by sandy strata tracts), and it is getting thicker towards south and east. The hilly areas in north covered by loess may not be regarded as a plain, if we consider its height (230–250 m). (These blocks are traditionally—and from a morphological point of view incorrectly—called "tables," e.g. Ercsi-tábla or Érdi-tábla; Fig. 7.6).

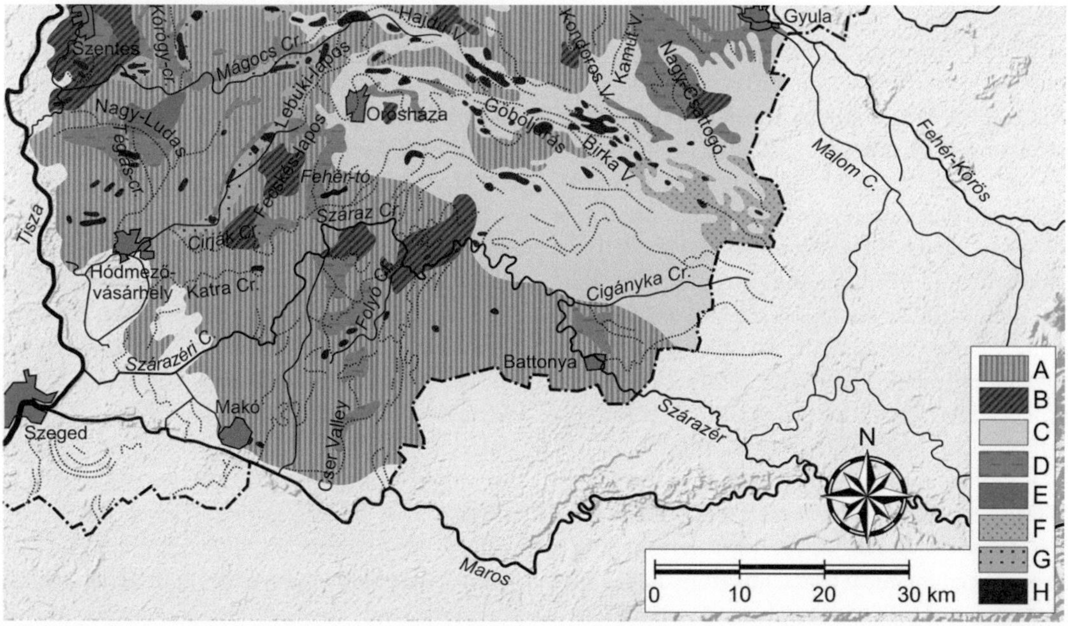


**Fig. 7.5** Surface formations of the Körös-Maros Interfluve (after Andó 1971, 2002; Gazdag 1964). *A* Pleistocene plain with a thick loess cover, *B* Early Holocene lacustrine areas covered by clay and loess, *C* central part of the alluvial fan of the Maros, with a loess cover, *D* salt-affected lacustrine and fluvial sediments covered by sandy-loessy mud, *E* surface covered by clay, *F* surface covered by fluvial sand, *G* subsurface blown sand layer, *H* sand dunes

**Fig. 7.6** Elevated loessy surfaces at Érd and Ercsi (after Ádám 1983, in: Pécsi 1967). *EV* Erosional valley, *S* subsiding area, *TVB* valley basin at Tárnok, *SPB* Pannonian block at Sóskút, *EAF* alluvial fan at Ercsi

Similar forms built of 10- to 60-m-thick loess could be found in the Duna-Sárvíz Interfluve (e.g. loess ridges of Pentele and Sárbogárd). Along the Danube, the Mezőföld has a well-marked border, an erosional escarpment (because of the lateral erosion of the Danube). Along this high escarpment, landslides cause a continuous hazard (e.g. in 1970 at Dunaföldvár, and in 2009–2010 at Dunaszekcső). Though the Mezőföld is 200–300 m high, it still belongs to the Great Hungarian Plain, partly because it is still covered by lowland-character loess, and partly because its hydrologic features, as the run-off is below the hydrological limit of plains (2.0 l/s/km$^2$).

Mezőföld is also the largest chernozem region of Hungary owing to the typical loess cover; thus, it is a region of one of the most advantageous physical geographical characters from the point of view of cereal production. The surface landforms are related to loess, e.g. derasional valleys, incised roads loess dolines and loess wells

**Hajdúság** is a low plain covered by loess, situated between the larger alluvial fans of the meso-regions of Hortobágy and Nyírség. (Hajdúság used to be more elevated than the latter two in the Pannonian; thus, it functioned as a geological barrier and was still not covered by fluvial sediments in the first part of the Quaternary.) The northern part of the Hajdúság is the Hajdúhát, where the blown sand landforms (blowouts, hummocks, residual ridges and large deflational flats) are covered by loess (to the north of Hajúvid and Hajdúböszörmény). Due to the loess cover, there was no blown sand movement in the Holocene. The southern part of the Hajdúság is higher by 20–30 m, and it resembles to a depositional plain, as it was covered by loess in 3–4 m thickness, and it is just slightly dissected by erosional forms (e.g. palaeo-channel of the Tisza, incised valley of Kösely or loess wells. The building material of the region contains sand towards the Nyírség. The alluvial fan of the Hajdúság is mainly built by rivers of the Bodrog, though rivers arriving from the direction of Érmellék also contributed in the development of its southern part. Owing to the loess cover chernozem is the typical soil type of the meso-region, which provides exceptionally good pedological conditions for cultivation; therefore, most of the landscape is plough land. The Keleti-főcsatorna (Eastern Main Canal) connects the Tisza and the Berettyó rivers supplying irrigation water for the areas of Hajdúság and Hortobágy during the dry summer months.

The northern edge of the Great Hungarian Plain is composed of a series of Quaternary alluvial fans of smaller rivers (e.g. Tápió, Galga, Zagyva, Tarna, Gyöngyös, Laskó, Eger, Sajó and Hernád Rivers). This constitutes the meso-region of **Northern Alluvial Fans**. The ongoing subsidence of the Great Hungarian Plain during the Holocene caused that these rivers dissected into the slightly sloping surface of the alluvial fans, creating low, flood-free ridges along the incised valleys. The surface of the meso-region has a sunny southerly exposure, and it is gently declining and covered by chernozem brown forest soil, creating excellent conditions for the cultivation of sugar beet, water melon and different kinds of grapes.

(B) *Meso-regions formed on floodplains and floodplain levels*

These landscape units formed on areas subsiding since the Quaternary, and perfect plain surfaces created by the rivers. The high floodplains, inundated by the highest floods, and the incised low floodplains involve tens of thousands of km$^2$ in the Carpathian Basin. The long-term inundations turned huge territories into marshland; thus, their flood relief necessitated the construction of a artificial levee system for flood prevention, and as a result, most of the territory became cultivated land. These floodplain units can mostly be connected to the Danube and the Tisza, and their classification also follows this division.

The **Danube Plain** is located along the Danube in a 20- to 30-km-wide strip stretching from the Pest Plain to the southern border of Hungary. Its relief can be clearly distinguished from the neighbouring areas. The surface of this series of plains and depressions was formed by the fluvial activity of the Danube. The fluvial evolution of the meso-region started when the Danube appeared here after an avulsion ant took its N–S flow direction. The channel shift was the result of the slip of the river from the alluvial fan of the Danube-Tisza Interfluve, and the renewed activity of the depressions of Baja and Kalocsa ca. 30–40 thousand years ago. The evolution of the region was determined by the erosional activity of the Danube, and it was influenced by NW–SE-oriented faults, along which earthquakes still happen (e.g. in 1956 at Dunaharaszti). The study of earthquakes is an important issue in connection to the safety and the enlargement of the Paks Nuclear Power Plant (Schweitzer 2005). According to Pécsi (1996), the N–S flow direction of the Danube is much younger, based on graves from the Avar Age found in the meadow soils on the floodplain near to Kalocsa.

Before the flood management works in the mid-nineteenth century the area used to be a marshland, characterized by natural cut-offs (e.g. oxbow lake: Lake Szelidi), fluvio-aeolian dunes and enclosed basins. These landforms indicate that the former courses of the Danube extended to the east, until the rim of Kiskunság, and the Danube thoroughly reworked the surface of the plain in 10–20 m depth; therefore, the plain is covered by Late Würm and Holocene sediments. The significant fluvial erosion spared the buttes at Solt and Tétel, which have a Pannonian basement and elevate 15–25 m above their environment (Fig. 7.7).

The Csepel Plain and the Solt Plain micro-regions of the Danube Plain are characterized by a dense river network, which received particular folkloric names referring to their important economic roles. For example the "fok" (crevasse) was a crevasse bed joining the Danube at mean water level, while the "ér" (brook) joined at flood levels, whereas the "turján" and the "őrjeg" refer to filled-up riverbeds that became marshlands (Strömpl 1930). The usage of the floodplain in the early Medieval times (Árpáds Age) is based on tributaries and crevasses joining the Danube (Andrásfalvy 1975). The enclosed basins and oxbow lakes farther from the Danube bed became marshlands during the cool Atlantic Phase, and a significant amount of peat accumulated in them, e.g. between Kecel and Császártöltés along the escarpment wall of Kiskunság (Image 7.3). As the result of river engineering and floodplain protection works almost all elements of the above-mentioned water system, the former Danube meanders and oxbows, and the salt-affected depressions lost their water supply and became dry. The area of the former wetlands was occupied by wet hayfields, and on the higher floodplain levels by plough lands. The particular floodplain conditions have been preserved just in the area of Sárköz and Gemenc.

The northernmost part of the Danube Plain meso-region is the Pest Plain micro-region

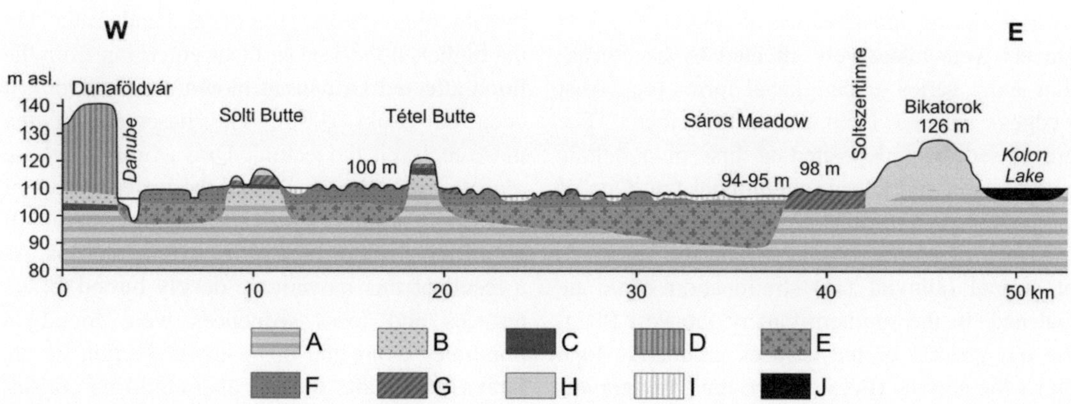


**Fig. 7.7** Cross-sectional profile of the Solt Plain (after Pécsi 2001). *A* Pannonian clay, *B* Pannonian sand, *C* red clay (Pliocene–Miocene), *D* loess with 3-4 palaeosols, *E* Danubian sandy gravel, *F* fluvial sand with small pebbles, *G* sandy loess and loessy sand, *H* aeolian sand, *I* alluvial sand, sandy loess, floodplain loess, salt-affected loess and clay in the basins and palaeo-channels, *J* Lake Kolon with peat

**Image 7.3** Peat filled the palaeo-channel at Császártöltés, Vörös-mocsár (photograph by Kiss, T.)

group, whose evolution could be explained in several ways. The erosional activity and the deposition of gravel sediment by the Danube arriving in the Great Hungarian Plain from the Visegrád Strait, created an alluvial fan which 4–5 terraces during the Quaternary (Fig. 1.29). The gravels were intensively affected by cryoturbation and a series of periglacial forms (e.g. frost wedges) can be identified within them. The gravel sediment deposited at first in a deltaic environment and later in an alluvial fan. One of the research questions is how the sediment sequence of terrace formation and the sequence of normal (alluvial fan) stratification could be matched. In the southern part of the Pest Plain, the old gravels of the terraces gradually slope under the surface (Pécsi 1959); thus, the gravels of terraces No. III and IV could be found several hundred metres below the surface (today it is more of a theoretical correspondence). Thus, the southern border of the Pest Plain is the inflectional zone where the terraces subside under the surface, where the terrace sequence turns into normal stratification.

The Hungarian part of the **Dráva–Danube Plain** is part of the Danube Plain. The region has two floodplain levels built of sand and gravel. On the higher, flood-free surfaces emerging from the flood-affected Ormánság blown sand movement occurred (Lóki 2000). The basement of the meso-region is the tectonic Dráva Graben, whose repeated subsidence affected the water system of a larger area in the Würm (e.g. the first capture of Zala River) and created a series of sub-basins. As a result of this movement, deeply buried Dráva terraces and loess sequences were found in boreholes along the downstream section of the Dráva floodplain; thus, it also could be considered as a lowland alluvial plain (Pécsi 1996; Lovász 2006).

The largest part of the floodplains in the Great Hungarian Plain belongs to the river-system of the Tisza. These riparian landscapes could be classified into three meso-regions, although the units do not separate from their environment as sharply as the Danube Plain. Before the nineteenth-century river regulations, the floods of the Tisza affected large areas, as the river had quite shallow channel; thus, the frequent floods almost turned its floodplain into a marshland. In the middle of the basin, the Tisza meandered in a 20- to 40-km-wide band, accumulating large amounts of mud, clay and sand; thus, the river continuously filled its floodplain and heightened the nearby areas. After the recession of floods, stagnant waters remained outside the heightened meander banks, and since the water cover changed even by a few centimetres' elevation difference, a diverse wildlife developed. Marshlands and bogs, riparian wetland forests, willow and poplar groves inhabited the landscape.

It is difficult to compose an integrated description of the **Upper Tisza Region**, as its Hungarian parts are not connected to each other, and they include only the smaller part of the meso-region, as it is mainly located in Ukrainian, Rumanian and Slovakian territories. The meso-region is covered by Holocene alluvial sediments arranged into two floodplain levels. The geomorphologic basement of the region is built of alluvial fans deposited by streams arriving from the Carpathian Mountains. This basement was segmented by fault lines; thus, the units subsided by different rates at the end of the Würm; thus, the meso-region became separated from the elevating Nyírség (Borsy 1989). The water system of the Tisza occupied the territory just about 16–20 thousand years ago, and since the low-lying and still sinking Szatmár and Bereg Plains filled up by the alluvium of the Tisza, the Szamos and its tributaries. Before the river regulations, these rivers changed their beds frequently on the perfect plains and created the surface forms seen still today: cut-offs, oxbow lakes, abandoned channels and natural levees. Levees are parallel to the river, and behind them, in the deeper-lying basins swamps developed (e.g. Ecsedi-láp) along the eastern edge of the Nyírség, though nowadays they are almost disappeared due to artificial draining; Lóki 2000). Unique forms in this landscape are the andesitic Miocene volcano ruins of Tipet-hegy at Barabás and Nagy-hegy at Tarpa, which are covered by loess.

The small Rétköz is situated north of the Nyírség. It was the floodplain of the Tisza covered by peat patches. The majority of its surface has been meadow and pasture since the water regulations. A similarly flat plain is the Bodrogköz formed by the Tisza and Bodrog rivers. Its evolution is connected to the fact that the rivers composing the Bodrog could not join the Tisza, so they got arranged into a new river course parallel to the Tisza, and their common confluence developed at Tokaj. The central part of the Bodrogköz was elevated and thus a drier surface; therefore, blown sand forms developed during the Pleistocene, though these floodplain islands were later partly destroyed by the lateral erosion of rivers (Lóki and Félegyházi 2007). The Tisza changed its flow direction in the territory of Bodrogköz, as it is indicated by the Karcsa Stream, which may have been a secondary channel of the Tisza (Fig. 7.8).

The Quaternary geological basement of the **Middle Tisza Region** is composed of alluvial sediments deposited by streams arriving from the North Hungarian Mountains. This evolutionary phase ended, when the Paleo-Bodrog and then Tisza appeared in the area about 15–20 thousand years ago, when these large rivers captured the former streams originally drained to the Körös Basin. The Tisza occupied the eastern part of the meso-region, and it shifted into its present location gradually, as a result of subsidences, but in the meantime, it reworked the surface by its huge meanders (e.g. Kadarcs, Üllő-Oktalan-lapos) to the Hajdúság (Borsy 1995; Lóki 2000; Pécsi 1996). South of the meso-region, the Nagykunság is located, characterized by a young loess sheets; however, on smaller surfaces, blown sand can also be found.

From the point of view of nature conservation, the most valuable area of the region is undoubtedly the Hortobágy, which is a low-lying, salt-affected, treeless, grassy steppe, lying at the same height above sea level as the floodplain of the Tisza. (This steppe has become

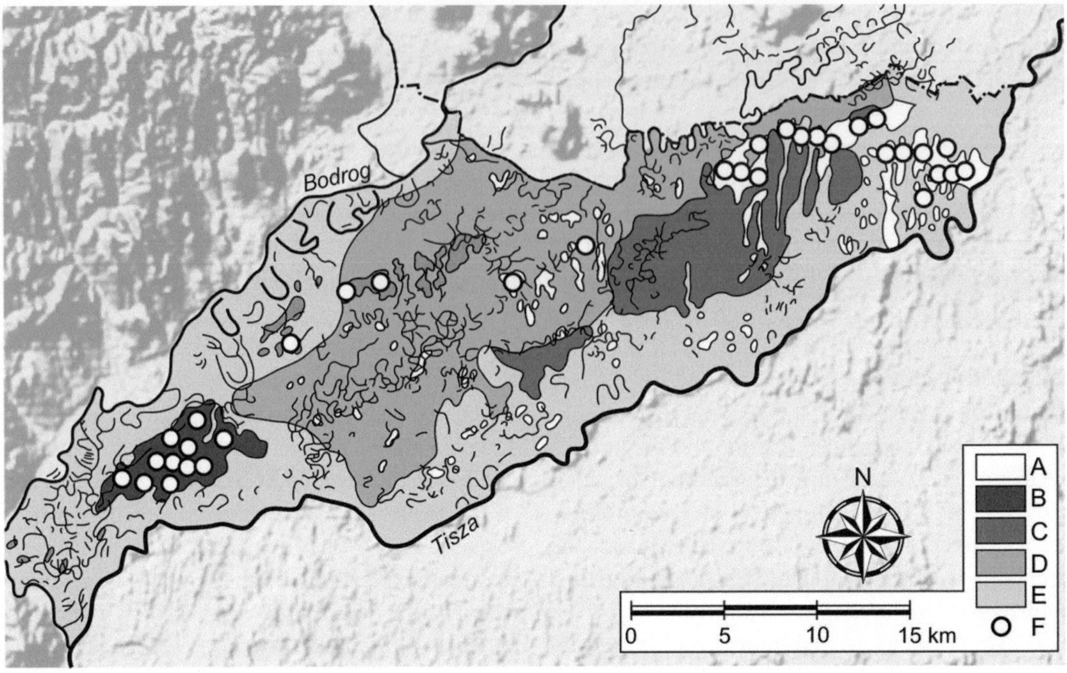


**Fig. 7.8** Geological and geomorphological sketch of the Bodrogköz (after Borsy et al. 1989 in: Marosi and Szilárd 1969). *A* Blown sand, *B* loessy sand, *C* marsh and peat, *D* clay, *E* alluvial clay and mud, *F* surface covered by blowouts

the symbol of Hungary and the Carpathian Basin, but several people argue that it is erroneous to associate Hungary only with the steppe; Csorba 2009). The Hortobágy Brook probably was a former meandering side-channel of the Tisza; thus, its surroundings are covered by young alluvial sediments, alluvial clay and salt-affected loamy soils. Since the water regulations of the Tisza and the flood control of the Great Hungarian Plain, the wet pastures of Hortobágy have become strongly salinized, and their character shifted from wetland to dry pastures. (The floods of the Tisza flowed across the Hortobágy and Nagykunság to the Körös River system through streams and crevasses of a comparatively large water discharge, 25–30 $m^3/s$. It explains the construction of the Kilenclyukú-híd (Nine Holed Bridge) at Hortobágy and a destruction of a similar bridge at Karcag. The salt-affected soils characterizing the meso-region are discussed in detail in Sect. 1.4.

The **Lower Tisza Region**, similar to the Körös Basin, is a sinking territory between fault lines, where the rivers of the Eastern Carpathian Basin flowed across since the Pliocene and deposited an approx. 1.5-km-thick (600–700 m in the Quaternary) fluvial sediments (Fig. 7.9). The floodplain levels (3) are located at 80–85 m asl, and they are confined by alluvial fans. Floodplain islands covered by loess and fluvio-aeolian dunes provide particular character of the landscape. These high surfaces are important from the archaeological point of view, since they have continuously been inhabited since the Neolithic—e.g. the world-famous Neolithic settlement of Szegvár-Tűzköves).

A large floodplain stretches towards the interior of the Great Hungarian Plain along the Berettyó and Körös rivers, between the alluvial fans of the Nyírség, Hajdúság and the Körös-Maros Interfluve. This is the perfect plain of the **Körös Region**, whose boundaries are difficult to determine from the landscape geographical point of view. It was a depression of great importance in the Great Hungarian Plain during the entire Pleistocene, where 500- to 700-m-thick mostly

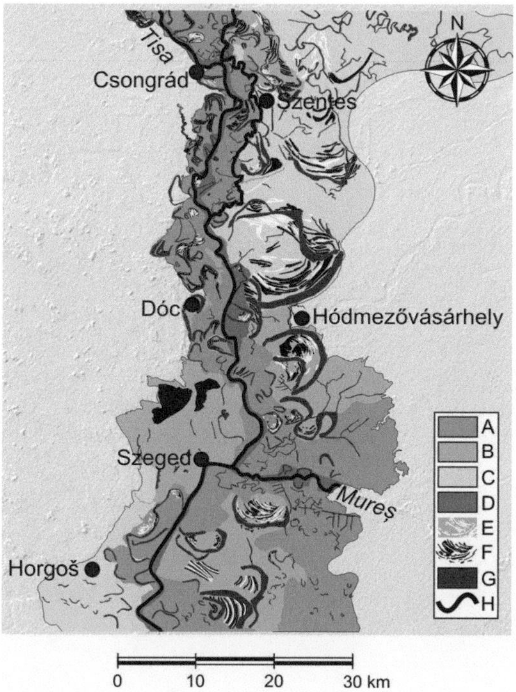


**Fig. 7.9** Main geomorphologic forms of the Lower Tisza Region (edited by Hernesz, P. using the data from Andó 1971). *A* Alluvium, *B* water, *C* natural levee, *D* artificial floodplain, *E* abandoned channel, *F* crevasse channel, *G* swale, *H* point-bar, *I* island, *J* floodplain island, *K* floodplain marshland, *L* settlement

fine-grained, Pleistocene fluvial mud and clay deposited. Before the water regulations, the meso-region was characterized by dense network of palaeo-channels (Fig. 7.10), low-lying marshlands and bogs, and in some places open waters. In the northern part of Nagy-Sárrét, in the neighbourhood of the Berettyó River, the rate of floodplain aggradation was high; therefore, there was no time for peat formation, though on the older surface of Kis-Sárrét (Körös Rivers) peat occurrences were frequent. In this wetland environment, only the elevated, flood-free system of levees was the natural surface where the first settlements could appear and served as the basis of kurgans and tells. The extensive water regulations of the nineteenth–twentieth centuries significantly transformed the surface; thus, its natural character ceased to exist, and almost all wetlands disappeared (and also the lifestyle connected to it).

## 7.3 Environmental Values and Hazards in the Great Hungarian Plain

(A) Main environmental values of the Great Hungarian Plain

In general, the Great Hungarian Plain is not rich in natural resources, except for hydrocarbon fields, the thermal waters and fertile lands, and its ecological, pedological, aesthetic and recreational values are also important from natural and social points of view. The water is rather an opportunity than a natural resource.

The renewable energy sources of solar and geothermal energy, the wind, water and biomass constitute more than 100 PJ potential utilizable energy annually, but its major contributors (90 %) are biomass and geothermal energy (Mezősi 2008). Most of the geothermal and one-third of the biomass energy of Hungary are connected to the Great Hungarian Plain. At the moment, nearly half of the biomass (31 PJ/year), and just a small proportion of the potential geothermal energy (3.3 PJ/year), are used for energetic purposes (Giber 2005). According to estimates, the total biomass production of the agriculture is 800–1000 PJ/year, and 40 % of this is provided from the Great Hungarian Plain. This refers to the fact that the soil is the most valuable natural resource of the country, whose share in the national wealth is estimated to be 22–23 % (Németh 2005), or 25 % (Lakatos 2003). (In the case of renewable energy sources, the currently low transformational efficiency must be considered; thus, they can satisfy the present need for energy only to a small extent.)

Among the non-renewable natural resources and raw materials, the natural oil and gas are the most significant. The largest natural oil reserves are located in the Lower Tisza Region, in the district of Algyő. The hydrocarbons usually reach marine sandstones, lacustrine and fluvial formations during their upwards migration; thus, the majority of the exploitable supplies can be found here. The natural gas of the Great Hungarian Plain occurs in similar structures in

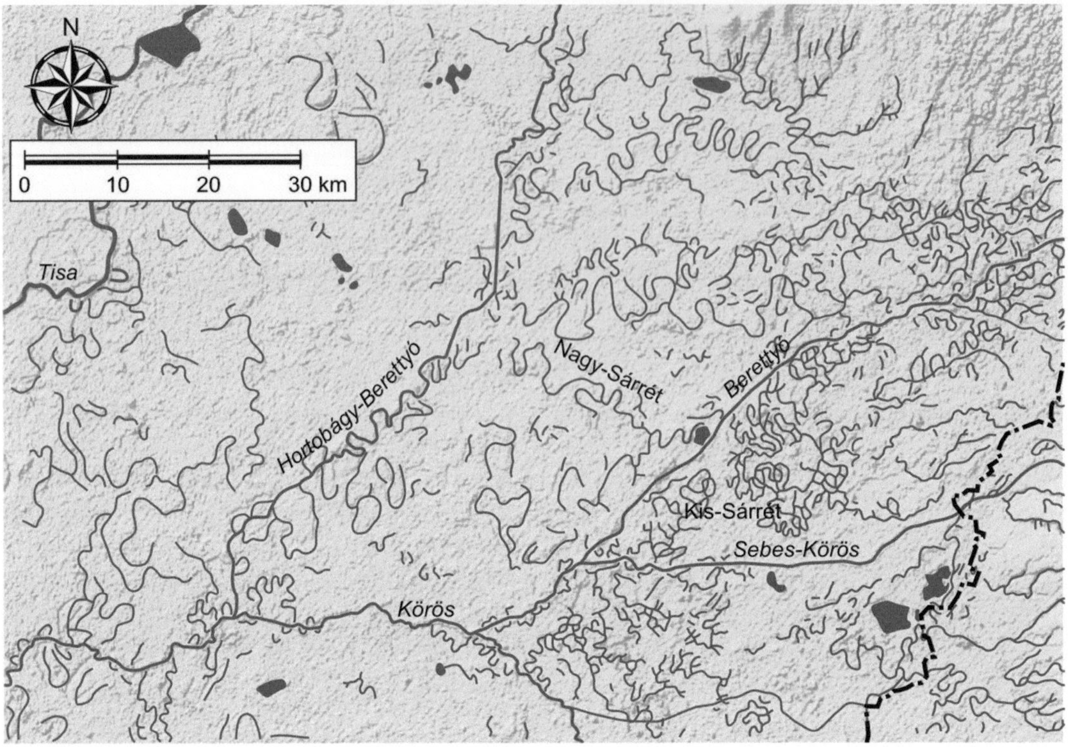


**Fig. 7.10** Network of the abandoned palaeo-channels in the Körös region (after Papp 1956)

various other locations. In the vicinity of Üllés, Pusztaföldvár, Hajdúszoboszló, Szeghalom and Szank could be found almost a half of the exploitable domestic natural gas reserves, apart from Algyő (Nagy 2001). Calculating with the current yearly production (ca. 1 million t/year in the case of natural oil and 3 million $m^3$/year in the case of natural gas), the reserves are sufficient for the next 15 years. In 2009, an underground gas storage with a capacity of 1.2 billion $m^3$ was constructed at Algyő, significantly increasing the security of gas supply.

The protection and management of the environmental values of the Great Hungarian Plain are directed by the headquarters of national parks. The Kiskunság, Hortobágy and Körös-Maros National Parks consist of several small units, since the natural values have been preserved only sporadically in the cultural landscape of the Great Hungarian Plain. The Hortobágy National Park established in 1972, possesses the largest contiguous protected area, preserving the special wildlife of the "puszta" and its traditional lifestyle, social memories and cultural landscape (e.g. nearly 1000 kurgans). The Hortobágy is also a World Heritage Site, a popular destination of the domestic and the international tourism (Fig. 7.11).

The Kiskunság National Park manages, protects and preserves the typical natural and landscape values (e.g. geomorphologic, pedologic, hydrographic and botanical) of the southern part of the Danube-Tisza Interfluve and the Middle and Lower Tisza Regions (Fig. 7.12). Its main task is to maintain the special farm life formed after the Ottoman invasion and to preserve the traditional farming methods. Its larger part has been a biosphere reserve since 1979. The Körös-Maros National Park was established in 1997 to preserve and manage the loessy landscapes and their typical natural values (alkaline steppe, marshland remains, palaeo-channels with special vegetation; Biró and Molnár 1998).

Apart from the above-mentioned three national parks, the Bükk National Park extends to the

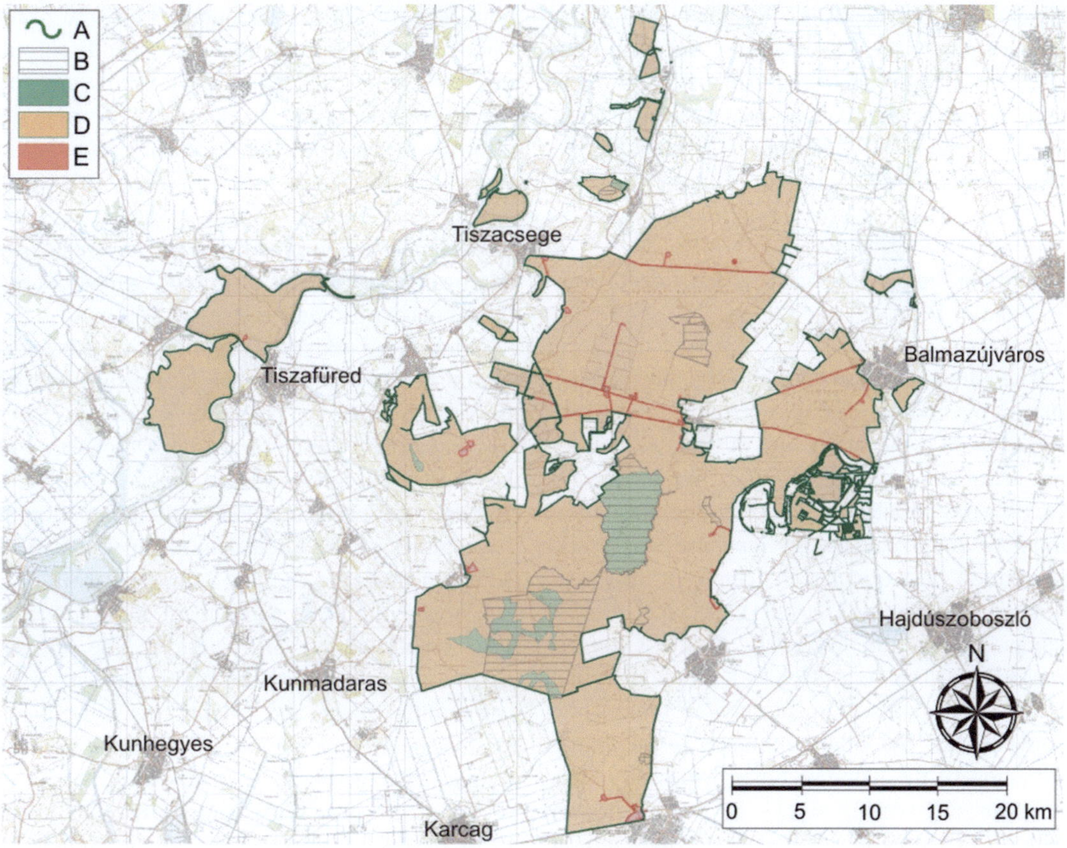


**Fig. 7.11** The zonal division of the territories of the Hortobágy National Park (after http://www.hnp.hu). *A* Border of the national park, *B* specially protected natural area, *C* natural zone, *D* nature-friendly utilization, *E* service zone

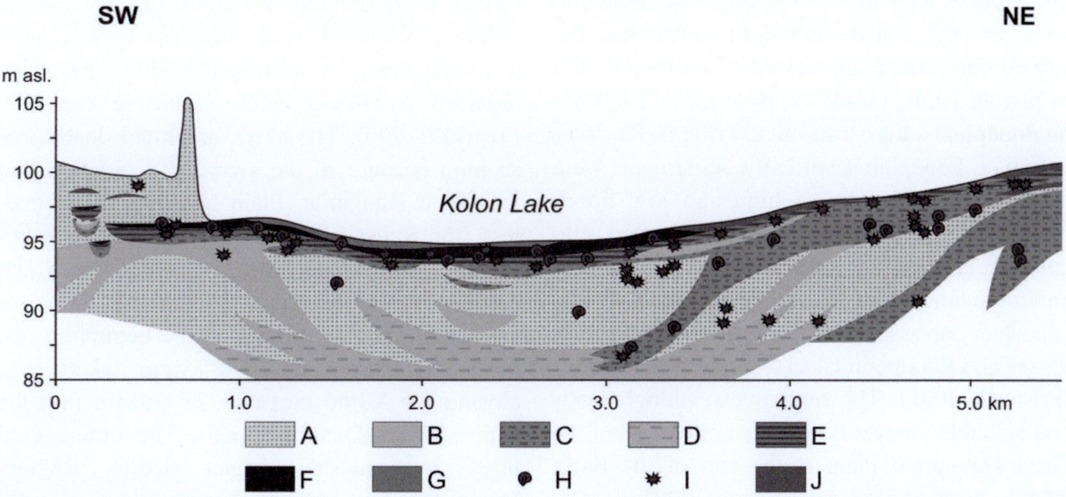


**Fig. 7.12** The geological cross section of Lake Kolon, at Bikatorok (after Molnár et al. 1979). *A* Small-grained sand, *B* fine sand, *C* silt of Danubian origin, *D* fine silt, *E* carbonate-silt *F* peat, *G* artificial filling, *H* Gastropoda-rich layer, *I* marshland-like formation, humic soil

Upper and Middle Tisza Regions, while the northern part of the Danube Plain, the Mezőföld, and the northern part of the Danube-Tisza Interfluve belong to the Danube-Ipoly National Park and the Danube-Dráva National Park manages the protection of the southern part of the

Danube Plain and the Dráva Plain (Fig. 5.7).

(B) Major human impacts on the Great Hungarian Plain

The Great Hungarian Plain has been transformed by two major anthropogenic influences. One of them is the succession of wars, especially the Ottoman occupation had great impact, resulting in population decrease and the abandonment of huge; thus, the land use was transformed too. In particular, the centre of the Great Hungarian Plain (Kiskunság) became almost depopulated at the beginning of the Ottoman invasion (lost battle at Mohács), although it had denser Cumanian settlement network than of the Nagykunság. The other influence is the transformation of the land-use types, especially after the river engineering works in the nineteenth century, as the application of newer agricultural technologies had smaller impact.

The anthropogenic transformation of the landscapes of the Great Hungarian Plain has taken place with different spatial and temporal rates. **In the tenth–thirteenth centuries,** the agricultural production was based on the rotation of plough lands, meadows, pastures and fallows on floodplains, loessy and sandy ridges. Hay was grown on floodplains, while the settlements were founded at the border of high and low floodplains. Certain surfaces were transformed into cultural landscapes; however, the sustainable environmental management is indicated by the so-called crevasse-management, where the crevasses and floodplain lakes were used as fisheries (Frisnyák 2001). The settlements, plough lands and valuable vineyards occupied only 5 % of the Great Hungarian Plain at the end of the Early Middle Ages (Árpád Age; Lóczy 2000).

**In the fourteenth–seventeenth centuries**, an economic shift took place between the plains and the hilly regions. Animal husbandry remained the main activity on the floodplains, but the grain production moved to the loessy and sandy ridges, vineyards were established on foothills and the forestry receded into the mountains. During the Ottoman occupation, the decreasing population moved to safer places, thus into large settlements and to the mountains, while a secondary (cultural) steppe was formed in the plains, enhancing significantly the spread of grassy vegetation and the decrease of forests (Frisnyák 1990).

After the Turkish occupation, **in the eighteenth century and at the beginning of the nineteenth century**, the recovery of the economy began and the cultural steppe ("puszta") of the Great Hungarian Plain began inhabited. (Between the 1720s and 1787s, the population of the counties of the Great Hungarian Plain grew from 125 thousand to 950; Somogyi 1994).

The geographical environment did not change considerably during the centuries; the applied technologies did not renew: the rotation of spring and autumn wheats and fellows was still typical. At the beginning of the nineteenth century, plough lands, gardens, vineyards and the built-in environments (i.e. settlements) occupied about 43 % of the territory of the Great Hungarian Plain (Lóczy 2000). The intensity of animal husbandry increased significantly; however, at the mid-eighteenth century, the droughts and the fodder shortage caused serious losses (Lakatos 1988). (Cattle and sheep were not kept in open anymore; thus, the traditional breeding had been replaced by newer, more intensive methods; Frisnyák 1990). The need for plough lands was so high because of the growing population that the Great Hungarian Plain became almost treeless (the forest cover reached a historical minimum of 4 %). The need for revitalization and development of the economy motivated the water engineering works, though at the beginning just local engineering measures were taken (e.g. closing the Mirhó crevasse, or constructing the Danube-Tisza-Danube Canal). The drainage of bogs and marshes began (Körös Region, Ecsedi-láp, etc.), but they were not successful due to their local character. Since the sandy areas began to be used intensively, blown sand movements started, and to stop the aeolian activity locust tree was introduced and planted, and vineyards became widespread.

The aim of the **nineteenth–twentieth-century** regulation works was to decrease the damage of flood all over the Great Hungarian Plain and to give way to the modernization of the economy. The flood control, the channel regulations and the irrigation–drainage canal constructions fundamentally changed the landscape of the Great Hungarian Plain. The land-cover difference between the floodplain levels (with high productivity) and the higher areas were eliminated, as all were transformed into plough lands. At the beginning of the twentieth century, 78 % of the Great Hungarian Plain was cultivated, while the extent of meadows and pastures decreased under 10 %. Several traditional land uses disappeared (e.g. floodplain orchards; Frisnyák 1990). Fishery and the related activities were also diminished. The economic role of sheep-breeding decreased by the appearance of Australian products in the market (Lakatos 1988). However, the real homogenization of the landscape and the intensive agriculture became relevant just after the Second World War, and mainly from the 1970s due to the socialist agricultural practices.

(C) Main environmental conflicts in the Great Hungarian Plain

We highlight those environmental problems of the Great Hungarian Plain, which are related to water. (The problems caused by deflation, transformation of the vegetation or drastic decrease of naturalness are not discussed.) These hydro-geographical problems originate from the abundance or the shortage of the water supply, occurring often within one year (see Sect. 1.3.3).

(a) Groundwater-level decrease in the Danube-Tisza Interfluve

The lowering of groundwater level can be detected in the entire Great Hungarian Plain, but it is the most apparent in the Danube-Tisza Interfluve (Fig. 7.13). It is primarily influenced by climatic changes (mainly by the significantly decreasing precipitation), the increased subsurface water withdrawal, land use and surface cover changes, though other minor factors also influence it. According to Pálfai's (1994) estimation, the role of climatic factors is 50 %, and anthropogenic influences account for the remaining 50 % of groundwater level decrease. Regarding just the human influence, the role of artesian water withdrawal is 25 % that of groundwater and hydrocarbon extraction is 8 %, of land-use changes is 10 %, and of water engineering is 7 %, respectively.

The list of triggering factors undoubtedly should start with the climate, since it seems to be the most important: as during the past century, consider the annual precipitation decreased by 100 mm. Researchers also draw the attention on evaporation, which has been previously disregarded, though it should be studied in the entire basin (Szesztay 2000).

The water withdrawal causes environmental changes, because the groundwater and the artesian waters compose a unified system, and there is no real impermeable zone between them. The artesian and groundwater extraction (without considering thermal waters) is about 500 thousand $m^3$/day. Its recharge was ensured until the mid-1990s, as the groundwater penetrated towards the artesian water bodies by ca. 15–25 mm/year rate (varying between 0 and 180 mm/year). However, nowadays, the artesian water level decrease is 30–40 cm/year in the Danube-Tisza Interfluve due to the intensive water withdrawal (Liebe 1994). All other influences (e.g. infiltration, evaporation) can modify the circulation in the aquifers. For example, the underground water infiltration of 1–10 m/year towards the lower lying peripheral areas of a positive gradient may attenuate the climatic influence in these regions.

The next group of factors leading to groundwater-level decrease is the change in surface cover, where the role of forests should be emphasized. The proportion of forests as a vegetation type has increased in land use. (The quantity of forested areas doubled in the past one hundred years in the Great Hungarian Plain.) The native oak and poplar forests have an extremely high water demand (400 mm/year and

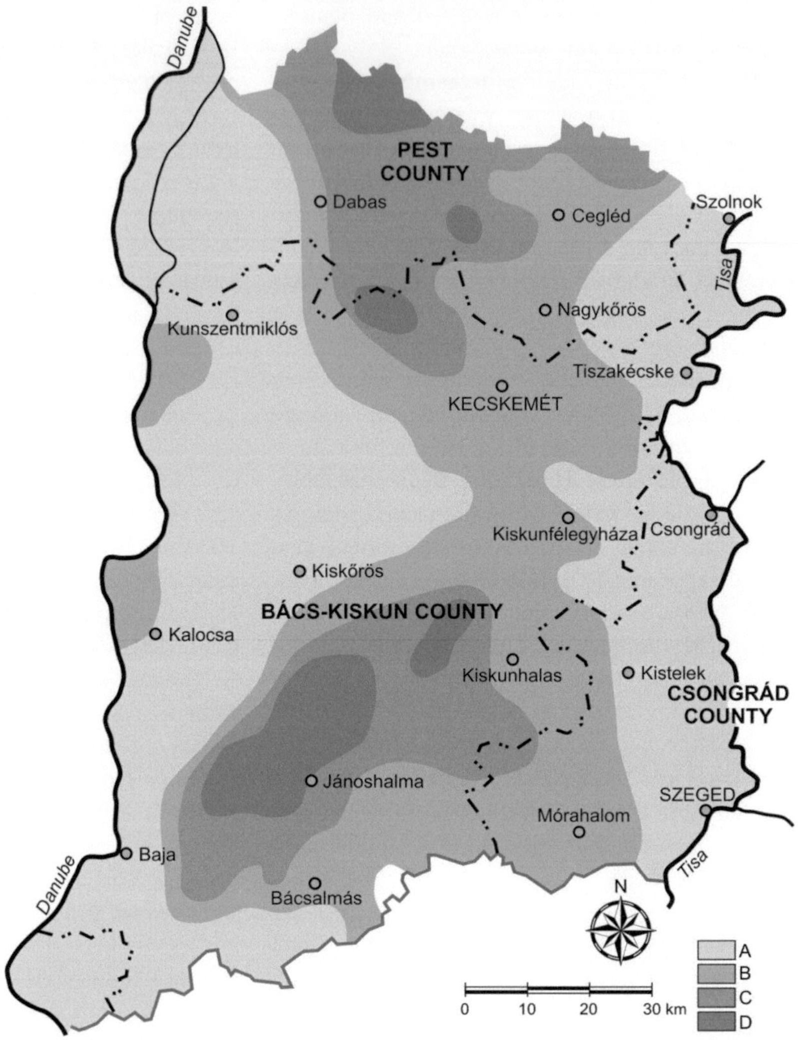


**Fig. 7.13** The decrease in groundwater level (m) in the area between the Danube and the Tisza (The figure shows the difference between medium ground water level in 1992 and 1956–1975, after Pálfai 1994). *A* Below 1 m, *B* 1–2 m, *C* 2–3 m, *D* more than 3 m

6–700 mm/year, respectively); thus, they have been replaced by pine and locust species, whose water demand is just a half or a quarter of the former. (The water demand of arable crops is 400–600 mm.) Thus, forests do not have a significant effect on groundwater-level decrease (especially not by "well-effect"). In the increasingly dry environment, the preservation of native tree species becomes highly questionable (Szodfridt 1994). In this approach, the transformation of forests is not the cause, but the consequence of aridification. (Mainly the areas too dry for agriculture have become forests.) Szesztay (2000) proved that the deforestation along the rims of the catchment resulted in 150–175 mm climatic water shortage (drought sensitivity), because of the changed conditions of evaporation, and the increased run-off (which poses flood and inland excess water hazards). It would also be worth to study the evaporation capacity of the changed land cover from this point of view.

The weight of the individual factors determining the groundwater-level decrease is difficult to estimate, because it may be different for different areas. Rakonczai and Bódis (2001) carried out a detailed analysis on the relationship of groundwater-level decrease and relief. According to their results, the decrease affected especially the higher surfaces, while in the low-lying periphery of the Danube-Tisza Interfluve, the decrease in groundwater level is much lower due to the recharge and the modified surface cover (Fig. 7.14).

(b) Reducing the flood risk on the Tisza

The floods in the Carpathian Basin are related to both climate and relief. The potential flood hazard is a logical consequence of the basin character and the lowland location. If a high amount of water prone to run-off is created in the upstream parts of the catchments, floods develop since neither the Danube nor the Tisza River is able to drain this amount of water without a flood. The serious difference between the two rivers from the point of view of flood hazard is that floods are shorter on the Danube, while longer (and consequently more dangerous) on the Tisza. (As it was mentioned before, floods can develop quickly on the smaller rivers in the hills and

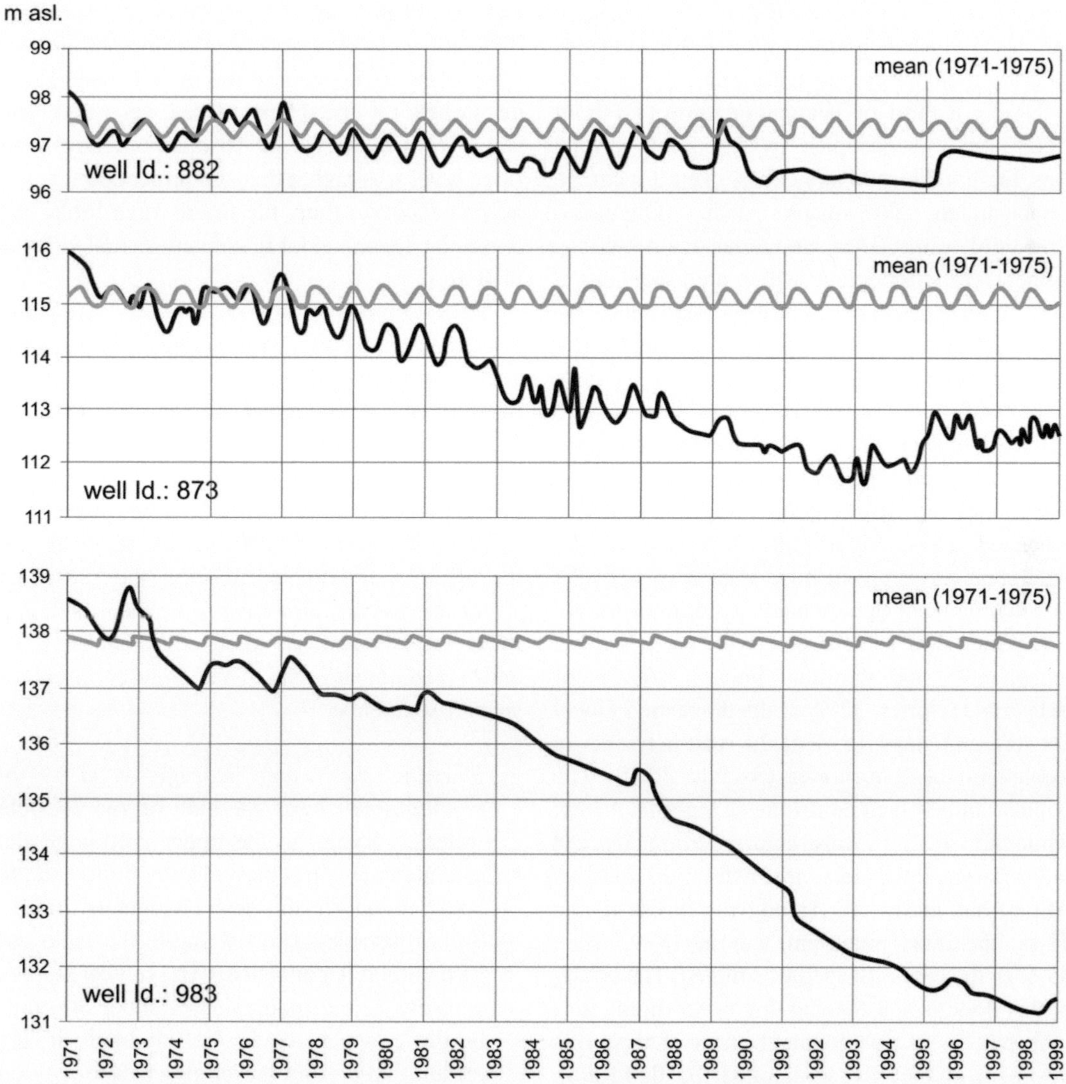


**Fig. 7.14** Groundwater level changes between 1971 and 1999 in the Danube-Tisza Interfluve in comparison with the average of 1971–75. The well No. 882 is at Kiskunfélegyháza, the No. 873 is at Kiskunhalas, and the No. 983 is at Jánoshalma (after Rakonczai and Bódis 2001)

mountains; thus, the maintenance of flood safety is a difficult task there.) Extreme amount of precipitation always plays in floods, but they can also develop when rain falls onto snow, frozen or water-saturated soil; thus, the run-off rate increases. Of course, the flood hazard had to be reduced to ensure the safe use of rivers and their environment (and it is still the main challenge of water management today). The issue of climate change was not raised in the nineteenth century, when flood control works were planned and done. Nowadays almost each scenario prognosticates the increase in winter run-off (and its decrease in summer), which also means more frequent floods (and droughts within the same year).

Nearly a third of Hungarian territory (about 28,000 $km^2$) is occupied by floodplain levels; thus, the flood hazard still exists, even if it can be demonstrated on maps with difficulties. Two-third of the Great Hungarian Plain belongs to the Tisza's floodplain, which are especially at risk. In the nineteenth century, floods used to endanger mostly agricultural areas. A paradoxical situation has evolved recently: the river regulation and flood protection works on rivers were designed to control floods; however, nowadays the flood hazard has increased due to higher flood waves and the denser population of the flood-protected areas. Significant proportion of the floodplain levels are in flood risk, despite of the fact that 97 % of the lowlands is flood-protected. Altogether 400 settlements (with ca. 3 million inhabitants), 300 thousand houses, 200 major industrial factories, 32 % of the domestic railway network and about 15 % of the road network are located in these territories (Szlávik 2002). The population involved constitutes a high risk value, while the value of infrastructure, settlements and industrial facilities (e.g. the natural-oil field of Algyő) has increased. Among the floods of the Tisza, the most important was in 1879, when 95 % of the city of Szeged was ruined. The oddity of this flood is that the ongoing water regulations took part in its development, since the cut-offs upstream of the city accelerated the flood conduction, while no cut-off was made downstream; thus, the flood wave occluded.

Apart from the one in 1879, several other, often devastating floods have evolved along the Tisza. A significant part of Szeged was ruined in 1712, and in 1816, according to written sources, thus travelling from Szeged to Kikinda (24 km far) was possible just on water (though, it was not rare at that time, because the majority of the floodplains was covered by floods for up to a half year; Réthly 1970). The flood in 1941 was also memorable (Szlávik 2002), when flood preparedness lasted for 210 days. More recently, large floods developed in 1970, 1999 and 2001, and the absolute water levels reached a record in 2006 at Szeged (Image 7.4). The flood of 2001—apart from 1970—caused the most damage, as it developed on the Upper Tisza very fast (the water level increased by 10 m in one day), and flood level was higher than the protection level of artificial levees; thus, the flood wave broke the levees (at Tarpa) and it inundated several villages in Bereg.

Various interventions of different scales and directions are known in mitigating flood hazard and controlling floods. One solution is to heighten the artificial levees, to provide flood safety even at higher flood levels. The other is to analyse the riverbed parameters in relation to human impact and to ensure the highest possible flood conduction efficiency of the channel and the artificial floodplain. Besides, the causes of rising flood levels also have to be identified, and flood management has to be comprehensive for entire catchment area. This strategy involves several segments:

1. *Increase the water storage capacity in mountainous areas*. A half of the feasible water reservoirs in the upper section of the catchment area has been already built, and the construction of the rest encounters mostly social obstacles. (For example, the valley to be dammed is populated.) The run-off storage capacity in the mountains is about 300 million $m^3$, which could be doubled, if the existing reservoirs are empty before a flood wave evolves. Altogether, the flood discharge of 1–1.5 days could be stored in the

**Image 7.4** The highest water level ever recorded at the embankment of Szeged (photograph by Kiss, T.)

catchment of the Upper Tisza. The change in land use (i.e. deforestation) and the resulted increased run-off are commonly mentioned as the cause of increasing flood development. However, Lóki et al. (2004) demonstrated that the moderate territorial changes in forests in the upper section of the Tisza's catchment area cannot reason the extreme floods at the end of the twentieth century. The limited water storage capacity of forests was also proven by Vágás (2001). The forest cover decreased by nearly 50 % in the past 2–3 thousand years (which anyway means less water storage and more run-off), but area of forest clearances during the past 30–40 years is lower than it could be according to the law (considering the regeneration rate of forests; Konecsny 2000); thus, today rather a reforestation is characteristic.

2. *Increase of the flood conveyance capacity of the river*. A theoretical solution for flood hazard could logically be if larger space (channel and floodplain cross-sectional area) would be provided for floods. However, the relocation of artificial levees further from the river channel would have very high costs (around a thousand billion HUF), because today the artificial levees are placed on elevated surfaces along the river (e.g. on higher floodplain levels, sand bars and fluvio-aeolian ridges). As the floodplains are convex, farther from the rivers, the elevation of the surface is lower; therefore, the relocated new artificial levees must be higher to retain even the present flood levels. The idea of the relocation of present-day artificial levees (in which case new systems must be constructed around settlements, roads, etc.) was also discarded

due to financial reasons, since its expenses could be the double of the former. Besides, the problem is not just the narrow floodplain, but also the narrowing channel, which is not able to conduct large discharge than 3600–4000 $m^3/s$.

Another possibility to reduce flood hazard is to increase the channel capacity of the artificial floodplains. At present, they drain just about 5 % of flood discharge, but the short-term goal could be to increase it to 10 %. To achieve this, the vegetational roughness of the artificial floodplain should be reduced, which could lower the flood level itself (the restoration of the original vegetation cover is illusorily). According to the data of the Central Tisza Region Directorate of Water Management, 1 billion $m^3$ sediment deposited between 1888 and 1945 on about 33 thousand ha large artificial floodplain of the Tisza (between Tiszafüred and Csongrád), which corresponds to 3 days' flood discharge of the Upper Tisza (Schweitzer 2004). The increased drainage capacity of the artificial floodplain could improve the flood conveyance of the system. It must be accepted that the primary function of the artificial floodplain is flood conveyance; thus during the planning of land use and buildings, it should be also considered. When the plans were made on water regulation in the nineteenth century, meadows and pastures prevailed on two-thirds of the artificial floodplain, whereas today it is occupied by invasive species (e.g. false indigo) and dense planted forests, which reduce the speed of the water and also the drainage capacity of the floodplain.

3. *Lowland storage*. Nowadays the water storage has been emphasized to decrease the speed and height of floods, which could be realized in the upper catchments or along the lower sections too. Simultaneously, there is a need to mitigate droughts by water storage in the basin. From the point of view of flood hazard, the aim is an approx. 1 m reduction in the flood level along the Tisza. To achieve it, new lowland reservoirs could be built along the Tisza and its tributaries, whose total capacity should be about 1.3–1.5 billion $m^3$. There is a limited space for building a large-capacity emergency flood reservoir in the Upper Tisza Region; however, there are more open territories in the Middle Tisza

**Image 7.5** Locks of the emergency flood reservoir at Tiszaroff, inaugurated in 2009 (photograph by Rakonczai, J.)

Region south of Tokaj. The series of emergency flood reservoirs is described in the New Vásárhelyi Plan. Although it is a logical concept, but many of its elements are not described/modelled in detail, though the works have started, and several reservoirs had been constructed (Image 7.5 and Fig. 7.15).

(c) River regulations on the Great Hungarian Plain and flood control of the macro-region

The regulation of the lowland rivers had several different aims beside flood control, as several considerations in economic policy supported and thrived to realize it. For example, the

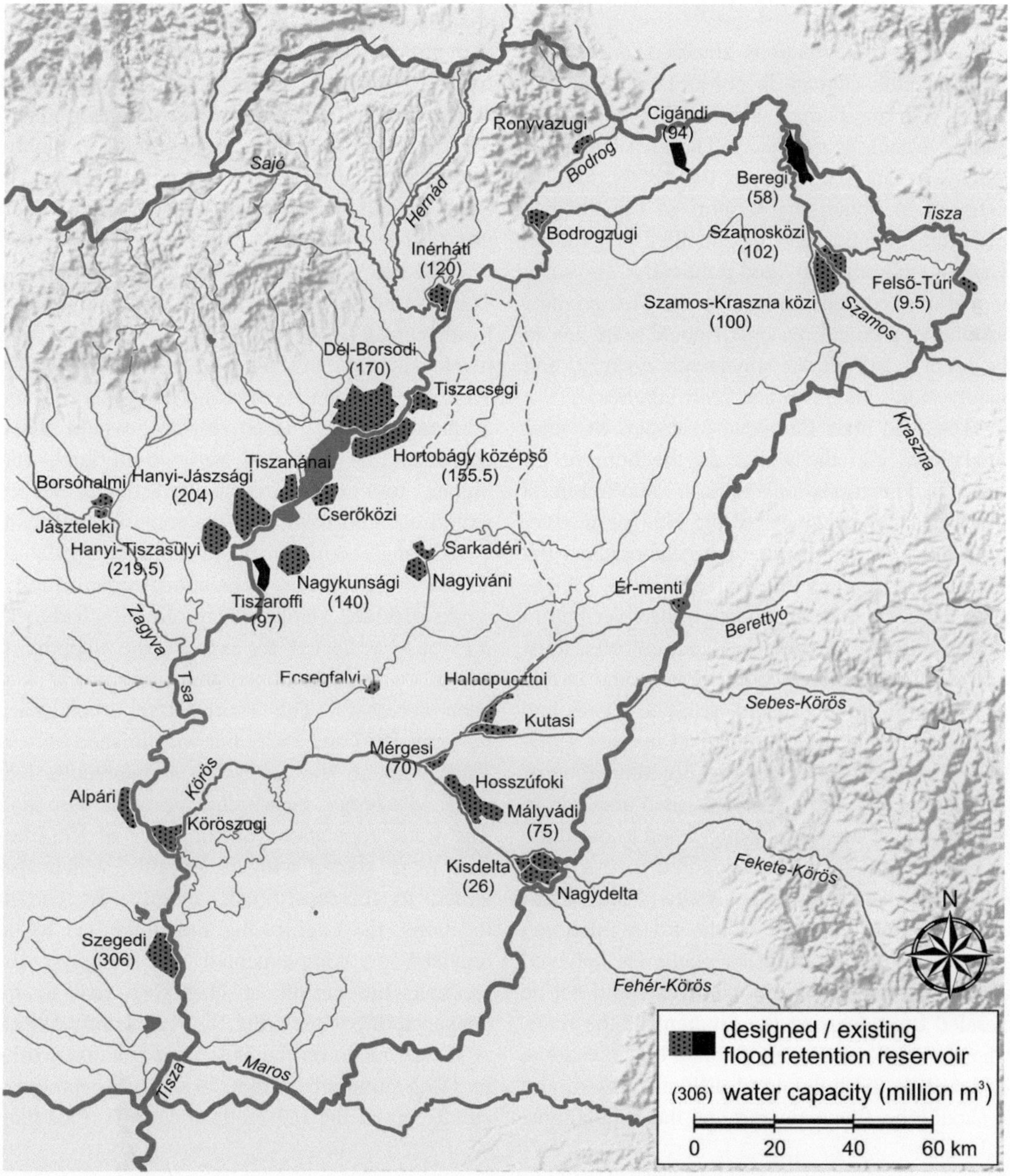


**Fig. 7.15** Proposed emergency lowland reservoirs along the Tisza and its tributaries (after Szlávik 2002)

growing land demand of the increasing population had to be satisfied, profit could be gained from the wheat boom caused by the Napoleonic wars, the flood protection had to be realized and safer shipping routes on rivers had to be provided (in order to complement the scarce road network on the Great Hungarian Plain). Besides, medieval type of agriculture had to be modernized, as it had a very low productivity after the Ottoman occupation.

The idea of regulations emerged as early as the eighteenth century in connection to several rivers (Sárközi-Duna, Sárvíz, Danube-Tisza-Danube Canal, Maros, etc.). The idea was worded by István Széchenyi in the 1830s, and he managed to obtain the support of the government, the counties and the country in general, which was strongly divided politically. The scale of works exceeded the one in the Netherlands, but due to political reasons it could start just in the second half of the nineteenth century, and mainly along the Tisza and its tributaries.

There had been European examples of water regulations, e.g. the works on the Sein or Po Rivers had centuries-long history. Pál Vásárhelyi was the leading engineer of the Hungarian river regulation works, and he elaborated the much-debated plans of the regulations of the Tisza. According to his concept, the river had to be shortened by series of artificial cut-offs; thus, a faster flood drainage could be triggered in this way, one on the other hand, artificial levees had to be built along the new channel in such a distance that the mean water depth/width ratios of the channel and of the whole flooded area had to be the same. Therefore, he suggested to build the artificial levees just a few hundred metres from the river. Apart from other experts, Pietro Paleocapa (leading engineer of the Po regulations) also gave an opinion on the plans. He believed that the overlapping of flood waves could not be avoided by increasing the gradient of the river, whereas the flood-safe agricultural lands could be achieved by the building of artificial levees, and it should be the main aim of the regulations. Therefore, he proposed just some cut-offs, and in this way the shortening of the Tisza could have been just by 20 % (instead of in Vásárhelyi's plan 40 % was aimed), and in addition Paleocapa suggested to build the artificial levees farther from the river (at 1–2 km distance), and he also recommended the construction of eight lowland reservoirs. However, this alternative plan was not supported widely by decision makers, due to smaller expansion of the territory of flood-free agricultural lands and the increasing costs of traffic (e.g. longer bridges, the improvement of shipping conditions would require greater effort). The biggest difference of the two plans was in the role of artificial floodplain and artificial levees in flood-wave drainage. The Vásárhelyi Plan would use 10 % of the flood protected area (15,000 $km^2$) as artificial floodplain (Fig. 7.16). The plan of the Italian Paleocapa included much fewer cut-offs (21) and wider floodplains. Both plans counted with the construction of 1500–1600 km artificial levee along the Tisza, but the artificial levees placed farther would have decreased the protected agricultural land, and higher, thus more expensive artificial levees would have been built on the convex floodplain.

The engineering works of the river regulation started in 1846, after tremendous organizational works. Besides, some basic information had to be previously collected, for example the mapping of the Tisza and its immediate environment had been necessary: The Tisza Survey took place between 1807 and 1825, but was finished only in 1840 with all the necessary supplements, e.g. water discharge measurements were essential. The water regulations on the basis of Vásárhelyi's plans started with the cut-offs, which also aimed to increase flood velocity. At certain locations, the engineering measured had to be repeated or complemented by dredging, for example the cut-off at Nagykörű had to be deepened three times (Fig. 7.17; Lászlóffy 1982). A survey made on the state of regulation works in 1872 summarized that 59 cut-offs were successful out of the 101 realized cut-offs (thus they

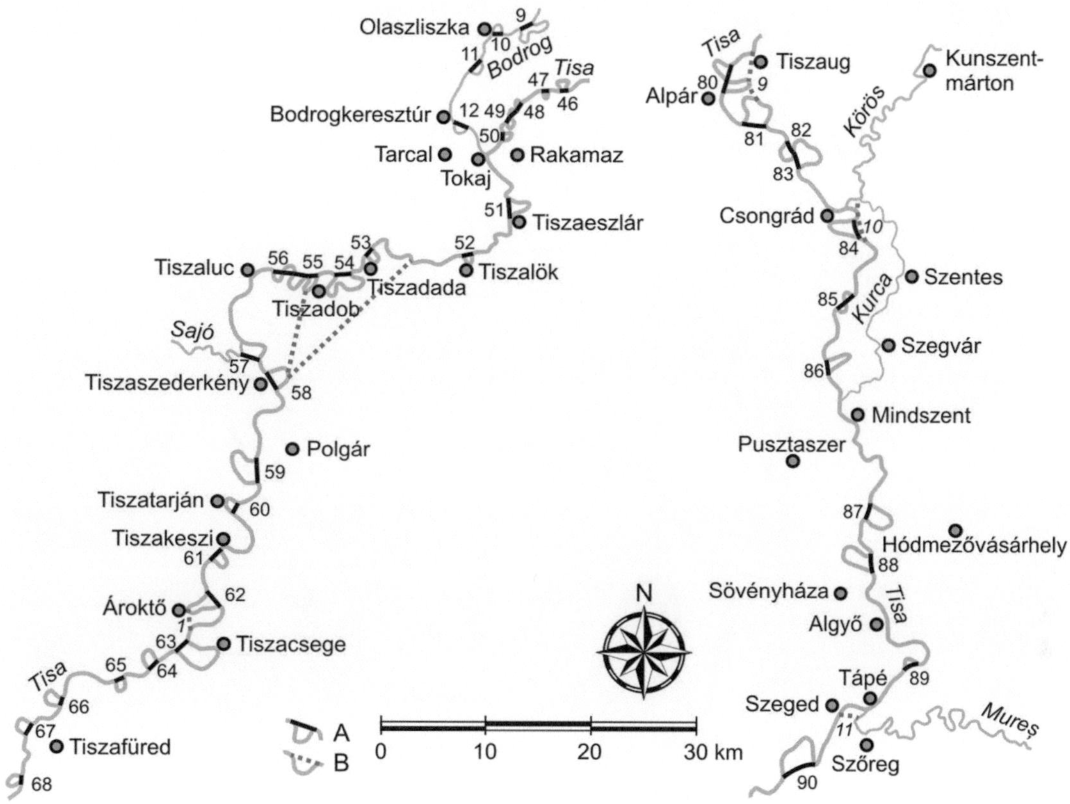


**Fig. 7.16** Cut-offs proposed by Vásárhelyi (*A*) and Paleocapa (*B*) along the Middle and Lower Tisza (after Lászlóffy 1982)

become main channels), 21 cut-offs were developing and 21 cut-offs were not successful, i.e. they had not become main channels. Finally, altogether 112 cut-offs were made, shortening the Hungarian section of the Tisza by approx 40 % (from 955 to 600 km). The shortening of the tributaries had a similar extent: the more than 1000-km-long Körös Rivers were shortened to 450 km, while on the Maros River 33 cut-offs were between Lippa/Lipova and shortening the river by 90 km.

The length of artificial in the river system of the Tisza is almost 4000 and 1400 km along the Tisza itself. The artificial levees were heightened several times (Fig. 7.18), for example after the flood of 1970 by 1.5 m on average. The structure and the increasing flood levels clearly show that the artificial levees need continuous heightening, which is usually carried out just after major flood events. The flood regulations started in 1846 and lasted in 1910, and its spatiality depended on the financial resources and the organizational background (Fig. 7.19). The higher gradient resulted in faster flood velocity; thus, during the years of the regulations in several cases, the quicker flood waves arriving from the regulated upper sections became blocked by the unregulated, highly meandering lower sections. Many people claim that this was one of the causes of the flood of 1879, which destroyed Szeged. It must be noted that water regime of the Tisza is influenced by the Danube and the tributaries too, due to the low gradient of the rivers; thus, their flood waves could be overlapped.

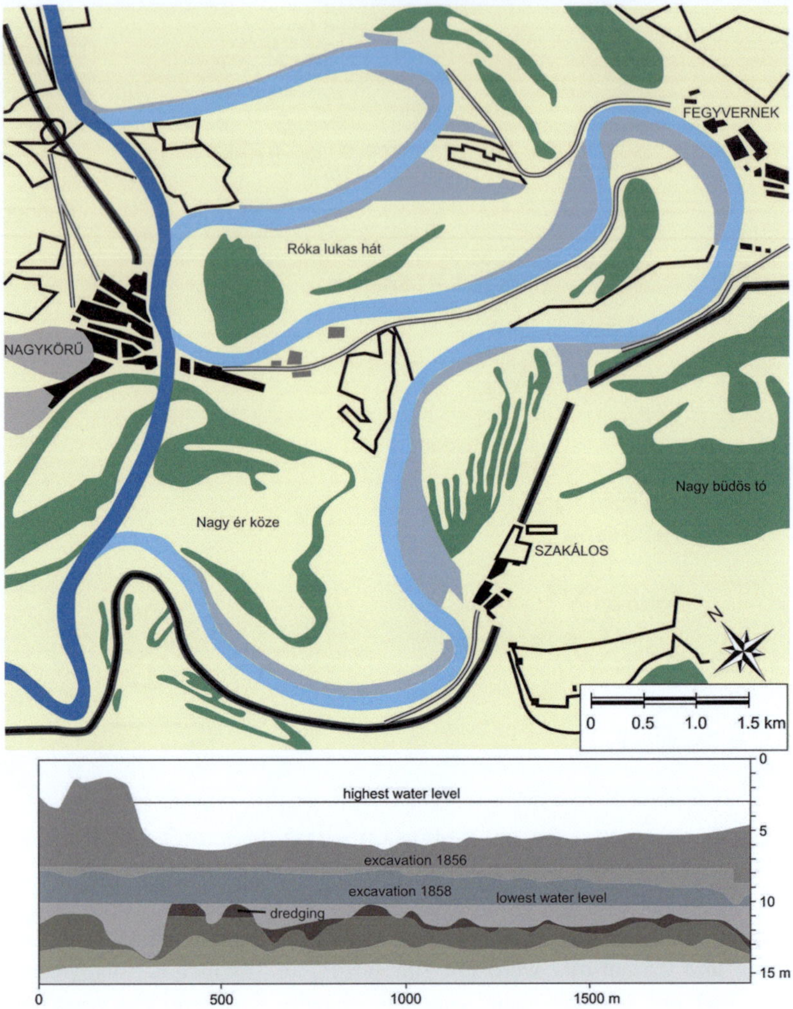


**Fig. 7.17** The development of the cut-off at Nagykörű, where artificial dredging was made in 1856, 1858 and 1861 to support the incision of the channel (after Lászlóffy 1982). *Dark blue* new channel created by the cut-off, *light blue* original bed of the Tisza, *green* wetlands (colour figure online)

### 7.3.1 Consequences of Channel Regulation and Flood Control Works

As an advantageous result of the nineteenth-century water regulation and flood control works is the increase of flood-free areas by 20 thousand km$^2$ on the former floodplains. The flood protection provided safety for the citizens, and it improved the possibilities of irrigation and shipping. The changes, which mainly took place in the Great Hungarian Plain, could be assessed to be favourable from the point of view of environmental ecology. Due to the large-scale economic and social benefits from the regulation works, it also mentioned as the "Second Conquest of Hungary." The unfavourable consequences mainly have a hydrological character, which is also manifested in other spheres (e.g. soil and vegetation). Somogyi (2000a) summarized the consequences of the regulation works, showing that on the Great Hungarian Plain 1300 km out of the total 2800 km river length have become partly or completely regulated. In the future, several

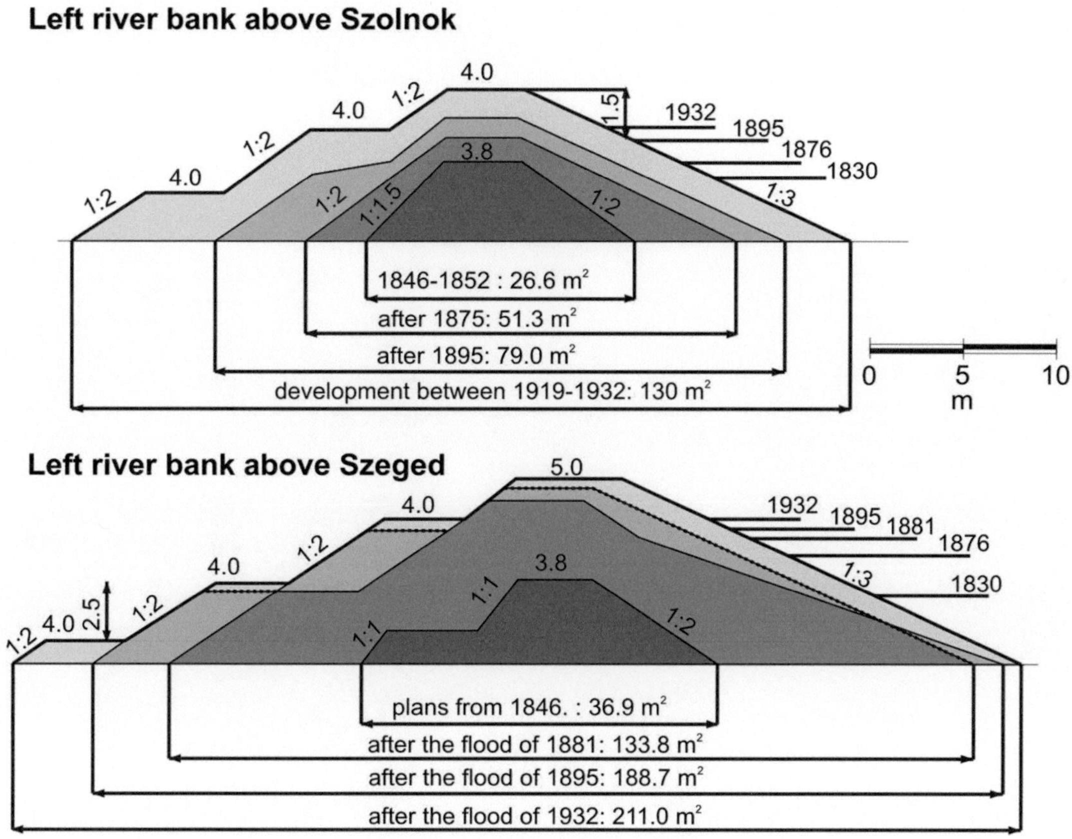


**Fig. 7.18** The layered structure of the artificial levees (after Lászlóffy 1982)

sections should be regulated (approx. 300 km) in the water system of the Tisza, though the Tisa itself is almost completely regulated.

(A) Hydrological consequences

In the nineteenth century, simultaneously with the river regulations, the extensive deforestation of the Carpathians intensified the surface run-off, and it was coupled with the increased gradient of the river caused by artificial cut-offs, which jointly led to the increase of erosional capacity of the river. Thus, the water system increased the amount of transported sediment and the intensity of channel incision, which finally resulted in channel pattern change. A good example of the latter is that the aggrading–meandering Tisza became a incising meandering, as its average depth grew by 2–4 m. The narrow artificial floodplain planned by Vásárhelyi (however, not consistently implemented) fastened the bank erosion. The altered hydrology affected all rivers and their floodplains, since they had been built of mainly organic materials of the rich vegetation and the inorganic material of floods. The two accumulation types separated as a result of the levee constructions; thus, more intensive the accumulation of inorganic materials on the artificial floodplains, while only biogenic material accumulates on the flood-protected floodplains (Somogyi 2000a). The increased sediment discharge resulted in intensive overbank

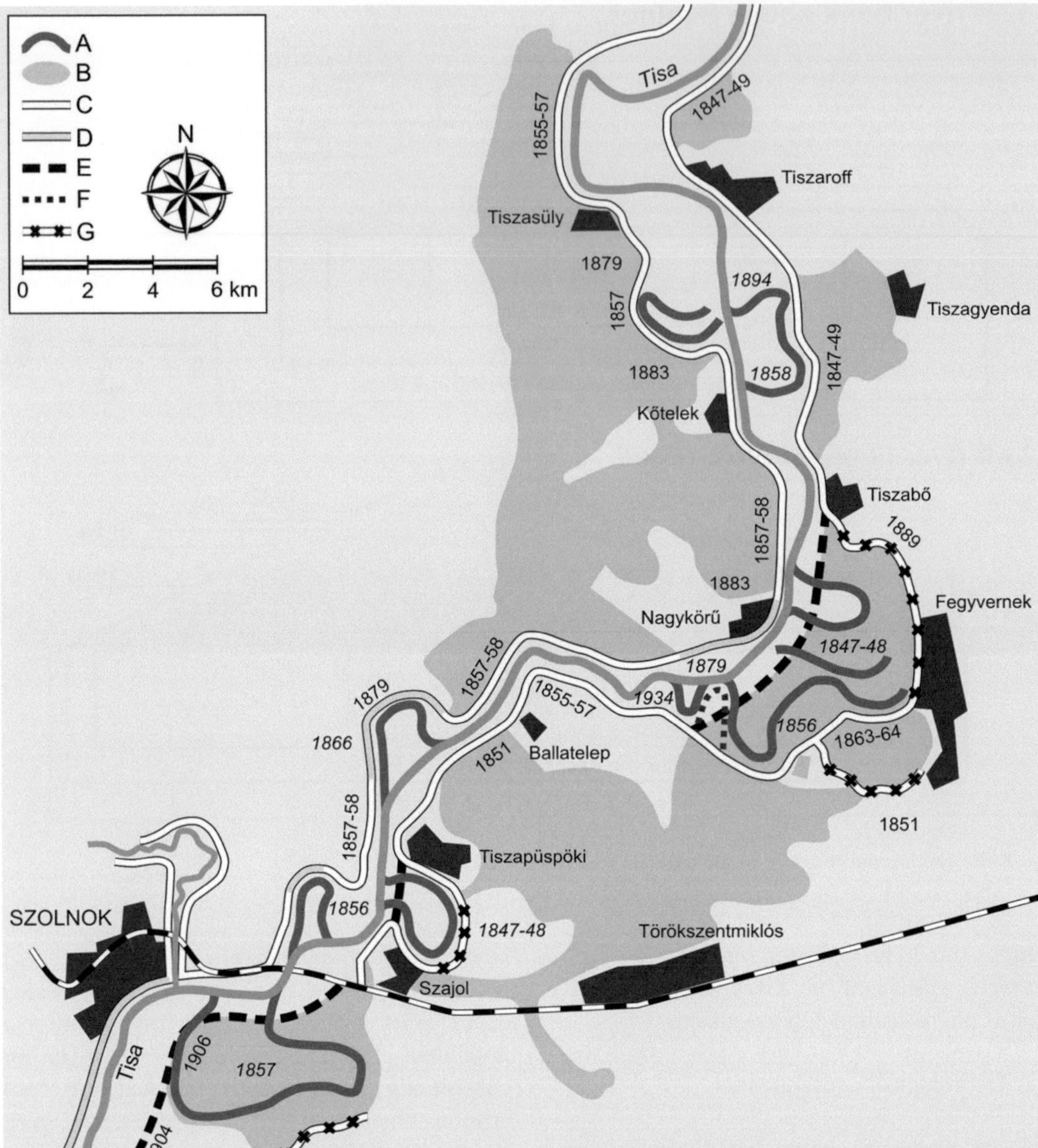


**Fig. 7.19** Changes in the location of artificial levees of the Tisza near Szolnok (after Lászlóffy 1982). *A* Cut-off meander with the year of the work, *B* flood-free area, *C* artificial levee finished by 1860, *D* artificial levee built between 1879 and 1889, *E* artificial levee built after 1891, *F* artificial levee abandoned after 1879, *G* abandoned artificial levee

aggradation on the floodplains and in the oxbows in the territory of the active artificial floodplains, while oxbow lakes on the protected side preserved their open-water character, which represents an important natural value nowadays (Pálfai 2001). The lowland sections of the Maros and the Körös Rivers were even more extensively regulationed; therefore, now the Maros has an almost straight river bed, and the system of the Körös Rivers has also shortened by 60 %.

(B) Increasing flood levels and further environmental impacts

The regulations resulted in narrower cross sections leading to 2- to 3-m flood level rise on average. The problem is that the floods have continued to rise, even if to smaller extents, while the low water levels have been decreasing (Szlávik 2002). At the beginning, the continuous flood level increase was the consequence of artificial levee constructions, but nowadays it is related to the increasing thickness of sediments deposited on the artificial floodplains, and to their high roughness. The drainage of marshlands, the flood control and the inland water regulations resulted in the decrease of the groundwater level, which also lead to the formation of a new surface cover and the alteration of vegetation. The loess and sand steppe, and oak forests used to represent the zonal, and the wetland meadows, marshes and bogs were the azonal natural vegetation. The loess and sandy steppes were transformed into agricultural area; in addition, the majority of the native riparian forests were felled. That is, on the flood-protected areas the former vegetation has remained only in patches. (Besides, before the regulations, 40 % of the territory used to be covered by hydrophile communities, whereas it decreased to 3–6 % afterwards.)

Originally the flood cover on the floodplains lasted for a half-year on the huge floodplains of the Tisza and its tributaries, but now the duration and the area of floods decreased dramatically. Therefore, some scientists claim that the flood control caused decrease in precipitation and they intensified the dryness of the area. The problem is that this cannot be proven (Antal at el. 2000), but the actual evaporation is undoubtedly smaller on the former floodplains, and that is why a heat energy surplus may appear above these areas, which is not significant in its extent; however, if the aridification intensifies further due to the climate change, it may become worthy to concern it. The above-mentioned effects may become emphatic at the meso-/micro-regional levels, in positive ways (e.g. irrigation, water retention) or in negative ones (e.g. disappearance of wetlands, groundwater-level decrease).

From the pedological point of view, the most important consequences of the river regulation works were the altered water and element household of the soils. This resulted in intensified salinization and salt accumulation in soils of the Great Hungarian Plain (Várallyay 2000). Since Treitz, several soil scientists have argued that one of the consequences of the drainage of wetlands is the drying of soils and the appearance of salt-affected soils (which are the product of the most recent past in the basin). The water regulations undoubtedly created the possibility of soil solution concentration and water-soluble salt accumulation (as the diluting effect of floods is missing) on the flood-protected areas. The stagnantly high groundwater leads to salt precipitation near the surface; thus, different types of salt-affected soils could develop. The other salt source causing salinization is the irrigation water. In particular, since the 1950s, salinization stared in an indirect way: the leakage of the irrigation canals increased the groundwater level containing salt, and this caused so-called secondary salinization.

It could be concluded that one of the most important elements of water management is the issue of artificial levees, and not only from the point of view of flood protection. The National Environmental Program also aimed at expanding the infrastructure of main protection lines from 62 to 75 %. In any case, we must be prepared for the challenges posed by drainage, since the abundance (flood and inland water) and the shortage of water (drought) seriously endangers the agricultural production, besides, they can occur within the same year, when the water has to be drained in the former case, and has to be resupplied in the latter. This also justifies the elaboration of a new and complex management system in connection to water reserves.

## References


Ádám L, Marosi S, Szilárd J (1959) A Mezőföld természeti földrajza. (Földrajzi Monográfiák 2.) Akadémiai Kiadó, Budapest

Ádám L (1983) A Dunántúli-középhegység alakrajzi jellemzése. Földrajzi Értesítő 32(3–4):413–420

Andó M (1971) Alsó-Tiszavidék. In: Marosi S, Szilárd J (szerk) A tiszai Alföld. Akadémiai Kiadó, Budapest, pp 142–165

Andó M (2002) A Tisza-vízrendszer hidrogeográfiája. SZTE TFT, Szeged

Andrásfalvy B (1975) Duna mente népének ősi ártéri gazdálkodása Tolna és Baranya megyében az ármentesítés befejezéséig. (Tanulmányok Tolna megye történetéből 7.) Tolna megyei Tanács Levéltára, Szekszárd, p 353

Antal E, Járó Z, Somogyi S, Várallyay Gy (2000) A XIX. századi folyószabályozások és ármentesítések földrajzi és ökológiai hatásai. Magyarországon. MTAFKI, Budapest

Biró M, Molnár Zs (1998) A Duna-Tisza köze homok-buckásainak tájtípusai, azok kiterjedése, növényzete és tájtörténete a 18. századtól. Történeti Földrajzi Füzetek 5:1–34

Borsy Z (1961) A Nyírség természeti földrajza. Akadémiai Kiadó, Budapest

Borsy Z (1977) Evolution of relief forms in Hungary in wind-blown sand areas. Földrajzi Közlemények 101 (25):3–16

Borsy Z (1989) Az Alföld hordalékkúpjainak negyedidőszaki fejlődéstörténete. Földrajzi Értesítő 37: 211–224

Borsy Z (1995) Evolution of the North-Eastern part of the Great Hungarian Plain in the past 50000 years. Questiones Geographicae, Special Issue 4

Borsy Z, Félszerfalvi J, Lóki J (1981) A jánoshalmi MÁFI alapfúrás homoküledékeinek elektronmikroszkópos vizsgálata. Acta Geogr Debr 20:35–50

Borsy Z, Félegyhézi E, Csongor É (1989) A bodrogköz kialakulása és vízhálózatának változásai. Alföldi Tanulmányok 13:65–85

Bulla B (1964) Magyarország természeti földrajza. Tankönyvkiadó, Budapest 424

Cholnoky J (1910) Az Alföld felszíne. Földrajzi Közlemények 38:413–436

Csorba P (2009) Mennyire összeilleszthető a hazai természet- és társadalom-földrajzi adatbázis? Földrajzi Közlemények 133(1):181–189

Frisnyák S (1990) Magyarország történeti földrajza. Tankönyvkiadó, Budapest

Frisnyák S (2001) Akultúrtáj kialakulása és terjedése az alföldön. 1. Magyar Földrajzi Konferencia, Szeged. www.geography.hu

Gábris Gy, Nádor A (2007) Long-term fluvial archives in hungary: response of the Danube and Tisza Rivers to tectonic movements and climatic changes during the quaternary. Quatern Sci Rev 22–24:2758–2782

Gazdag L (1964) A Száraz-ér vízrendszere. Földrajzi Értesítő 13:367–374

Giber J (2005) Megújuló energiák szerepe az energiaellátásban. B+V, Budapest

Justyák J (2002) Magyarország éghajlata. Kossuth Egyetemi Kiadó, Debrecen

Kertész Á, Mika J (1999) Aridification—climate change in South-Eastern Europe. Phys Chem Earth (A) 24 (10):913–920

Konecsny K (2000) Az országhatáron túli tájalakítás hatása az Alföld vízviszonyaira. In: Rakonczai J (szerk) A víz szerepe és jelentősége az Alföldön. Nagyalföld Alapítvány, Békéscsaba

Lakatos V (1988) Krónika a kun pusztákról. Isis, Kecskemét

Lakatos D (2003) A termőföld mint alapvető erőforrás. http://www.date.hu/szervez/agrvid/kozgaz/04.htm (2008–03)

Lászlóffy W (1982) A Tisza. Akadémiai Kiadó, Budapest

Liebe P (1994) A rétegvízkészletek és a nyomásszintek változása a Duna–Tisza közi hátságon és azok hatásai a talajvízszintekre. In: Pálfai I (szerk) A Duna–Tisza közi hátság vízgazdálkodási problémái. Nagyalföld Alapítvány, Békéscsaba, pp 25–29

Lóczy D (2000) Az alföldi tájak változó hasznosítása és értéke. In: Frisnyák S (szerk) Az Alföld történeti földrajza. Nyíregyházi Főiskola, Nyíregyháza, pp 221–228

Lóki J (2000) Az Alföld általános képe. A dunai Alföld. A tiszai Alföld. In: Karácsony D (szerk) Pannon enciklopédia. Kertek, Budapest, pp 296–301

Lóki J, Félegyházi E (2007) ABodrogköz kialakulása és mai felszíne. http://geogis.detek.unideb.hu/Kutatas/Felszinfejlodes/Bodrogkoz.pdf

Lóki J, Szabó J, Konecsny K, Szabó G, Szabó Sz (2004) Az erdősültség és az árhullámok kapcsolata a Felső-Tisza-vidéken. 2. MFK, Szeged. www.geogaraphy.hu

Lovász Gy (2006) A pleisztocén–holocén hordalékkúpok fejlődéstípusai Magyarországon. Tiszteletkötet Hahn Gy.-nek. Egyetemi Kiadó, Miskolc, pp 117–121

Marosi S (2000) Síkságok és dombságok. In: Karácsony D (szerk) Pannon enciklopédia. Kertek, Budapest, 285–287

Marosi S, Szilárd J szerk (1969) Tiszai Alföld. Akadémiai Kiadó, Budapest

Mezősi G (2008) Magyarország környezetföldrajza. JATEPress, Szeged

Miháltz I (1967) A Dél-Alföld felszínközeli rétegeinek földtana. Földtani Közlöny 97(3):294–307

MNA (1989) Magyarország Nemzeti Atlasza. Carthographia, Budapest

Molnár B (1977) ADuna–Tisza köz felső pliocén (levantei) és pleisztocén földtani fejlődéstörténete. Földtani Közlöny 107:1–18

Molnár B, Iványosi Szabó A, Fényes J (1979) A Kolon-tó kialakulása és limnogeológiai fejlődése. Hidrológiai Közlöny 59(12):549–560

Nagy B (2001) Energiavagyon és kiaknázhatóság. Magyar Tudomány 48(11):1274–1279. http://www.matud.iif.hu/01nov/nagy.html

Németh T (szerk) (2005) A talaj vízgazdálkodása és a környezet. MTATAKI, Budapest
OMSZ = Mersich I (szerk) (2003) Magyarország éghajlati atlasza. OMSZ, Budapest
Pálfai I (szerk) (1994) A Duna–Tisza közi hátság vízgazdálkodási problémái. Nagyalföld Alapítvány, Békéscsaba
Pálfai I (szerk) (2001) Magyarország holtágai. KVM, Budapest
Papp A (1956) ANagy-és Kissárrét vidékének régi vízrajza. Acta Univ Debr 3:1–7
Pécsi M (1959) A magyarországi Duna-völgy kialakulása és felszínalaktana. Akadémiai Kiadó, Budapest
Pécsi M (1996) Geomorphological regions of hungary. MTAFKI, Budapest
Pécsi M (2001) Geomorfológiai felszínek képződése a lepusztulás, a felhalmozódás és a lemeztektonika tér- és időbeni változásának hatására. Földrajzi Értesítő 50 (1–4):33–48
Pécsi M, Somogyi S (1967) Magyarország természeti földrajzi tájai és geomorfológiai körzetei. Földrajzi Értesítő 16:285–302
Péczely Gy (2002) Éghajlattan. Nemzeti Tankönyvkiadó, Budapest
Rakonczai J, Bódis K (2001) Ageoinformatika alkalmazása a környezeti változások kvantitatív értékelésében 1. Magyar Földrajzi Konferencia, Szeged, p 18. www.geography.hu
Réthly A (1970) Időjárási események és elemi csapások Magyarországon 1701–1800-ig. Akadémiai Kiadó, Budapest
Schweitzer F (2004) A tiszai hullámtér geomorfológiai vizsgálata: víztározók, hullámtérbővítés, gátépítés. Jászkunság 2004(1):1–12
Schweitzer F (2005) Atomerőmű Magyarországon. História 27(6–7):22–23
Somogyi S (1994) Az Alföld földrajzi képe a honfoglalás és a magyar középkor időszakában. Észak- és Kelet-Magyarországi Földrajzi Évkönyv, 61–75
Somogyi S (2000a) Atermészeti változások és a társadalmi-gazdasági folyamatok kölcsönhatása az Alföldön a Honfoglalás előtt. In: Frisnyák S (szerk) Az Alföld történeti földrajza. NYF, Nyíregyháza, pp 7–25
Stefanovits P (1981) Talajtan. Mezőgazdasági Kiadó, Budapest
Strömpl G (1930) Helyneveink vízrajzi szókincse. Hidrológiai Közlöny 9:113–128
Sümeghy J (1944) A Tiszántúl. (Magyar tájak földtani leírása 6.) Magyar Királyi Földtani Intézet, Budapest
Sümegi P (2005) Loess and upper paleolithic environment in hungary. Aurea, Nagykovácsi
Szesztay K (2000) Az Alföld vízháztartása. In: Pálfai I (szerk) A víz szerepe és jelentősége az Alföldön. Nagyalföld Alapítvány, Békéscsaba, pp 7–16
Szlávik L (2002) Árvízvédelem. In: Somlyódi L (szerk) A hazai vízgazdálkodás stratégiai kérdései. MTA, Budapest, pp 205–244
Szodfridt I (1994) Az erdők és a talajvíz kapcsolata a Duna–Tisza közi homokhátságon. In: Pálfai I (szerk) A Duna–Tisza közi hátság vízgazdálkodási problémái. Nagyalföld Alapítvány, Békéscsaba, pp 59–66
Vágás I (2001) Az ezredforduló árhullámai a Tiszán. Magyar Tudomány 8:958–966
Várallyay Gy (2000) Magyarország talajviszonyai a folyószabályozások előtt. In: Somogyi S (szerk) Folyószabályozások és ármentesítések. MTAFKI, Budapest, pp 109–125
Várallyay Gy (2001) A talaj vízgazdálkodása és a környezet. Magyar Tudomány 46(7):799–815

# 8 Physical Geography of the Little Hungarian Plain


**Abstract**

The Little Hungarian Plain is situated in the north-western part of the Carpathian Basin; it is similar to the Great Hungarian Plain, as it is also a sinking area, though with a much smaller territory. Its material consists of several 100-m-thick loose fluvial sediments, and along the rim of the region aeolian material also deposited. The main character of the region is determined by the fluvial activity of the Danube. The alluvial fans are closely spaced determining the landscape forms, and influencing the utilization of the region and its potential environmental conflicts. In a meso-scale, a detailed description is given on the landform evolution of the area, its main and most characteristic geomorphological forms and general physical geographical characteristics. Besides, the exploitable resources (e.g. pebble and gravel bodies, $CO_2$ gas, ecological network of natural parks) comprehensive analysis is given in connection with the Bős/Gabcikovo Hydropower Plant, focusing on the very intensive landscape alteration and its consequences. This debate is the origin of an international conflict between Hungary and Slovakia, which is in connection with the discharge, water quality and ecological (partially social) issues of the Danube and the region.


## 8.1 Location and the Main Evolutionary Periods of the Macro-Region

The Little Hungarian Plain is the counterpart of the Vienna Basin of a similar character. It is situated in West Hungary, along the Danube and the Rába rivers. In geomorphological point of view, the entire Little Hungarian Plain is composed of a flat alluvial fan in the centre of a basin and (older) alluvial fans in the boundary regions of the basin.

According to the Hungarian landscape division still in use today (Marosi and Somogyi 1990), the Little Hungarian Plain comprises three meso-regions: the Győr Basin, the Komárom-Esztergom Plain and the Marcal Basin (Fig. 8.1). Several micro-regions are distinguished within the three meso-regions (the updated version (2009) of this division will be applied in the followings). A different division has tried to adjust these units to the borders of the neighbouring countries, and concerning the Little Hungarian Plain and the West Hungarian Border




G. Mezősi, *The Physical Geography of Hungary*, Geography of the Physical Environment,
DOI 10.1007/978-3-319-45183-1_8

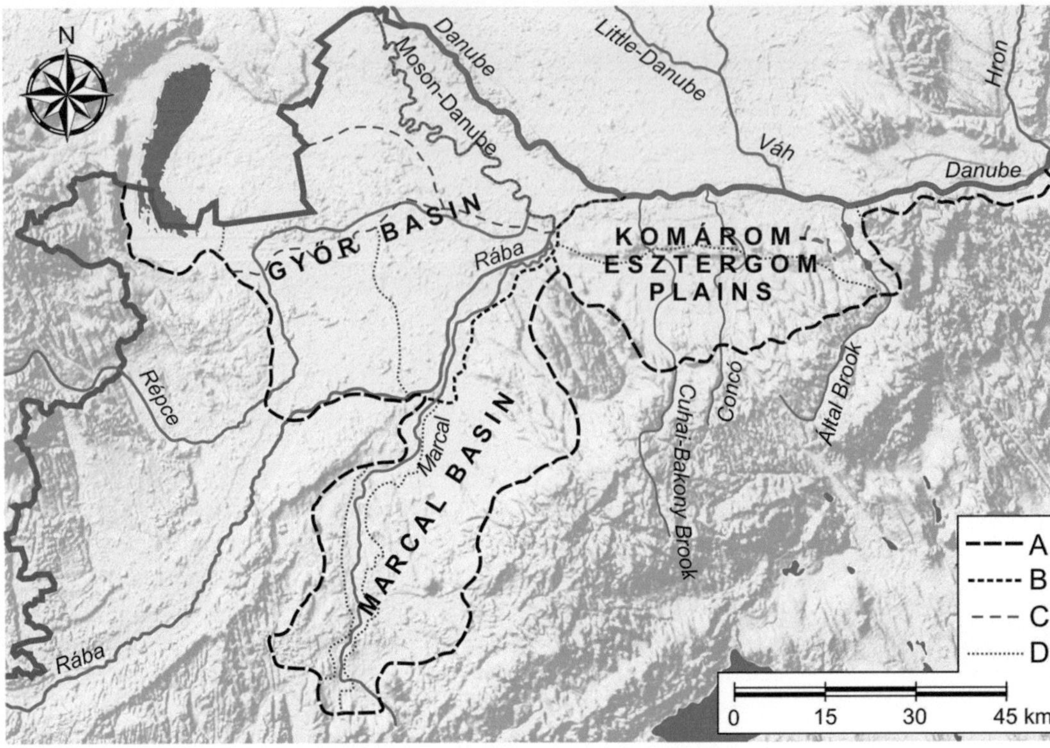


**Fig. 8.1** Natural landscapes of the Little Hungarian Plain (Marosi and Somogyi 1990). *A* Macro-region border, *B* meso-region border, *C* border of groups of micro-regions, *D* micro-region border; see the list of regions in Appendix A

Region as a unity to a certain extent (which is justifiable from a landscape ecological point of view), it defined 7 meso-regional units belonging to different levels of the previous division (Hajdú-Moharos and Hevesi 2000).

The evolution of the Little Hungarian Plain is similar to the one of the other large depressions of the Carpathian Basin within the Great Hungarian Plain. The deep structure, divided by the Rába fault line (Sect. 11.2) into two different parts, is not visible on the surface. The basement to the north-west of the line is composed of Paleozoic crystalline schist, which can be seen on the surface of the Sopron Mountains. The basement to the south-east consists of the Mesozoic, mostly carbonated ranges of the Transdanubian Mountains, though at its bottom Permian red sandstone can be found in a fold, and this rock is on the surface near the Lake Balaton. In connection with the formation of the water system, it was previously mentioned that the Győr Basin started to sink dynamically in the middle of the Miocene (almost until the depth of 3000 m). As a result of the subsidence, the surface schist of the Sopron Mountains can be found 2000 m deep at Répcelak. The carbonated zone to the east of the Rába fault line began to sink at the beginning of the Tertiary. The most significant cover material was provided by Pannonian sediments, similar to other regions.

The marine sedimentation gradually transformed into fluvial aggradation. The environment at the end of the Miocene can be described by the Baltavár Fauna Complex. The wildlife was characterised by ancient giraffe, gazelle, and some antelope species, ancient three-toed horse, deers and roes of the savanna woodlands and scrublands (Pécsi 1993). There were periodically reoccurring dry climatic periods; however, the climate was semi-arid. In this environment, the physical weathering resulted in sand, which was transported from the mountains towards the

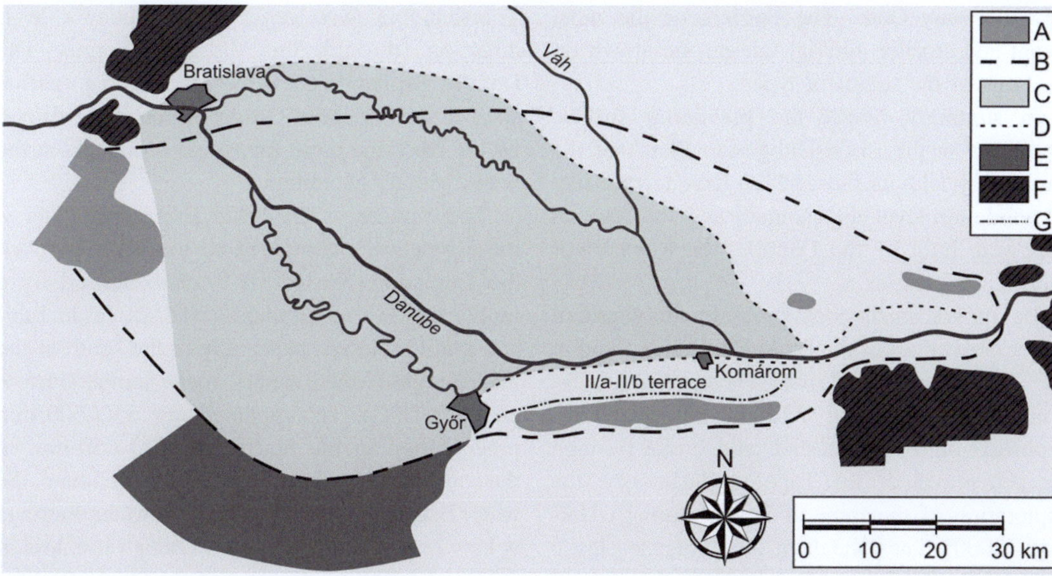


**Fig. 8.2** The supposed extension of the older and the younger alluvial fans in the Little Hungarian Plain (after Pécsi 1996). *A* Surface remains of the older alluvial fan, *B* extension of the older alluvial fan, *C* surface sediments of the younger alluvial fan, *D* supposed extension of the younger alluvial fan, *E* alluvial fans of the Rábca and the Répce Rivers, *F* Paleozoic and Mesozoic mountains, *G* edge of terraces No. II/a, II/b and III

basins by intermittent semi-desert waterflows and wind during the dry season. Meanwhile, the rivers with large sediment load formed pediments at the edges of the mountains. (When the climate became more humid, savanna areas with woodlands, scrubs and grasslands appeared in patches, and red soils and red clay were formed on the surface colonized by the vegetation.) Therefore, as the Danube and its tributaries arrived to West Transdanubia in the Pliocene, then to the Little Hungarian Plain ca. 2.5–3 million years ago, these rivers accumulated large amount of sand and gravel (originated from the physical weathering of Alpine crystalline rocks). The mostly sandy sediment deposited on Upper Pannonian layers, which filled the major part of today's Transdanubia in several hundred metres' thickness in some places.

The formation of the alluvial fan(s) of the Little Hungarian Plain probably started at the beginning of the Quaternary. Several researchers identified two surfaces joined and built on each another. The formation of the older alluvial fan lasted until the Mindel (the age of the Danubian terrace No. IV.), while the younger one was formed in the subsequent 400 thousand years (Fig. 8.2). The remnants of the older and orographically higher alluvial fans could be found between Bana and Bábolna (in a form of a "terrace island") and on the Parndorf Plateau. Their nomenclature and origin need some clarification. In general, the higher terraces are older (with old terrace sediments), while in the basins "normal" depositional sequence is characteristic; thus, the older materials can be found deeper. The mentioned two areas are older and situated higher than the young alluvial fans; however, they are not the results of fluvial terrace formation. The surfaces of alluvial fans deposited on each another and next to each other resemble to terraces in their appearance: the younger and lower alluvial fans merged with the higher, older and more eroded remnants, like terraces do, and appeared as new geomorphologic levels. The other problem is the name of the Parndorf "Plateau", since it refers to a gravel (and partly loessy) surface at only 180–200 m asl. The apex of the older alluvial fan (and according to Pécsi (1996), the Ancient Danube too, may have joined the Bruck Gate to the south, and not the

Devín/Dévény Gate. The borders of the most recent and smaller alluvial fan can be drawn on the basis of the sediment types.

In summary, today the prevailing surface landforms in the Little Hungarian Plain are the Quaternary alluvial fans, which have terrace-like or floodplain-level appearance; and the alluvial fans were built by the Danube, the Rába River and their tributaries.

Basalt volcanism took place in the southern and eastern edges of the Little Hungarian Plain in the Pliocene and the Lower Pleistocene (ca. 3–5 million years ago). (Basalt tuff and lava deposited onto Kemeneshát, and broke through the pea gravel of the Tapolca Basin—see the exploration of the mine at Lesencetomaj.) Harangi (2000) dates the lava and volcano-clastic rocks of the Ság Mountain located at the edge of the Marcal Basin to 5.5 million years (Fig. 8.3). The Kissomlyó and Somló developed under similar circumstances, and the latter was basically not affected by mining activities. The basalt volcanism terminated 3.5 million years ago (Harangi 2000), while others claim that it ended as early as 5.3 million years ago (Kovács 2008).

Considering the facts above, it could be assumed that basalt volcanism—apart from the subsidence of the Győr Basin and the elevation of the Vardar Ridge—may have been one of the causes of the flow direction change of the Danube, that is, the former N-S flow direction in Transdanubia was replaced by today's W-E direction (through the Visegrád Gorge). The Danube probably occupied the existing marine strait at the Visegrád Gorge, which is confirmed by the Miocene coral limestone sediments at the Szent Mihály Mountain.

The climate of the Little Hungarian Plain is more balanced (compared to the other parts of the Carpathian Basin): it is characterised by a moderately warm summers (20.5–21 °C in July) and a mild winters, especially in the south of the macro-region (the January mean temperature is above −1 °C), accompanied by 550–600 mm precipitation in the north, and 600–650 mm in the southern part. Despite of the mild climate, the water balance is negative, and the water shortage is between 50 and 100 mm, although it is less in the southern part. The zone of the Devín/Dévény Gate is one of the windiest areas of the country. Similarly to the Great Hungarian Plain, this macro-region also belongs to the forest steppe zone with low forest cover; however, the forests are extensive along the large rivers. Hardwood gallery forests characterize the banks of the Danube, and the proportion of hydrophile communities is high in the Szigetköz and the Fertő-Hanság Basin.

The geographical situation of the macro-region is mainly determined by its basin character, which is responsible for the high proportion of loose sediments, the plain surface, the

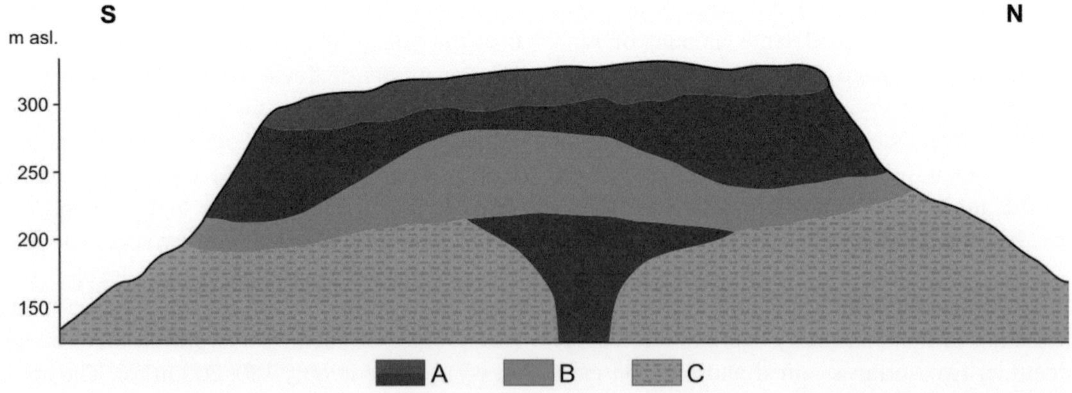


**Fig. 8.3** Mining reshaped the Ság Mountain; therefore, the figure only implies the structure created by the Strombolian-type cinder layers and the subsequent Hawaii-type lava flows resembling to (after Harangi 2000; Somogyi 1975). *A* Basalt, *B* basalt tuff, *C* sandy clay

fluvial processes and the near-surface groundwater level. A series of hydromorphic soils evolved in the low-lying territories of the basin (e.g. peat soils in the area of the former bogs of Hanság, and the alluvial and the highly decomposed peat soils of Szigetköz), while meadow soils (often with a shallow fertile layer) and chernozem soils developed in the orographically higher territories, whereas brown soils characterize the highly elevated areas (similarly to the Komárom-Esztergom Plain; Góczán 2000).

## 8.2 Overview on the Evolutionary Types of the Regions of the Little Hungarian Plain and Their Physical Geographical Characteristics

Most of the Little Hungarian Plain—and especially the **Győr Basin**, which includes the major part of its central area—is composed of floodplains of alluvial fans, which extend from Bratislava to Komárom in a length of more than 100 km and a width of 60–80 km, considering their Hungarian extension. The Szigetköz, the Moson Plain, Rábaköz and the micro-regions of the Fertő-Hanság Basin represent this geomorphological form. (The majority of the landscape belongs to Slovakia, e.g. Csallóköz). The gravelly sediment of the Danube is almost 250 300 m thick in the Győr Basin, and it stores large amount of water. The sandy coarse gravel layers probably started to deposit at the beginning of the Pleistocene in the normal stratigraphic order. Nowadays, the Danube flows along the elevated axis of the alluvial fan (the Danube has an aggrading or equilibrium character; Fig. 3.6), the Moson Danube flows in the southern, while the Little Danube in the northern edge of the alluvial fan. The two typical and well analysed micro-regions of the Győr Basin are Szigetköz and Hanság in the Fertő Basin (e.g. Kogutowicz 1930; Barczi et al. 2004).

The geological structure of the Szigetköz micro-region (Fig. 8.4) resembles to a series of interleaved plates, as its surface is covered by a dense system of paleo-channels and abandoned meanders. In accordance with the alluvial fan character, the central (younger) part is situated 4–5 m higher than the multichannel system at the edge (i.e. Moson Danube, Little Danube). The problems of shipping originating from the aggrading channel of the Danube have been analysed in detail in the hydrological chapter about the Upper Danube (see Sect. 3.1.2). The water regulations of the river (the Old Danube) attempted to solve the problem of shipping in the nineteenth and twentieth centuries (Fig. 8.5); however, there had been previous works too with a modest success in the Roman times. The hydropower plant system on the Austrian section of the river (and the deposition of bedload in the reservoirs) resulted in incisions in the lower section.

The evolution of the Fertő-Hanság Basin has been analysed in detail in Sect. 3.1.3. Its geographical units are uniformly water-affected—the presence of peat indicates it in both micro-regions—and situated in enclosed basins; however, they differ in water supply and traditional land use. In the Fertő Basin, water was important for the

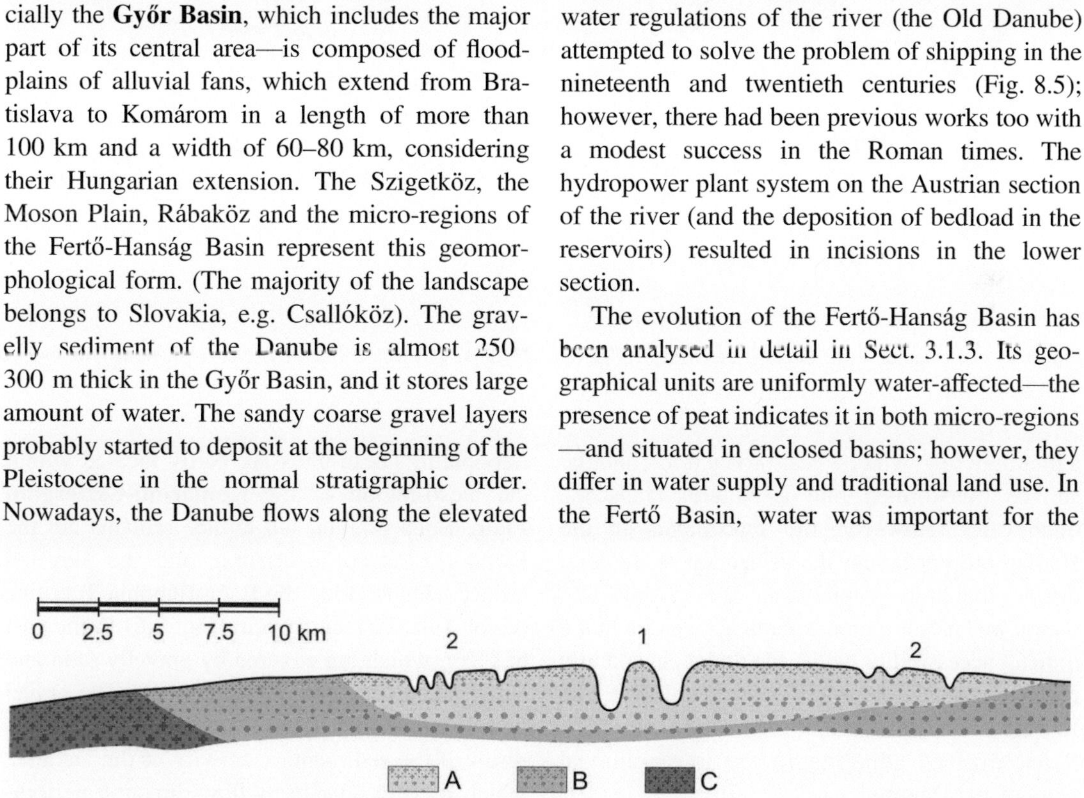


**Fig. 8.4** The profile of the alluvial fan of the Little Hungarian Plain (after Pécsi 1996). *1* main channel with bars (their environment was elevated due to alluvial sedimentation), *2* meandering side channel, *A* Holocene gravel and alluvial fan material reworked by channel migration and sedimentation, *B* Würm cryoturbated gravel, *C* Riss-Würm interglacial gravel with traces of cryoturbation

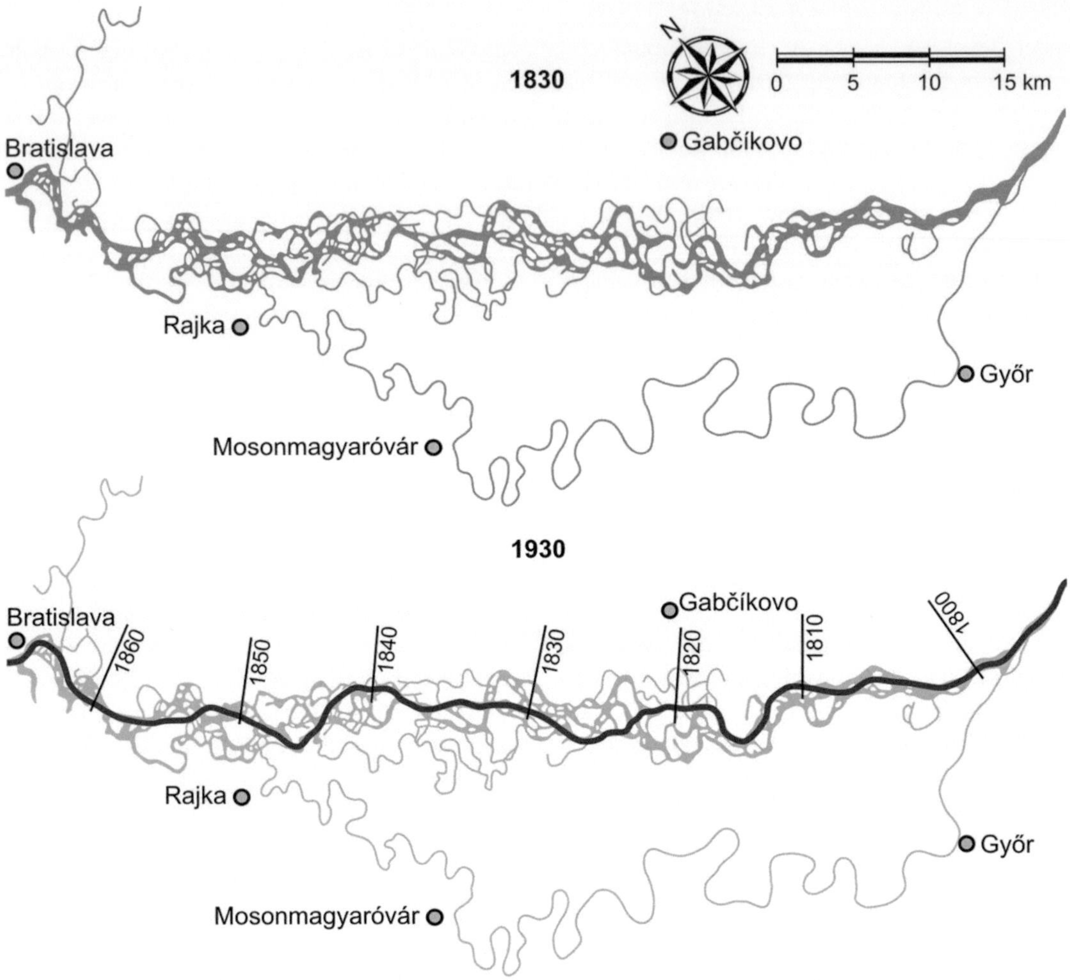


**Fig. 8.5** The channel of the Danube between Bratislava and Gönyű was transformed by river regulations (Ádám and Marosi 1975)

agriculture (e.g. wine, wheat production), and the surface transformed into a cultural landscape quite early. However, the marshlands in the Hanság did not favour the utilization of the land due to the high groundwater levels, thus, the region was preserved in its natural state due to the difficult accessibility before the drainage, and as a result, a particular way of lifestyle developed. The Rábaköz is an alluvial fan formed by the Rába River, evolved adjusting to the lower alluvial plain of the Danube.

The other units of the Little Hungarian Plain constitute flood-free alluvial fans, which are elevated by 50 m over the present-day floodplain. These units are partly located within the meso-region of the **Komárom-Esztergom Plain**. Such alluvial fan terrace remains are the Parndorf Plateau in Austria, and the elevated terrace islands along the Bana-Bábolna-Tata line (Győr-Tata Terrace System; Fig. 8.6) to the east of Győr, which are covered by gravelly sand and early Pleistocene alluvial fan gravels. The gravel field is the remain of the alluvial fan, and it is the result of the sedimentary activity of the Danube, which used to change its flow direction periodically in the Pliocene and the Lower Pleistocene. Remains of alluvial fans of different ages and heights in the terraces and fluvial landforms can

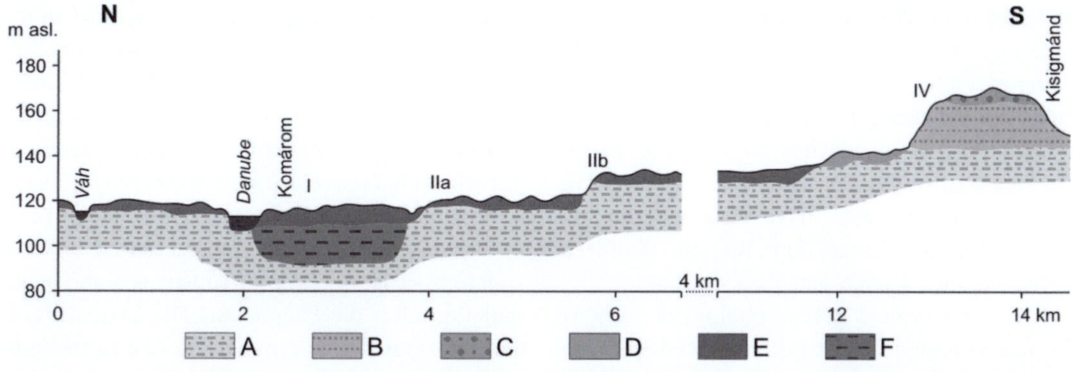


**Fig. 8.6** A geomorphologic profile of the Győr-Tata Terrace System between Komárom and Kisigmánd (after Pécsi 1996). *A* Pannonian clay and sand, *B* (Pliocene) cross-bedded sand, *C* terrace gravels (terraces No. I–IV.), *D* blown sand, *E* alluvial sandy mud, *F* fluvial sand and gravel

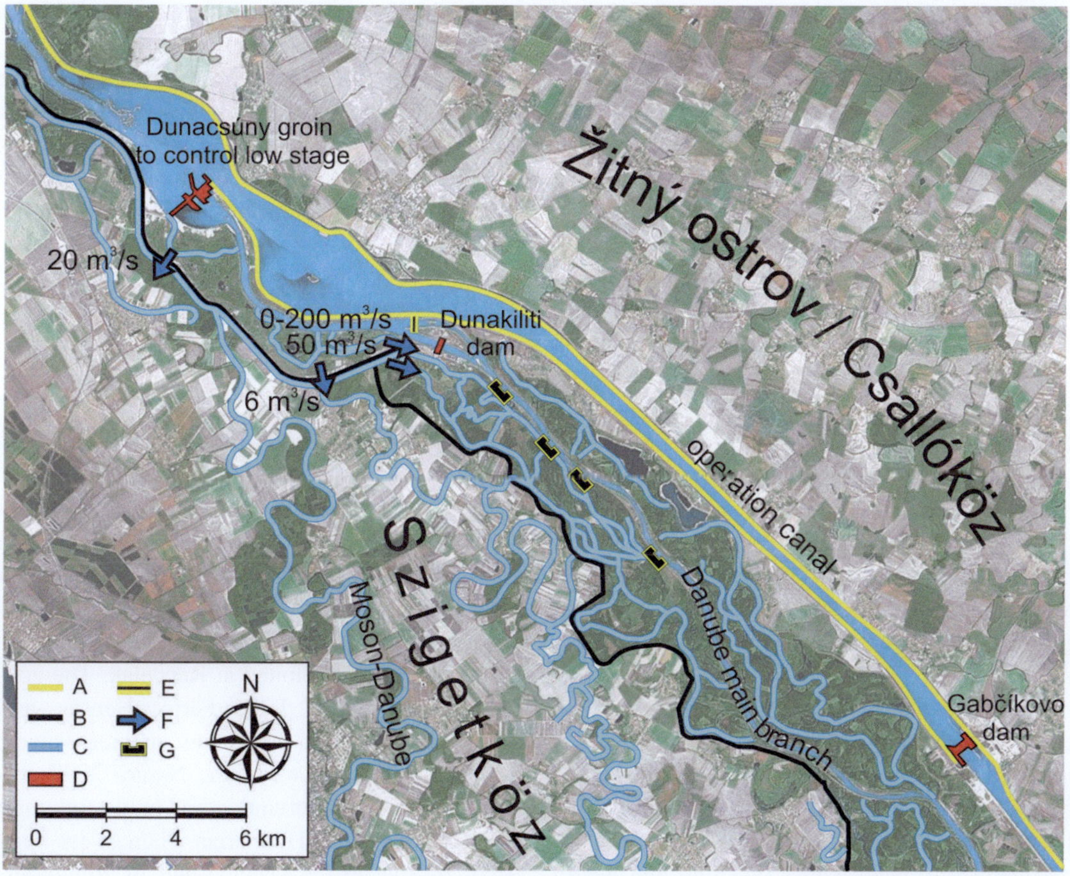


**Fig. 8.7** The facilities of water regulations on the Danube between Dunacsúny/Čunovo and Bős/Gabčíkovo. *A* Power canal, *B* artificial levee, *C* side channels in Szigetköz, *D* dam and lock, *E* groynes, *F* supply point of the side channels, indicating the water discharge at a given point, *G* side-channel lock

be found here. This micro-region has a particular situation, and it is has more connections to the Parndorf Plateau if we consider the natural features, but according to the landscape geographical divisions it is attached to the east, to the fluvial terraces of the Danube. The rivers arriving form the Bakony Mts. have dissected the formerly unified alluvial fan remain (Pándzsa, Cuhai-Bakony, Concó and Által Brooks).

The third typical geomorphological form of the macro-region is the system of terraced valleys, which also belongs to the meso-region of the Komárom-Esztergom Plain. The older terraced alluvial fan stretches into the Danube Bend, and younger terraces joined to the foreground of the valleys.

A series of terraces consisting of 7–8 levels evolved at the northern foot of Gerecse Mountains along the Danube between Komárom and Esztergom. The sequences of alternating layers of fluvial terraces, alluvial fans, foothill surfaces and delta gravel at the edge of Gerecse Mts. are covered by travertine, which protect the underlying forms from erosion. The travertine was deposited from thermal springs upwelling from the karstic rocks of the basement. The thick loess cover also took part in the protection of the layers. The geysers transported the underlying old quartz gravels along their chimney and funnel-shaped upwellings into the travertine, where the gravel was deposited (e.g. at the edge of Gerecse Mts. at Dunaalmás or Kőpite).

Similar terrace systems developed along the tributaries (or more exactly, they joined to certain terrace levels) as the terrace sequence of the Danube. The best known is terrace No. V. of the Által Brook at Tata, 50 m above the present-day floodplain level. The terrace is covered by a thick travertine, and within the limestone Pécsi (1973) identified the cultural layer of the prehistoric man of Vértesszőlős. A former thermal spring and lake provided unique ecological conditions for the prehistoric man, whose dwelling and hearth were found, and the flora and fauna of that time can also be reconstructed by the findings (Kretzoi and Dobosi 1990). The prehistoric man findings of Vértesszőlős represent one of the oldest sites in Europe.

The fourth typical geomorphologic form of the Little Hungarian Plain is the basin, which is represented by the meso-region of the **Marcal Basin**. The streams arriving from the Bakony Mountains and the Alps (mainly from here) built a series of gravel alluvial fans close to each other in the basin. In contrast with the Győr Basin, the alluvial plain of the Marcal Basin was characterised by erosional processes instead of subsidence and accumulation after the Pannonian. The Marcal River and its tributaries became active due to the subsidence of the Győr Basin. In geomorphological point of view, the plain surface of the Marcal Basin is interrupted by the Somló Mt. (423 m asl) with its lava-topped butte, which proves that the erosion by wind and rivers has eroded at least 200-m-thick sequence from the Pannonian layers under the basalt top (Image 8.1).

The Igmánd-Kisbér Basin also belongs to the Komárom-Esztergom Plain meso-region. It was formed by erosional–derasional processes, mass movements, wind erosion and the fluvial activity of streams flowing from the Bakony Mts. to the Danube.

## 8.3 The Most Important Natural Values and Some Environmental Conflicts of the Little Hungarian Plain

The Little Hungarian Plain is poor in traditionally utilizable minerals. Its specific mineral sources are the peat, which represents a conservation asset, the significant amount of gravel deposited by the Danube (e.g. at Hegyeshalom) or the carbon dioxide gas exploited at Répcelak. The ecological systems of Lake Fertő, Hanság and Szigetköz represent important natural values; moreover, they belong to the Natura 2000 network. The most important natural (and economic) value of the Little Hungarian Plain is its huge water reserve in the Danube and its tributaries, and in the Pleistocene–Holocene gravel layers of the floodplains and alluvial fans. At the moment, there is no surface water withdrawal from the Danube, and the extraction of subsurface water by bank-filtered wells is also partial. The water

**Image 8.1** The view of Somló (photo by Kiss, T.)

energy resources and the transport possibilities of the river are not completely exploited. Between 1969 and 1985, three or four times more gravel was extracted from the alluvial fan and the river bed than what the river transported from its Austrian section.

The regional development plans may focus on the above-mentioned environmental resources and values, and the related nature management tasks in the macro-region, which has a typical agricultural character.

The conservation of the valuable wildlife, the traditional land use and lifestyle was also supported by the establishment of the Fertő-Hanság National Park (1991) and by enlisting the "Fertő-Neusiedler Cultural Landscape" as an UNESCO World Heritage Site after a joint Austrian–Hungarian proposal. The territory of the Fertő-Hanság National Park contains the Lake Fertő, the reeds around the lake, which are several kilometres' wide zones in some places, the saline lakes along the shores, and the marshes and the fens. The basin of the Hanság—which used to constitute a joint water system with Lake Fertő—has to be highlighted because of its special wildlife. The lower, unregulated section of Répce River also belongs to the National Park.

One of the conflicts between society and nature was the drainage and drying out of the area of Lake Fertő and Hanság, whose adverse effects are still perceived through the constructed drainage system, contributing to the further decrease of the remaining natural wetlands. Wetland habitats are especially endangered, and their territory is diminishing countrywide. The process is the result of an earlier ideology, in which the arable land is more valuable than the state of the environment. In the area, approx. 1000 $km^2$ may have been covered with bogs, but its 97 % was drained during the water regulation works of the nineteenth–twentieth centuries (Lájer 1998). (Apart from the Hanság, the Ecsed Swamp and the Sárrét were also drained.) The wetland habitats of the floodplains have also been diminished all around the country for similar reasons, since the river regulations have practically terminated the functioning of floodplains in the ecological sense, which constituted a quarter of the surface of Hungary. The water management problems are indicated by the spread of forests on the valuable marsh meadows in the Fertő-Hanság National Park.

The other group of environmental conflicts is related to the operation of anthropogenic facilities. The dense network of roads, railroads and electric lines in the landscape disturbs the species, results in habitat fragmentation, and on the whole, endangers the natural values. It is further influenced by the "uncontrolled" tourism in protected areas.

The environmental hazards are primarily connected to the social exploitation of the above-mentioned resources and assets. The problems outlined in the following are related to the availability of water and mainly refer to the extensive potable water resources of the central part of the Győr Basin.

The main anthropogenic impact in Szigetköz was the establishment and operation of the Bős/Gabčíkovo hydroelectric power plant. From the 1950s, the basic idea was to utilize the energy provided by the Danube in Hungary (jointly with Czechoslovakia at the time) in lowland conditions, similarly to the several small Austrian hydroelectric power plants. The erroneous investment project, the Gabčíkovo-Nagymaros Dam, had a disputable environmental impact and has brought up still unresolved environmental and political issues (Szigetköz 2006). The subject of the treaty was the establishment of a hydroelectric power plant producing about 720 MW energy, which would have been supplied by a reservoir stretching to Bratislava. The power plant was joined to a separate power canal at Bős/Gabčíkovo, constructed for this single purpose, which did not appear in the original plans, though (Láng 2002). Since the intention was to operate the power plant at peak running, the 5–6 m high floods would be created twice a day. To level the flood waves off, a further power plant was planned in the Danube Bend at Nagymaros.

The construction started in the 1980s, and the related environmental conflicts came to the light at the end of the decade. Besides these, profitability and financial problems emerged too, which were further encumbered by political and geopolitical conflicts related to the change of the political system of the countries. Owing to all this, the Hungarian party terminated the treaty, and withdrew from the realization of the power plant of Nagymaros, which also meant the renouncement of peak running. Therefore, the Czechoslovak party built an independent power canal (Fig. 8.7) and started to operate the Gabčíkovo power plant in 1992. The two parties still could not agree on the use, because Hungary requires (more) water in the original channel of the Danube instead of obtaining electric power. The International Court of Justice found both parties responsible and called on both states to reach an agreement.

The water supply of the alluvial fan is an urgent environmental issue. At present, 80 % of the mean water discharge of the Danube (about 2000 $m^3$/s) flows in the artificial concrete canal (Image 8.2); thus, the water supply of the former main channel and the side channels is not appropriate, and thus the natural wildlife is declining. Furthermore, the groundwater level has decreased by 1–3 m in the whole Szigetköz and especially in its central region. Nowadays, the groundwater table approaches the original level just in the side channels. The vegetation and the wildlife of the main channel have to adapt to this change.

The resupply (infiltration) of the alluvial fan and the significant drinking water reserves is ensured by an artificial channel system built for this purpose and operated at great precision by experts, which utilizes the former side channels on the floodplain of the Danube. A great groin has been installed into the main branch to elevate the water surface and to facilitate the operation of the side channels (Image 8.3). The water discharge of the original main branch decreased drastically to 60–120 $m^3$/s, which is not enough to maintain the earlier level of infiltration. The infiltration of the alluvial fan used to be ensured by the main channel having 10–20 times as much discharge. The present system of side channels is blocked by locks to obstruct water drainage towards the main branch, which is located 4–5 m lower. In this way, water can infiltrate to the alluvial fan of Szigetköz (Fig. 8.7).

The created large water reservoir behind the Bős/Gabčíkovo Hydropower Plant entails a conflict related to water quality. Pollution accumulates in the reservoir (Vargha 2000), and the

**Image 8.2** The satellite image of the investments realized—the lock at Dunacsúny/Čunovo and the dam at Dunakiliti

**Image 8.3** The groin in the former main channel of the Danube at Dunakiliti

potential use of potable bank-filtered waters becomes limited. Several issues must be resolved, since the different needs of forestry, conservation (proper water levels), traffic (navigability of the main channel) and flood control conflicted as early as the planning phase; thus, the facilities were constructed with a transitional character.

Water quality problems affect the central part of the Győr Basin to the greatest extent, since the areas of subsurface water quality protection can be found here, and the flood hazard is also present. In this sensitive environment, settlements with high population and with hazardous waste production over the average are located (VÁTI 2005).

Two hazards have to be mentioned in connection with the soils of the Little Hungarian Plain. One is the processes of degradation (salinization and hydrogeological hazard), and the other is the soil contamination. Nowadays, degradation means the loss of expected soil functions (e.g. fertility) due to certain factors (Németh et al. 1997). In the case of the Little Hungarian Plain, the heavy mechanical composition is the factor most inhibiting fertility.

## References


Ádám L, Marosi S (szerk) (1975) A Kisalföld és a Nyugat-magyarországi-peremvidék. Akadémiai Kiadó, Budapest

Barczi A, Lóczy D, Penksza K (2004) Magyarország tájai. Kisalföld, Győri-medence. Környezetvédelmi és Környezetgazdálkodási Felsőoktatásért Alapítvány, Gödöllő

Góczán L (2000) A Kisalföld. In: Karátson D (szerk) Pannon enciklopédia. Kertek, Budapest, pp 302–305

Hajdú-Moharos J, Hevesi A (2000) A Kárpát-Pannon térség természeti tájbeosztása. In: Karátson D (szerk) Pannon enciklopédia. Kertek, Budapest, pp 274–285

Harangi R (2000) Bazaltvulkánok a Kisalföldön. In: Karátson D (szerk) Pannon enciklopédia. Kertek, Budapest, pp 306–308

Kogutowicz K (1930) A Dunántúl és a Kisalföld írásban és képekben I. M. Kir. Ferenc József Tudományegyetem, Szeged

Kovács I (2008) Újabb adatok a Somló felszínfejlődéséhez. Földrajzi Értesítő 57(3–4):257–271

Kretzoi M, Dobosi TV (szerk) (1990) Vértesszőlős Site, Man and Culture. Akadémiai Kiadó, Budapest

Lájer K (1998) Bevezetés a magyarországi lápok vegetációökológiájába. Tilia 6:84–238

Láng I (szerk) (2002) Környezet-és természetvédelmi lexikon I. Akadémiai Kiadó, Budapest

Marosi S, Somogyi S (1990) Magyarország kistájainak katasztere. I–II. Magyar Tudományos Akadémia Földrajztudományi Kutatóintézet, Budapest

Németh T, Stefanovits P, Várallyay G (1997) Talajvédelem. Spácium, Budapest

Pécsi M (1973) A vértesszőlősi ópaleolit ősember telephelyének geomorfológiai helyzete és abszolút kora. Földrajzi Közlemények 97(21):115–119

Pécsi M (1993) Negyedkor és löszkutatás. (Elmélet–Módszer–Gyakorlat 54) MTAFKI, Budapest

Pécsi M (1996) Geomorphological regions of Hungary. MTAFKI, Budapest

Somogyi S (1975) Felszínfejlődés és domborzat. In: Ádám L, Marosi S (szerk) A Kisalföld és a Nyugat-magyarországi-peremvidék. Akadémiai Kiadó, Budapest, pp 413–423

Szigetköz (2006) http://szigetkoznet.extra.hu/asajtotema/duna.htm; http://szigetkoznet.extra.hu/cikktema.htm

Vargha J (2000) Bős-nagymarosi vízlépcső környezeti kockázatai. In: Karátson D (szerk) Pannon enciklopédia. Kertek, Budapest, pp 476–477

VÁTI (2005) Környezetbiztonság. http://www.vati.hu/static/otk/int/jelentes2005178.pdf

# Physical Geographical of the West Hungarian Border Region

# 9

**Abstract**

It is difficult to sum up the main characteristics of the West Hungarian Border Region, as it constitutes of meso-regions with quite different characters and origin: Palaeozoic mountain ridges of the Alps stretch into the basin and pebble sheets connected to them. The meso-scale description of the region gives detailed analyses on this variegated region. The foothills of the Alps, the pebble and gravel sheets of the Alpine streams, and the alluvial fan of the Rába River (Kemeneshát region) are well-distinguishable units with special characters. The affiliation of the Zala Hills located SW of the region is the topic of debates, as the main characteristics represent a shift between two macro-regions (i.e. West Hungarian Border Region and Transdanubian Hills). From the point of view of natural resources, the Zala Hills are the most important because of its natural oil fields. Nowadays, in the macro-region, the ecological values and the conflicts in connection with them are the most important: for example the water quality of the rivers and streams, or the factories producing hazardous waste are located on environmentally sensitive areas.

## 9.1 Physical Geography and Landscape Development

The West Hungarian Border Region, macro-region is part of a larger macro-region divided by four countries. If the whole system is considered, a different name may be more justifiable, for example the West Transdanubia, which was suggested by Hajdú-Moharos and Hevesi (2000). From Hungarian point of view, the word "marginal" seems logical (Fig. 9.1), since it refers to the fact that the core of the landscape is the low, crystalline foothills of the Eastern Alps (Sopron Mts, Kőszeg Mts and Vas Mts), divided by Tertiary sediments. These foothills of the Alps in Hungary are not connected to Little Hungarian Plain, since they differ significantly also in their geographical and geomorphological structure and evolution. Thus, it is reasonable to treat this landscape as a separate macro-region, even if only a small part of it can be found in Hungary. A high level of heterogeneity also characterizes the whole macro-region.

The fundamental difference in the bedrock is indicated by the Rába Line (and partly the Hahót fault line) (Figs. 9.1 and 1.2). The fault line is



G. Mezősi, *The Physical Geography of Hungary*, Geography of the Physical Environment,
DOI 10.1007/978-3-319-45183-1_9

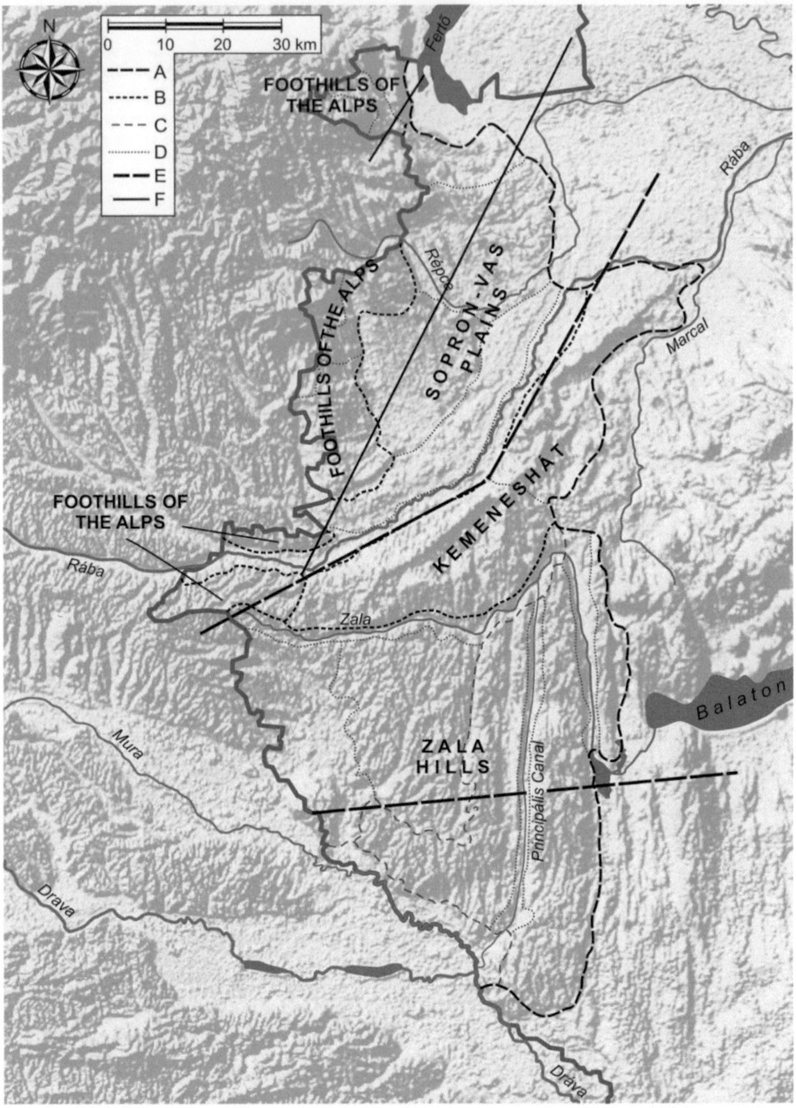


**Fig. 9.1** Micro-and mezo-regions and fault line structure of the West Hungarian Border Region (after Marosi—Somogyi 1990). *A* Macro-region boundary, *B* meso-region bondary, *C* micro-region boundary, *D* micro-regional border, *E* fault line (Rába Line dividing large structural units and *F* secondary fault line

located deep underground. The Palaeozoic crystalline sediments are situated west to the Rába Line (and appear also on the surface e.g. Sopron Mountains). The Palaeozoic formations are the foothills of the Alps and to these foothills other foothills are connected (e.g. Pinka, Gyöngyös), which are lowering towards the broad floodplain of the Rába River. This geological structure (the sediments of the Eastern Alps continue until the Rába Line in the basement of the Neogene sediments) underlie that the Sopron-Vas Plain is part of this macro-region.

After the Upper Cretaceous-Paleogene basin environment, the territory of the macro-region used to be a dry surface, until the Middle Miocene. After that period, marine/lacustrine sediments deposited in the whole area, except for the Kőszeg and Sopron Mountains e.g. the Badenian (Sect. 1.2) Leitha limestone at Fertőrákos, or the older Brennberg brown coal. (At that time in the Zala Hills, more than 3000 m thick Neogene sediments were accumulated.) This surface was covered by Pannonian lacustrine and Plio-Pleistocene fluvial sand and gravel in notable and uniform thickness. The fluvial sediments were probably deposited by the Danube, which flowed in the last 5 million years formerly to the southern direction, and later towards the Győr Basin. The alluvial fans of the streams coming from the Alps covered this surface. The largest one is the Kemeneshát, built by the Rába River, which is an independent landscape unit, separated from its environment due to its island-like relief character (Sect. 1.3).

The Zala Hills fit into the macro-region because of its sediments, partly Mesozoic and Neogene (storing petroleum), mostly marine sediment sequence, but also because of its geomorphologic, lithologic, ecologic and economic features. The Muravölgy Plain is fit more into the Dráva-Mura system, only its small extent in Hungary is justify its classification to the Zala Hills. The high landscape dissection is also indicated by the fact that the notion of a landscape in everyday life ("experienced") has manifested only in the cases of Őrség, Hetés, or Vendvidék, and local specificities of separating the landscapes have not been associated with the meso-region of the Sopron-Vas Plain.

It is difficult to provide a comprehensive, macro-regional assessment because of the high landscape dissection and the small Hungarian extension. The Hungarian part of the macro-region includes the pieces of the central, crystalline range of the Alps, having a fragmented, blocky structure on the surface, and the series of the adjoining foothill surfaces. The oldest, Old Paleozoic metamorphic (gneiss, phyllite, mica and Devonian dolomite), formations are represented by the pieces of the Sopron Mountains with tertiary faults. Here, the dominant forms were created by a series of tectonic effects. Whereas, in the Kőszeg Mountains, which consist of Mesozoic phyllite and slate, the different rock qualities created the dominant (e.g. karst) forms.

The macro-region is characterized by a more balanced and moderately warm climate with mild winters (with −1 to +2 °C mean monthly temperatures in January) and cool summers (with 18–20 °C mean monthly temperatures July). An Alpine climatic influence also effects slightly the Lower Alps. The annual sunshine duration (1800 h) is the lowest in the country due to the highest cloud cover. The yearly precipitation sum exceeds the 800 m in major part of the region (at the slopes of Írott-kő it also exceeds the 900 mm); thus, the water balance shows a 50–150 mm surplus, only the periphery close to the Little Hungarian Plain is drier (Péczely 1975).

The positive water balance, the fragmented relief and the low permeability sediments led to the formation of a dense water network as part of the Rába River, the Mura River and the Zala River systems. The formation of the water network was determined by the Quaternary tectonic movements, starting in the Pliocene: the subsidence of the Győr Basin and the uplift of the Vardar Ridge (on the territory of the Zala Hills). The main river of the macro-region is the Rába River, which collects the water flows arriving from the Alps, and its asymmetric, terraced valley has become a characteristic element of the landscape.

The original (natural) vegetation of the landscape is the closed oak, beech and extensive Scotch pine in the humid parts of mountains and plains. The present forest cover is 30 % of the region in spite of the anthropogenic impact and fragments of the natural vegetation has been preserved (perhaps) in the forms of pine species and peat bogs (Károlyi and Pócs 1968). The eastern part of the region is typically a cultural landscape, where natural vegetation could be preserved only in small patches, and the woody habitats are represented by turkey oak forests. The elements of the Alpine flora often appear in the highest regions and in the northern slopes [e.g. peat moss species, pine species and fern species (e.g. ostrich fern)].

Various soil types developed due to the diverse relief, climate and vegetation. Several types of the zonal brown forest soils can be found, the brown forest soil with clay illuviation —often in podzolic versions—occurs mainly in the Sopron and Kőszeg Mountains. It is moderately eroded, unfavourable soil moisture regime and low fertility.

Several other soil types also occur from the more eroded, medium fertility brown earth (mainly in the west of the Zala Hills and the

Kerka Region) to the acidic, non-podzolic brown forest soils (e.g. Kőszeg Mountains). The chernozem types occur on the loess-covered, gravelly alluvial fans. Moreover in the basins, valleys and the edges of foothills azonal soils are also occur (meadow and peat soils, soils of slope sediments).

## 9.2 Geographical Units of the West Hungarian Border Region

The name **Lower Alps** comprises two neighbouring blocks, belonging to the range of the Alps: the Sopron Mts. and the micro-region group of the Kőszeg-Vas Mts. (the existence of the block is questionable in the case of the latter). The Sopron Mts (557 m asl) are mostly built up of Palaeozoic metamorphic mica and coarse-grained gneiss. The slopes of the mountains are covered with Tertiary gravel layers. In earlier periods, the top region was also covered by this gravel sediment; however, here, this layer was eroded, thus surface formation could start on the crystalline basement. The geomorphological records of the Miocene tectonic movements are the morphological forms adjusting to the fault lines (Fig. 9.2), uplifted horst block, and valley incisions along the fault lines, main and side ridges and a staircase piedmont towards Fertő Lake. The Lajta limestone and the Lower Miocene coal depositions are evidences of the Upper Miocene marine and coastal environment. (A large quarry was established in Fertőrákos to exploit the Lajta limestone, which was used even in the Roman period. Today it serves cultural purposes.) The first coal mine of the country was opened to exploit the 10–15 m thick coal layer in Brennbergbánya in 1759 (Makádi 2000).

The slightly elevated crystalline Balfi block to the west of Fertő Lake and to the north of Sopron Basin is also a part of the foothill of the Eastern Alp. The channel of the Ikva River is following a paleo-channel of the Danube, and the remnants of the Lower Pleistocene gravel cover of the former

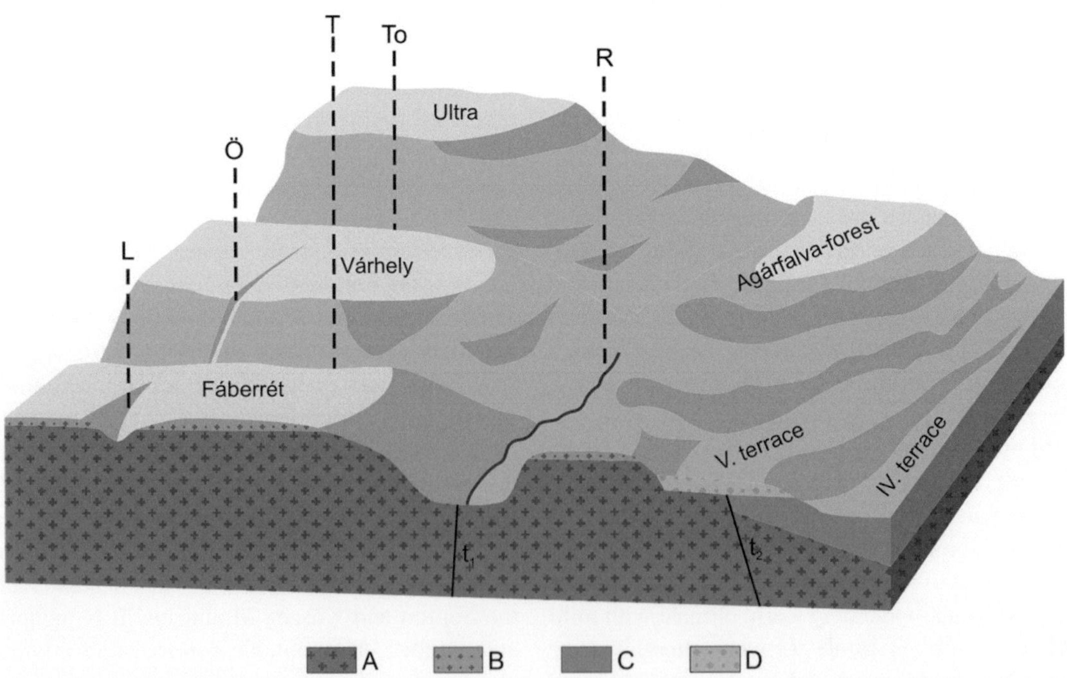


**Fig. 9.2** Surface forms of the Sopron Mts adjusting to the fault lines (after Ádám and Kárpáti 1975). *A* Crystalline slates, *B* remnants of gravel cover, *C* Miocene layers, *D* Pleistocene terrace gravel, *R* Rák Stream, *T* Tacsi Trench, *To* Tolvaj Trench, *Ö* Ördög Trench, *L* anonymous stream

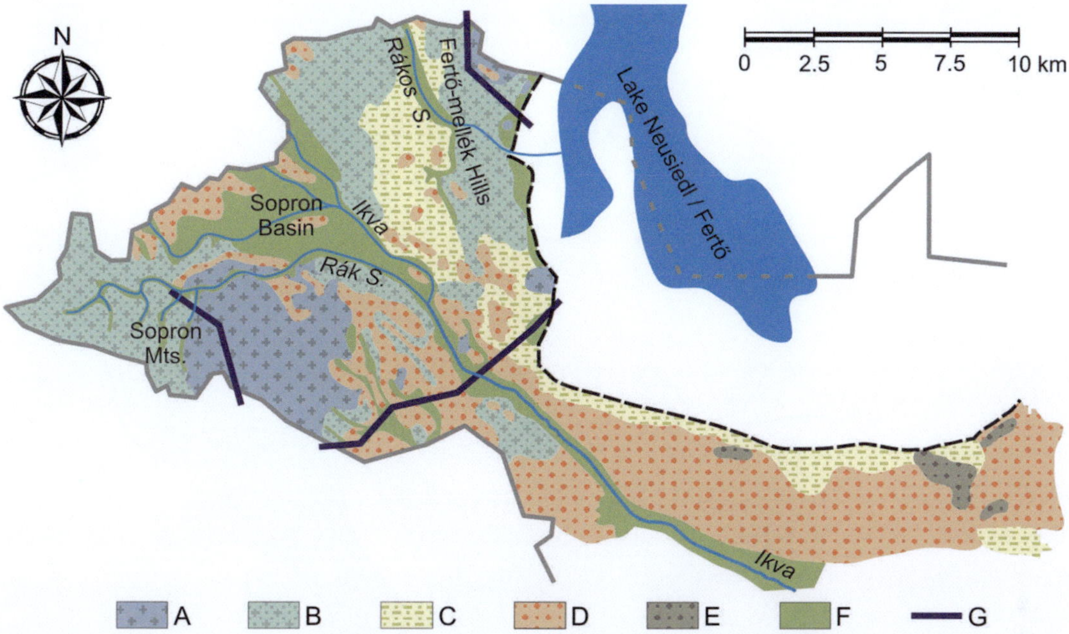


**Fig. 9.3** Geologic and geomorphologic map of the Sopron Mts and the Ikva Valley (after Ádám and Kárpáti 1975). *A* Crystalline slates, *B* Miocene sediments morphologically in a staircase piedmont position, *C* Pliocene clay, sand and gravel, morphologically the remnant of the foothill of Fertő Hills Fertő-melléki-dombság, *D* alluvial fan and gravel cover remnant of the Ikva River, *E* gravel cover wreck of the Danube and *F* Pleistocene sediments (clay, sand, loess), *G* main fault directions

river can be found on the surface at Fertőd (Fig. 9.3). The mountains are covered with closed beech and pine forests. The surroundings of the Sopron Mots are famous for grape and wine production, which is cultivated on the southern slopes and the on foothill of the Mountain.

The Kőszeg Mts and the Vas Mts represent the easternmost surface occurrence of the Variscan crystalline bedrock of the Alps. The plateau-like block of the mountains represents the highest base-level of Transdanubia (Írott-kő 882 m asl). Its rock material is still younger than the Sopron Mountains, because they are connected to the development of the Alps (in the Jurassic-Cretaceous). The calcareous and quartz phyllites, chlorite schists and serpentinites near Fertőcsatár have similar origin (Makádi 2000). Surface forms are the result of the diverse rate of erosion influenced by the erodibility of rock types, complemented by solifluction and derasion at the edges.

**The Sopron-Vas Plain** is an alluvial fan, built up of lowland-type gravel sediments, named on the basis of different gravel covers in the literature (e.g. West Hungarian Gravel Cover). The landscape development of this homogeneous meso-region is strongly related to the development of the gravelly alluvial fans, which deposited on the cross-bedding fluvial sand of the second half of the Pliocene. Before the formation of the alluvial fans, from the Middle Miocene, the area was a sedimentary basin, where more than 1000 m sediment was deposited. Rivers had deposited their alluvium on the surface of this sediment, and their flow direction had been defined by the sinking of the Dráva Basin, and later the Győr Basin (see Sect. 1.3). The development of the alluvial fans, the main structural forms of the meso-region was started with the flow direction shift of the Danube into the Győr Basin and with the appearance of the Rába River in the landscape at the beginning of the Pleistocene, although these flow directions had already existed before the Quaternary. Three larger alluvial fans were developed in south-north direction (Fig. 9.4). The oldest belongs to the Pinka River, which joined to the

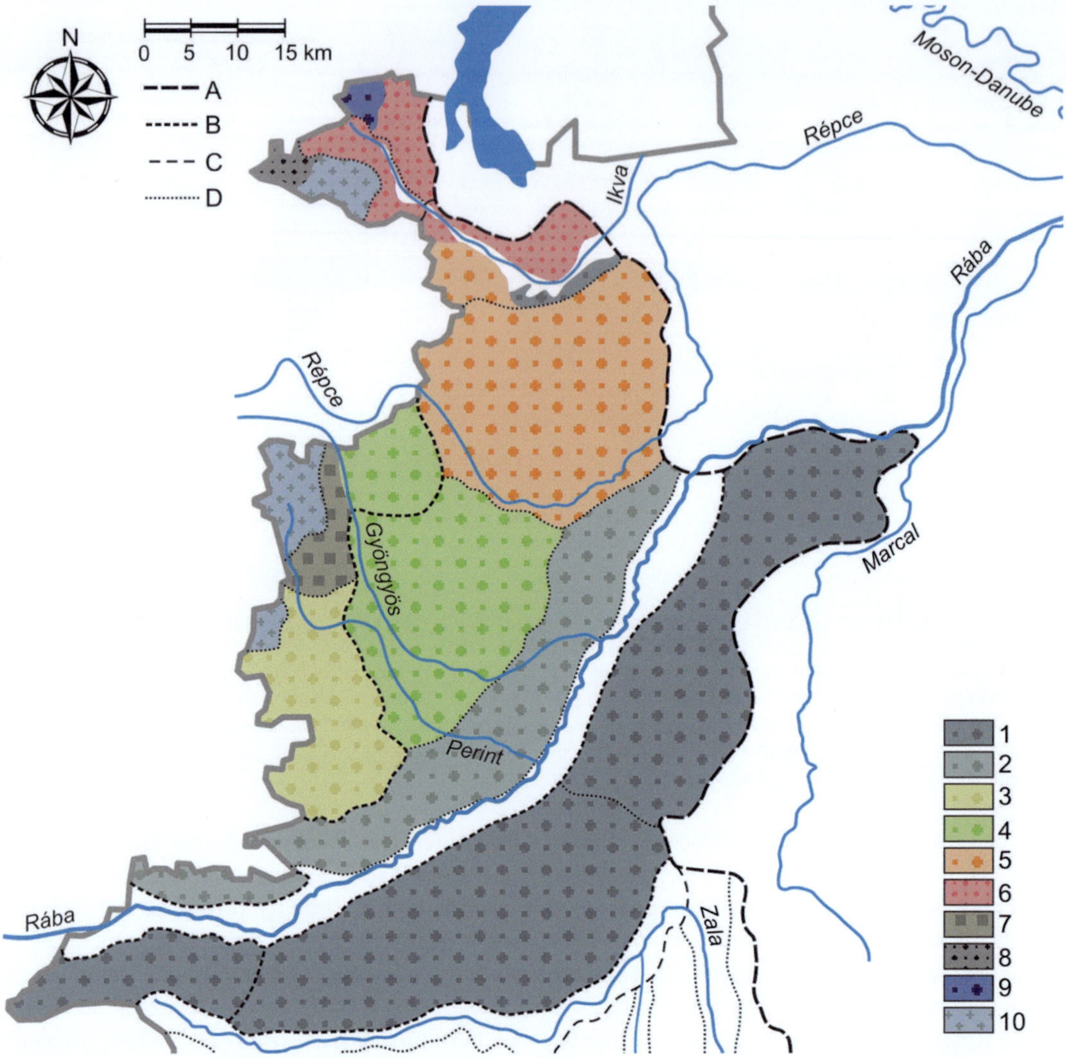


**Fig. 9.4** The genetic map of the Sopron-Vas Plain and Kemeneshát (after Ádám 1975). *A* Macro-region border, *B* meso-region border, *C* border of micro-region groups, *D* micro-region border, *1* gravel cover of the Rába River at the right bank side, *2* gravel cover of the Rába River at the left bank side, *3* the gravel cover of Pinka River, *4* gravel cover of the Gyöngyös River, *5* gravel cover of the Répce River, *6* gravel cover of the Ikva-Vulka River, *7* thin layer of angular gravel of the Kőszegi Mts, *8* the Miocene gravel cover and conglomerate of the Sopron Mts, *9* Pliocene gravel and *10* Variscan crystalline faulted mountains

Kemenshát, built by the Rába River, from the Lower and Middle Pleistocene. The alluvial fan of the Gyöngyös River was formed in the Middle Pleistocene, while the youngest in this sequence is the gravel cover of the Répce River, formed in the southwestern edge of the Győr Basin (heading towards the Hanság, and later towards the Győr Basin). The gravel layers deposited by the Répce were built from the beginning of the Pleistocene, and they become younger towards the south. They deposited on the sediments of the Danube (and later the Ikva River) (Fig. 9.5).

The eastern border of the alluvial fans is the Rába Valley which has a connection with the Kemeneshát. The valley has an asymmetric shape, with terraces on the western side. The Rába River started to form the valley, located in a NNE-SSW oriented structural line, after the river

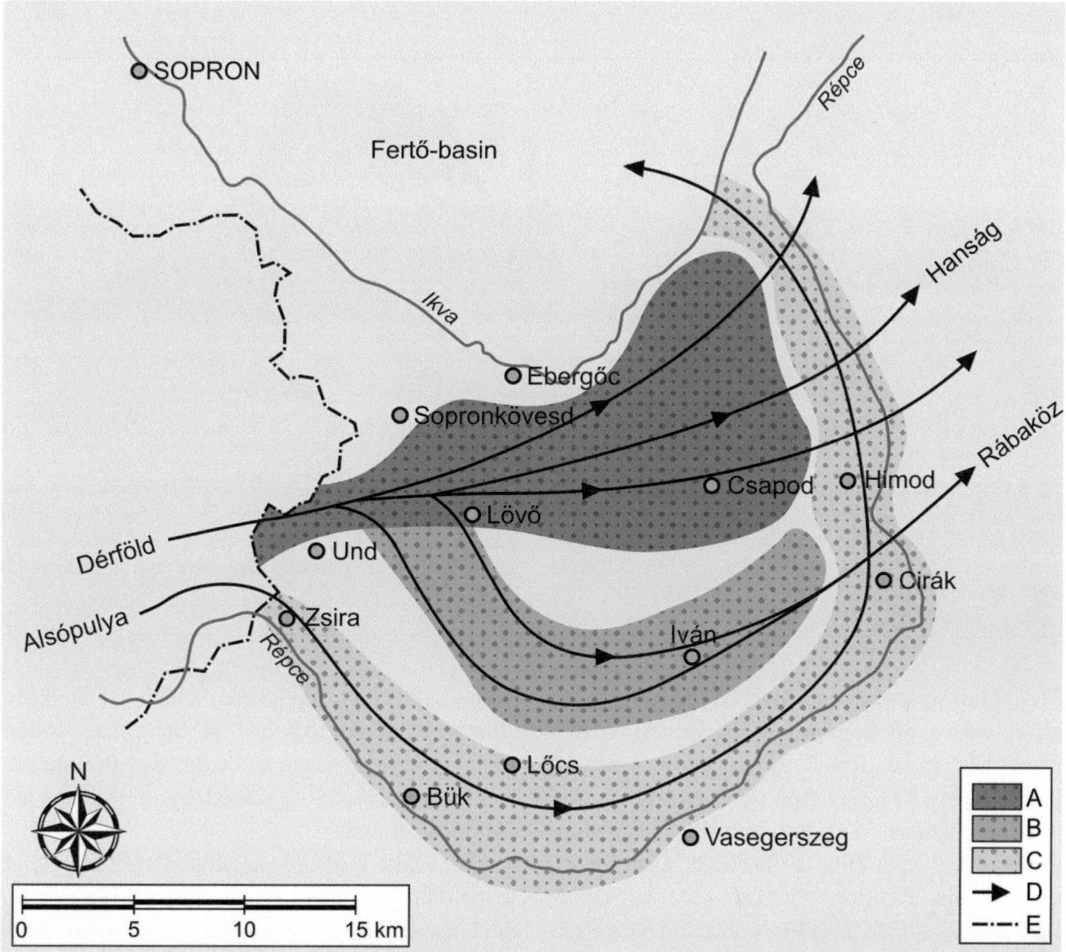


**Fig. 9.5** Gravel covers the Quaternary flow directions of the Répce River and (after Ádám 1975). *A* gravel cover from the end of the Lower Pleistocene, *B* Middle Pleistocene gravel cover, *C* Late Pleistocene gravel cover, *D* Pleistocene flow directions of the Répce River, *E* state border

left the Kemeneshát (the alluvial fan). The slope of the Rába River is relatively high, thus its energy was used. Seventeen watermills were closed in 1877–78 under Sárvár (Károlyi 1962), because they hindered flood protection. Because of the floods, the channel also had to be regulated. In the case of the Rába, these interventions had greater environmental impacts than the construction of the power plant at Ikervár, opened in 1900 (the first in the country).

The basement of **Kemeneshát** is composed of Upper Pannonian clay, which characterizes the entire Transdanubia, covered by cross-bedded fluvial sand. The sand is 20 m thick on average, and is related to the Danube (Fig. 9.6). The Rába River played the most important role in the formation of Kemeneshát. At the beginning of the Quaternary, the Rába River flew to the north-east towards the dynamically subsiding Little Hungarian Plain, to the foot of Bakony Mts., and built a gravelly alluvial fan, larger than its present extent. This accumulation phase was followed by erosion, when the 3–3.5 million-year-old basalt volcanoes formed on the Pannonian-Pliocene sediments (here and also in the Marcal Basin) were weathered out.

The more intense subsidence of the western part of the Győr Basin in the Middle Pleistocene forced the Rába River to slightly change its flow direction, thus the river shifted down from its

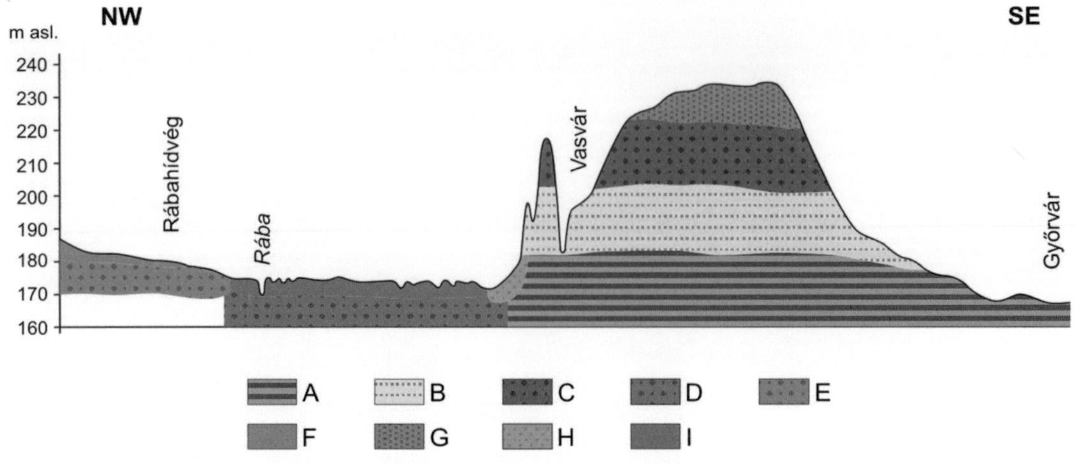


**Fig. 9.6** Schematic geomorphologic profile across the Kemeneshát formed on Pannonian clay (the alluvial fan of Rába), and the asymmetric Rába Valley (with merged terraces on the left side), between Rábahídvég and Vasvár (after Somogyi 1975, 2000). *A* Upper Pannonian clay, *B* upper Pliocene cross-bedded sand, *C* lower Pleistocene gravel of the Rába River, *D* late Pleistocene and older gravel of the Rába River, *E* late Pleistocene gravel of the Rába River, *F* loamy loess sediment, *G* allogenic sandy, loessy sediment, *H* sandy sediment and *I* present-day floodplain sediment

alluvial fan westward and reached the edge of the gravel cover of Sopron-Vas Plain. After this phase, the Rába River turned to its present north-easterly flow direction by the subsidence of the Rába Trench and the uplifting of Kemeneshát. (The separation of the older alluvial fan between the Parndorf Plateau and the Győr-Bana-Bábolna-Tata fault line can also be dated to this period.). The Rába River continuously downcut the Kemeneshát in the tectonically preformed section, consequently, a slope with 40–60 m relative height and 40–50 % steepness was formed on its western side, and a more gentle slope in the eastern side.

The occurrence of young river terraces on the west bank of the Rába River and the missing of terraces younger than terrace III form the east side can be explained by the differences in the surface formation. The intensive downcutting of the right side explains the lack of terraces. The separated plateau-like north-eastern part of the Kemeneshát is called "Cser" in Hungarian, after the characteristic tree species (oak) of the region. The gravel cover is sporadically covered by loessy–loamy sediment and brown forest soil and high quality chernozem brown forest soils. Terraces started to develop on the left bank after the sliding from the alluvial fan, which are therefore young (three terraces can be identified—Ádám 1975), however, they can be dated with difficulty due to the dissection caused by significant tectonic movements.

The **Zala Hills** in an alluvial fan-like landscape. The basement of the alluvial fan—in the Zala Basin—is a several thousand metre thick sandy, clayey, marly Pannonian sediment that is covered by cross-bedded (fluvial) sand, which sand is characteristic in the whole region (Juhász 1994). The development of the alluvial fans of the Rába and the Mura started on this basement W-E direction. The development of the distinct Zala Hills is connected to the uplift of the area (the uplift of the Vardar Ridge in the late Pliocene, early Pleistocene), causing the dissection of the 200–300 m asl fan surface by the rivers.

The recent character of the Zala Hills is determined by the system of meridional valleys and hilly uplands. In the past hundred years, a key question is the origin and evolution of the north-south oriented meridional valleys (e.g. Válicka, Principális, Zala, Sárvíz Stream) (Cholnoky 1918; Lóczy 1913). The wind erosion, as an explanation of the valley formation (connected to the climate of the Pliocene, thought to be arid at that time), is dated back to Cholnoky. This theory is based on observations in arid and semi-arid deserts that the straight, longitudinal valleys are formed by unidirectional winds, and parallel ridges (so-called yardangs) remain between them. Bulla (1964) and Sümeghy (1951) found that the meridional valleys of Transdanubia were mostly formed by stream erosion, however, some fault lines predicted the direction of the erosion (this may be regarded as an early stage of the tectonic explanation). Lovász (1975) was the first to point out that these valleys are complex forms and various processes took part in their evolution. Pécsi (1986) also confirmed this theory, and later it was also confirmed that the deflation (in the period of the Messinian Salinity Crisis) cannot be excluded, because these valleys had existed before the deposition of the cross-bedded sandy sediment (Schweitzer and Szöőr 1992; Pécsi 1996).

Nowadays, in the south-north oriented valleys, marshes and bogs are common (e.g. the 12 km long peat mine of Pötréte is located in this of environment). Thus, the valley did not favour the establishment of settlements, which is why the settlements were established at the footslope and ridges of the hills (especially on the characteristic "szeg" or "szer" in the Őrség and Göcsej regions). The valley bottoms were used as meadows or pasture, and the gentle slopes as plough land. Forests (especially beech) are typical on the brown forest soils and brown earth owing to the relatively high precipitation sum.

Another key geomorphological feature is the evolution of the double river capture of the Zala River. Góczán (1960) provided a comprehensive explanation (based on Lóczy, Cholnoky and Kéz) that the Zala River flew in a NE direction into the Győr Basin until the Middle Pleistocene, the terraces until the Riss can be identified in this direction (terrace IV is the youngest existing terrace, terrace III. cannot be identified; Gábris and Mari 2007). The faster subsidence of the Dráva Trench, and partly the Felső-Kapos-Kalocsa Depression caused that the Lower Zala River through headward erosion (today's Rinya Stream) captured the Zala at Türje, and led its water into the Dráva Trench. The age of the capture corresponds with the evolution of terrace III (Gábris and Mari 2007). The second river capture (at Zalavár) can be linked with the development of the Balaton Basin (in the Middle Würm).

## 9.3 Some Natural Values and Environmental Conflicts of the West Hungarian Border Region

The mineral resources of the macro-region are interesting from the point of view of geoscience, rather the economic point of view: copper, antimonite, graphite (Velem), bentonite (Fertőmellék) and talc (which was extracted in an underground mine in Felsőcsatár until 2007, but the quality of the material was not appropriate for even cosmetic industrial use, and the mine was closed down due to unprofitability—Rokob 2008).

Coal played a traditional role, as an energy source. The brown coal of the abandoned Brennbergbánya have already been mentioned, but the Pliocene lignite site of more than 1 billion tons near the settlement of Torony has to be also mentioned. The question of lignite mining is a good example of an environmental conflict of the region. On one hand, coal mining has economic-energetic interest (and the mining activity here was active until the 1980s characterized by low capacity). On the other hand, the region possesses archaeological and architectural values (Roman relics, Ják Abbey) and damaging the landscape may reduce the touristic attraction; thus, the intended mine did not get an exploration license in 2006.

The hydrocarbons used to have a major role in the economy after the Second World War. The petroleum and mineral gas resources of the Zala Hills are partly connected to the caverns of the karstified (and buried again with 8–10 % porosity) limestone (Nagylengyel), and partly, to the Miocene coarse clasolite. Petroleum production started in the territories of Budafa and Lovászi in the late 1930s. These sites have been exhausted by now, due to the intense exploitation. Later there were attempts to use carbon dioxide for the extraction before the turn of the millennium (Bíró 1999). Thermal waters of a significant amount can also be found in the Neogene sediments (e.g. Bük, Sárvár, Zalakaros), which were often discovered by the 1000 m deep exploration drillings for petroleum. On the balneologic use of thermal waters a touristic industry developed.

Metamorphic phyllite constitutes the majority of the Kőszeg Mountains, composed of marine sediments, clay and sand, which has already been mined by the Romans. The material was largely used in the buildings of Szombathely, in the construction of the water system, or the Castle of Kőszeg, similarly to the Leitha limestone, which was used for the building of the Roman castle in Sopron (Rokob 2008). The region is also rich in gravel and fluvial sand due to the alluvial fans and terraces. (It is a significant environmental problem that several former gravel pits are used as dumping sites, both legally and illegally—e.g. Sótony).

The surface waters usually have medium quality (class III), however, on the basis of nutrient balance and microbiologic pollutants, they belong to class IV. This is also true for the Marcal River, however, at one section, water quality is the worst in nutrient balance point of view (HKA 2005). An important environmental conflict between Hungary and Austria was the foaming of water in Rába River in the first decade of the twenty-first century. The reason was the wastewater of the factories operating in Austria, which was treated, however, the excessive amounts of salt and the naphthalene sulphonate exceeding the contamination limit values in the water created a critical situation. Due to the installation of new filtration equipment in water treatment process, less surfactants, which cause foaming, might enter the river. The solution of the water quality problems was an important issue in the first decade of the twenty-first century.

Two types of environmental conflicts have to be highlighted considering the entire macro-region. One is connected to the water quality protection areas of surface catchments, since they may be located in areas where wastewater treatment has a low ratio (e.g. Zala catchment or the water quality of Lake Balaton, in connection to which the situation of Small Balaton was mentioned). The other one is that there are some significant plants that produce hazardous waste in the sensitive protected areas of subsurface water quality (VÁTI 2005).

The Őrség National Park along the Austrian, Slovenian and Hungarian state borders was

**Image 9.1** The National Park preserves the traditional land-use and cultivation types, and some dwelling areas also preserves the traditional settlement structure in the Őrség (source: http://www.origo.hu/i/0709/20070910orseginem.jpg)

**Image. 9.2** Sporadic settlement in the Vendvidék (photo by Farsang, A.)

established to protect and preserve the special hilly environment, characterized by subalpine climate, valuable water surfaces, meadows, forests, and the relict Pleistocene bogs. The National Park preserves the traditional land-use and cultivation types and some dwelling areas also preserves the traditional settlement structure; thus, the "szeres" settlements in the Őrség (image 9.1) and the sporadic settlements in the Vendvidék (image 9.2) (KvVM 2008).

## References


Ádám L (1975) A Nyugat-magyarországi-peremvidék természetföldrajzi jellemzése. In: Ádám L, Marosi S (szerk) A Kisalföld és a Nyugat-magyarországi-peremvidék. Akadémiai Kiadó, Budapest, pp 68–81

Ádám L, Kárpáti L (1975) Alpokalja. In: Ádám L, Marosi S (szerk) A Kisalföld és a Nyugat-magyarországi-peremvidék. Akadémiai Kiadó, Budapest, pp 354–362

Bíró Z (1999) Kőolajkihozatal-növelő eljárás: a szén--dioxidos művelés. Magyar Tudomány 6:692–696

Bulla B (1964) Magyarország természeti földrajza. Tankönyvkiadó, Budapest 424

Cholnoky J (1918) A Balaton hidrogeográfiája. In: A Balaton tudományos tanulmányozásának eredményei. I. kötet, II. rész. MFT Balaton Bizottság, Budapest

Gábris G, Mari L (2007) A Zala lefejezése. Földrajzi Értesítő 55(1–2):39–55

Góczán L (1960) Közép-Nyugat-Dunántúl felszínfejlődési problémái. A Magyar Földrajzi Társaság XIV. Vándorgyűlése. Zalaegerszeg, pp 27–30

Hajdú-Moharos J, Hevesi A (2000) A Kárpát-Pannon térség természeti tájbeosztása. In: Karátson D (szerk) Pannon enciklopédia. Kertek, Budapest, pp 274–285

HKA (2005) Hazánk környezeti állapota. KvVM, Budapest. http://www.kvvm.hu/cimg/documents/0104_magyar.pdf

Juhász G (1994) Magyarországi neogén medencerészek pannóniai s. l. üledéksorának összehasonlító elemzése. Földtani Közlöny 124(4):341–365

Károlyi Z (1962) AKisalföld vizeinek földrajza. Földrajzi Közlemények 86(10):157–174

Károlyi Á, Pócs T (1968) Délnyugat-Dunántúl flórája I. Acta Acad Paed Agriensis 6:329–390

KvVM (2008) http://www.nemzetipark.gov.hu/

Lóczy L (1913) A Balaton környékének geológiai képződményei és ezeknek vidékek szerinti telepedése. MFT Balaton Bizottsága, Budapest

Lovász G (1975) Zalai-dombság. A domborzat kialakulása és mai képe. In: Ádám L, Marosi S (szerk) A Kisalföld és a Nyugat-magyarországi-peremvidék. Budapest, Akadémiai Kiadó, pp 442–454

Makádi M (2000) Az Alpokalja. In: Karátson D (szerk) Pannon enciklopédia. Kertek, Budapest, pp 309–310

Marosi S, Somogyi S (1990) Magyarország kistájainak katasztere. I-II. Magyar Tudományos Akadémia Földrajztudományi Kutatóintézet, Budapest

Pécsi M (1986) A zalai meridionális völgyek, dombhátak kialakulásának magyarázata. Földrajzi Közlemények 55(1–2):3–12

Pécsi M (1996) Geomorphological regions of Hungary. MTAFKI, Budapest

Péczely G (1975) Az éghajlat általános jellemzése. In: Ádám L, Marosi S (szerk) A Kisalföld és a Nyugat-magyarországi-peremvidék. Akadémiai Kiadó, Budapest, pp 61–63
Rokob K (2008) Kőbányászat a nyugati végeken. Élet és Tudomány 5, 144–147. www.mafi.hu/microsites/Foldeve/ET/fold_eve2.pdf
Schweitzer F, Szöőr G (1992) Adatok a Magyar-medence száraz-meleg klímájához a mogyoródi„ sivatagi kéreg" alapján. Földrajzi Közlemények 106(3–4):105–123
Somogyi S (1975) Felszínfejlődés és domborzat. In: Ádám L, Marosi S (szerk) A Kisalföld és a Nyugat-magyarországi-peremvidék. Akadémiai Kiadó, Budapest, pp 413–423
Somogyi S (2000) Ősvízrajz. In: Karátson D (szerk) Pannon enciklopédia. Kertek, Budapest, pp 238–242
VÁTI (2005a) Az ország környezeti állapota. http://www.vati.hu/static/otk/int/jelentes2005154.pdf (2008.03)

# Physical Geography of the Transdanubian Mountains 10

**Abstract**

The Transdanubian Mountains mainly consist of Mesozoic carbonate sedimentary blocks, but young volcanic elements also appear in the region. The main character of the macro-region is determined by limestone and dolomite covered by younger sediments, as they influence the rate of (weak) karstification and the type of vegetation, which is also influenced by elevation and lithological changes. The main natural resource of the area is bauxite with declining exploitation, manganese and brown coal with already closed mines. Nowadays, they rather constitute a reserve. The mining activity already terminated, but it caused very complex environmental problems, and the relaxation period of the landscape is very long, as the extreme lowering of the karst water table caused several problems. The most well-known consequence of the mining was the limited utilisation of Hévíz Lake, as the discharge of the hot springs supplying the lake decreased dramatically. The problem of decreasing water supply appeared in several lakes, as karst water withdrawal was very intensive during the mining. Probably, the regeneration and relaxation of the hydrological system will last for decades after the termination of mining. In the eastern and north-eastern meso-regions, the greatest problems are in connection with dense urbanisation (region of Budapest) and with intensive industrial activity along the Danube in the northern foreland of the Gerecse Mts. (e.g. at Almásfüzitő).

## 10.1 Location and the Main Evolutionary Periods of the Macro-Region

The Transdanubian Mountains is a low (average: 300–400 m asl) mountain range stretching from SW to NE. It is mainly composed of carbonated sedimentary rocks and some Tertiary volcanic materials. The horst range is segmented by grabens and valleys perpendicular to the main direction (e.g. Mór Graben, Tata Graben, Váli-víz Valley and Pilisvörösvár Graben). The macro-region consists of three meso-regions: the Bakony Mts., the Vértes-Velence Mts. and the Dunazug Mts; the latest is a group of several members (Gerecse, Pilis, Budai Mts.; Fig. 10.1).



G. Mezősi, *The Physical Geography of Hungary*, Geography of the Physical Environment,
DOI 10.1007/978-3-319-45183-1_10

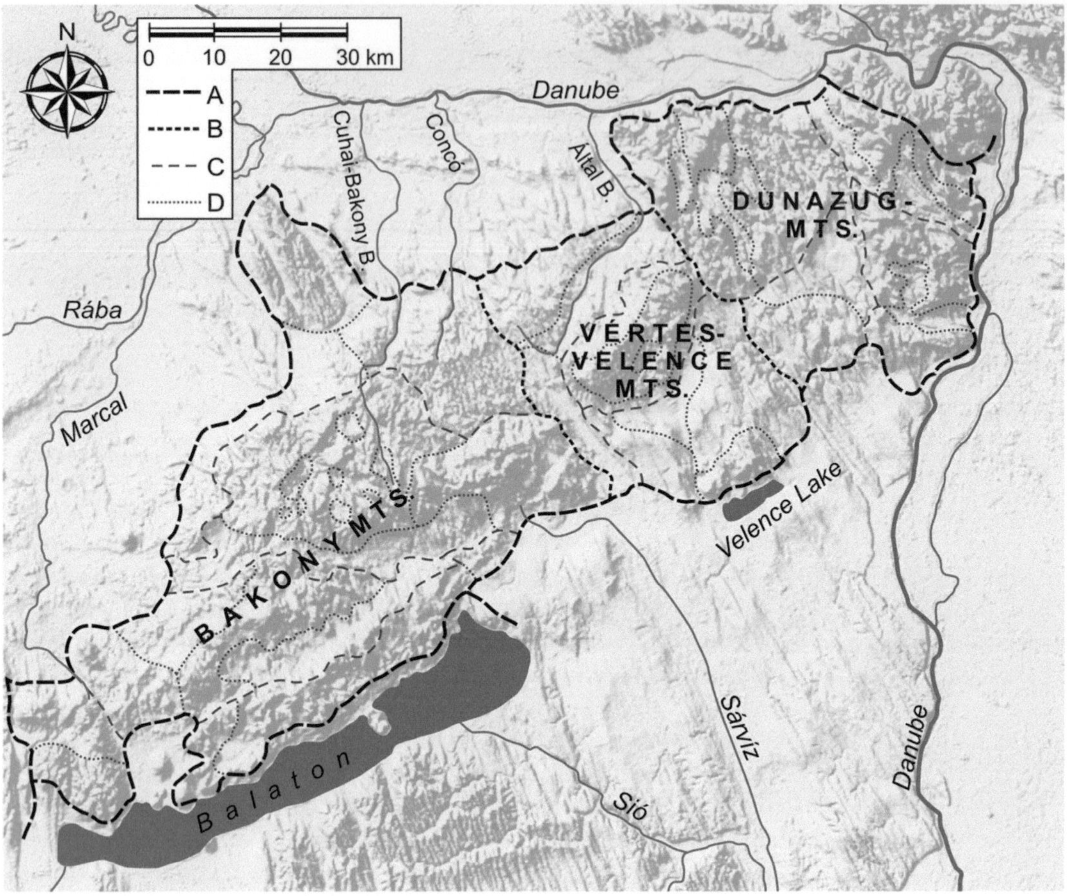


**Fig. 10.1** The landscape division of the Transdanubian Mountains (after Marosi and Somogyi 1990) *A* macro-regional border, *B* meso-regional border, *C* border of groups of micro-regions, *D* micro-regional border

The mountains and their foregrounds contain various natural resources (e.g. bauxite, brown coal, thermal water, Lake Balaton).

The macro-region evolved on the ALPACA plate, which merged of smaller crust plates since the late Mesozoic. Until the end of the Cretaceous, they located on the southern platform of the Tethys; thus, their material is mainly composed of carbonated sediments (see Sect. 1.1). These deposits constitute the main bodies of the mountains of the Bakony, Vértes, Gerecse and Pilis (Budai et al. 1999). In the second half of the 70-million-year-long Cretaceous, the surface became terrestrial; thus under the tropical climate, the karstic (cone karst) surfaces became planated, and as the result of the weathering, bauxite was formed. Karstification is not only indicated by the cone karsts discovered during the bauxite mining (see the protected geological site of the mine at Darvastó), but also some paleo-karst forms were explored by manganese mining (see Sect. 1.1). The bauxite was covered by Eocene marine sediments, which become thinner towards the north-east, from a thickness of several hundred metres to only 10 m (see Sect. 1.2). (The Eocene sediments were eroded during the Oligocene, and most of the coal formations were destroyed; Juhász 2000.) Since the end of the Eocene, the relief units composing the mountain range shifted to their present location after a few hundred kilometres of drift towards north-east (see Sect. 1.1). These units may have got buried, and then elevated and exhumated several times during their shift; thus, their sediment was eroded from the karstic blocks (Pécsi 1996).

After the north-eastern drift, the units became stable since the Middle Miocene arranged to similar directions. The terrestrial surface development was diversified in few places by a marshland environment, where coal could form (Hámor 2001). The predominantly terrestrial state was characteristic in the Upper Miocene too, when the elevated surfaces were formed by fluvial processes. The exceptions were mainly the southern peripheral areas, where marine and later Pannonian lacustrine sediments deposited. In the close vicinity of Lake Balaton, some remnants of this abrasional activity were preserved.

Since the Miocene the relatively uniform structure has become elevated from the environment, and finally the present form of the mountain range with horsts and grabens was created as a result of Pliocene–Pleistocene tectonic movements. The segmented structure of the mountains developed since the Cretaceous, and it characterizes the macro- and micro-regions too. For example, the Bakony and Vértes Mts. are separated by a graben. The horsts of the Gerecse Mts. are separated by N-S grabens and fault lines, whereas the faults and the connecting basins have NW-SE directions in the Pilis and Buda Mts. The Pleistocene uplift by 200–300 m (especially in the Gerecse Mts.) is indicated by the levels of travertine occurrences arranged vertically (Scheuer and Schweitzer 1988) and the similar vertical segmentation of the Danubian fluvial terraces (Ság and Pécsi 1987). An intensive erosion accompanied the increase of the relative relief, which significantly contributed to the surface formation of the mountains. Basalt volcanism connected to the Pliocene–Pleistocene tectonic movements resulting in large basalt surfaces (e.g. Agártető) and buttes shaped by erosion (e.g. Badacsony, Szent György-hegy, Haláp). The second half of the Quaternary used to be a period of karstification on covered and uncovered carbonated surfaces, and it was also the period when alluvial fans developed along the foothills (small ones due to the chemical weathering of carbonates). Nowadays, karstification is active on covered karst surfaces; consequently, the number of surface karst forms is low (an exception is the karrenfeld of the Tés-fennsík), and the number of larger and visitable caves is also small (e.g. Lóczy Cave, Lake Cave at Tapolca formed in young, Miocene limestone). The low rate of active karstification is also explained by the high proportion of dolomite and the extensive segmented structure of the mountains.

The climate of the Transdanubian Mountains is determined and influenced by various geographical features. The geographical situation may modify the subcontinental character, which is manifested as a sub-Mediterranean climatic effect, for example, along the northern coast of the Lake Balaton (creating favourable conditions for grape and fruit production). The climate is also modified by the direction of the mountain range, since it is perpendicular to the dominant NW and the SE winds, the latest are the most frequent during warm fronts (Péczely 1987). This geographical setting results in precipitation surplus, which leads to a positive water household. As the precipitation infiltrates into the karst, it will result in abundant springs, and especially in the northern units, the discharge of streams is increased by karst water inflow (Marosi 1987).

The modifying factors have manifested at the level of climatic elements as well. The mean temperature of the summer half-year of the Transdanubian Mts. is 17.9 °C; it is 19.5 °C in July at 300 m asl and 21 °C in the periphery, respectively; while the mean temperatures of January vary between −2 and −3 °C in the areas above 300 m asl. It is difficult to record regional differences, since the effect of the altitude on temperatures overrides them significantly. According to the calculations of Péczely (1987) in the macro-region in the summer half-year, the vertical temperature decrease is 0.58 °C/100 m, which value is higher than the average of the entire Carpathian Basin. A modest precipitation decrease can be observed from west to east in the Transdanubian Mts., and a vertical trend also can be detected similarly to the temperatures (approx. 35 mm/100 m increase rate). Thus, the highest amount of precipitation (approx. 800 mm) is measured in the Bakony, while the average value is only 550–600 mm in the foreground of Dunazug Mts. Therefore, the entire Bakony, and most of the Vértes and the Dunazug Mts. are mostly humid; the 1.0 contour line of aridity index is at the southern edge of the macro-region. However, the intramountain basins are drier.

The hydrography of the mountains determined by the geological structure of the horsts and grabens, the positive water regime and the hydrological features of the mostly karstic rocks (they are responsible for the poor surface water network). The streams of the Bakony Region belong to the catchment of the Marcal River (e.g. Gerence) of the Danube (e.g. Sárvíz) and Lake Balaton; while the streams of the Vértes are collected by the Császár-víz and Által Brooks, and the local erosional base of the Dunazug Mts. is the Danube. The quality of surface waters had been deteriorated by industrial and mining activities—reduced or terminated by now—and by the intensive agricultural cultivation, which increased the sediment transport and accumulation, and sometimes river bed regulations were needed (Somogyi 1987c). In the macro-region, lakes with highly different origin can be found: the karst spring lakes of Hévíz (see Sect. 3.1.3) and Tapolca, the volcanic lake of Belső-tó in Tihany and there are smaller lake basins on the basalt cover on top of the buttes (e.g. Bika-tó on Fekete-hegy). The other group of lakes are connected to depressions in the foreground of the mountains, which were created by erosion or subsidence, and their former basins have already been filled up. The groundwater plays a significant role only in the small intramountain basins, in terraces and in larger valleys. Karst water represents a regional value among the subsurface waters, both from qualitative and quantitative points of view, which wells up as springs in the peripheral areas of the mountains (there are more than 100 springs just in Budapest.).

The zonal soils of the region are brown forest soils. The diversity of soils is caused by altitude, precipitation and lithological differences. As the humid-cool climate is replaced by drier-warmer one, the brown forest soils with clay illuviation are replaced by brown soils, and then by chernozem forest soils. If the fertile layer remained thick, chernozem forest soils developed on the carbonated rocks; otherwise, redzinas have been formed on karst surfaces and leptosols on volcanic ones. The redzinas and leptosols have low quality due to their bad water regime, while the brown soils and brown chernozem soils are more suitable for agricultural purposes (Góczán 1987). The soil erosion is the most severe in the northern hill slopes of the Bakony and Vértes Mts., and on the south-western, cultivated slopes of the Buda Mts.

The natural vegetation of the Transdanubian Mts. shows high diversity, partly because of the existence of sub-Mediterranean, subcontinental and subatlantic climatic influences, and because of the relief and lithological effects (Marosi 1987). The zonal plant communities are beech, hornbeam–oak forests and turkey oak forests, which could be modified by human impact (Jakucs and Fekete 1987, in: Ádám et al. 1987). At higher elevations in the Bakony and Vértes Mts., beech and hornbeam–oak forests developed adjusting to the subatlantic climate effect, while turkey oak forests (mostly on forest soils) are widespread in the Dunazug Mts. owing to the subcontinental climate influence which gets stronger towards north-east. Calciphilous oak forests, karstic shrub forests and grasslands appear on karstic surfaces (Martonné 2006).

## 10.2 Overview on the Evolutionary Types of the Regions of the Transdanubian Mountains and Their Physical Geographical Characteristics

The central part of the macro-region with the largest extension is the **Bakony Region**, where the characteristic features of the whole mountain range occur, usually territorially separated. These are carbonated and volcanic rocks; the typical landforms are the Mesozoic horsts, tectonic basins and volcanic buttes.

The westernmost part of the Bakony Region is the Keszthely Mt. Its horst has clearly defined borders from each direction by structural lines and grabens. The mountain is composed of Triassic limestone and dolomite, which were dissected and forming a plateau characterised by traces of karstification of different ages (Juhász 2000). Besides, some subsurface forms also occur (e.g. the “well-cave” at Cserszegtomaj). The karst water represents a special asset (from drinking water to medicinal water), whose anthropogenic affectedness will be analysed later in this chapter. The Vadleány Cave is a special form within Hungary, as it was formed by Pannonian abrasion.

North of the Keszthely Mts is the Tátika Group, and to the east the Tapolca Basin is located. The members of Tátika Group are characterised by basalt dykes, while the unique sight of the Tapolca Basin is provided by basalt buttes with basalt columns at their sides (Fig. 1. 20). According to the landscape division made in 1990 (mostly followed in the present volume as well), as micro-regions, they belong to the Balaton Basin and the Transdanubian Hills, based on their characteristic relief and land use, whereas in a later landscape division (by Hajdú-Moharos and Hevesi 2000), they are part of the Bakony Region. The volcanic material deposited on Pannonian sediments, protecting them from erosion. The volcanoes elevate from the basement of the basin by 300 m (e.g. Haláp, Csobánc, Badacsony, Szent György-hegy); besides, volcanic material occurs in other regions of the Bakony Mts. too, which may support the classification of the micro-region as a member of the mountains. Taking this into consideration, the Tapoca Basin is a specific unit of the Bakony Mts., and the land use (e.g. vineyards, orchards) that evolved on volcanic soils underlines this. The postvolcanic activity accompanying volcanism affected various other territories of the Bakony Mts, as it is proven by the acidic springs in the Káli Basin (e.g. Kékkút) or the 2–3 million-year-old geyser cones of Tihany Peninsula. (Nowadays, only about 50 thermal water cones or geyser cones remain; the most spectacular among them is the Aranyház; Image 10.1).

Along the northern shoreline of the Lake Balaton is located the Balaton Uplands, being a part of the south-eastern wing of the Transdanubian Mountains. Its larger eastern part is composed of limestone and dolomite arranged into horsts and plateaus of different heights (e.g. Veszprém Plateau), though its minor part is built of Paleozoic red sandstone (along Lake Balaton) and Tertiary volcanic material (Ádám 1990). From the Miocene this micro-region was separated from its environment, thus it was an island in the Pannonian Sea and Lake; and along its southern shoreline, abrasional levels and sandy-gravelly beaches developed, which were identified by the famous geomorphologist, Cholnoky. The volcanism created natural values not only in Tihany Peninsula (approx. 7 million years ago), but also in the tectonic basins (e.g. Káli and Pécsely Basins) segmenting the micro-region. The basalt columns of Hegyestű (Fig. 10.2), the caves in Mesozoic limestone (e.g. Lóczy Cave), the karst springs or the block field of Kővágóörs on Miocene formations are all geological treasures of the Balaton Uplands National Park.

The main mass of the Bakony consists of Triassic-Cretaceous limestone and dolomite, which are arranged to horsts and plateaus of different heights. The largest unit is the North Bakony. It is characterised by structure of grabens and horsts (the Veszprém-Devecser Graben separates the North and the South Bakony Mts.), and its formerly uniform plateau was dissected by NW-SE structural lines; thus intramountain basins, trenches and plateaus are the most significant landscape elements (Ádám 1990; Martonné 2006). Several karstification periods took place on the horst plateaus (the highest is the Kőris-hegy: 704 m asl) from the end of the Cretaceous until the Pleistocene. The typical Hungarian occurrences of covered karst can be observed, where the Tertiary or Quaternary sedimentary cover has remained (Veress 2004). The slopes of plateaus were formed by periglacial processes during the Quaternary, while their edges got dissected by dense valleys in the humid climatic periods (e.g. Cuha Stream, Gerence Stream, Séd).

The South Bakony also consists of a series of horsts, basalt lava cones (e.g. Kabhegy,

**Image 10.1** Thermal springs welled in the form of geysers subsequent to basalt volcanism, forming the cone of Aranyház in the Tihany Peninsula. The calcareous and siliceous mud of hot lakes solidified gradually (photo by Kiss, T.)

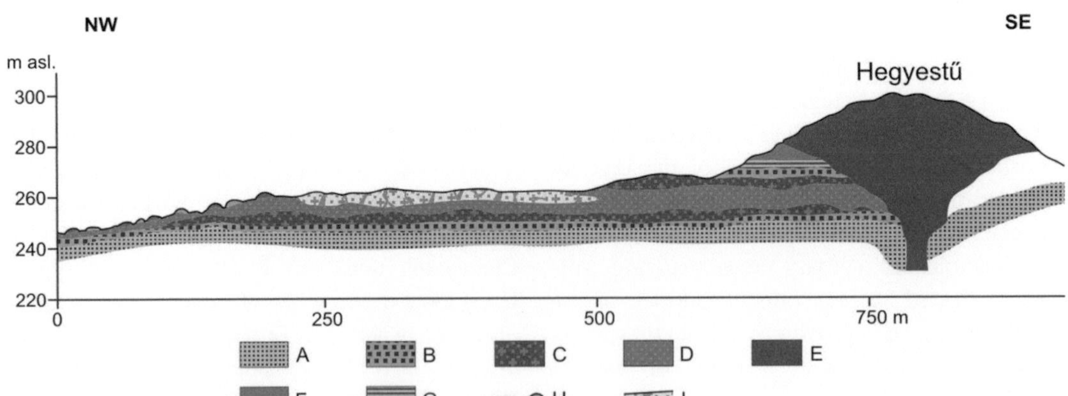


**Fig. 10.2** Geological sketch of the block field at Kővágóörs in the foreground of Hegyestű (after Juhász 2000) *A* Permian red sandstone, *B* Pannonian abrasional gravel, *C* Pannonian beach sand and gravel, *D* Pannonian fine-grained sand, *E* basalt, basalt tuff, *F* red clay, *G* Pannonian reddish clay, *H* sandstone blocks, *I* contiguous sandstone cover

Agár-tető) and intramountaine basins. Bauxite accumulated in the Cretaceous karst holes of the carbonated surfaces. The slopes also evolved by periglacial processes. Loess used to cover the alluvial plains of the northern foreground of the Bakony in Bakonyalja (e.g. the triple range of Pannonhalmi Hills); thus their forms have corrasional and deflational origin.

The NW-SE-oriented Mór Graben separates the **Vértes** and the **Velence Mountains** from the Bakony Mts. The Vértes Mts. are mainly composed of dolomite and limestone, and it is characterized by horsts, similarly to the entire macro-region, bordered by faults from each side (Fig. 10.3). Its evolution is similar to that of the Bakony: a planated, karstified surface has been formed on the carbonate sediments deposited on a Palaeozic basement, and bauxite was deposited at the end of the Cretaceous (Csillag et al. 2006). The edges of the block were covered by a shallow sea in the Eocene, where brown coal sites were formed in the bays and along the north-western foreground of the mountains (it was mined at Tatabánya and Oroszlány). The block of the Vértes Mts. became dissected as a result of tectonic movements between the Eocene and the Miocene. The separate blocks were buried and exhumed various times (Pécsi 1996), while they elevated into different heights and created smaller basins and hills, following the structure of horsts (Juhász 1988a). Their present height is the result of the Pleistocene uplift. The Eocene cover eroded from the surface, and the sediments were preserved just over the cryptoblocks in the bottom of basins (e.g. Gánt Basin).

The Velence Mts. are unique in the macro- and also in the meso-region, regarding their material, situation and landscape. The Palaeozoic granite batholith that constitutes the core of the mountain (see Sect. 1.1) was located in the southern hemisphere, near to the tropics at the beginning of its evolution. It used to be covered by Paleozioc limestone and shale, which were mainly eroded by intensive planation. The repeated erosional processes resulted in weakly segmented 250–300 m high mountains (Meleg-hegy is at 352 m asl, elevating from a granite environment), whose forms are rounded (Ádám 1993). Granite corestones (boulders) became exhumed on the slopes of foothill surfaces in

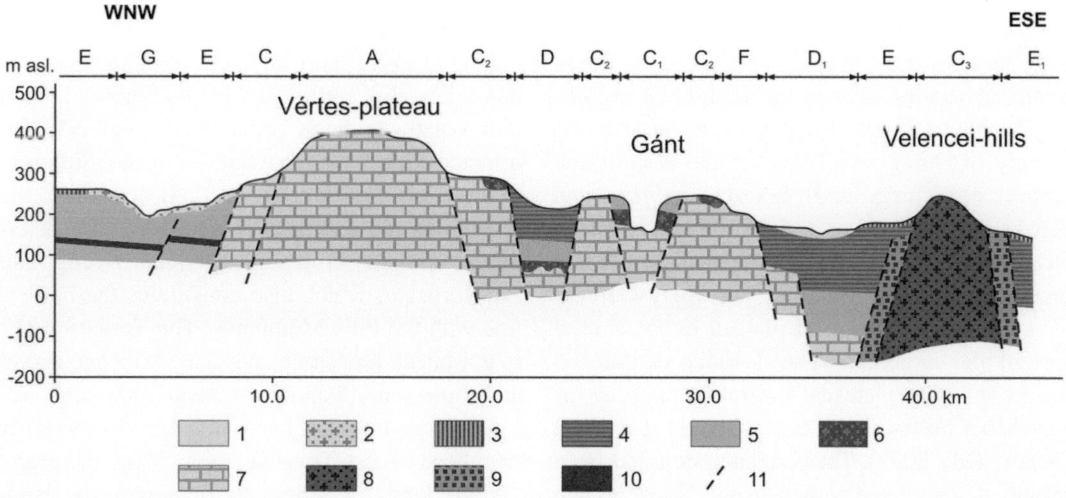


**Fig. 10.3** Main relief units of the Vértes (after Pécsi 1996 and MT 2002) *1* recent alluvium, *2* foothill debris cone, *3* loess, *4* Pannonian sandy-clayey sediments, *5* Tertiary basin sediments, *6* bauxite, *7* Triassic limestone and dolomite, *8* Paleozoic granite, *9* metamorphic crystalline rocks, *10* lignite layer, *11* faults *A* horst in hilltop position, *C* horst in foothill position, $C_1$ buried horst in foothill position, $C_2$ horst in low position, exhumed or semi-exhumed, $C_3$ exhumed horst in hilltop position, *D* buried erosional surface in graben position, $D_1$ intramountain graben filled by loose sediment, *E* terraced foothills, $E_1$ foothill surface on solid sediment, *F* abrasional terrace, *G* foothill basin

connection with subtropical planation processes. (The woolsacks near Pákozd are protected). Andezite lava (hardly traceable on the surface) filled the fault lines related to the Eocene subsidence, and volcanism was accompanied by postvolcanic activity too (e.g. Pátka: mineral ores). The faulted surface is covered by granite gravel in the west and loess in the east, providing highly different ecological conditions.

The three independent mountains (i.e. Gerecse Mts., Buda Mts. and Pilis Mts.) are integrated by the name **Dunazug Mountains**. The mountains are similar in their geological structure, dissected character, landscape forms and utilization. They are built of Mesozoic limestone and dolomite. These characteristics also make them fit in the macro-region (Leél-Őssy and Pécsi 1988). The difference between them is the different thickness of Tertiary sediments, which play different roles in landform development and the limited occurrence of raw materials (i.e. brown coal, base materials for the cement industry).

The most extensive member of the mountain group is the Gerecse which is characterized by N-S and NW-SE oriented fault lines, which is indicated by the spatial distribution of erosional valleys, small plateaus and basins between them (Juhász 2000). The typical limestone and dolomite surfaces are covered by Eocene carbonaceous formations only in the peripheral regions (at Tatabánya in the West Gerecse and in the vicinity of Dorog and Tokod in the eastern and north-eastern foreground). Loess cover deposited between the horsts (i.e. in the grabens), covering the Pannonian sand and clay layers, even sometimes above 300 m altitude. The Nagy-Gerecse (634 m asl) is the highest plateau in the central part of the mountain. The Tertiary carbonated blocks of the plateaus uniformly appear as island-like horsts in the East, Central and West Gerecse (MT 2002). The plateau system has been exhumed since its elevation in the Neogene and has been able to preserve its form due to the structure of the karstifying rocks, since the erosion and landform evolution are slow processes under such circumstances. The result is a comparatively rich surface and subsurface karst landscape (e.g. ponors, spring caves, Szelim Cave of Tatabánya; Leél-Őssy 1979). Several streams flow from the elevated karst surfaces towards the erosional valleys (e.g. Által Brook in the west, Danube in the north) building travertine pools. Their springs of these streams have got into higher altitudes due to the continuous elevation of the mountains since the Pliocene.

In the East Gerecse, Scheuer (1999) identified six, successive travertine formation phases since the Pliocene, while in the West Gerecse, he attached 19 springs to travertine formation phases (Fig. 10.4). According to Scheuer and Schweitzer (1988), the springs of the Central Gerecse lost their continuous water supply in the Middle Pleistocene, and the West Gerecse became more dominant in the process. The older travertine deposited on the slopes by the streams, protecting the former Pannonian abrasional terraces and the Pleistocene fluvial terraces from erosion, and preserving valuable archaeological finding in its material (see Sect. 8.2). The amount of travertine is sufficient for the purpose of construction; for example, the Romans used the travertines in the vicinity of Dunaalmás for road constructions (today the Roman mine is a protected site.)

The geographic features of the macro-region can also be recognised in the Buda Mountains: they are mostly built of Mesozoic limestone and dolomite; the carbonated formations arranged into horsts, grabens and valleys; and NW–SE oriented fault lines have had an outstanding role in the landscape development. At north-east, the mountains are bordered by a structurally determined graben stretching between Dorog and Pilisvörösvár, which also constitutes the borderline with the Pilis Mountains. The Tertiary rocks (e.g. Eocene limestone and marl, Miocene gravel and limestone, Pannonian sand and clay) are much more frequent here compared to the other members of the Transdanubian Mts.; therefore, various kinds of sediments determine the landform evolution (Juhász 2000). Another difference is that the horsts are much more segmented (Pécsi 1996).

The genetics of the planated horsts of the Buda Mountains can be analysed (MT 2002), but regarding their situation, they can be classified

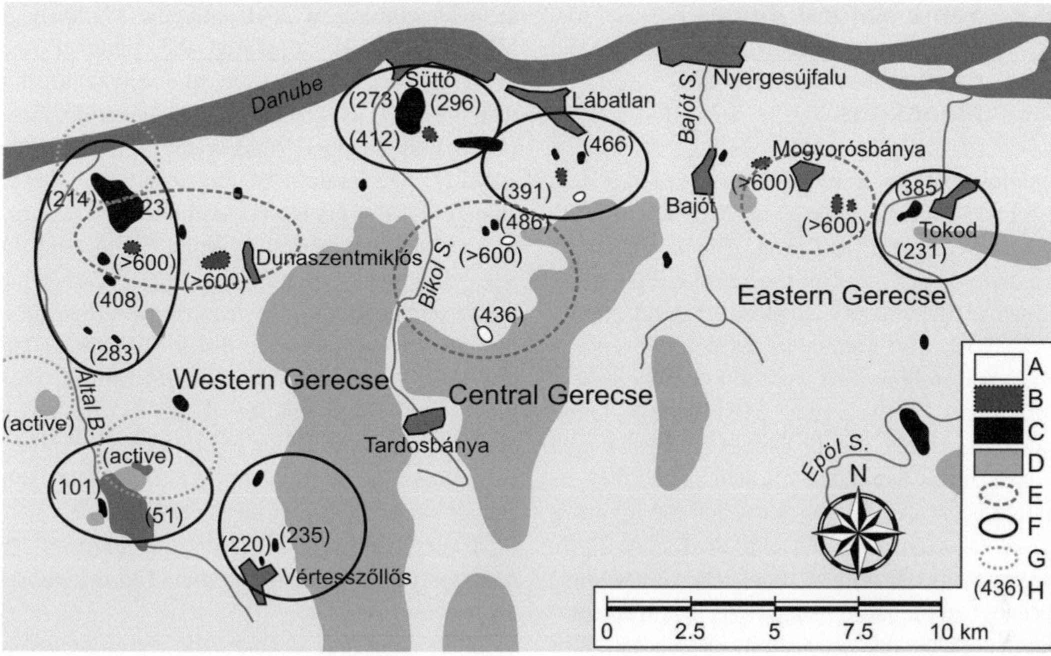


**Fig. 10.4** Location of karst springs in the Gerecse Mts., depositing travertine from the Upper Pannonian until the Holocene (after Kele 2008) *A* Upper Pannonian springs and travertine, *B* Upper Pliocene springs and travertine, *C* Quaternary springs and travertine, *D* surface appearance of Mesozoic carbonated rocks, *E* spring field older than 600 thousand years, *F* Middle and Upper Pleistocene spring field, *G* Holocene spring field, *H* altitude of spring

into two groups (Juhász 2000, Martonné 2006). The first group comprises the conical karst horsts of Farkas and Csíki Hills, which are covered by younger sediments, e.g. Hárshegy Sandstone, and the travertine is even much younger (it covers the Castle Hill of Buda, which is the terrace No. IV. of the Danube). The second group includes the smaller horsts of the Hármashatár Hill, which are rich in surface and subsurface karst forms. The tectonically preformed thermal water caves (e.g. Ferenc-hegyi Cave, Mátyás-hegyi Cave and Pálvölgyi Cave) constitute special asset, which require engineering geological attention.

The elevated horsts of the Pilis Mountains are dissected by NW–SE oriented fault lines, and some horst broadens into plateaus (e.g. Hosszúhegy), one of them includes the highest point of the macro-region (Pilis-tető: 757 m). Similarly to the other parts of the macro-region, the Pilis Mts. are built of karstifying limestone and dolomite. Caves are typical in the region. The 5 km-long cave system of the Legény Cave, Leány Cave and Ariadne Cave is known for almost a hundred years; it is archaeologically explored, and the cave system itself joins the Pilis-tető at 500 m asl (Székely 2003). The strong tectonic influence also characteristic in the Pilis Mts., as the region is bordered by tectonic faults from each side, therefore the Mesozoic formations north of the mountains are deeply sunken. Along the steep edges of the horsts erosional landforms, e.g. gullies and rock towers appear. (For instance, the rocks of Oszoly are excellent sites of rock climbing.) The younger, Tertiary sediments are also common here, similarly to the Buda Mts., and the thick Eocene brown coal of the Dorog–Piliscsév Basin, which had been mined for a long time, must be mentioned. It is common that both the horsts and the younger sedimentary strata are covered and preserved by a travertine cover (MT 2002).

## 10.3 Some Natural Values and Environmental Hazards of the Transdanubian Mountains

Significant energy sources and ore reserves can be found in the Transdanubian Mts. The Eocene brown coal basins (at Dorog, Tatabánya-Mány, Oroszlány basins, and in the north-eastern foreground of the Bakony) dispose of good quality brown coal. The amount of exploitable coal is about 800 million tons, and the production in 2007 was 1.4 million tons; MBFH 2008. Other important reserves are the Cretaceous brown coal at Ajka (it has not been exploited since 2004, its amount is ca. 270 million tons) and the lignite at Várpalota occurrence (not exploited since 1996, its amount is 190 million tons). The Cretaceous–Eocene bauxite and manganese mining represents a considerable opportunity. Nowadays, four bauxite mines are in operation (Halimba, Bakonyoszlop, Sándor-bánya and the open cast mine of Óbarok), producing 0.5 million tons/year; thus calculating with this rate, the bauxite is sufficient for 100 years. The manganese ores at Úrkút will probably also last for a century (MBFM 2008).

The preservation and maintenance of the special natural vegetation, the geographical and geological natural values (e.g. buttes, numerous caves), and the great hydrographic values (e.g. springs and lakes) are the tasks of the Balaton Uplands National Park and the Danube–Ipoly National Park.

In the topic of environmental conflicts in the Transdanubian Mts., those effects of former brown coal and bauxite mining will be highlighted that have involved the quality and quantity of karst water reserves. The human-induced landscape changes would deserve a separate chapter. (Therefore, the highlighted task of national parks is the protection of traditional landscape structure and husbandry.)

The exploitation of minerals was the most intensive in the 1980s, when more than 2.500 mines were in operation in the territory of the country. Their number seriously reduced after the change in the political system. (For example, at the beginning of 2010 only the Márkushegy Mine functioned, supplying the Vértes Power Plant by coal.) However, the environmental damage (e.g. degradation of vegetation, land acquisition, changes in the water regime, surface stability deterioration of the abandoned mines, changes of the landscape aesthetics due to mining, soil contamination by mine dumps and the need for their recultivation; Juhász 2000) has partly remained, and the recultivation (e.g. mine dumps) has been realized just incompletely. The abandoned non-metallic mineral mines (e.g. gravel and sand) can be found in the whole country, not only in this region. They involve a continuous, large areas in the South Great Hungarian Plain and along the motorways.

*Karst water level decrease induced by anthropogenic activity in the Transdanubian Mountains*

In the previous chapter, the development of Lake Hévíz was described (see Sect. 3.1.3). The lake, which is unique from various points of views, was affected by human impact through its recharging springs. The problem originated from the good quality, deep-level bauxite of Nyirád, which was intensively mined since 1965–1970. Underground mining also characterizes coal (at Tatabánya, Mány, Ajka, Balinka) and manganese mines (at Úrkút), but deep-level bauxite mines were also common (at Halimba, Fenyőfő, Kincsesbánya). These mines were accompanied by a considerable amount of water extraction in the 1970s and 1980s; thus, the total amount of extracted water in the Transdanubian Mts. reached 800 $m^3$/min (13–14 $m^3$/s; Fig. 10.5), which is the double of the mean water discharge of the Zala (6 $m^3$/s). In the 1980s, the water withdrawal at Nyirád was about 5 $m^3$/s, and it exceeded the recharge of the karst water on the karstic catchment of the Transdanubian Mts. west of the Mór Graben.

The most important consequence of the extreme water withdrawal was the significant sinking of karst water level, which entailed a significant discharge decrease of the karst springs. The unfavourable phenomenon created a critical situation for thermal springs; for example, the balneologically ideal surface temperature of

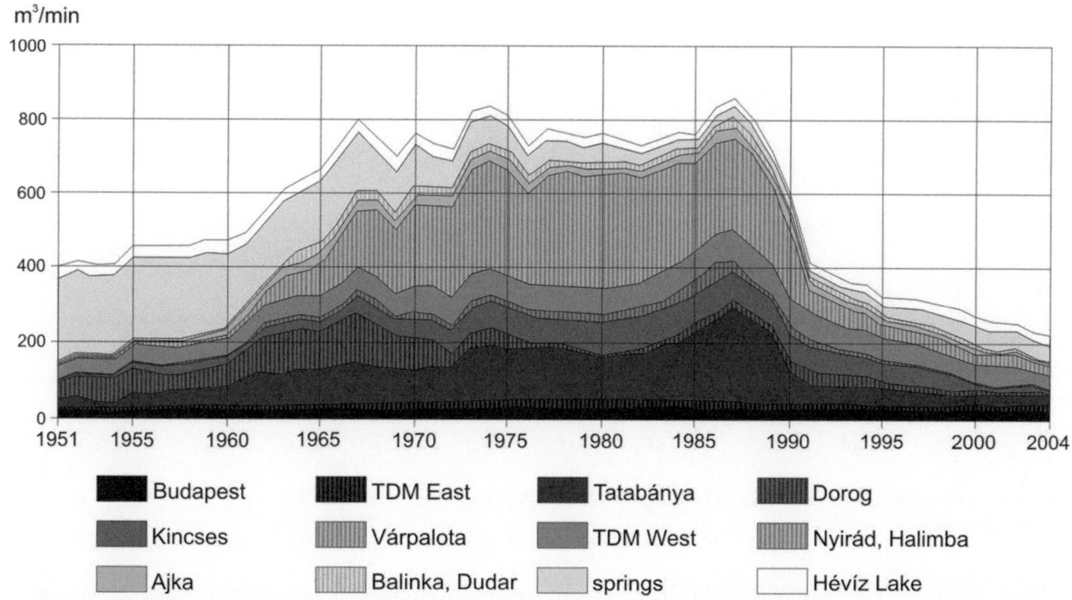


**Fig. 10.5** The amount of water extraction mostly for mining purposes in the Transdanubian Mountains (after Alföldi and Kapolyi 2007)

Lake Hévíz was maintained just for 1–2 months a year. Other regions have also experienced serious discharge loss and spring depletion. For example, because of the water extraction of the coal mines in Tatabánya within the framework of the Eocene Programme, the water from the Lake Cave at Tapolca disappeared, the discharge of the springs in Tata decreased considerably and even the discharge of thermal springs of Buda dropped. The absolute level of karst water decreased considerably by 1990, as the total amount of water-level drop was 100 m in some places, for example, around the most active Nyirád site (Alföldi and Kapolyi 2007).

Balogh and Lovász (1988) estimated the discharge of the deep springs in Hévíz to be 710 l/s in 1950, 559 l/s in 1967, only 263 l/s in 1978 and just about 300 l/s in 1984. Finally, a decision was made in 1990 to terminate the water withdrawal for mining purposes, simultaneously with the closure of mining in Nyirád, as a reaction to the negative environmental effects described above and to the disappearance of the political pressure to mine bauxite extensively. In 2004, the discharge of the springs in Transdanubia was already 72 $m^3$/min (1200 l/s). Although the thermal water recharge started to improve from 1990–1995 in Hévíz and at other sites too (Fig. 10.6), the models predict that another 15–20 years are needed for the total restoration of the karst water (Csepregi 2007). The water supply of the Lake Cave of Tapolca has also been restored, and the springs of Tata have also started to well, etc. If water extraction is limited around Hévíz and the wells are closed to support the positive water balance, a 500 l/s discharge can be achieved in the long run, which is necessary to recharge the karst water. The limitation is justified also because several researchers suppose that the discharge decrease of thermal springs has not only been caused by mining, but also by the water extraction of the wells along the Lake Hévíz to utilize the region for recreational and communal purposes.

Similar processes have taken place in other regions of the Transdanubian Mts. as well. Thus, for example, the discharge of the Fényes Springs of Tata is expected to be 22.2 $m^3$/min by 2020 and 30.6 $m^3$/min until 2040. The recovery of the estimated former water discharge (78 $m^3$/min) is not probable in the near future, even if the present growth rate could be maintained; however,

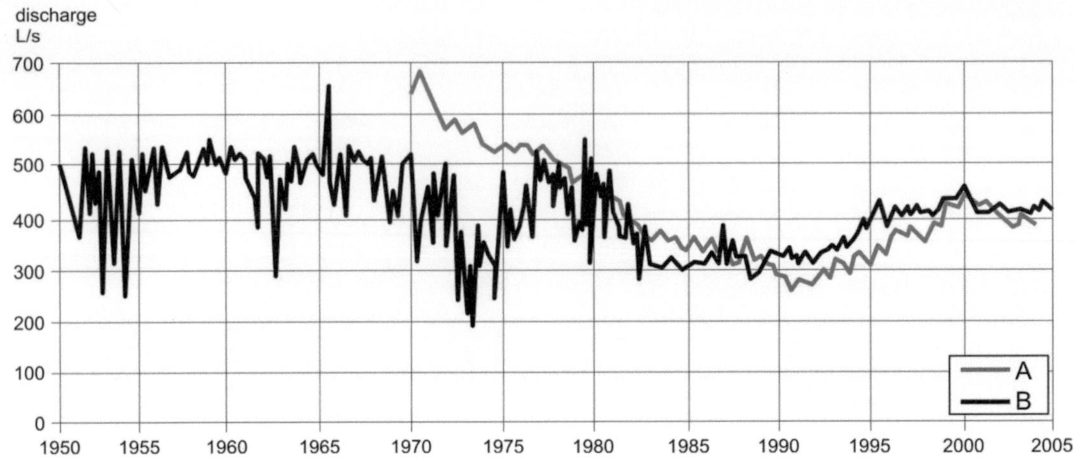


**Fig. 10.6** Discharge changes of thermal springs between 1950 and 2005 (after Csepregi 2007) *A* measured discharge (l/s), *B* calculated discharge (l/s)

significant upwelling of karst water is still probable (Ballabás 2004).

The overexploitation of springs has been stopped by the decrease of the water discharge in the largest part of Transdanubia. Several formulas are known to calculate water discharge of springs, which depends on the infiltration coefficient (which is about 35 % in the case of the Hungarian karsts), precipitation and water withdrawal. The anthropogenic impact may involve withdrawal (for example, the 40,000 l/min discharge of the spring of Miskolctapolca has decreased to its third, due to water extraction), and infiltration as well, which can be modified by the alteration of surface cover and land use (Mádlné 2008).

## References

Ádám L. (1990) A Balaton medencéje, a Balatonfelvidék és részmedencéi In: Marosi S, Somogyi S (szerk.) Magyarország kistájainak katasztere. MTAFKI, Budapest, pp 483–513

Ádám L (1993) A Velencei-hegység fejlődéstörténete és felszínalaktana. Földrajzi Értesítő 42(1–4):93–110

Ádám L, Marosi S, Szilárd J (szerk) (1987–1988) A Dunántúli-középhegység. A–B. Akadémiai Kiadó, Budapest

Alföldi L, Kapolyi L (szerk) (2007) Bányászati karsztvízszint-süllyesztés a Dunántúli-középhegységben. MTAFKI, Budapest

Ballabás G (2004) Visszatérő karsztforrásokkal kapcsolatos településfejlesztési és környezetvédelmi lehetőségek és veszélyek Tata város példáján. http://geogr.elte.hu/TGF/TGF_Cikkek/ballabas2.pdf

Balogh J, Lovász G (1988) Vízföldrajzi-vízföldtani erőforrások. In: Ádám L, Marosi S, Szilárd J szerk. (1987–1988) A Dunántúli-középhegység. B. Akadémiai Kiadó, Budapest

Budai T, Császár G, Csillag G, Dudko A, Koloszár L, Majoros Gy (1999) A Balatonfelvidék földtana. MÁFI, Budapest

Csepregi A (2007) A karsztvíztermelés hatása a Dunántúli-középhegység vízháztartására. In: Alföldi, Kapolyi (szerk.) Bányászati karsztvízszint-süllyesztés a Dunántúli-középhegységben. MTAFKI, Budapest, pp 77–112

Csillag G, Fodor L, Selmeczi I (2006) A Vértes és környéke neogén geomorfológiai fejlődéstörténete. 3. MFK előadás, Budapest. www.geography.hu

Góczán L (1987) Talajok. In: Ádám L, Marosi S, Szilárd J (szerk.) A Dunántúli-középhegység.A. Akadémiai Kiadó, Budapest, pp 354–382

Hajdú-Moharos J, Hevesi A (2000) A Kárpát-Pannon térség természeti tájbeosztása. In: Karátson D (szerk.) Pannon enciklopédia. Kertek, Budapest, pp 274–285

Hámor G (2001) Magyarázó a Kárpát-medence miocén ősföldrajzi és fáciestérképéhez. MÁFI, Budapest

Juhász Á (1988a) A Bakonyvidék. In: Pécsi M (szerk.) A. Dunántúli-középhegység. B. Akadémiai Kiadó, Budapest, 31–101

Juhász Á (2000) A Dunántúli-középhegység. In: Karátson D (szerk.) Pannon enciklopédia

Kele S (2008) Édesvízi mészkövek vizsgálata a Kárpát--medencéből: paleoklimatológiai és szedimentológiai elemzések. PhD-értekezés, Budapest

Leél-Őssy S (1979) A Dunazug-hegység genetikus felszíntipológiai térképe. Földrajzi Értesítő 27(1–2):39–50

Leél-Őssy S, Pécsi M (1988) A Dunazug-hegyvidék. In: Ádám L, Marosi S, Szilárd J (szerk.) A Dunántúli-középhegység. B. Akadémiai Kiadó, Budapest, pp 333–341 és 360–369

Mádlné SJ (2008) http://applied.geology.elte.hu/oktatas/alkhidro/2008_alkhidro%204ea.pdf

Marosi S (1987) A Dunántúli-középhegység főbb táji jellemzői. In: Ádám L, Marosi S, Szilárd J (szerk.) Dunántúli-középhegység. A. Akadémiai Kiadó, Budapest, pp 11–24

Marosi S, Somogyi S (1990) Magyarország kistájainak katasztere. I–II. Magyar Tudományos Akadémia Földrajztudományi Kutatóintézet, Budapest

Martonné EK (2006) Magyarország tájföldrajza. KLTE, Debrecen

MBFH (2008) http://www.mbfh.hu/home/html/index.asp?msid=1&sid=0&hkl=72&lng=1

Glatz F (szerk) (2002) Magyar Tudománytár I. Föld, víz, levegő. Kossuth, Budapest

Pécsi M (1996) Geomorphological Regions of Hungary. MTAFKI, Budapest

Péczely G (1987) Éghajlat. In: Ádám L, Marosi S, Szilárd J (szerk.) Dunántúli-középhegység. A. Akadémiai Kiadó, Budapest, pp 243–281

Ság L, Pécsi M (1987) Kainozoikum – földtani képződmények. In: Ádám L, Marosi S, Szilárd J (szerk.) Dunántúli-középhegység. A. Akadémiai Kiadó, Budapest, pp 75–121

Scheuer G (1999) A gerecsei mészképző hévforrások összehasonlító vizsgálata. Hidrológiai Közlöny 79 (3):152–156

Scheuer G, Schweitzer F (1988) A Gerecse és a Budai-hegység édesvízi mészkőösszletei. Akadémiai Kiadó, Budapest

Somogyi S (1987c) Vizek. In: Ádám L, Marosi S, Szilárd J (szerk.) Dunántúli-középhegység. A. Akadémiai Kiadó, Budapest, pp 282–336

Székely K (2003) Magyarország fokozottan védett barlangjai. Mezőgazda, Budapest

Veress M (2004) A karszt. BDF, Szombathely

# Physical Geography of the Transdanubian Hills

# 11

**Abstract**

The Transdanubian Hills in the southern-central part of Hungary consists of hills built of fluvial and aeolian material, though some mountain blocks built of Palaeo- and Mesozoic marine sediments (Mecsek Mts., Villányi Mt.) are also represented. The latest ones have small area, but they have great importance as their uranium and black coal reserves were mined for a long time. The environmental consequences of the mining activity should be solved in the future. In a meso-regional scale, the most important characteristics of the region (without considering the mountainous carbonate blocks) are connected to the loessy and (aeolian) sandy material of the hills, and to the Quaternary evolution of the river system. The Lake Balaton and its basin belong to this macro-region, being the largest lake in Central Europe and one of the most important touristic destinations. Therefore, the sustainable management of its water household and the protection of its water quality are very important issues. The environmental problems and conflicts in connection with the hydropower plant and reservoir series of the Dráva River should be solved in the future too.

## 11.1 Location and the Main Evolutionary Periods of the Macro-Region

The Transdanubian Hills consist of the hilly areas between the Transdanubian Mountains and the Mura-Dráva Valley, and the Mecsek Mts. The units of the macro-region are very diverse from a geological point of view, as they originated from different locations and submerged only later. The so-called South Transdanubia landscape unit contains parts belonging to the Tisza Unit geologically, as their evolution started on the shelf area of the European plate These old Palaeozoic and Mesozoic formations can be found on the surface in the Mecsek Mts., the Villány Mts. and the Geresd Hills in several places, and they represent different landscape types (Fig. 11.1). These surfaces built of older rocks elevated like islands form the Pannonian Sea at the end of the Miocene (the abrasion levels are the evidences on it), therefore Pannonian clayey and sandy sediments used to cover the surface up to ca. 400 m asl. Afterwards, during the Pliocene (cross-stratified) sand, and in the Pleistocene, alluvium were deposited,





G. Mezősi, *The Physical Geography of Hungary*, Geography of the Physical Environment,
DOI 10.1007/978-3-319-45183-1_11

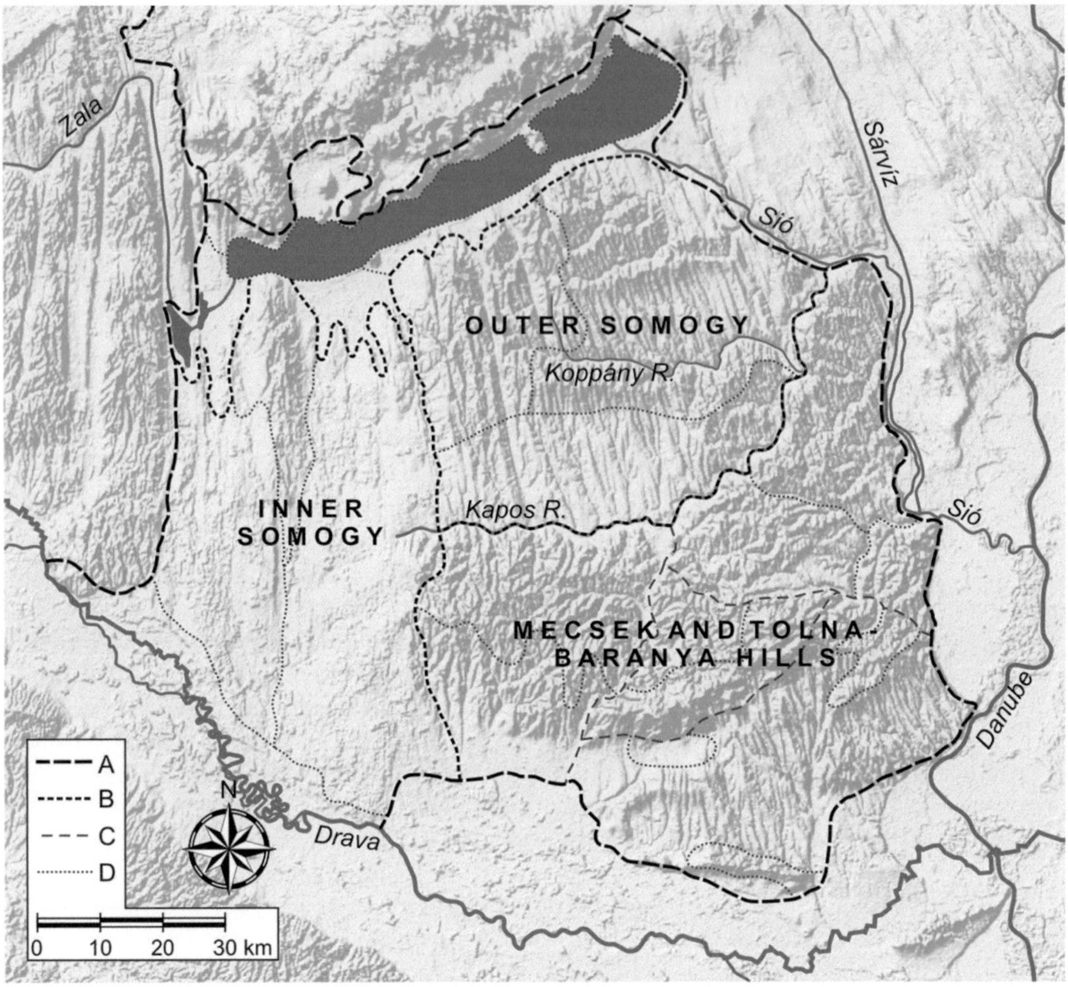


**Fig. 11.1** The landscape division of the Transdanubian Hills (after Marosi and Somogyi 1990) *A* macro-regional border, *B* meso-regional border, *C* border of groups of micro-regions, *D* micro-regional border

constituting the basement of the hills in the lower areas. This surface, dissected by Pliocene and Pleistocene tectonic movements, served as the basis of valley formation.

The originally flat alluvial fans became arranged into hills as a result of the Pliocene and Pleistocene tectonic movements (as a result, the Pannonian sediments were exhumed by erosion). The movements can be linked to a triple generation of depressions according (Ádám et al. 1981). The oldest one is the depression of the Croatian-Slavonian Lake, which is several million years old, and it was reactivated in the form of the Dráva Graben in the Quaternary, sinking at different rates. The second one is the Felső-Kapos-Kalocsa Graben, where a 150-m-thick fluvial sediment sequence was deposited by the Middle Pleistocene (Marosi 1969). The youngest is the 10- to 50-thousand-year-old depression of Lake Balaton (supposedly not trenched) and the mountain foreground depressions with similar origin (e.g. Lake Velence, Tikacs, Sárrét in Fejér County).

The alluvial fan surface was segmented by the Pliocene and Pleistocene tectonic movements, creating a dissected surface for valley development, thus a unique system of meridional valleys (and the ridges between them) developed in Inner

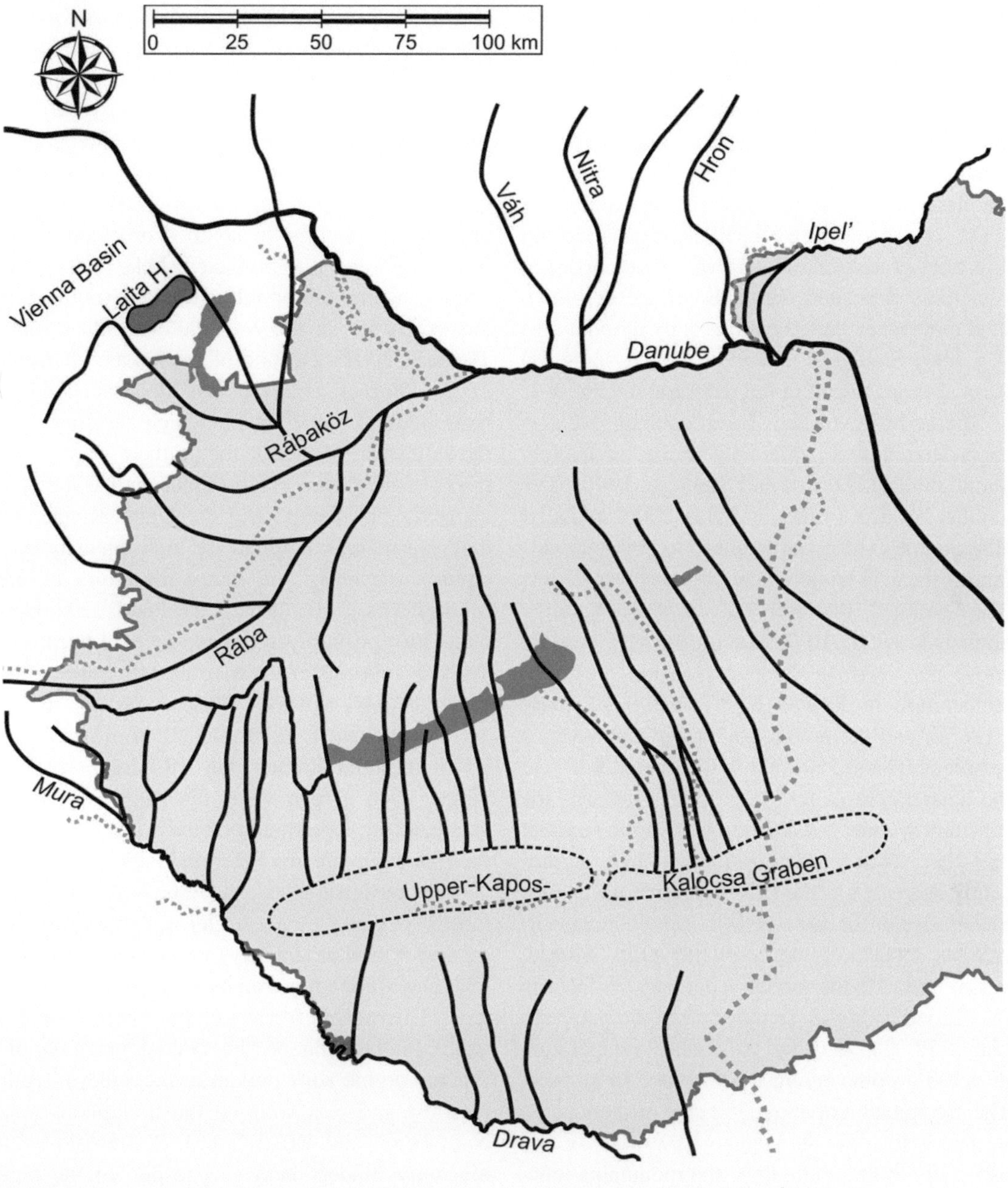


**Fig. 11.2** Reconstructed river network of Transdanubia in the Middle Pleistocene (after Ádám 1981)

and Outer Somogy. According to Cholnoky (1926), they were yardangs (ridges formed by wind erosion) at the beginning, similarly to the Zala Hills; however, their evolution was explained by stream erosion and tectonic impacts later (Fig. 11.2). However, based on some research in the 1990s, a (semi-)desert deflational relief formation phase cannot be excluded (Schweitzer 1993). The valleys, having NNW–SSE directions, were thought to be as much as 5–6 million years old, and they could be buried (by Baltavár Sand) and exhumed later (Pécsi 1996; MT 2002). The meridional valleys drained the waters of the Transdanubian Mountains into the

Felső-Kapocs-Kalocsa Graben, which was active until the middle of the Quaternary, but it was gradually filled up by fluvial sediments.

Apart from the meridional valleys, the Quaternary uplift of the Mecsek Mts. (and the Transdanubian Mts.) by about 200 m played a part in the dissection of the hills and the formation of a dense water network. By the Pleistocene, the main directions of the streams and rivers developed, and later loess deposited onto the surface or blown sand movement started on the sandy alluvial fans (e.g. Inner Somogy). In the Quaternary some terraces also developed in the meridional valleys.

The territory of the Transdanubian Hills is characterized by a moderately warm, moderately humid climate (Marosi and Somogyi 1990). The number of sunny hours is 1900–2000 annually, although this value can be modified by local differences, e.g. in winter the air of the city of Pécs is often polluted, therefore the duration of sunny hours is lower by 10 % than on the Misina peak, above the town. In the macro-region the mean temperature in July is 21 °C, and in January −1.5 °C, and the amount of precipitation varies between 600 and 750 mm. In the winter half-year the temperature pattern gets milder towards the west and warmer towards the east in the summer half-year. The amount of precipitation significantly decreases to the E-NE. In contrary to this, the climate of the Mecsek Mts. can be characterized as moderately humid with mild winters, where the Mediterranean character is clearly manifested by higher summer mean temperatures (July: 22 °C), and by a smaller temperature decrease accompanying the increase in altitude. The precipitation of about 750 mm pronouncedly decreases towards the edges of the Mecsek and especially to the south, thus, the mountains function as a precipitation divide (Martonné 2006). The favourable climatic features, for example, the spring arrives the earliest in the whole country on the sunny slopes, sheltered from the northern winds, facilitated the establishment of several settlements and fruit and grape production on the foothills of the Mecsek Mts.

The water network of the hills has adapted to the meridional valleys, the tectonic preformation and the local erosional bases of Lake Balaton, the Dráva Valley and the Felső-Kapocs-Kalocsa Graben. The uplifts in the Würm and the loose, loessy sediment cover results in the development of a dense and complex water network. The streams flowing to the north (to the direction of Lake Balaton) have often been artificially transformed; therefore, natural lakes mainly appear in the valleys running to south. For example, in Inner Somogy the protected Lake Baláta (at Szenta), which preserved its bog environment, is located in one of the N–S valleys (Sümegi and Töröcsik 2007). Despite of the karstic character of the Mecsek and Villány Mountains, dense river network developed on their territory, and the valleys are longer on the northern slopes. The rivers of the macro-region drain their water typically to the Danube and the Dráva River, and they are mainly supplied by high-intensity precipitation events. The water reservoirs of the Mecsek Mts (Lake Orfű, Lake Pécsi, Lake Herman Ottó) primarily serve recreational purposes. The karst springs of the Mecsek Mts. support the water supply of Pécs: Tettye Spring (0.25–40 $m^3$/min) and Orfű Spring (0.35 $m^3$/min) can be highlighted among the 18 larger springs, which, even if their water discharge is volatile, attenuate the extremities of the water regime of the mountain streams (Martonné 2006).

The dominant soil types of the hills, which are mostly covered by loess, loess derivatives and sand in a smaller territory, are brown forest soils with clay illuviation, and its subtype in the sandy areas. Chernozem forest soil and forest soils also appear, depending on the rate of precipitation. Hydromorphic soils prevail in the valleys, while meadow and peat soils on the floodplains (e.g. Nagy-berek and Kis-Balaton). Soil erosion significantly hinders land use despite of the high forestation of the meso-region, and especially in areas where the forest cover is missing. In the Mecsek and in the Villány Mts., redzinas developed on carbonated surfaces, otherwise, the forest soils with clay illuviation prevail, which have a thin fertile layer and a bad water regime (Rajkai 1990).

The natural plant communities of South Transdanubia are beech forests mixed with linden; turkey oak and hornbeam forests; the sand

and sand steppe communities appear on dry surfaces; and the alder marsh forests developed on wetlands (Lehmann 1981). The original linden–turkey oak forests used to occupy 20–35 % of the territory (Galambos 1990), but the cultural steppe has displaced them (e.g. Outer Somogy).

The Mediterranean influence is reflected in the natural vegetation of the Mecsek and Villány Mts., especially on the slopes with southern exposure. Here, in the plant associations of downy oak-dominated karst shrub communities ca., 30 % of the species have sub-Mediterranean and Balkanic origin, and many of them are protected (MT 2002). On the karstic southern slopes calciphile oak forests, and on the northern slopes Illyrian beech forests developed, while on the non-carbonated southern slopes we can see turkey oak and lime trees (with chestnut forest secondarily), and on the non-carbonated northern ones calcifugous oak and beech forests have grown.

On the whole, the macro-region is characterized by (1) mountain landscape types used for the purposes of forestry, mining (terminated, thus it needs recultivation today) and recreation; (2) hilly landscape types are the most common, they are dissected up to a different degree and used for agriculture and forestry; and (3) plains characterized by loessy cultural steppe, and blown sand areas covered by forests and pastures (Ádám et al. 1981).

## 11.2 Overview on the Evolutionary Types of the Regions of the Transdanubian Hills and Their Physical Geographical Characteristics

The **Balaton Basin** meso-region contains the Lake Balaton, which is the largest lake of Central Europe (for detailed development history, see Chap. 3), with great natural values (e.g. Kis-Balaton), recreational and cultural historical assets (e.g. the first Hungarian written words form the deed of endowment of the Abbey of Tihany from 1055). Lake Balaton constitutes an independent meso-region with its characteristic abrasion levels, bays and buttes.

The geological basement of the region is composed of several hundred metres of thick Pannonian sediments, which occur near the surface in the north of the lake, in the Riviéra of the Balaton. According to some opinions, the deposited Pannonian sediments were reshaped by Pannonian abrasion, and Pliocene and Pleistocene planation (Cholnoky 1936; Ádám 1990). On these Pannonian sediments, 3- to 4-million-year-old basalt lava were deposited in some places (Fig. 1.20), which preserved the sediments underneath; therefore, basalt-covered buttes could develop after a period of fluvial erosion (e.g. Fonyód and Boglár Buttes in the southern lakeshore). The loose sediment near the northern shoreline was also stabilized by Pannonian travertine found at ca. 200 m asl. in a plateau position today. Several landforms have connected to and been preserved by postvolcanic activity in the Balaton Basin. Such are the maars on the Fekete-hegy in the Káli Basin, the geysers in the Tihany Peninsula, and thermal, often medicinal springs at various places (MT 2002).

On this Pannonian surface the rivers flowed from the Bakony Mts. to SSE, towards the Felső-Kapos-Kalocsa Graben, covering the area of today's Balaton by sediments. On these alluvial fan surfaces, meridional valleys of Somogy developed, between them the meridional ridges existed for a long time in the area of the Balaton Basin, resulting in distinctive sub-basins of the Balaton, which began to join in the middle and at the end of the Würm (Ádám 1981), though nowadays it is dated to 15–16 thousand years. Several professional hypotheses were born about the evolution of the Balaton basin (see Chap. 3; Marosi and Szilárd 1981b). The most widely accepted interpret the basin as a tectonic depression (though the latest seismic studies have not proven the existence of fault lines at the edges of the lake examinations; Magyari 2000). At the end of the Quaternary the basin was uplifted, although to a smaller extent than the areas in Somogy, thus, the evolution of the basin could start by burial by Pannonian marine sediments and a subsequent erosion after the regression (Marosi and Pécsi 1999).

In close connection to the erosional basin of the lake bays developed, which have transformed into groves (e.g. Nagy-berek, Kis-Balaton) and are separated from the lake by bars and barriers. In these bays stagnant water existed for a long time, thus, peat and bogs were created. The barrier bars created by the NW winds and the ridges between the meridional valleys took part in the segmentation and separation of the sub-basins of Lake Balaton. The (lacustrine) abrasion also played a role in the formation of the southern foreshore. The steep, eroded coasts of the meridional valleys of northern exposure suggest tectonic influences.

The Somogy Hills is composed of the different morpho-lithogenic units of Inner Somogy and Outer Somogy. **Outer Somogy** is built of Late Miocene, Pliocene sandy and, in a smaller extent, clayey sediments. Its 200- to 300-m-high surface is dissected by erosional and derasional valleys, which in the west part of the region join the NNW-SSE-oriented meridional valleys, and in the east part they run perpendicular (e.g. Koppány, Jaba, Kis-Koppány Rivers). Thus, the systems of the meridional and the perpendicular valleys created a chessboard pattern of some typological units.

The evolution of the ridges remaining between the valleys is connected to the fact that the area used to constitute the flat foothills of the Transdanubian Mountains, stretching until the Felső-Kapos-Kalocsa Graben. In the Upper Pleistocene, the uplift of the central part and the subsidence of the depression of Lake Balaton created obsequence in the meridional valleys. The SSE declination of the young terraces could be explained by the original flow direction: it was preserved to the south of the divide developing in the middle of the meridional valleys, while in the valley sections to the north of the divide the direction of the streams turned to north, thus in an opposite direction as before (Szilárd 1967).

The northern slopes of the landscape unit are steep due to the asymmetrical uplift, thus here mass movements and the development of gullies and ravines are typical. The ridges have a terraced layout in the E–W direction, but their edges are at risk of mass movements. The region is dominantly covered by easily erodible, thick loess; therefore, its land-use planning requires special attention. The loessy layers are mostly of aeolian origin, but slope loess (derasional loess) also appear because of the orographical situation (Szilárd 1983).

**Inner Somogy** appears more like a plain landscape (Marosi and Szilárd 1981b). Its basement is composed of sandy sediment from the Late Miocene and the Pleistocene, which was covered by fluvial sediments gradually thickening towards the south, towards the Felső-Kapos-Kalocsa Graben. At the beginning, the formation of the fluvial sediment sequence was connected to the Palaeo-Danube (its erosional base was the Croatian–Slavonian Lake), then to the rivers of the mountains stretching from the Lower Zala to the Felső-Kapos-Kalocsa Depression. This alluvial fan sequence had such a low gradient in some places that lacustrine sediments also occurred (Marosi 1970). The Marcali Ridge in the axis of Inner Somogy got into a higher orographic situation due to Quaternary uplift, thus in the Würm loess could cover it. In the next phase of the development, an intensive blown sand movement started at the edges of the dry alluvial fans composed of fluvial sand. This may have started at the beginning of the Würm (Marosi 1967), but according to the latest results, the present-day aeolian forms developed in the Late Pleistocene and Holocene. Shallow deflational flats and frequent blowouts characterize the majority of the region (Lóki 1981), though large accumulational forms (e.g. hummocks, parabolic dunes) also occur in the southern part of the meso-region.

The **Tolna-Baranya Hills** consist of three main units: the Tolna Hills (including Völgység, Tolnai-Hegyhát and Szekszárd Hills), the Zselic and the Baranya Hills (which connect the relief of the Mecsek and Villány Mts.). The landform evolution was highly similar in these landscape units:

- The Pannonian sediments were covered by fluvial deposits of rivers flowing towards the Felső-Kapos-Kalocsa Graben (the alluvium is thin in the Zselic and Szekszárd Hills). After the surface became dry, loess sheets covered the region.

- The area was affected by significant tectonic influence afterwards (Ádám 1981; Martonné 2006), leading to the formation of large loess ridges. (For example, the Hegyhát was asymmetrically uplifted, and the Szekszárd Hills got separated from their environment, though unlike the other units they had been higher originally, and they were uplifted in the Quaternary rather than had a basin character.)
- Significant valley development started as the result of the uplift. (The Szekszárd Hills have a deep, highly segmented valley system; the Völgység got its name after the dense valley system; and the surface of Hegyhát is also segmented by more than 700 derasional valleys; Ádám 1981).
- Fault scarps were formed, and they have become the delimiting elements of the landscape (for example, the northern side of Hegyhát, or the steep edge of the Szekszárd Hills towards the plain of Sárköz).
- In the humid periods, landslides became active (for example, on the western and northern edges of Hegyhát).
- Typical loess erosional forms can be identified, e.g. loess dolines deep roads and loess ravines (Szekszárd Hills, Hegyhát).

The Zselic also belongs to the meso-region. It has a similar geological structure as it was described above, and it is highly segmented by valleys. In the northern part, the deep valleys have a north-south direction, whereas the eastern part was affected by E–W-oriented faults followed by the valley system and the ridges in between. The fluvial sand, deposited onto the Pannonian clay and sand, is covered by loess, which is 10 m thick on average, though it is 30–40 m in the south. In the area of the highest point of the Zselic (Hollófészek: 358 m), the rocky basement is closer to the surface. The sediments covering the surface are in close relation with the landforms: landslides are extremely frequent, making agricultural cultivation difficult, and along the above-mentioned directions, derasional valleys, loess ridges and canyons developed.

The Baranyai-Hegyhát is basically the northern foothills of the **Mecsek Mts**. It is similar to the Zselic in its structure and landforms. The geological structure of the landscape is determined by the Mecsek and the Villány Mountains, and their foothills. Both mountain horsts are built of mainly Mesozoic carbonated sediments, and they elevate like islands from the Pécs Plain and the South Baranya Hills. The body of the mountains used to be located in the shoreline of the Tethys and Vardar Oceans, in the foreground of the European crust piece, which fact defined their sediment sequences, as well (see Sect. 1.1). At present, the sediments are getting younger in the west-east direction, thus the connecting landforms are also different (Fig. 11.3). The Upper Permian sandstone situated in the south-western foreground of the Mecsek Mts. contains uranium (which was exploited) and indicates a terrestrial environment. (The sandstone of Jakab-hegy covering the above-mentioned layer is already Triassic and contain cemented gravel layers, whose particular surface erosional forms are the "Babás szerkő"; Image 11.1.). The Triassic limestones refer to deep-sea environment in the West Mecsek and the Villány Mts, on them a rich surface and subsurface karst form-assemblage was formed: e.g. the Abaliget Cave or the hydrothermal shaft cave of Beremend (Ördöglyuk), which was explored by quarrying. The palaeo-karst of Nagyharsány-hegy (Szársomlyó) demonstrates the karstification under tropical climate of the Cretaceous and contains a modest amount of bauxite. In the Middle Mesozoic, in coastal marshes and bogs, the vegetation was rich, thus (Liassic) black coal formed in the Lower Jurassic.

In the Lower Cretaceous, volcanism became active in the East Mecsek. (The product of submarine volcanism is the phonolite, which constitutes the large mass of the Zengő Peak.) Some erosion levels older than Eocene were formed on sandstone and marl, as a consequence of the tropical climate (Misina-tető, Jakab-hegy, Dobogó; Fig. 11.3). The Villány Mts. are bordered by fault lines on each side, and along them medicinal water wells spring, enriched by sulphur and hydrogen sulphide (at Harkány, Siklós, Beremend).

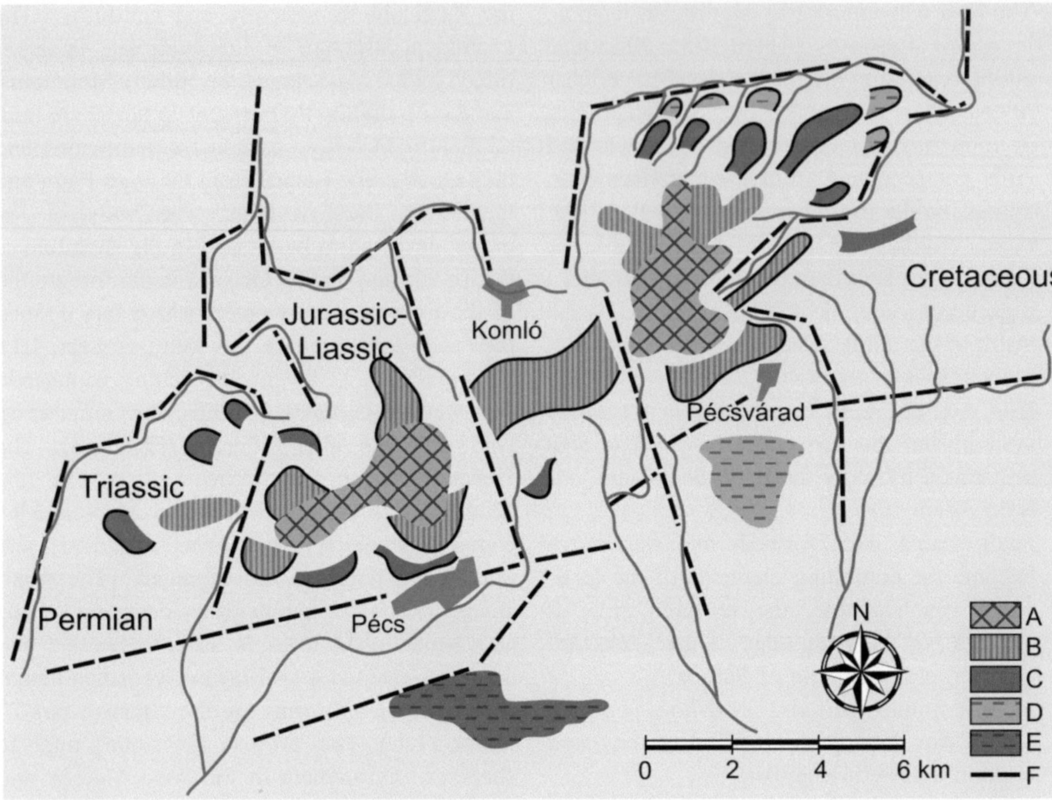


**Fig. 11.3** The main structural lines, erosional surfaces and lithological structure of the Mecsek Mts. (after Lovász 1981; Lovász and Majoros 1994; MNA 1989) *A* Eocene–Oligocene surface, *B* Carpathian (Miocene) abrasional surface, *C* Lower Pannonian abrasional surface, *D* pediment, *E* buried foothill debris cone, *F* main structural line

Since the Middle Miocene, the central and peripheral parts of the Mecsek became subsided, thus gravel debris deposited onto them, and then it was followed by a short marine transgressional period. It is proven by Leitha Limestone, brown coal at Hidas, and some Miocene clay in the northern edge. (These clay sediments provide good conditions for active mass movements; therefore, the land use is still hindered by them, for example the community development of Komló.) Abrasion terraces were formed on the surfaces which were situated above the Pannonian Sea (Lovász and Majoros 1994; MT 2002).

Between two horst ridges, in the south-eastern foreground of the Mecsek, the hills (Geresd Hills) are covered by loess. Within this hilly region, the granite block of Mórágy emerges to the surface. It is a Palaeozoic, 300-million-year-old structure (Pécsi 1996). The Pécs Plain is a peripheral member of the broad trench system of the Dráva, which was formed in the Pleistocene in the SW front of the Mecsek Mts. The alluvial fans of the Pécs Plain filled the depression, and nowadays they play an important role in the water supply of Pécs. The South Baranya Hills are the pediment remains between the Mecsek and the Villány Mountains, where the erosion and aggradation of valleys repeated several times, created the present landscape of the hills (MT 2002). Loess only accumulated along its edges in thin layers.

**Image. 11.1** Babás-szerkő in the Mecsek Mts. *Source* http://www.panoramio.com/photo/10760969

## 11.3 Some Natural Values and Environmental Issues of the Transdanubian Hills (with the Collaboration of Bata, T.)

The macro-region is rich in natural resources on Hungarian scale. Apart from coal and uranium, several other utilizable raw materials represent a significant value, but they are not exploitable at the moment because of environmental and economical reasons. Such are the gravel deposits, bauxite and natural gas occurrences in the Dráva Valley. The ecological and recreational features also represent significant value (e.g. Mecsek Mts., Lake Balaton).

The Jurassic black coal in the Central Mecsek Mts. is one of the greatest values of the region. The exploitable coal is almost 2000 million tons (MBFH 2008), which is sufficient for a hundred years counting with the extraction of the past years (when mining was active); however, its exploitation depends on economical and environmental conditions. Black coal mining started on the surface in the 1790s in the Mecsek Mts, applying simple tools. It was transformed into industrial subsurface mining to satisfy the needs of steam boats on the Danube. Open cast mines were also created due to the quick development of Pécs (Szirtes 1994). Coal mining was stopped in 2004, because the local power plant switched to the utilization of biomass and natural gas.

The mining of the uranium started in 1955 from Permian sandstone in the south-western foreground of the Mecsek. The mines produced about 20.8 million kg of uranium metal (Lóczy 2006b), but the extraction terminated in 1997, due to low profitability. This was the time when mining and ore processing lost their strategic importance in Hungary. Environmental restoration and landscape rehabilitation of the mines and spoil heaps started afterwards (Lendvainé Koleszár et al. 2003).

Two national parks are responsible for the management of the important geological, pedological and botanical values of the Transdanubian Hills. The Danube–Dráva National Park was established in 1996 involving the floodplains of the Danube and the Dráva River, on a territory of about 50 thousand ha. Thus, the majority of the managed natural values are connected to rivers, for example the oxbow system of Gemenc and Béda, or the valuable oak-ash-elm riparian forest of the floodplains. A traditional usage of oxbow lakes, crevasses and natural wetlands was typical along the Danube until the end of the eighteenth century, as the yearly floods refreshed the floodplain wetlands and lakes, providing good opportunity for fishery, for pastures and hayfields. Typical landforms of the Dráva are the sand and gravel bars, and the abandoned channels. Lake Baláta, which has plant relics form the Ice Age, is also specially protected. The Mecsek Protected Area with its particular geomorphological forms, the botanical and faunistical values of Szársomlyó Mt. and the picturesque, undisturbed Zselic belong to the national park too. The territory of the Balaton Uplands National Park extends to the southern and south-western part of Lake Balaton, thus, to the area of the Transdanubian Hills and Mountains macro-regions too.

Two of the human-induced environmental conflicts in the macro-region will be highlighted. The first is connected to the former mining of coal and uranium in the Mecsek Mts., and the second is related to the water usage of Lake Balaton and Dráva River.

One of the significant effects of mining was the 46 million tons of rock transported to the surface and piled in spoil heaps, where the soils were mixed with uranium, radium, acids, manganese and limestone. The subsurface mining created a significantly high volume of cavity system (18 million m$^3$), which had to be filled up and rehabilitated after the end of mining activity (Lóczy et al. 2007). The mine pits and spoil heaps deteriorate the landscape aesthetically, and also damaged forests, meadows and alluvial soils under them. The depressions and the water level decrease caused by subsurface mine blasts constituted further environmental conflicts. One of the most serious consequences is that the flow directions and patterns of subsurface water changed; thus, the recharge of the Pannonian aquifers decreased from the direction of the Mecsek Mts., though they play an important role in the water supply of Pécs. The reclamation and rehabilitation of the landscape after mining involve the burial of the subsurface caverns and waste heaps, and covering them by quickly growing tree species (Lóczy 2006a).

Lake Balaton is an important element of the Transdanubian Hills. Also taking into consideration the quantitative and qualitative aspects outlined in Sect. 3.1.3, several concepts were born after the unusually dry year of 2003 to supply water from the Rába and Zala rivers (Fig. 11.4). Fortunately, it was decided that the changes in the natural water regime should not be altered, and the natural environment and processes should be preserved. (As it was mentioned, the Balaton originally belongs to the slightly salt-affected lakes, and a freshwater resupply would disturb the chemical balance of the lake.) The water level of the lake reached the desirable level in the first decade of the twenty-first century, what is more, it inundated the lower shoreline areas in 2010, as it was an unusually humid year.

The Lake Balaton had a drinking water quality, but it has changed mostly due to the dissolved or mineral nitrogen and phosphorus. The water supply of the lake is partly provided by streams flowing into the lake, and the Zala River is the most significant among them. Intensive agriculture is used to characterize the catchment area of the Zala River in the second part of the twentieth century; thus, the river transported chemicals into the Lake Balaton, which resulted in eutrophication. Eutrophication means the accumulation of plant nutrients, especially nitrogen and/or phosphorus compounds. The problem is posed by its consequence, since the higher nutrient supply means the acceleration of the growth of algae and more complex plant life forms. The increased oxygen consumption of the algae, due to the more intense decomposition of biomass, leads to the decrease in the amount of dissolved oxygen in the water, creating unfavourable conditions for all the aquatic living organisms. All these change the quality of the water into an undesirable direction. Eutrophication can cause serious problems in the western basin of Lake Balaton; however, it affects most of the shallow lakes of Central Europe. The problem can be solved by decreasing the amount of wastewater and organic material transported into the system, which is also applied in the case of Lake Balaton.

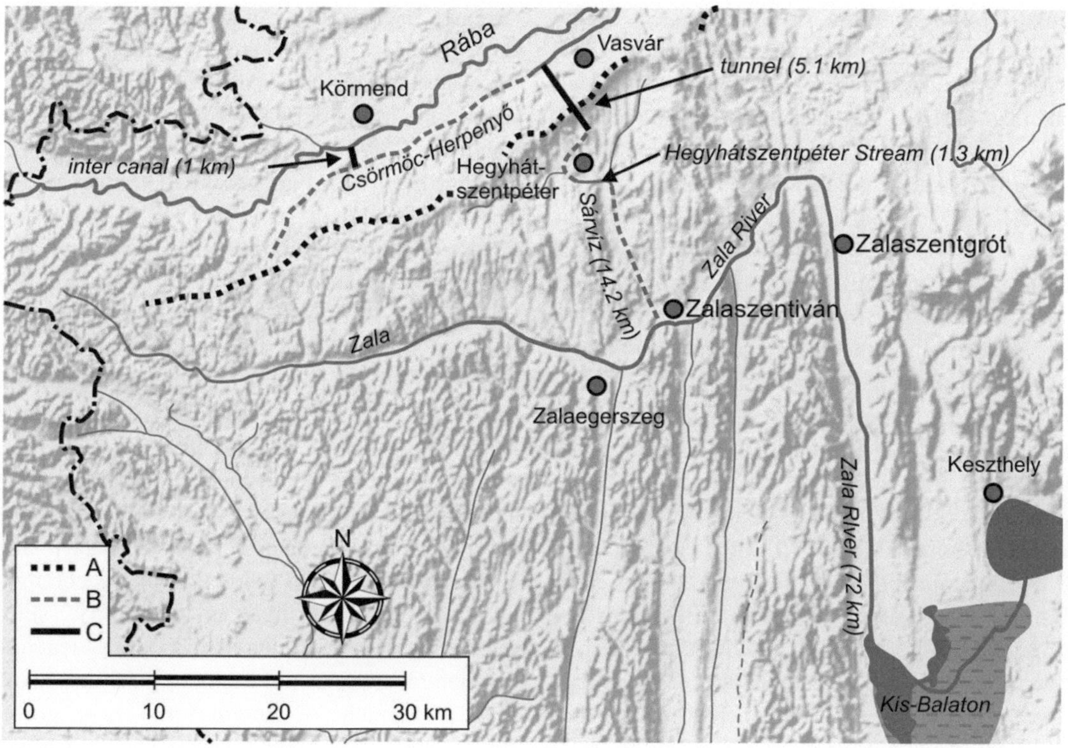


**Fig. 11.4** Proposed plan to supply water to Lake Balaton from the Rába River (after Szabó 2003) *A* divide, *B* lake-bed regulation, *C* artificial channel

In the 1980s and 1990s, the Zala River is used to drain water polluted with nitrogen and phosphorus, considerably deteriorating the water quality of the Keszthely Bay. The water reservoir system of Kis-Balaton (at the confluence of the Zala River) was built to solve this problem. Its first phase has been operating since 1985, so in the last decades, a high amount of suspended solids, nearly 500 tons of phosphorus and 2650 tons of nitrogen have been deposited in the reservoir, instead of being transported into the Lake Balaton. (The water quality of the reservoir of Kis-Balaton is bad due to an intense biological production, and the high organic content.) The quality of water flowing into the lake became better due to nutrient retention, which was also supported by the fact that the number of livestock farms had decreased on the catchment area of the Zala River, and the use of fertilizers had also decreased. As the result of the filtering function of the Kis-Balaton system, the growth of algae declined in the Keszthely Bay. However, this system was closed at the beginning of 2000s, because the deposited mud, containing sodium and phosphorous nutrients, was stirred up during the floods of the Zala flowing through. Therefore, it is not advisable to drain high discharges through the filters (Mezősi 2008). Nowadays, the construction of the second phase of the water reservoir system is being planned in order to improve water quality, and also because of some complex conservational and hydrological aims. Though the water quality of Lake Balaton is good, it deteriorates from the east to the west, however, intensive algae growth has not been observed lately.

The regulation of the Balaton having sensitive water regime has been intended for thousands of years. In the third century AD, Emperor Galerius's lock was built to regulate the outflow of water into the Sió River. In 1827, the Parliament declared the necessity of draining the water of

Lake Balaton, aiming for a decrease in water level. Thus, a lock was built at Siófok in 1863, with a drainage capacity of 50 $m^3/s$, which was enlarged in the 1900s. Depending on the amount of precipitation, it makes possible the water exchange of the whole lake within 1–5 years (MT 2002). The lake and its ecological conditions were directly affected by the interventions related to the increasing utilization of the catchment area and the shoreline from the eighteenth century (e.g. deforestation, land-use changes and the use of chemicals in cultivation in the twentieth century). Lake Balaton became a fashionable recreational destination about 100–120 years ago: the first steam boat started to operate in 1846; the railway was constructed in 1861 along the southern shore, and in 1910 in the northern one.

A government decision declares that Lake Balaton is a national treasure, which must be protected and developed, maintaining the permanent ecological balance of the environment. The problems of the environmental and ecological state of Lake Balaton are connected to water quality and the sewer system (Márkus 2001). Water quality alternates depending on the amount of precipitation and the intensity of human impact in the catchment area. In the past years, good water quality and quantity has characterized the lake (KvVM 2005). The waste-water sewer system (Bíró 1996) has been realized (40 out of the 43 shoreline settlements have joined the system; KvvM 2005), and modern facilities for wastewater collection and treatment have been built (at Balatonfüred, Keszthely), which helped a lot to manage the environmental problems. The purpose of building the facilities for wastewater treatment and storage was to stop the growth of algae in the lake by reducing the external phosphorous load. There is a connection between water quantity and quality in the case of all our lakes, and the appropriate water quantity is also important because the shallow lakes become more concentrated due to the loss caused by evaporation, which is more disadvantageous from the biological point of view (planktonic algae growth), than from the chemical one (Magyarics 2003). Among the environmental burdens affecting Lake Balaton, the intensive recreational use must be also mentioned. (According to a statistical survey, the tourists spent 4.1 million nights in the region in 2009.)

The other hydro-geographical conflict of the Transdanubian Hills is connected to the Dráva River. The construction of a dam–reservoir system in Croatia and Hungary was planned in 1977 (VIZITERV–ELEKTROPROJECT 1977), to improve shipping and flood safety, and to produce energy. On the upper, Austrian–Slovenian section, already more than 20 hydroelectric power plants operated on the Mura and the Dráva Rivers. There are three finished power plants in Croatia working at full capacity; however, there is an ongoing debate between Hungary and Croatia about the construction of a common hydroelectric power plant at Novo Virje, since the ecological conditions of the Dráva, which is in a natural state, would change significantly, and the development of natural forms (e.g. gravel bars, islands and oxbow lakes) would slow down or terminate, though these forms are important for the protected wildlife (Remenyik 2004). The utilization water and the sustainable landscape management became an important topic along the Hungarian–Croatian section of the Dráva, also because of the involvement of the Danube–Dráva National Park.

## References

Ádám L (1981) Tolnai-dombság. In: Ádám L, Marosi S, Szilárd J (szerk.) A Dunántúlidombság. Dél-Dunántúl. Akadémiai Kiadó, Budapest, pp 117–124

Ádám L (1990) A Balaton medencéje, a Balatonfelvidék és részmedencéi In: Marosi S. –Somogyi S. (szerk.) Magyarország kistájainak katasztere. MTAFKI, Budapest, pp 483–513

Ádám L, Marosi S, Szilárd J (szerk) (1981) A Dunántúli-dombság. Dél-Dunántúl, Akadémiai Kiadó, Budapest

Bíró P (szerk) (1996) A Balaton és vízgyűjtőjének ökológiai állapota. MTA Limnológiai Kutató Intézet, Tihany

Cholnoky J (1926) A földfelszín formáinak ismerete (Morfológia). Királyi Magyar Egyetemi Nyomda, Budapest

Cholnoky J (1936) Magyarország földrajza. Franklin, Budapest

Galambos J (1990) A Dunántúli-dombság sajátos táji adottságai. In: Marosi S, Somogyi S (szerk.) Magyarország kistájainak katasztere. I–II. Magyar Tudományos Akadémia Földrajztudományi Kutató Intézet, Budapest, pp 483–596 (kistájanként)

KvVM (2005) Ezerarcú Balaton – tervek és tények. http://www.kvvm.hu/cimg/documents/1002_onkormanyzati.ppt

Lehmann A (1981) A Dunántúli-dombság növényzete. In: Ádám L, Marosi S, Szilárd J (szerk.) Dunántúli--dombság – Dél-Dunántúl. Akadémiai Kiadó, Budapest, pp 211–248

Lendvainé Koleszár Z, Hideg J, Csáki F (szerk) (2003) Szilárd ásványbányászati alprogram. Uránbányászat. (Kármentesítési Füzetek 9.) Környezetvédelmi és Vízügyi Minisztérium, Budapest

Lóczy D (2006a) A Nyugati-Mecsek Tájvédelmi Körzet vízkészleteinek minősítése. In: Kovács F. – Hevesi A. (szerk.) Tiszteletkötet Hahn György 70. születésnapjára. Egyetemi Kiadó, Miskolc, pp 319–330

Lóczy D (2006b) Esettanulmány: Az uránbányászat hatásai a domborzatra a Mecsekben. In: Szabó J, Dávid L (szerk.) Antropogén geomorfológia. Egyetemi jegyzet. Debreceni Egyetem, Debrecen, pp 164–167

Lóczy D, Gyenizse P, Pirkhoffer E (2007) Veszélyforrás a város peremén: felhagyott bányaterületek Pécsett. In: Kerekes A, Fazekas I, Szabó G (szerk.) Települési környezet konferencia abstract kötete. Debreceni Egyetem Tájvédelmi és Környezetföldrajzi Tanszék–Kossuth Egyetemi Kiadó, Debrecen. http://geo.science.unideb.hu/taj/dokument/telkonf/dokument/loczy_d_et_al.pdf

Lóki J (1981) Belső-Somogy futóhomok-területeinek kialakulása és formái. Acta Geogr Debr 18–19:81–111

Lovász G (1981) Baranyai dombság, a Mecsek és a Villányi-hegység. In: Ádám L, Marosi S, Szilárd J (szerk.) Dunántúli-dombság – Dél-Dunántúl. Akadémiai Kiadó, Budapest, pp 124–136

Lovász G, Majoros Gy (1994) Magyarország természeti földrajza. JPTE, Pécs

Magyari O (2000) Vizeink szeizmikus „röntgenképe". In: Karátson D (szerk.) Pannon enciklopédia. Kertek, Budapest, pp 242–244

Magyarics A (2003) A Balaton vízminőségi- és vízszint-szabályozási problémáinak kezelése. Kézirat. http://www.gwpmo.hu/pdf/Balaton1.htm

Márkus F (2001) A Balaton betörése. http://ujsagcikkek.freeweb.hu/2001/nepszabadsag/NS3758.htm

Marosi S (1967) Megjegyzések a magyarországi futóhomok-területek genetikájához és morfológiájához. Földrajzi Közlemények 15(3):231–255

Marosi S (1969) Adatok Belső-Somogy és a Balaton hidrogeográfiájához. Földrajzi Értesítő 18:419–456

Marosi S (1970) Belső-Somogy kialakulása és felszínalaktana. Akadémiai Kiadó, Budapest

Marosi S, Pécsi M (1999) Gemorphological evaluation of the Balaton Basin. In: Pécsi M (ed) Landform evolution studies in Hungary. (Studies in geography in Hungary 30.) Akadémiai Kiadó, Budapest, pp 62–65

Marosi S, Somogyi S (1990) Magyarország kistájainak katasztere. I–II. Magyar Tudományos Akadémia Földrajztudományi Kutatóintézet, Budapest

Marosi S, Szilárd J (1981b) A Dunántúli-dombság domborzatának kialakulása és általános jellemzése. In: Ádám L, Marosi S, Szilárd J (szerk.) Dunántúli--dombság – Dél- Dunántúl. Akadémiai Kiadó, Budapest, pp 93–103

Martonné EK (2006) Magyarország tájföldrajza. KLTE, Debrecen

MBFH (2008) http://www.mbfh.hu/home/html/index.asp?msid=1&sid=0&hkl=72&lng=1

MNA (1989) Magyarország Nemzeti Atlasza. Carthographia, Budapest

Mezősi G (2008) Magyarország környezetföldrajza. JATEPress, Szeged

Glatz F (szerk) (2002) Magyar Tudománytár I. Föld, víz, levegő. Kossuth, Budapest

Pécsi M (1996) Geomorphological regions of Hungary. MTAFKI, Budapest

Rajkai K (1990) A Dunántúli-dombság kistájainak talajtani viszonyai. In: Marosi S, Somogyi S (szerk.) Magyarország kistájainak katasztere. MTAFKI, Budapest, pp 483–596

Remenyik B (2004) Vízerőművek építése vagy természetvédelem a Dráván. http://geography.hu/mfk2004/mfk2004/phd_cikkek/remenyik_bulcsu.pdf

Schweitzer F (1993) Domborzatformálódás a Pannóniai--medence belsejében a fiatal újkorban és a negyedidőszak határán. MTA, doktori értekezés, Budapest

Sümegi P, Törőcsik T (2007) Hazánk növényzete az éghajlatváltozások tükrében. Természet Világa 138 (7):1–9

Szabó G (2003) Duzzadó tervek. HVG 6(28):54–57

Szilárd J (1967) Külső-Somogy kialakulása és felszínalaktana. (Földrajzi Tanulmányok 7.) Akadémiai Kiadó, Budapest

Szilárd J (1983) Dunántúli és Duna-Tisza közi löszfeltárások új szempontú litológiai értékelése és tipizálása. Földrajzi Értesítő 32(1):109–166

Szirtes B szerk. (1994) A mecseki kőszénbányászat. I–II. Kútfúrás Kft., Pécs

VIZITERV–ELEKTROPROJECT (1977) Közös érdekű Dráva-szakasz komplex hasznosításának gazdasági elemzése. Budapest–Zágráb

# 12 Physical Geography of the North Hungarian Mountains

**Abstract**

The members of the North Hungarian Mountains are built of mainly Miocene andesitic volcanoes. They were strato-volcanoes with single volcanic vents, and during their evolution calderas developed. The only one exception is the Zemplén Mountains in northeast, where centro-labial volcanoes developed. In the macro-region the Bükk and Aggtelek Mountains built of carbonates, thus they have very rich surface and subsurface karstic form assemblage. These morpho-lithological characteristics of the meso-region determine its natural values and potential environmental conflicts too. In the volcanic mountains the hydro-thermal activity resulted in the development of precious and non-ferrous metal ores, but nowadays they are not exploited. In the intra-mountain basins and in the forelands of the mountains considerable amount of brown coal (its mining terminated) and lignite deposited. The exploitation of the latest is in progress, and the need for reclamation is continuous. In the twentieth century the macro-region was the industrial centre of Hungary, therefore several environmental conflicts should be solved in the future.

## 12.1 Physical Geography and Landscape Development

The macro region, which involves the mountains and hills in the northern part of the country, got its present name as a result of the designation of the state borders after World War I. The name of the region was Mátra Forest, and it used to spread onto Western and Eastern Carpathians in Slovakia, but, as opposed to the categorization today, did not involve the Visegrád Mountains which was part of the Transdanubian Mountains (then called Bakony Forest). Most of the physical geographical features (surface evolution, relief, climate, water supply, soil and wildlife) support the attachment of the Visegrád Mountains to the North Hungarian Mountains (Fig. 12.1). According to some categorizations, the Gödöllő Hills appear as a separate unit (Hajdú-Moharos and Hevesi 2000) and its attachment to the highly diverse Cserhát has ethnographic reasons. The affiliation of the basins in this macro region is very intricate, since they form landscape geographical units with their Slovak parts. The Hungarian classification considers these unconnected fragments as one unit out of necessity



G. Mezősi, *The Physical Geography of Hungary*, Geography of the Physical Environment,
DOI 10.1007/978-3-319-45183-1_12

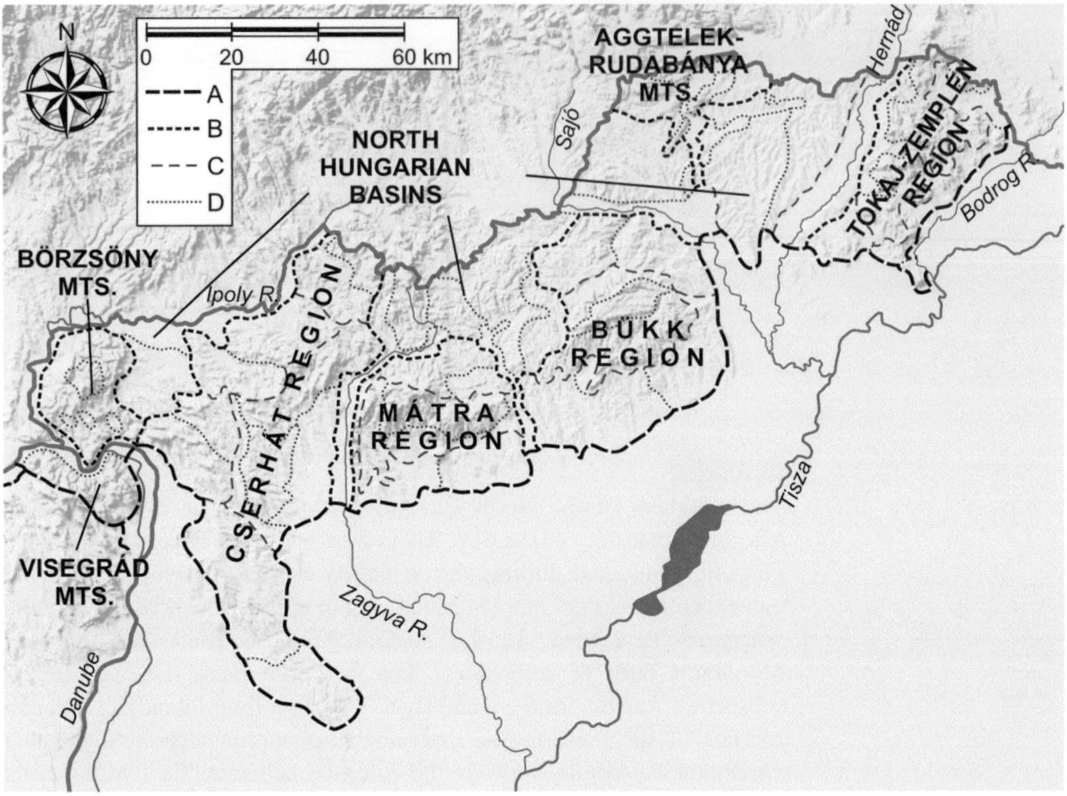


**Fig. 12.1** The landscape geographical division of the North Hungarian Mountains (according to the 1990 landscape division) *A* macro region border, *B* meso region border, *C* border of groups of micro regions, *D* micro region border

(called the North Hungarian Basins. Within the recent Hungarian borders, the Rudabánya, the Szalonna and the Aggtelek Mountains, which belong to Western and Eastern Carpathians in Slovakia in landscape geographical point of view, has to be considered as an individual unit.

The basis of the macro region consists mostly of carbonated formations that evolved on the southern platform of Tethys. Today, some of the remains are on the surface or located near the surface (see Sect. 1.1). From a physical geographical aspect, the area is characterised by different morphology and lithology. On the one hand, the units of the Aggtelek and Rudabánya Mountains, and also the Bükk Region, basically consist of Mesozoic carbonated sediments. On the other hand, there is a sequence of mainly Miocene volcanoes, which is highly characteristic of the region (Fig. 12.2). Based on these different parent rocks diverse landscape units evolved throughout centuries or millennia.

Horst mountains are located between the volcanic mountains, for example, the Bükk Mountains, and the Aggtelek and Rudabánya Mountains. The average height of the Bükk Mountains is 800 m, and its eroded upland is concentrically surrounded by a lower eroded surface and a wide, but dissected foothill surface. The younger sediments from the Tertiary in the area of the Aggtelek and Rudabánya Mountains eroded almost completely. Thus, the most typical Hungarian karstic forms (dolines, uvalas, limestone pavements, sinkholes and spring caves) evolved on the surface of both meso regions, though the mountain regions are very distinct concerning their origins and landscape geographical relations.

A number of stratovolcanoes (or their remnants), built of lava and Miocene tuff, adjoin the

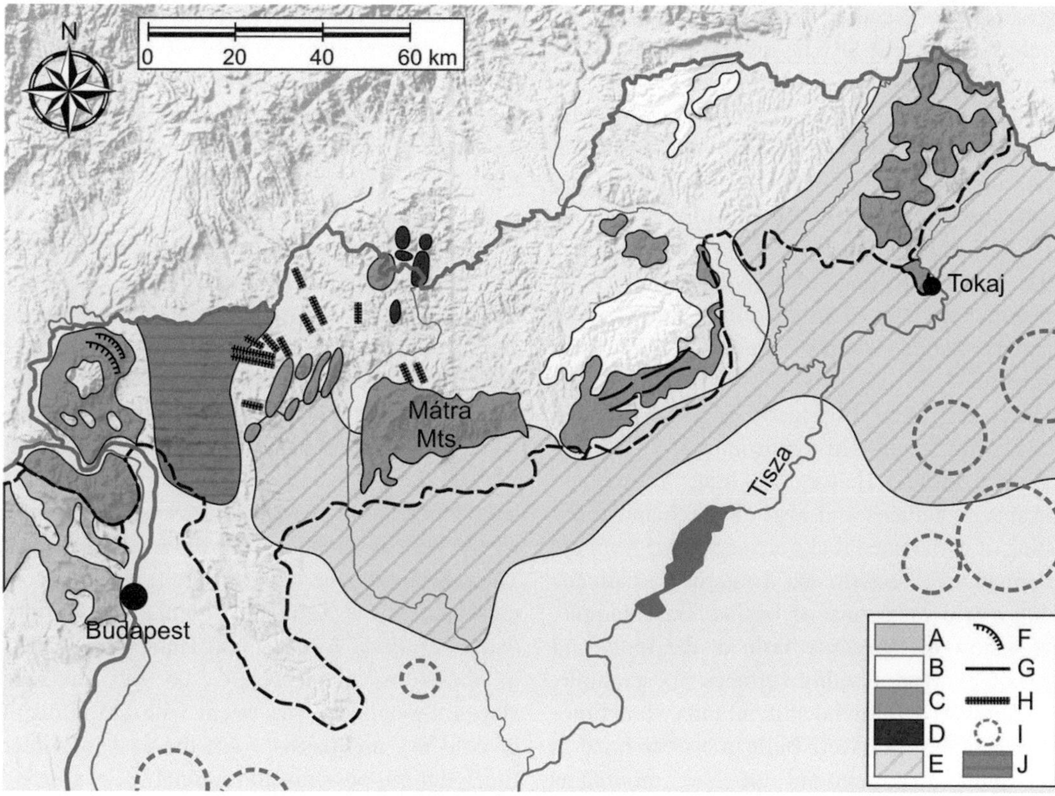


**Fig. 12.2** Mountain types and surface formations of the Inner Carpathians in the Hungarian territory (after Székely 1989, 1993) *A* series of sedimentary horsts from the Mesozoic, *B* highly folded limestone mountains from the Mesozoic, *C* surface volcanic formations from the Miocene, *D* basalt from the end of the Pliocene and the beginning of the Pleistocene, *E* Miocene volcanic formations covered with postvolcanic and depressed sediments, *F* the remains of the estimated outer caldera rim, *G* ignimbrite staircase, *H* weathered out andesite dyke, *I* assumed buried eruption centres, *J* pyroclasts deposited between the formerly huge volcanoes

North Hungarian Mountains. The mountain range, which is younger and more active heading from the west to the east, is one of the largest volcanic mountain wreaths of Europe. In the last 20 million years, the region was characterised by lessening volcanic activity resulting in various processes and forms. Lava flows were interrupted by massive tuff ejections, thus, volcanic tuff depositions predominate in several areas. The Visegrád Mountains, the Börzsöny, the Cserhát, and to the east, the Mátra and most parts of the Zemplén Mountains are built of andesite, and, in a smaller proportion, dacite lavas, tuffs and their agglomerates. Volcanism was the most active between 10 and 16 million years, coinciding with a significant thinning of the lithosphere. The change in activity was recognised a century ago indicated by the different presence of rhyolite and dacite tuffs (Mauritz 1910). Based on this, Miocene volcanic activities can be divided into three phases, and dated to around 19 ± 1.5, 15 ± 1.5, 13 ± 1.5 million years with the help of K/Ar measurements (Balla et al. 1981; Hámor et al. 1987; Korpás 1998; see Sect. 1.2). The Inner Carpathian volcanic wreath extends as far as the Harghita Mountains, where volcanic activity was present during the Pleistocene, as well (Harghita Madadaras Mountain, Kakukk Mountain).

At the time of their formations, the volcanic mountains of the North Hungarian Mountains were high, layered cones of the size of Mount Vesuvius. There were several other smaller volcanoes active in their surroundings. All these characteristic forms had significantly eroded and degraded by the end of the Miocene Period and the thinner volcanic lava layers and dykes around the centre had got heavily truncated. For example, in some locations in the Cserhát, only the paths of the lava lodes are visible on the surface. According to Pécsi (1996), the position of the Leitha limestone on surfaces at different heights (100–150 m asl) suggests that the volcanic cones eroded almost to their erosion base (sea) level rather quickly (in about 1 million years), then emerged and started their multifactorial formation.

Hills and basin hills made of loose material evolved in the northern foreground of the former volcanic centres. Hence, the Inner Carpathian volcanic mountains and also the highlands consisting of carbonated rocks are separated from the mountainous Western and Eastern Carpathians along a wide sequence of basins. For example, that is how the spacious basin of the Ipoly and the Sajó evolved. Foothill surfaces (for example, the Cserehát) and fluvial alluvial fans, sometimes covered with loess, were built in a wide band in the southern foreground of the mountains towards the Great Hungarian Plain.

The climate of the North Hungarian Mountains is a mountain climate, and shows slight vertical segmentation. The areas above 300 m have moderately cool and moderately wet climate, while the areas above 700 m have cool and wet climate (Péczely 1979). Irradiation is below average in the island-like elevated landscape segments, on a Carpathian Basin scale. (It is only lower in the Lower Alps.) Due to the relief characteristics, the length of the irradiation is very diverse in the mountains. (Here the average value is less than 50 % of the astronomically possible sunshine duration, which is 4100 MJ $m^{-2}$.) This means almost 50 % more sunny hours on average in winter, since the region has a higher relief, thus, during inverse positions some areas rise above their foggy surroundings. However, summers are more cloudy and rainy, therefore the number of sunny hours is less. The convection of the macro region is mostly controlled by orographic impacts. As a potential feature, we may point out the high wind-energy of the southern slopes.

The annual mean temperature of the mountain range as a whole is 7.0 °C (MET1 2009). The mean temperature in January it is −3 °C, and in July it is +20 °C. The vertically changing temperature in the summer is characteristic of the Earth's mean (its average value is 0.65 °C/100 m —the value of 0.58 °C/100 m given in Chap. 10 apply to the Transdanubian Mountains and the summer half-year). However, during the winter half-year, when the occurrence of inversive air stratification is more frequent, the value is around 0.2–0.4 °C/100 m (Péczely 1987). In the mountain areas there are 120–140 days of frost annually and, as mentioned earlier in Chap. 2., max. 70–100 days of snow cover. The annual mean temperature is the lowest in Kékestető (5.5 °C), where we can only count with 6 summer days. The most extreme temperature values can be measured in the Zemplén Mountains. Heat prevails—similarly to the Great Hungarian Plain—in the summer, especially on the southerly slopes, though, we can count with the intrusion of cold dry air masses from the east in winter. Precipitation, on a macro regional scale, shows a decreasing tendency as we head from the west to the east, which cannot be compensated even by the height of the mountains. Therefore, the amount of annual mean precipitation is between 550 and 800 mm. To demonstrate the extremes, the wettest month in Hungary was measured in Dobogókő (June, 1958–444 mm).

From a hydrographical point of view, the volcanic and the carbonated areas of the macro region are clearly separated. The surface waterflows often adjust to the morphological framework in the volcanic areas. Radial (on the sheet of volcanic cones), branched (in craters) and arched (in calderas) water networks are frequent, but the most complex is the one in the Zemplén Mountains due to its intense relief segmentation. The Ipoly conducts its water into the Danube, while the Zagyva, the Galga, the Tarna, the Sajó, the Hernád and the Bodrog rivers—sometimes following structural paths (for example, the Lókos Stream in Cserhát)—conduct their water into the River Tisza. Great floods related to these rivers are statistically rare (MT 2002), although the destructive runoff floods of 2010 suggest otherwise. The mountains are rich in springs with smaller discharges. Springs and waterflows are rare on the karstic surfaces of the Bükk and the Aggtelek Mountains, however, at the edges of the mountains, there are several very abundant surfacing karst springs. Their unreserved waters join the Tisza. Some of these springs are temporary and others are permanent (for the hydrographical analysis of the springs, see Sect. 1.3). Karstwaters are important from the aspect of their use as drinking water, and their protective zones are determined by law (Mezősi 2008). Karstwaters at the southern edge of the Bükk often combine with postvolcanic impacts, and surface as lukewarm springs (for example, Eger, Miskolctapolca).

The zonal soil cover of the macro region is the brown forest soil—its types vary according to the relief and the impact of rocks. In the basins and at the southern edges, with the enhancement of the lowland character, chernozem brown forest soils become typical (Stefanovits 1981). The Mátra and the volcanic mountains in general are covered with brown forest soils with clay illuviation. Erubase soils and brown earth are characteristic on volcanic detritus, which turn into reddish brown forest soil near the edges. In the wetter areas of the Zemplén Mountains, grey podzolic forest soil is present, while in most parts of the mountain, the slightly leached loamy brown forest soil is characteristic. On the carbonated surfaces of the Aggtelek and Rudabánya Mountains, the Bükk, etc., rendzina soils prevail, but apart from that, we may encounter loamy brown forest soils in the Bükk, and in most parts of the macro region, we can see non-podzolic brown forest soils related to shale in the lower ridges and valleys.

The North Hungarian Mountains have a rich vegetation, and botanists refer to it as one unit (*Matricum;* Soó 1964). Naturally, the vegetation consists of deciduous forests, where oak and beech forests are common. If their vertical positions are regular, oak and turkey oak forests can be found up to 600 m on the southern side, and up to 400 m on the northern side, whereas on the northern side and above the previously mentioned heights, beech forests are characteristic (Hortobágyi and Simon 1981). Hornbeam, ash and (in the Mátra) even maple forests mingled with oak forests. All of the volcanic mountains were covered with thick forests. The inhabitants who had settled into the valleys and basins thinned the forests heavily and used the edges as arable land (for example, in the Cserhát and the Mátra). The situation of the Zemplén Mountains is special, since in that area, it is not only the lower foothills covered with loess that are precious agricultural territories, but also the higher, 200–300 m hill country of southern and south-eastern exposure, which were formed on volcanic rocks (for example, Tokaj-Hegyalja wine region). The potential forest coenoses in the Bükk and the Aggtelek Mountains are the montane beech, the karst shrub forest and on the lower levels, turkey oak and hornbeam oak, which are enriched by swards and steppe meadows (Soó 1964).

## 12.2 The Genetic Overview and Some Physical Geographical Issues of the North Hungarian Mountains

From a landscape geographical point of view, the volcanic areas of the macro region mostly form individual units. The remains of volcanoes as individual units had been highly specifically classified by Székely (1983, 1997) (Among others, he determined consistencies such as ruin, wreck, remnant, etc.). The categorisation above is difficult mostly because the majority of the volcanoes is recognizable, however, they are significantly eroded.

The young Tertiary volcanic range of Hungary begins with the Visegrád Mountains in the Danube Bend. The Börzsöny, which is similar to the Visegrád Mountains concerning its morphology and origin, is connected to the latter from the north, at the gapped Visegrád gorge.

The meso regions of the Visegrád Mountains and the Börzsöny have a very similar physical geographical structure, form and process. In the Oligocene and the beginning of the Miocene, sandy and clayey sediments got deposited onto the carbonated base in several hundred metres' thickness. (In the case of the Börzsöny, the base to the north of the Diósjenő Line, which is part of the Carpathian Vepor range, is already Paleozoic). The volcanoes started their activity in 100–200 m deep sea.

(A) Volcanic activity in the **Visegrád Mountains** can be divided into two phases. The first resulted in volcanoes of 2–6 km in diametre, which were sometimes accompanied by small subvolcanic bodies (bulging cones—the best known being Hill Csódi in Dunabogdány) and lodes (Korpás 2000). The volcanoes sorrounded by structural lines (Fig. 12.3) emerged like islands, and turned into calderas by explosions rather than ruptures (with dendric water supply inside). Volcanism in the mountains took place between 14.5 and 15.5 million years (Hámor et al. 1987). The material for the greater cones was provided by the 6–12 m somma volcano in the area of Visegrád—which was recognized by Cholnoky (1936). The material of the stratovolcanoes mainly consists of andesite and andesite tuff. The stratovolcano structure can be observed in Vadálló-kövek (belonging to the somma in Dömös) or in Rám Canyon (at the northern side of the mountains). To demonstrate the size of the volcano, Hill Szent Mihály on the northern bank of the Danube was also part of the formation. The estimated original height of Hill Keserű and Visegrád stratovolcanoes was 1100 m (Korpás 1998). The towering Thirring Rock, which stands at the edge of the central part, is a relic of the exposived caldera (Fig. 12.1). The size of the caldera was also determined by remote sensing data obtained from river network reconstruction (Székely 1997; Nemerkényi 1987), although the existence of a double caldera was questioned (Karátson 2007). Szabó (1996) depicted that here, and also in the Mátra, landslide areas developed (under appropriate tectonic and lithological conditions) in the peripheral zones, which may have highly contributed to the effective erosion of the volcanoes in warm-temperate climate (Image 12.1).

(**B**) **The Börzsöny** also belongs to this volcanic group. Its base was formed on schlier and also on thick layers of clay and sand, which developed in marine, seaside, lacustrine and fluvial environments as a result of a prolapse that took place in the Oligocene and lasted until the beginning of the Miocene (Fig. 12.3). Its volcanic activity took place in its present location, but took longer than that of the Visegrád Mountains (13.5–16.5 million years ago—Balla et al. 1981). Based on reconstruction, the mountain was approx. 1800 m (Balla 1978), or 1300–1400 m high (Karátson 2000), consists mainly of andesite and andesite tuff, and it is described as a stratovolcano, which was complemented by subvolcanic formations in its later phase (Kubovics and Pantó 1970; Balla et al. 1981; Korpás 1998; Karátson 2000).

The spectacular central caldera (Fig. 1.14) may have dilated with the implosions of the steep volcanic conduit, as a result of explosions. The great explosions ejected large amounts of material, thus the volcano gaved away—perhaps several times—into the magma chambers below and developed a caldera. However, the central caldera that determines the shape originates from several volcanic centres. Fekete-víz creek developed a water network in it, which led to fluvial erosion. The truncatedness (lack) of the south-western part of the central caldera is explained by the explosions caused by subvolcanic intrusion, which is characteristic of the mountains (Korpás 1998), while according to others, the reason for the hiatus lies in slope mass movement (Karátson 2007). Mineralization related to volcanism is discussed in Sect. 1.2.

Almost a dozen smaller parasitic craters settled onto the cone mantle of the Upper Börzsöny. Concerning the forms, there is asymmetry between the eastern and the western side of the mountain. On the eastern side, we can observe volcanic cones (for example, Nógrádi-Várhegy), emerging layer heads and laccoliths, while the western side has a gentler slope, with layer slabs on its surface, and the lava stocks that had been formerly stuck under the surface reached surface as a result of the emergence of the mountains (for example, Mount Csák near Szob, or the earlier mentioned Mount Csódi near Dunabogdány). The formations found on the rim of the caldera

**Image 12.1** The Thirring Rock in the Visegrád Mountains show the edge of the former caldera. *Source* http://www.panoramio.com/photo/4905703

were formed by the wind and periglacial processes, for example Szabó Rocks and Oltár Rock on Csóványos.

In the last century there were several kinds of theories explaining the evolution and the date of formation of the Visegrád Gorge on the Danube Band. The results acquired about its surface development or its age are very different from each other, and due to the lack of further information, the question is still not settled today. According to one group of the assumptions, the Danube only occupied this valley section in the second half of the Quaternary or in the last ice age, about 120 thousand years ago (Ruszkiczai-Rüdiger et al. 2006). According to the other group (which are supported by several data), the Danube was present in the Visegrád Gorge as early as the Miocene (if it existed at the time), although it may have been temporarily blocked, buried and then exhumed (Bulla 1964; Schweitzer 1993; Pécsi 1959, 1996 and for details see Sects. 1.3 and 1.4). In the regions of Nagymaros and Visegrád, on both sides at about 330–360 m asl, we can find coral limestones, which may refer to the presence of a shallow Miocene gorge. Based on this, the presence of the Ancient Danube as early as the Pliocene seemed to have been proven (MT 2002).

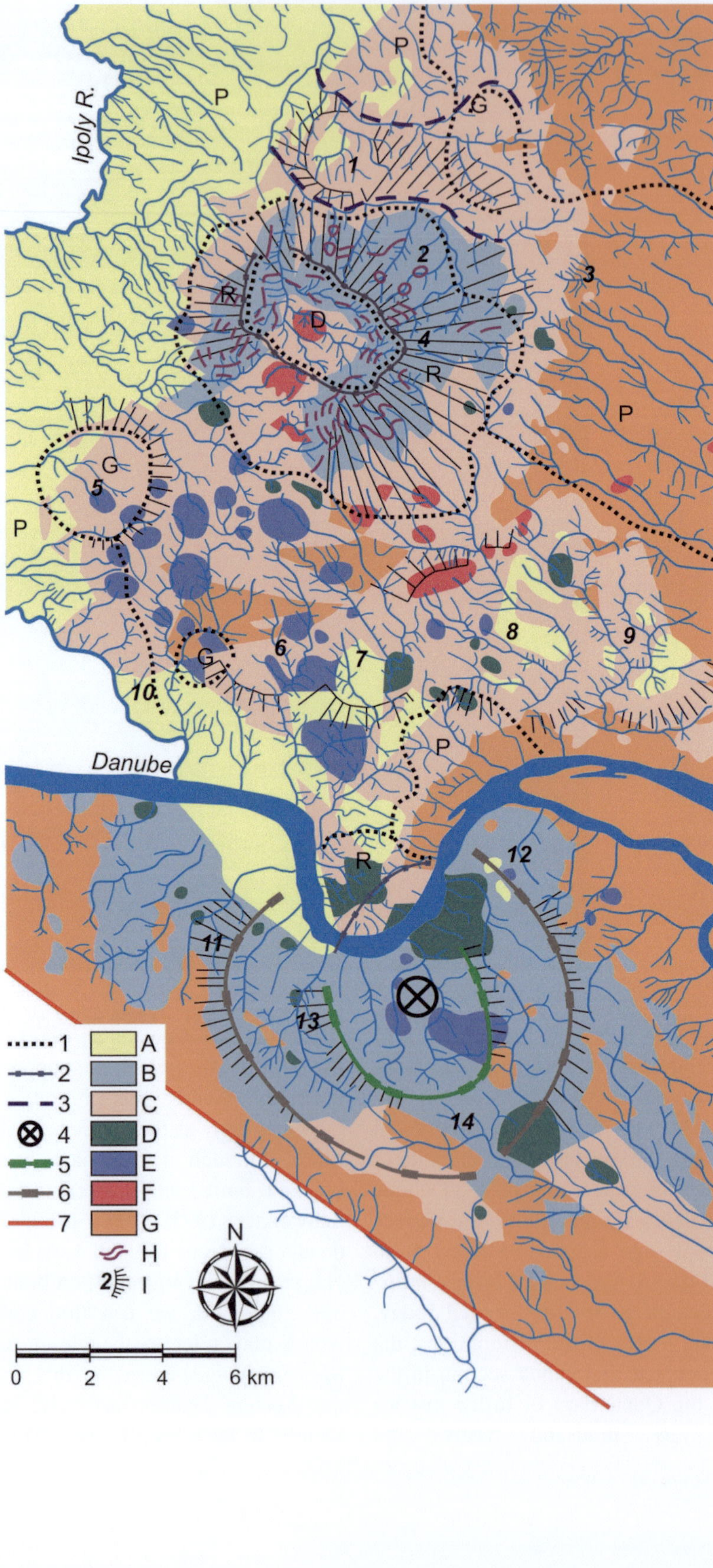
Ipoly R.
Danube
P
G
R
D
1
2
3
4
5
6
7
8
9
10
11
12
13
14
A
B
C
D
E
F
G
H
I
N
0
2
4
6 km

◀ **Fig. 12.3** A schematic geological map of the Börzsöny and the Visegrád Mountains and some of their characteristic form groups (after Korpás 1998, Gábris 1983, Székely 1997) *1*—border of valley system types, 2—inner rim of the caldera, *3*—border of the assumed outer rim of the caldera; the Visegrád Mountains, *4*—assumed eruption centre in the inner caldera, *5*—rim of inner caldera, *6*—rim of outer caldera, *7*—structural line, border between Mesozoic sediments and volcanic mountains; *A*—cover sediments of volcanoes, *B*—andesite stratovolcanic formations, *C*—andesite-dacite stratovolcanic formations, *D*—pyroxene andesite, *E*—biotite-amphibole pyroxene, *F*—biotite-amphibole andesite, dacite, *G*—vulcanite bedrock sediments, *H*—lava terraces, *I*—remnants of volcanic cones; Inscriptions: *D*—branched water supply inside the caldera, *R*—radial, outward valley network, G—ring-shaped water network on slope of former side-cone, *P*—parallel consequent water network on former slope of lava flow. *1* Kemence stratovolcano, *2* Király-kút stratovolcano, *3* Hill Kámor stratovolcano, *4* Upper Börzsöny stratovolcano, *5* Hill Közép stratovolcano, *6* Magasrét stratovolcano, *7* Pusztatorony stratovolcano, *8* Börzsönyliget stratovolcano, *9* Magyarkút stratovolcano, *10* Ipolydamásd stratovolcano, *11* Visegrád stratovolcano, *12* Nagyvillám stratovolcano, *13* Hill Keserű stratovolcano, *14* Pilisszentlászló stratovolcan

The Danube is the main landscape forming factor in the valley between the volcanic mountains. In fact, it does not only separate the two mountains from each other, but also connects them. The Visegrád Gorge has not only been a space for living to the local people for a thousand years, but also has a major interregional economic and political role. The role (the assessment or function of their features and their exploitation) and use of the landscape units can vary from age to age. In our case the highly similar meso-regions with primarily economical main functions (forestry, mining) could be taken as one large unit. Their history has slightly changed in the last century, mainly due to the slightly different development of social and aesthetic functions of lower hierarchy.

**(C) The Cserhát Region** is relatively clearly separated from the Börzsöny (to the west) and from the Mátra (to the east), and shows the sight of a hill country rather than a middle mountain range. It includes micro regions of different genetics. Smaller volcanic or carbonated formations that determine the character of the smaller landscape units emerge from hill countries, often divided by valleys, consisting of generally Tertiary shallow marine sediments. Castles built on these high grounds provided protection for the people of the valleys who had created a dense network of settlements. They stand somewhat isolated from each other, still, some of them have kept their old folk traditions and costumes.

The West Cserhát is the continuation of the Mesozoic mountains of the Transdanubian Mountains. Here, the carbonated horsts that were covered with Hárshegy sandstone (Naszály, Csővári and Romhányi Mounts) sank deep. This group of mountains only emerges sporadically, in an island-like way from the declivous hill ridges built from Tertiary clay and sand. In the vicinity of Nézsa, the bauxite remnants of the former dolines on top of the limestone block refer to Paleozoic karstification (Korpás 1998; Horváth 2000). The limestone horst of Naszály got heavily truncated due to intensive limestone mining. In this group of micro regions, volcanic formations and their form groups have a subordinate role.

The image of the East Cserhát (or Volcanic Cserhát), which lies between the valleys of the Galga and the Zagyva, is determined by volcanic formations. Lava erupted mainly along the NE-SW and perpendicular cross faults during the Mid-Miocene Period (Noszky 1940 mentions andesite lodes). Three trench rows, and among them horsts (or rather, eroded row of lodes), separated along the faults as a result of elevations that took place from the end of the Pliocene (Székely 1997). The system of trench rows can be recognised in the running direction of the volcanoes, and the Zagyva also occupied such a trench (Láng 1967). Apart from the andesite range of Mounts of Tepke and Szanda (Fig. 12.4, Image 1.5), the lodes adjoining them—which were shaped by erosion—and the lode traces are also characteristic on the surface. Székely (1983) examined the direction of the row of lodes and came to the conclusion that the centre of the lode directions point to the centre of the Mátra (Fig. 12.5). That is how he explained that the East Cserhát belongs to the Mátra.

The lava under the surface sometimes solidified in the form of lakkolite, which could have

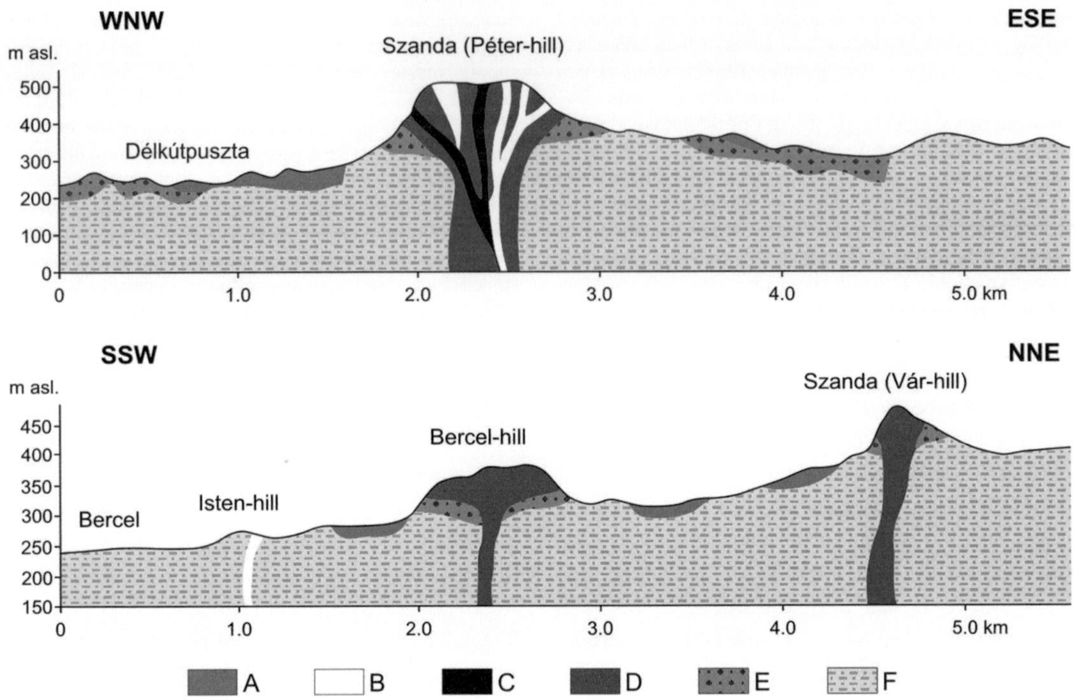


**Fig. 12.4** The schematic geological structure of the andesite volcano of the Volcanic Cserhát (after Noszky 1940) *A*—loess, *B*—hydrothermal lode, *C*—hydroandesite, *D*—pyroxene andesite, *E*—lower Helvetic gravel, *F*—katti clayey sand

surfaced due to later tectonic elevation and effective erosion. Mount Karancs (the largest lakkolite in Cserhát - Horváth 2000) built of Mid-Miocene andesite evolved in this manner in the North Cserhát and the Karancs Region. The whole surface has a gentle slope towards the valley of the Ipoly, to the north.

The Gödöllő Hills and the Monor-Irsa Hills are also claimed to be parts of the Cserhát Region. They partly consist of Pannonian sandy-clayey sediments, and are related to the sediments of the Ancient Danube and the waterflows coming from the north. The surface which emerged from the middle of the Quaternary is covered with loess, with blown sand around the edges, thus, the prevailance of loess-derasional forms is obvious.

**(D) The Mátra** are the highest mountains in the North Hungarian Mountains and also have the highest point in Hungary (Kékestető—1015 m). The valley of the Nagy-patak in Gyöngyös, which is the deepest regressing waterflow in the ridge, morphologically divides the mountains into western and eastern sides. Apart from this, the Upper Mátra is also mentioned as a common name for the higher regions (from a landscape geographical view, these regions are jointly called Mátra). Using the landscape subdivision of the upper, eastern and western micro regions of the Central Mátra (which is followed by Hajdú-Moharos and Hevesi, as well) is more advisable than the use of the categorisation of 1990 (also used in Hungary—see Appendix).

These units are connected (similarly to other akin Mátra Forest units) by the form that evolved as a result of Mid-Miocene volcanism (which took 5–6 million years, between 13 and 19 million years) and the material it produced. The difference between the eastern and western sides does not actually lie in the difference between their volcanic activities, but in the difference between the forms. Earlier, the West Mátra was categorized as caldera and the East Mátra was a

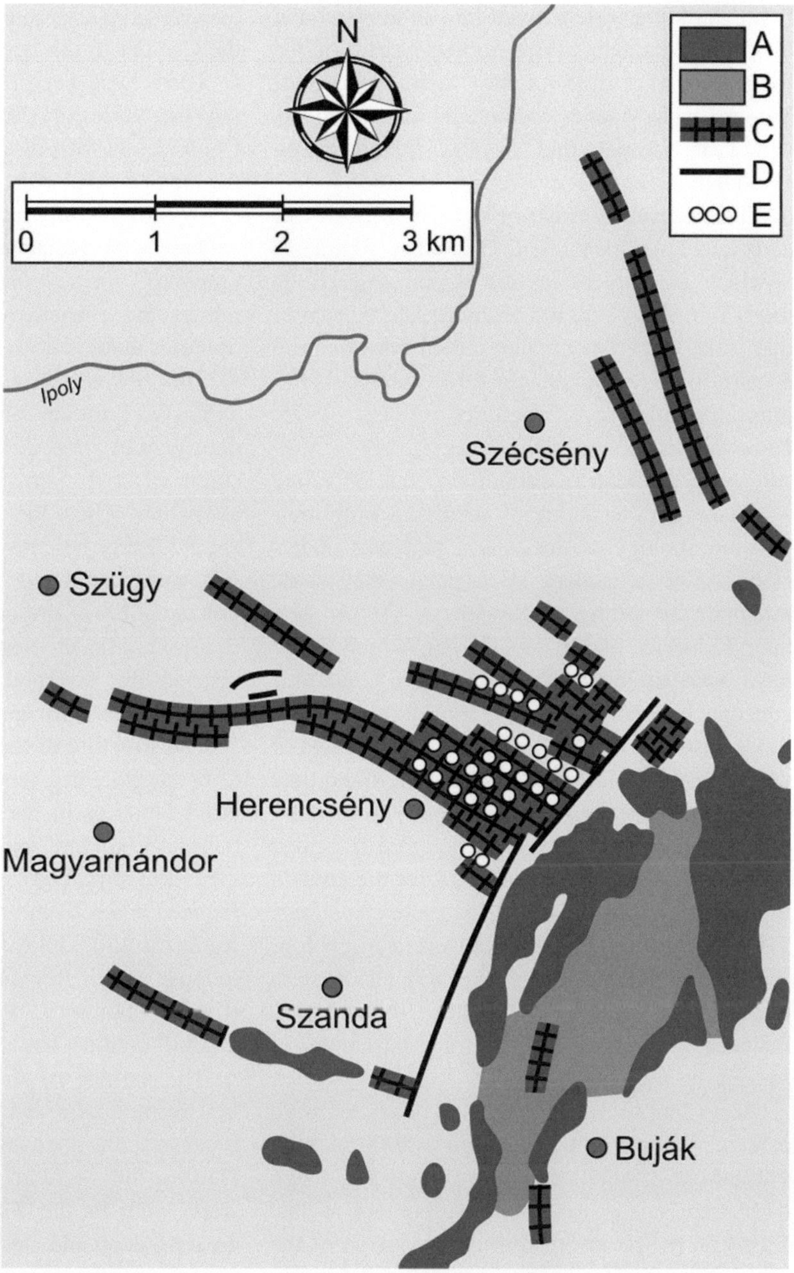


**Fig. 12.5** The types of volcanic remnants in the East Cserhát (after Székely 1983a) *A*—horsts from Mid-Miocene stratovolcano remnants, *B*—trench basins between horsts filled with postvolcanic sediments, *C*—andesite lodes, *D*—fault lines, *E*—basins between lodes

fissure volcano type, but today, as we will see, they are both categorized differently. The geomorphologic difference can be explained by the change in surface evolution and orography during the Pleistocene. (Due to the different opinions of a dozen experts, the problem of surface evolution is the main professional issue of the landscape unit.)

The Mátra is separated from the Cserhát by the wide valley of the Zagyva from the west, while from Bükk, from the east, it is separated by the valley of the Tarna. Partly due to the nearly 2000 m sinking of its southern side, it is connected to the Great Hungarian Plain with a dip slope, therefore the streams running to the south dissected the slopes with long, deep valleys (while building wide alluvial fans in the East and the West Mátraalja). The northern edge of the mountains, that descend into hills, consists of Tertiary clayey-sandy sediments and involves slides, is steeper due to the forced tilting (Mátralába).

The basement consists of Eocene volcanic and carbonated sediments, and Oligocene sediments —which are typical in this region (e.g. Buda marl, Tard clay)—on top of them. The volcanoes that make up the mountains settled onto them in a maritime environment, in three clearly distinguishable phases of activity (Varga 1975; Kubovics and Pantó 1970; Székely 1997). Volcanism in the Mátra—similarly to the Börzsöny —consisted of a series of recurring eruptions. The morphology of the central part was understood to be a caldera as large as the whole mountain according to Szádeczky (1938), and later Kubovics and Pantó (1970), which could have been similar in size to today's Etna. This concept has since been reconsidered several times based on more recent data. The remains of the original volcanic forms cannot be found here either, since the surface had been significantly degraded by external forces, therefore it gained its present height and landscape after the erosion as a result of Pliocene and Pleistocene elevations. The latter process can be well traced in geological cross sections. The entire volcanic system tilted to the south, consequently, the volcanic material got embedded into the basement of Jászság on the southern part.

According to Baksa et al. (1981) an extensive caldera ruptured into the approx. 3000 m high volcano of 250–270 $km^3$ in volume. The range of Mátrabérc is assumed to be a remnant of this. Based on the reconstruction of the divide of the Mátra (Fig. 12.6) a caldera of almost 13 km in diametre, and a volcano of an approx. 30 km in diametre can be drawn (Gábris 1983, Székely 1997). The existence of the half caldera structure (its southern side tilted and subsided) is accepted by most (e.g. Pécsi 1996; Karátson 2000; Hámor 2001); in addition, as we have pointed out earlier, Székely (1997) even connects the western part of the Cserhát to this area.

There have been other explanations concerning the issue of the Central Mátra structure. Cholnoky (1936) discovered eruption points on the main ridge (on Mátrabérc). Nemerkényi (1987) could only accept the existence of the caldera with great constraints, based on morphometric measurements. Karátson (2007) dismisses the existence of the caldera, as in his opinion, there is no real proof of the rupture, we are not aware of any material related to an explosive eruption, and the presence of the caldera cannot even be determined with the measurement of gravitational and magnetic anomalies. Thus, the arch identified as a caldera should rather be determined as a row of smaller volcanoes arranged in an arch (Fig. 12.7). Zelenka (1997) noticed several potencial eruption points (to the south of Abasár and Markaz) next to the assumed (main) caldera based on aerial magnetic images.

The structure of the East Mátra was presumed to be centrolabial, where volcanoes lie adjoining cross faults as in the case of middle mountains (similarly to the Zemplén Mountains). Today, they are identified as stratovolcanoes built of terraced andesite and rhyolite tuff (Székely 1997; Karátson 2007). Several traces of hydrothermal (copper ore in Recsk) and postvolcanic activity (Csevice Spring in Parád) can be observed on this hill country landscape.

The surface of the mountains was strongly transformed as a result of the different climatic processes in warm-wet and cold-dry periods. During the warmer, interglacial periods, the waterflows in the Mátra dissected the surface creating deep and steep valleys as a result of the elevation of the mountains and the prolapse of the Great Hungarian Plain, and evoked significant erosion (Székely 1983; Kerényi 1991). The relative growth of the relief reinforced the transformation of the slopes during the periglacial period, as well. With the freezing-out of rocks, the movement of stone blocks and block fields and

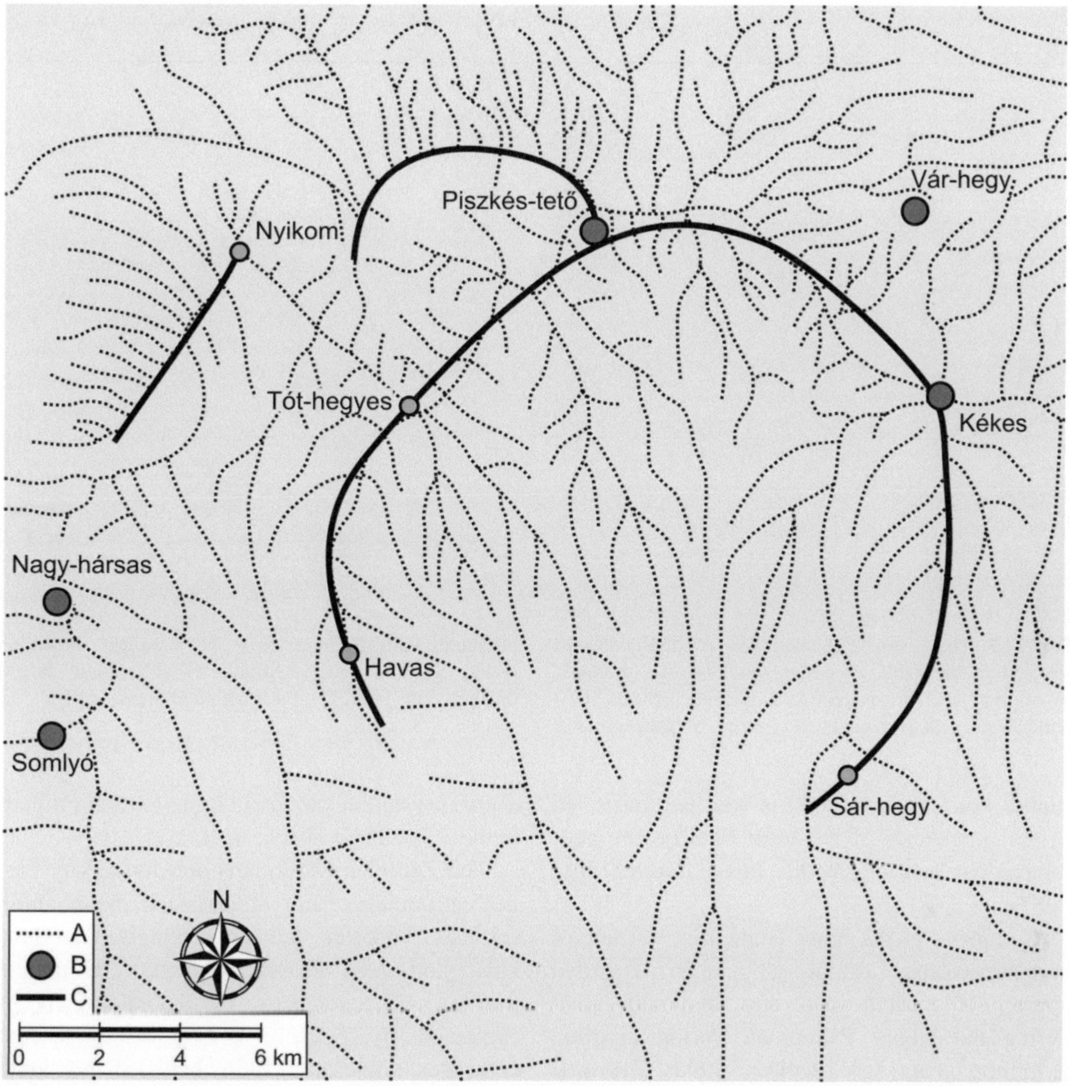


**Fig. 12.6** The reconstruction of the divides in the Mátra, based on aerial and satellite pictures (after Gábris 1983; Székely 1997) *1*—divides, *2*—volcanic domes, *3*—volcanic structures

also solifluction, the periglacial processes may have caused a degradation of approx. 50 m.

E) The name of the **Zemplén Mountains** came to existence after World War I. In geology, the name Tokaj-Eperjes Mountains is used for the complete NE-SW oriented, connected mountain range. In physical geography, the terms Tokaj-Zemplén Mountains or Tokaj (Zemplén) Mountains (Székely 1997), and also Tokaj Mountains (which seems the most adequate) are used (e.g. Pinczés 1998a; Gyarmati 2000) for the Hungarian parts. The volcanic characteristics of the mountains correspond to the macro region, although they differ in many ways from the mountains described earlier. We cannot find one large, or a few stratovolcanoes joining a volcanic centre, since the volcanic cones are set along a NE-SW oriented and some perpendicular rift systems, and also in their cross sections. The reason for this is explained with the individual subduction of a crust fragment (Balla 1978). The centrolabial system that lacks the presence of a

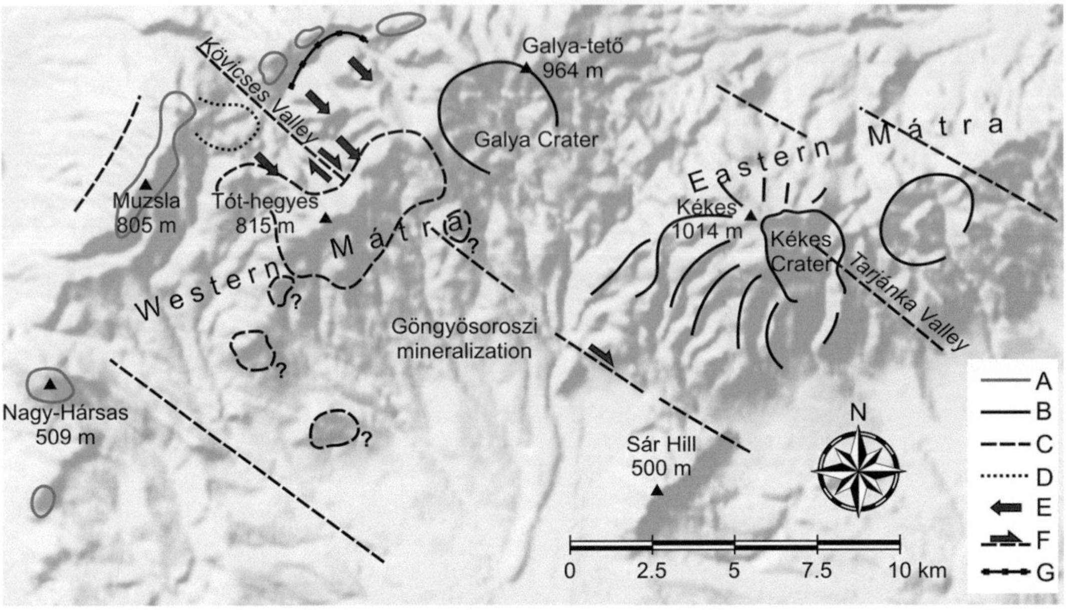


**Fig. 12.7** The main characteristics of the volcanic structure of the Mátra (after Karátson 2007) *A*—remnants of original lava floods and lava pouring centres, *B*—remnant of volcanic crater, *C*—areas of the assumed mountain slides (uncertain), *D*—assumed mountain slides, *E*—direction of slides, *F*—subsequent lateral displacement fault, *G*—faults of slide areas

central volcano (Fig. 12.8) is well demonstrated in the appearance of the form both on the geological map and the satellite Image (Nemerkényi 1987).

Compared to the other Hungarian volcanoes, here the volcanic activity is younger: it started approx. 15 million years ago, and still lasted during the Upper Pannonian Period, approx. 9 million years ago (Székely 1997; Gyarmati 2000). The activity, which can be divided into three phases, was accompanied by violent eruptions, which produced volcanic rocks of varied composition (andesite, rhyolite, trachyte, dacite and their tuffs, and basalt at the time of the younger eruptions, which is known from deep drilling in Sárospatak.) In the beginning, the volcanic substance accumulated in a marine environment, but later it continued to do so in a thickness of thousand metres in a dry environment (Gyarmati 1977). The accumulation zone was 30–50 km wider than today and the southernmost area of subsidence was the basement of the Nyírség (Pinczés 1998a). Heading towards the west, the volcanic substance settled on top of gradually deeper crystalline mountain bases (which surfaces in the Vilyvitányi Block; Sect. 1.1).

The Zemplén Mountains are divided by bigger and smaller parts of dissected mountaintop surfaces, lakkolite remnants, longitudinal lava lodes and deep tectonic valleys. The unique position of Sátor-hegy in Sátoraljaújhely is explained by Székely (1997) with adjacent lakkolites. Cholnoky (1936) believed this could be compared to Mount Vesuvius, and included the individual mountain parts into the "somma" section (for example, Sátor-hegy, Magas-hegy—Fig. 12.9). Based on morphological elements, this categorisation is once more accepted today (see Láng 1953; Pinczés 1998a, b; Gyarmati 2000). The age of the "somma" was measured with K/Ar method and estimated to be 11 million years by Pécskay et al. (1986). The stratovolcanic surfaces have all the complete landscape of periglacial forms, from cryogenic blocks through rock streams, block fields to debris cones and derasional valleys (Pinczés 1998a).

Hydrothermal transformations were connected to the slowly ceasing volcanic activity, which are

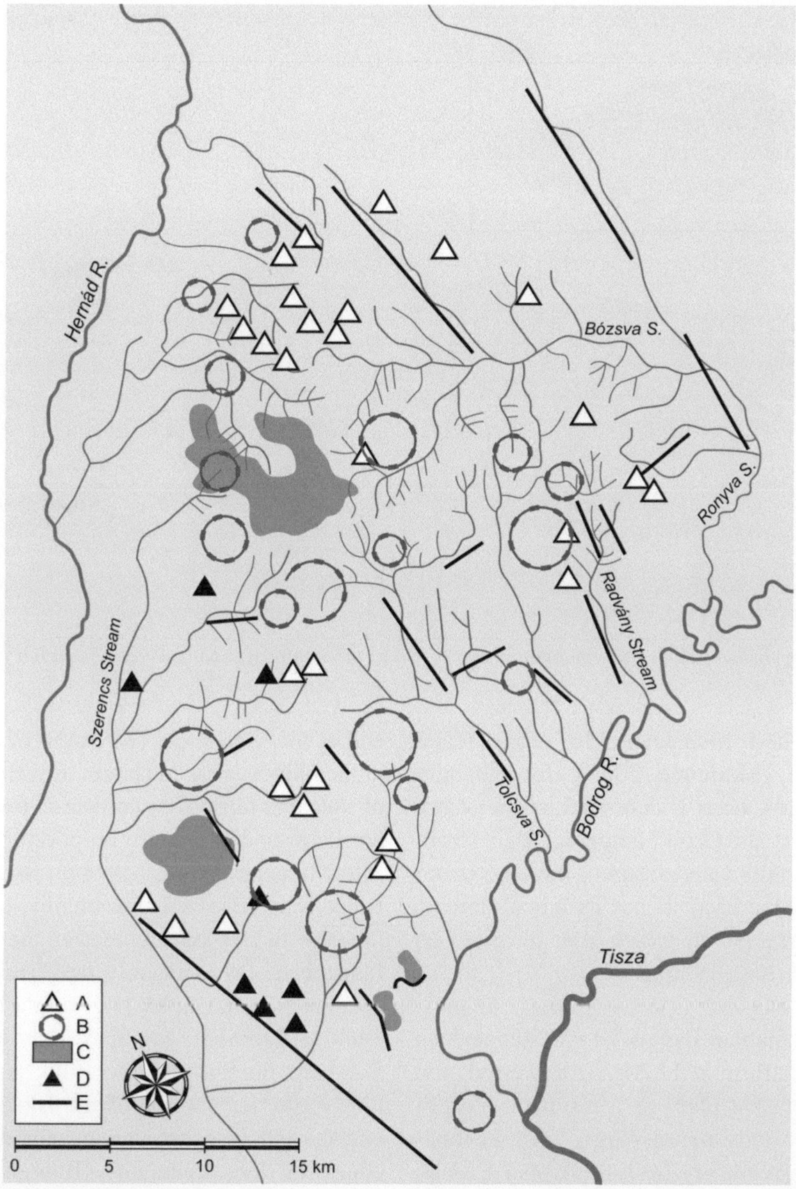


**Fig. 12.8** The volcano-morphological scheme of the Zemplén Mountains (after Nemerkényi 1987; Pinczés 1998b) *A*—exhumed subvolcanic stocks, lakkolites, *B*—eruption centre based on geological map and satellite Image, *C*—lava cover, lava flood, *D*—limnoquartzite surfaces, *E*—structural lines

indicated by smaller mineral occurences and lodes (Telkibánya). The appearance of non-metallic raw materials, such as the Pálháza perlite and the Mád bentonite, which can be related to rhyolite volcanism or its weathering, deserve special attention. Other perlite occurrences in the contact zone at the place of subvolcanic intrusion in rhyolite tuff evolved as a consequence of heat impact (for example, in Erdőbénye). The Zemplén Mountains are also rich in special minerals, for example, jasper (Komlóska), opal (Erdőbénye) and fossils that silicified as a result of volcanism.

**(F) The Bükk Region** is the largest unit of the North Hungarian Mountains, and also the one

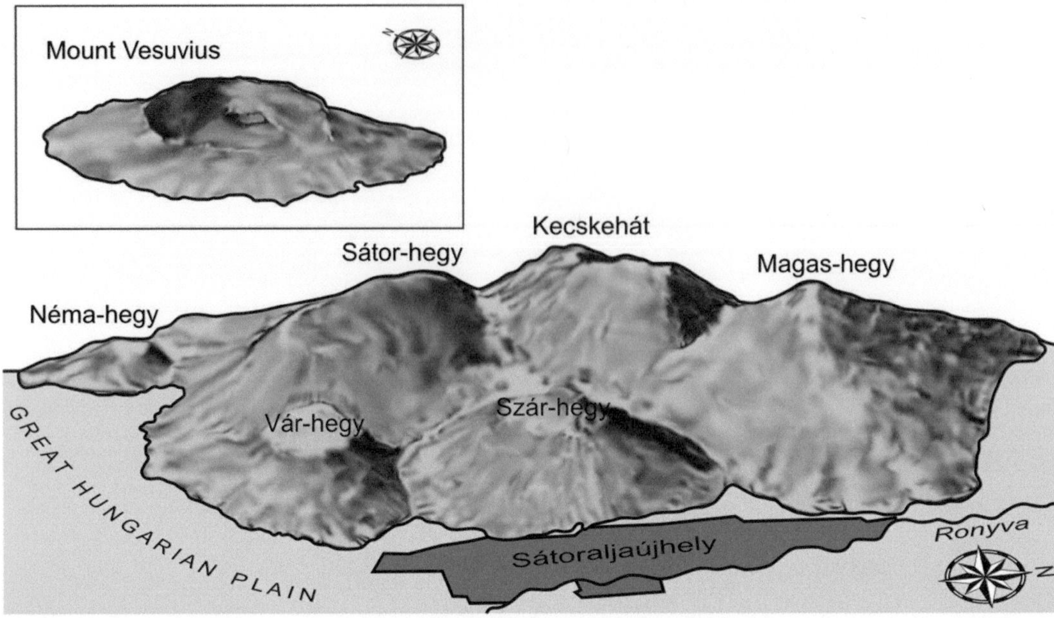


**Fig. 12.9** Comparison of Sátor-hegy in Sátoraljaújhely and its environment to Mount Vesuvius (after Cholnoky 1936)

with the highest mountains: 50 peaks reach above 900 m (Martonné 2006). Its typically carbonated mass, horst is delimited by the Mátra (from the west), the Great Hungarian Plain (from the south), and the valley basin of the Sajó (from the north and the east). It got its name after its typical tree vegetation, which was recorded as early as 1240 (Kiss 1980).

From a geological aspect—similarly to parts of the Transdanubian Range—it was situated on the African platform in the Mesozoic Period, and got into its present place as a result of complicated tectonic movements (Sect. 1.1). It functioned as a sedimentary from the Upper Carbon until the Jurassic; calcic and clayey layers subsided in it, which got enriched with andesites (porphyrites) and basalts (diabase) (Fig. 1.4). Later, as a consequence of tectonic movements (the closing of the Tethys), the Bükk became a folded mountain consisting of inclined foldings and overlapping folds (Balogh 1964; Csontos 1999). Since the crust that holds today's Bükk was situated much more to the south—around the northern part of the Tropic of Cancer—Hevesi (2000) assumed the presence of cone karst, towerkarst and peneplaned karst surfaces at the end of the Mesozoic. The extent of transgression in the Oligocene is uncertain as well as the extent of volcanic tuff coverage, which originated from northern and southern eruption points in the Mid-Miocene Period. Although most parts of the Bükk were involved presumably, since most of these parts may have eroded in the Pliocene and Pleistocene. We can only find traces of Eocene sediments in the Bükk (e.g. on the Kis Plateau), while Oligocene sediments only occur in the Southern Foothills of the Bükk (Bükkalja) and the Northern Foothills of the Bükk (Bükklába), and the volcanic remains are thicker around the edges of the mountains (Balogh 1964). The north-eastern movement of several units in the basin, the same way as that of the Bükk, started in the Paleogene (as a result of the closing of the European and African continental crusts), consequently, the Bükk occupied its present position (Sect. 1.1).

The sea started to retreat in this area from the end of the Badenian (Mid-Miocene), from the Sarmatian (approx. 10 million years ago), and the surface started to emerge parallelly. Several conditions of karstification characteristic of the mountains came to existence, and this was the time when the present framework of the water network—the almost radial water system of the mountains (with eastern, north-eastern and south-western flow directions) evolved. These waterflows formed large Pliocene foothill surfaces in the foreground of the mountains (Hevesi 1980). At the end of the Pliocene, the determining landscape elements of the middle mountains separated with the growing intensity of elevation (which could have been 370–620 m asl at the time). On the increasingly exhumed limestone we could identify different generations of dolines (presently in mountain top positions) and sinkholes, and also a sequence of spring caves (e.g. Vidróczki Cave, Szeleta Cave). The waters of the spring caves created the valley system that later got inherited onto the limestone surfaces of the central and southern areas (Hevesi 1978). The dry and icy periods of the Pleistocene hindered intensive karstification, on the one hand, due to the decrease of infiltration, and on the other hand, the accumulation of loess derivatives which also hindered infiltration (not only around the edges, but in greater heights as well; Tóth 1984). As a result of the continuing elevation, the surface water system gradually ceased to exist, dry valleys and a series of ponors evolved and the water continued to flow in subsurface systems.

The central unit of the mountains is the Bükk Plateau, which is divided by the Garadna Valley. The lower Kis Plateau is 200–300 m lower than the 600–800 m high Nagy Plateau (Fig. 12.10). The continuous surface—20 km long in a west-east direction and approx. 6–7 km wide—, mostly eroded from its former cover sediment to

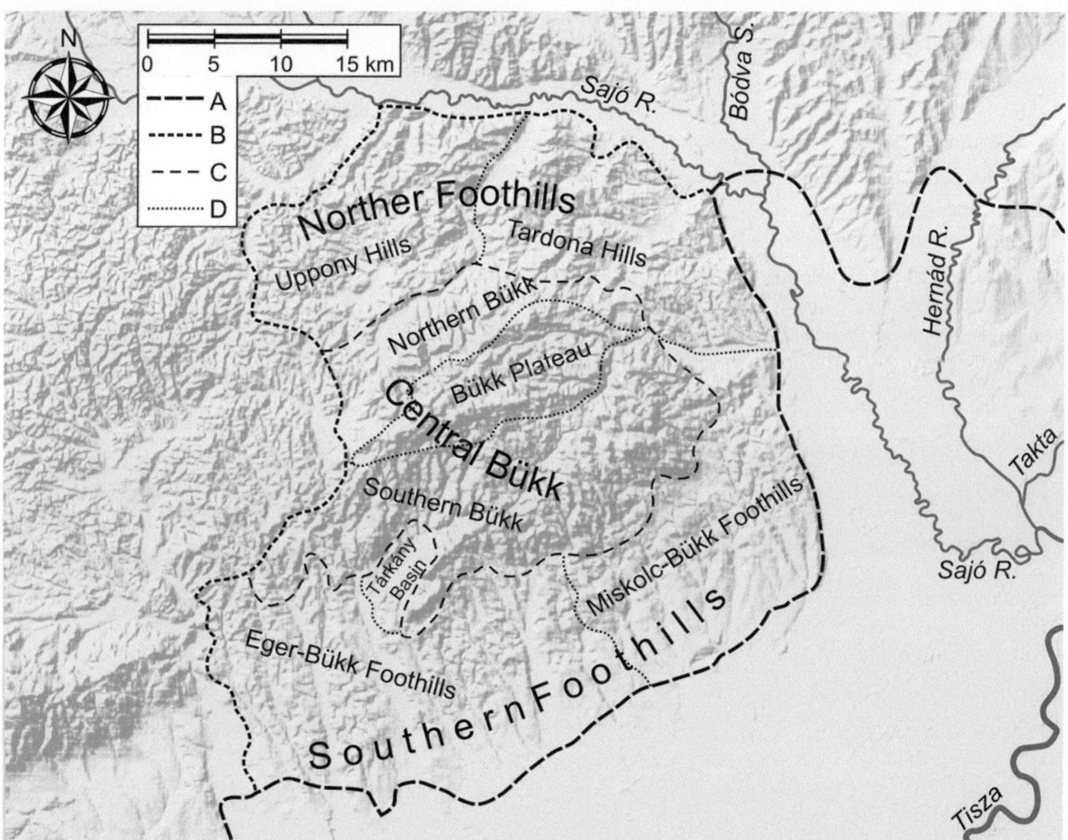


**Fig. 12.10** The geographical landscapes of the Bükk Region (according to Marosi and Somogyi 1990) *A*—macro regional border, *B*—meso regional border, *C*—border of groups of micro regions, *D*—micro regional border

the limestone, thus, there are no larger water systems on it, and the forms are of karstic origin: karrenfelds, ponors, shaft caves (the deepest one is Kiskőhát vertical—shaft-cave). However, smaller permanent and temporary waterflows are not only found connected to karst edge sinkholes, but they can also be found on the Nagy Plateau (e.g. on Csipkéskút, Jávorkút, Létrás) and on the Kis Plateau as well, for example, Csókás and Kaszás Meadow sinkholes. The dolines inherited the former flow directions of streams, therefore they may be arranged into series of dolines (Hevesi 1986). Apart from the lapas of two or three doline series, the largest karst hollows of the Bükk Plateau are the poljes. Nagymező, which was formed at the meeting point of valleys of several ponor series, is the most spectacular part of the Bükk Plateau with the capricious depth and alignment of its ponors and uvalas (Martonné 2006 and Image 12.2). Due to its untouched character, it gives home to block-fields, rich in species.

The characteristic forms include crags on plateaus and also on their rims, which hold peneplane remnants from the Cretaceous and the Mid-Eocene, and have the appearance of domes. The ones found on the rims called "Bükk rocks" (e.g. Istállós Rock, Tar Rock, Örvény Rock) owe their forms to periglacial processes and the different extent of erosion of limestone and shale rock (Hevesi 1986, 2000).

In the middle part of Bükk the scarce natural resources were not in favour of establishing settlements. In the mid-eighteenth century, Slovak settlers came for forest use, and founded so-called huta villages (for example, Óhuta, Répáshuta). The inhabitants of these villages took part in iron production mostly from the 1800s (Veres 2002).

The exhumed tropical planation limestone plateau of the Central Bükk is surrounded by the extensive, lower (300–700 m high) double ring of hills and mountains (Tóth 2001). The northern and southern inner rings are divided into mountain ridges, hillsides and declining ridges with valleys by a system of radial streams running down from the edge of the plateau, supplied by abundant karst springs. The large southern part (Southern Bükk) is further divided by the Hór Valley, based on morphological and petrological differences. The south-western part consists of weathered out volcanites (gabbro and basalt) and non-karst rocks (mainly Lower Jurassic shales) mostly covering Triassic limestone. A good example of the volcanites is the Szarvaskő Canyon of the Eger creek, which is additionally

**Image 12.2** Nagymező http://www.panoramio.com/photo/10857677

surrounded by pillow lava (Hevesi 2000). The surface presence of limestone is indicated by the karst formation of the Hór Valley, its caves and other, mostly Transdanubian types of karst forms (Ódorvár, Hajnóczy Cave). In the south-eastern part, between the Szinva and the Hór Valley, we can find a series of karst and non-karst tracts. The water collected from shales in limestone areas created caves. We must highlight the sinkhole of Pénzpatak creek for its water abundance, and Suba-lyuk Cave for its archeological values. Waters coming from the Nagy Plateau created one of the largest travertine cavity system of Europe (Anna Cave), which evolved inside a travertine barrier built in the Szinva Valley (Hevesi 1986). The northern part of the ring surrounding the Bükk (Northern Bükk)—similarly to the Southern Bükk—is strongly divided by valleys and appears in the form of a low and middle mountain area. Its petrological structure is varied. The western part mostly consists of shales, while in the rest of the areas, carbonated Triassic and Jurassic formations predominate. Travertine formation is characteristic here as well; we can find 11 out of the 26 travertine deposits of the mountains here (e.g. the travertine series of the Fátyol Waterfall in the Szalajka Valley).

The southern part of the outer ring, the Southern Foothills of the Bükk (Bükkalja), is a large foothill surface divided by valleys. Due to the NE faults running between the edge of the Great Hungarian Plain and the middle mountains, large amounts of rhyolite tuff and dacite tuff (ignimbrite), and subordinately, lava came to the surface (Balogh 1964; Császár et al. 1998). The so-called beehive stones, which were deepened into volcanic tuffs between the 11th and 14th centuries, are valuable cultural and historical relics. (They seemed to have had cultic purposes, but an apiary purpose is more probable.) Along the faults, we can find abundant thermal springs, for example, Eger, Bogács, Kács, Miskolctapolca. The northern part, the Northern Foothills of the Bükk (Bükkléba), is from the Tertiary, and consists of brown coal and marine sediments in some places. Its surface was formed by slope mass movements (Pinczés 1986).

(G) **The Aggtelek-Rudabánya Mountains** is only the southern foothill of the large, Slovak fault block (Gömör-Szepes Ore Mountains and Gömör-Torna Karst), which is covered with Mesozoic limestone and is rich in minerals. The Hungarian parts consist of the low mountains of the Aggtelek Karst and the Rudabánya-Szalonna Mountains. Their rocks mainly consist of limestones (Wetterstein, Gutenstein and Steinalm limestones), which evolved in the Mid and Upper Triassic—their evolutions are discussed in details in Sect. 1.1. Tropical karst formation must have taken place on their surface which had become dry land during the Cretaceous (Borger – Beck 2000). Despite the strong erosion, there are still some remnants indicating tropical karst formation on the Pelsőci Plateau in Slovakia. From the second half of the Miocene, the entire area obtained a significant amount of sediment cover as it was part of the Gömör-Szepes Ore Mountains foothill surface. Its gravelly material played a role in the widening of cave passages in later karst formations. Terra rossa, which was a product of weathering processes, remained filling the hollows of dolines. From the end of the Pannonian, a significant elevation took place in the area. As a result, on the one hand, the foothill surface elevated to different heights and underwent segmentation (e.g. Haragistya, Nagy-oldal); on the other hand, the level of karst water decreased and large shaft caves evolved (e.g. the 236 m deep Vecsembükk vertical—shaft-cave introduced in Sect. 1.7). Highly diverse karst landscape evolved on the stabilized, elevated surface and also beneath the surface, which determines the character of this landscape unit. Intensive karst formation is characteristic all around the Gömör-Torna Karst Region—the words "töbör"(dolina) or "zsomboly" (vertical or shaft cave) are of Slovak origin (Jakucs – Móga 2000). The intensity of karst formations was measured in lengthy sequences by Zámbó (1998). Based on these data, he also provides assessments on the speed of karst formations. The subsurface forms of allogenic karst type "B" described in the area (Jakucs 1971) created not only the Hungarian, but also (evolving from the

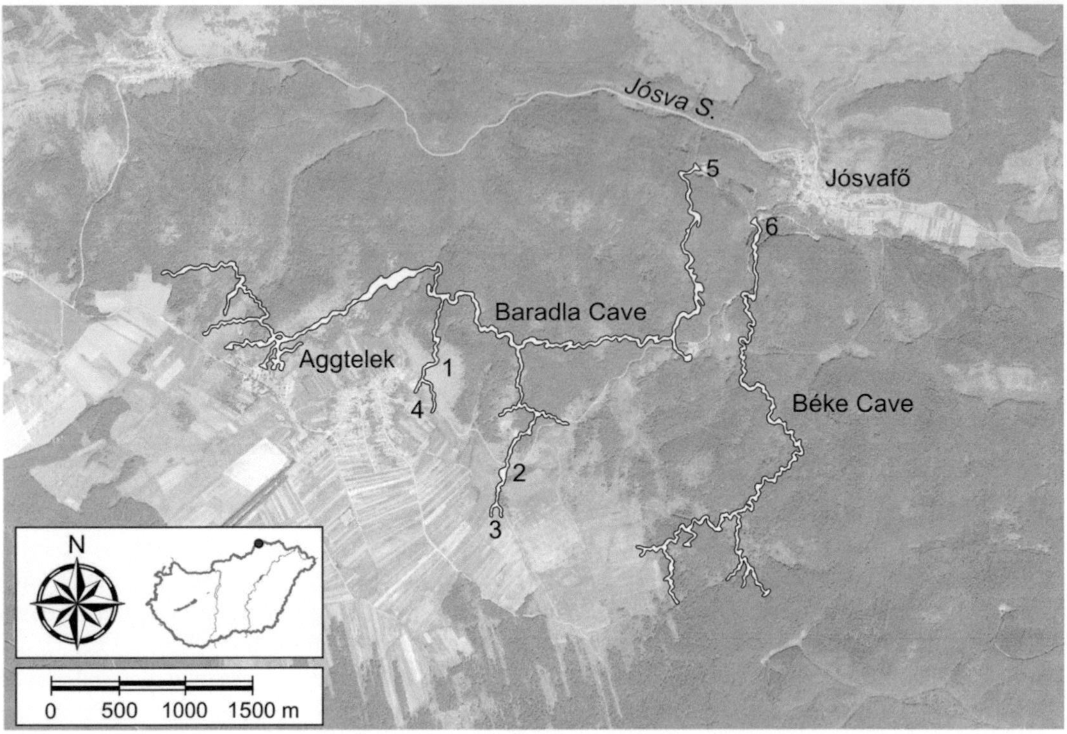


**Fig. 12.11** The branch systems and the significant surface connections of Baradla Cave and Béke Cave (based on googlemaps) *1*—Törökmecset Branch, *2*—Retek Branch, *3*—Zombor-lyuk sinkhole, *4*—Ravasz-lyuk sinkhole, *5*—Jósva spring, *6*—Komlós spring

second half of the Pleistocene) one of the largest and richest cave systems of Europe (Fig. 12.11). The approx. 24 km long Baradla Cave was inhabited by the prehistoric man, as well (Kordos 1984). Its large passages and halls of 400–800 thousand air cubic metres were broadened by water and flint entering the cave through sinkholes (Image 12.3). Among the caves in the area, it is worth mentioning the 7 km long Béke Cave, which was explored in 1952, and the 3 km long Szabadság Cave in Égerszög. We can find rich karst landscape on the surface, as well. Row and plateau dolines (Mezősi 1985), karren, different generations of sinkholes and abundant karst springs at the edges of mountain blocks (for example, Jósvafő spring, or the Lófej spring on Haragistya Plateau, which is rare because of its intermittent character) can be found here.

Limestone horst ranges emerge from the sedimentary cover in the Rudabánya Mountains and the Szalonna Mountains (which are divided by the River Bódva). Structurally, these two mountains differ from the Aggtelek Mountains, but they went through similar surface development. The evolution of a relatively poorer karst landscape was mainly caused by lithological and orographical differences (the emergence was smaller and the exhumation was slower). The limestones of the Rudabánya Mountains were criss-crossed by hydrothermal solutions in some places, and the evolving siderite (which became limonite reaching the surface) sheet enabled the existence of the only serious Hungarian iron ore mine in Rudabánya. (Though, it has been closed down.) The greatest occurence of hominides, who lived 8–12 million years ago, and the discovery of a new species (Rudapithecus) is also connected to Rudabánya. The anhydrate occurences at Alsótelekes (introduced in Sect. 1.1) could replace the lost jobs of the closed iron ore mines, although they are only available in limited amounts.

**Image 12.3** We can find abundant dripstone formations in the erosional passages of the Baradla Cave (Photo by Rakonczai, J.)

(**H**) The different landscape classifications claim that the sequence of valleys lying in an E-W direction among the mountains mentioned earlier make up the **North Hungarian Basins** meso region. The aggregation, which contains numerous micro regions, can be grouped around two greater units. According to the classification of 1990, the units of the western part are the Ipoly Valley, the Nógrád Basin and the Upper Zagyva-Tarna hills, which sum up the Hungarian parts. The units of the eastern part are the Gömör-Heves Hills, the Borsod Basin, the Cserehát and the Valley Basin of the Hernád. The meso region is mainly delimited by terraced river valleys found in valley basins and depressions (for example, the Hernád Valley, the Mid-Ipoly Valley, the Zagyva Valley, the Valley of Tarna).

These non-connected geographical units have several shared characteristics. One of them—connected to surface evolution—is the shared geological basement. The surface is covered with schlier, marl and sandstone (in great thickness in the western parts of the basins) originating from the end of the Oligocene and the beginning of the Miocene. Miocene volcanic tuff, which settled onto these, can be found in several locations; its most beautiful emergence is demonstrated in Image 12.4. The Darnó Line created a sedimentological border in the micro regions of the eastern part. To the east of it, the area is basically devoid of Paleogene deposits. The occurrence of coal formations from the Badenian stage is characteristic in both large areas of the meso region (for example, Zagyva Valley, the Putnok Hills, the Ózd-Egercsehi Basin), which implies a marshy shoreline environment (Image 12.5).

The homogeneous structural-morphological development of the meso region basically lasted until the Badenian volcanism. From a morphological point of view, both the eastern and the western parts can be regarded as basin hills. However, after the Badenian volcanism, the evolution of the western part continued with the

**Image 12.4** The erosive (badland) surface of the Kazár rhyolite tuff (Photo by Karancsi, Z.)

**Image 12.5** The Castle of Somoskő, standing on Pliocene basalt (Photo by Karancsi, Z.)

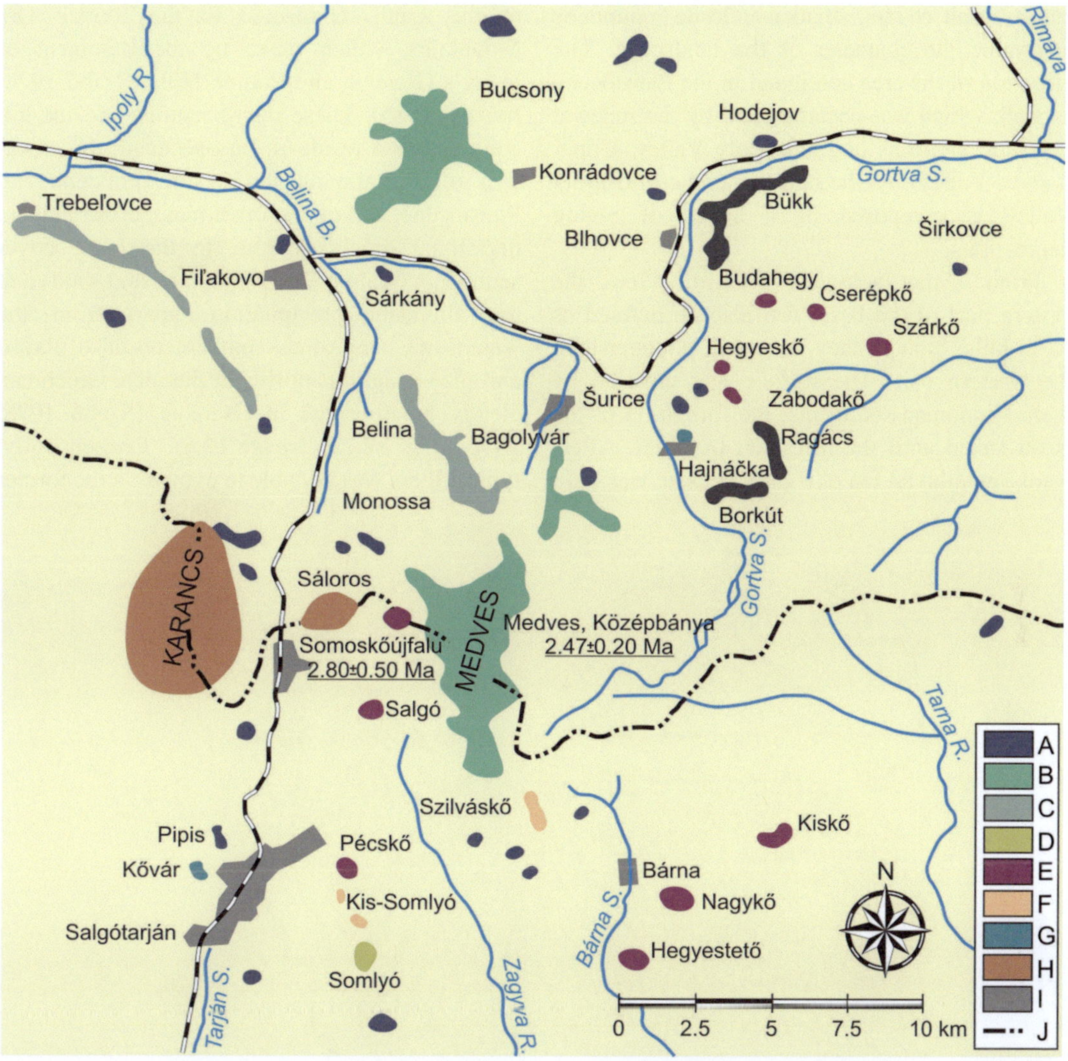


**Fig. 12.12** The main geomorphological formations of the Medves Region and the broader environment of Karancsság (after Jugovics 1940; Székely 1993, 1997) *A*—basalt mountains in general, *B*—basalt plateaus, broader cover ruin and debris, *C*—basalt ridge and narrower basalt tongue ruin and debris, *D*—remains of basalt cover, *E*—basalt cone, *F*—basalt intrusive dome, *G*—basalt and agglomerate cone remnant, transformed by human activity, *H*—eroded volcanic conduit trunk, *I*—eroded Mid-Miocene andesite lakkolite, *J*—state border

slight elevation of the basin area. As a result, the further surface development of the area was fluvial. For example, the surface of the Nógrád Basin was eroded by the Ancient Ipoly, its gravelly remains can be found on the surface (Horváth et al. 1997). This sort of evolution of the hill country was interrupted by basalt volcanism in the second half of the Pliocene. Pliocene (2–4 million years old) basalt covers (Medves—the largest such formation in Hungary), basalt cones (Salgó, Somoskő) and, in some places, maars are characteristic in this area (Image 12.5 and Fig. 12.12). 47 individual basalt occurrences have been mapped in the Hungarian and Slovak areas at the crossings of NE-SW and perpendicular faults (Jugovics 1940). Apart from larger basalt covers, smaller andesite mountains determine the character of the landscape. The elevation of the area continued in the Pleistocene as well, which was accompanied by the renewal of valley incisions (e.g. the Ipoly Valley, Upper Zagyva Valley). At the same time, the Mid-Ipoly Valley is composed of a series of young depressions.

From a morphological point of view, the eastern part of the basin can also be defined as basin hills, though, they are much younger than the western part. The surface was covered by thick Pannonian sediments; the formation of the basin lasted until the recession of water. Afterwards, an alluvial fan of significant size was built to the south, connected to the Slovak Ore Mountains, which make up the basement of today's Cserehát and Putnok Hills (Szabó 1978; Mezősi 1985). These micro regions became hill countries as a result of, on one hand, the elevation of the alluvial fan in the Pliocene and Pleistocene, and on the other hand, erosional and derasional processes. Due to the loose cover sediments (mainly Pannonian clay and sand) and the abundant precipitation, apart from the waterflows, slope mass movements also played and play a significant role in the area, which can clearly be observed in Cserehát (Szabó 1978, 1996—Fig. 12.13, Image 12.6). Consequently, typical loess was not able to evolve on the former

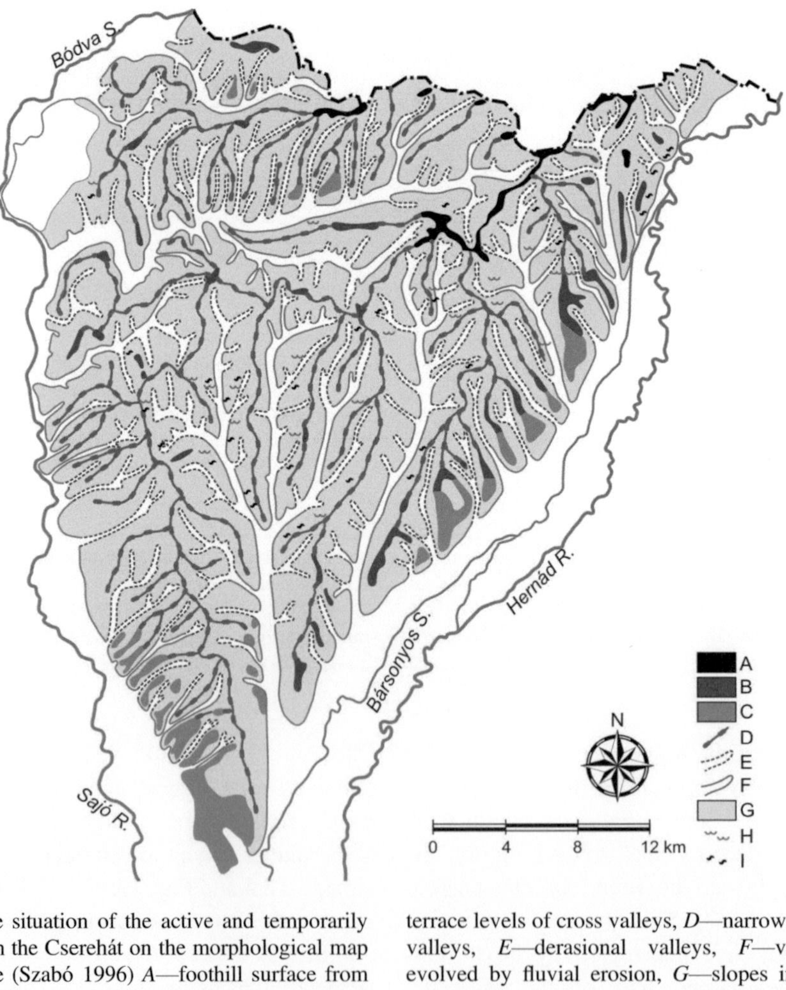


**Fig. 12.13** The situation of the active and temporarily inactive slides in the Cserehát on the morphological map of the landscape (Szabó 1996) *A*—foothill surface from the end of the Pliocene, mainly with remains of gravel cover, *B*—broad, roof-like ridges between valleys, *C*—terrace levels of cross valleys, *D*—narrow ridges between valleys, *E*—derasional valleys, *F*—valleys, mainly evolved by fluvial erosion, *G*—slopes in general, *H*—active slides, *I*—temporarily stabilized, and eroding slide forms

**Image 12.6** Surface covered with slide forms in the Cserehát (Photo by Kiss, T.)

alluvial fan surfaces. The mass movements make traditional agriculture and the construction of a transport network difficult, and in some places they even endanger some parts of the settlements (MT 2002).

The peneplaned Szendrő Mountain, which consists of Paleozoic, metamorphic sandstone, limestone and shale, emrges like an island in the hill country that mainly consists of Pannonian sediments (see also Chap. 1.1). Most of its surface was covered with Pannonian-Pliocene sediments. A great alluvial fan of the Ancient Bódva and (to the east) the Ancient Hernád was built on the Pannonian marine sediments throughout the eastern part of the basin. In the Cserehát and also the Putnok Hills, as a result of the Quaternary elevation, fluvial erosive processes became active and turned the alluvial fan surface into a dissected hill country. In Szendrő Mountain on the incision sometimes created canyon-like valley sections (e.g. Bódva). This erosive process exhumed the peneplaned landscape, as well. The two ancient rivers created terraced valleys as they slid from the alluvial fans to their present locations, similarly to other great rivers in the region, for example, the Sajó (Szabó 1995; Mezősi 1984).

## 12.3 Some Natural Values and Environmental Hazards of the North Hungarian Mountains

Due to the countless natural values of the macro region, a detailed presentation would be quite lengthy. This time we would like to draw a thumbnail sketch, and we will only discuss the effects of mining on the environment in detail.

From a natural point of view, one of the connecting elements of the macro region is the presence of energy resources. Such is the occurence of the Mid-Miocene brown coal. Its mining near Salgótarján (mining here began in 1868) and in mines on the side of the broad valley basin of the Sajó (mining in Borsod gained momentum between 1786-1812 – Borsod 1986) ceased due to the low quality of coal and the unfavorable site conditions. The exploitable amount of coal assets is 100 million tons (Kovács 2005).

It is worth highlighting the other energy resource, lignite, which is available in the largest quantity. Lignite is mined in Southern Foothills of the Mátra and Bükk (Mátralja and Bükkalja) even today (8 million t/year), and is used in the power plant of Visonta to produce electricity. The developmental historical description of the topic can be found in Sect. 1.2 (Fig. 1.19 and Image 1.8). The mined lignite layer is approx. 50 m thick (the explored amount is 2 or 3 times as much), but the depression of the mountain foreground, which was repeatedly interrupted, resulted in the evolution of several thin lignite zones (Jaskó 1981; Pécsi 1996). As a result of this, the lignite-waste heap ratio is 1:5 in Bükkábrány, while in Visonta it is 1:9. The large open cast mines cause agressive transformation on the surface (as part of drainage works, 12 $m^3$/min is being extracted in Bükkábrány), which is mitigated by extensive recultivation.

Partly related to these features, some typical industrial areas have developed (mostly in Upper Zagyva—Tarna Hills and in the Sajó Valley). The intensive landscape use has significantly transformed the surface, which is covered with waste heaps in several locations and (under) mining has caused mass movements (Karancsi 2002; Horváth et al. 1997; Sütő 2001; Szabó et al. 2004). In some areas, considerable surface depressions are common; surface depressions have been measured up to 20 m in the Mecsek, and collapse may happen at a speed of 1–3 m/day. Karancsi (2002) showed that 268 depressions out of the 543 evolved as a consequence of mining activity in the area (Fig. 12.14).

Hydrothermal activities during the Tertiary volcanism, and subvolcanic intrusions resulted in occurences of precious metal and non-ferrous metal (Sect. 1.2; Zelenka 1997). Gold, silver, copper, lead and zinc ores used to be mined in Nagybörzsöny, Recsk, Gyöngyösoroszi and Telkibánya. Modern mining started in 1927 in the Mátra, and restarted after World War II. near Gyöngyösoroszi (the mine was closed down in 1986). The majority of the lead, zinc, and copper ore stocks are related to low level occurrences in Recsk. Their extraction awaits higher raw material prices to come. (The copper mine in Recsk is presently in the state of preservation.) The hydrothermal processes that took place in carbonated rocks played a major role in the evolution of low quality iron ore in Rudabánya. The extent and importance of ore mining in the Middle Ages is indicated by the facts that Telkibánya got a mining town right for its gold, Rudabánya got one for its silver and copper ore, and Nagybörzsöny was founded on gold and silver mining in the 14–15th centuries (Mattányi 2006).

Non-metallic mineral raw materials are becoming more and more important, since many of them are used to decontaminate pollutants, or used in the field of soil improvement, and they can be important construction materials in heat and sound insulation as well. There are considerable zeolite sites (Mád) or kaolinite, bentonite and perlite sites (in the Tokaj Mountains). Their evolution is connected to rhyolite tuff, and has been described earlier from a developmental historical point of view. Fireclay (in Felsőpetény), which is related to Hárshegy sandstone in the Cserhát, belongs to this group and can be dated to the same age, and is an important material in the production of ceramics in Romhány.

The climatic features of the middle mountains (for example, the favourable effects of the mountain climate on our health in the Mátra, and some related sports facilities) were discussed in Chap. 2. The hydrography of the macro region was also discussed in detail in Chap. 3. The water stored in karsts and the great number of mineral and thermal water springs is an especially important feature. Together with the vegetation, they form the basis of recreation in the macro region. In the long list of protected natural values and unique landscapes in Hungary, the North Hungarian Mountains are represented by several entries: from the Fátyol Waterfall in Szilvásvárad through Lake Arló to the andesite in Bér (Tardy 1996; BNP 2009).

Three national parks serve to present and protect the natural values. The mission of the Bükk National Park, established in 1977, is mainly to maintain the karst occurrences, the geomorphological forms, and the rarities of the flora and fauna in the mountains. Here we must highlight the seven forest reserves established in the area. In the Aggtelek National Park, the Baradla Cave System has crucial—and also international—importance; it has been a part of the UNESCO World Heritage since 1995. The Aggtelek National Park was established in 1985 to preserve the forms on and below the surface, though its karst vegetation and cultural values (e.g. sustainable village of Gömörszőlős) are also characteristic of the landscape (Tardy 1996). The

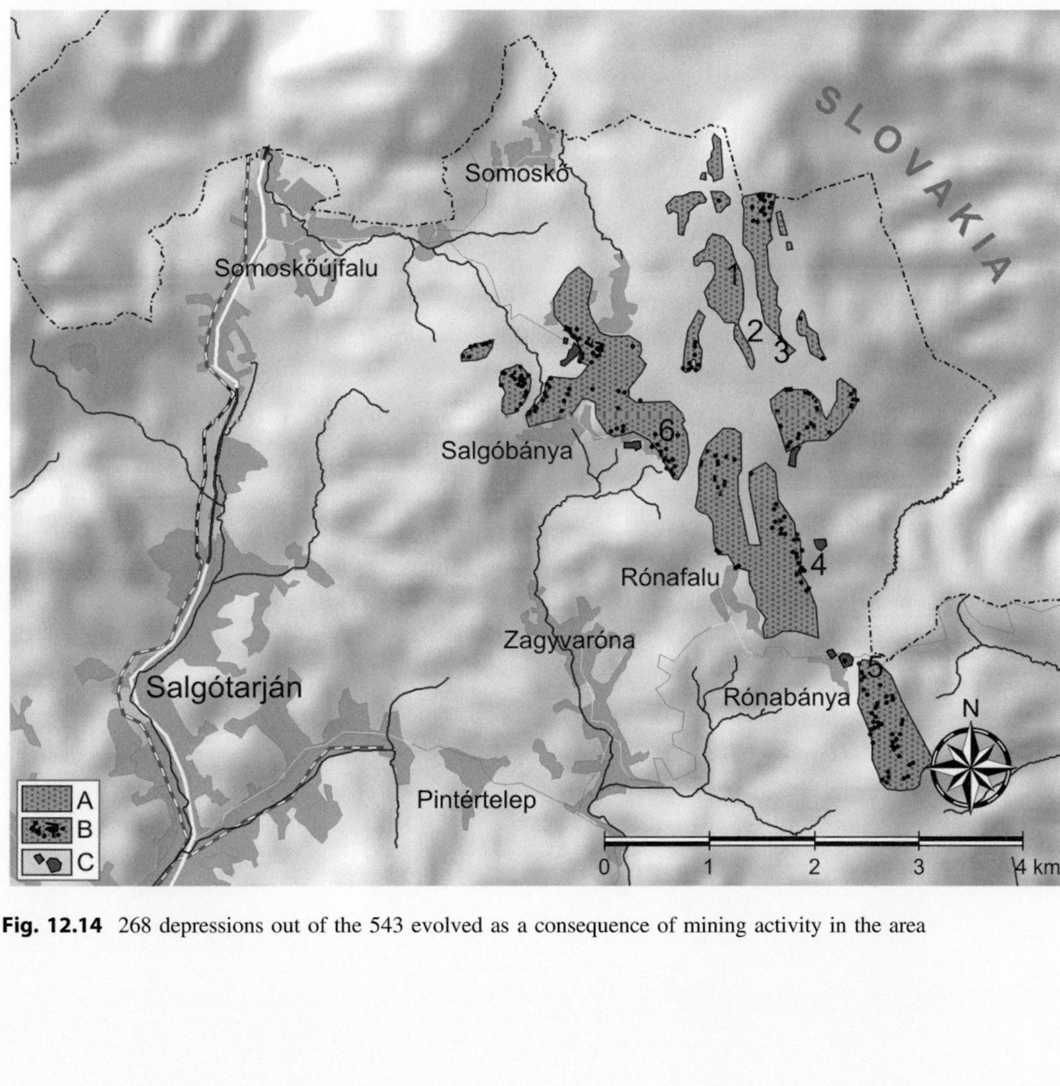


**Fig. 12.14** 268 depressions out of the 543 evolved as a consequence of mining activity in the area

**Image 12.7** Spoil heap at Salgótarján (photo by Kiss, T.)

aim of the Duna-Ipoly National Park since its foundation in 1997 has been to preserve and protect the characteristic picture of the landscape, its untouched nature, greenfield sites, isolatedness, richness in species and its cultural values (DIPNI 2009).

The environmental conflicts of the landscapes also have a historical aspect. The conflicts between economic and natural factors (rather than solely natural ones) are mostly characteristic in this macro region, as well. Some of these are related to non-revitalized, unmodernised industrial areas (e.g. the Sajó Valley) and the improvement of abandoned mining areas (e.g. the Sajó Valley, and partly the Borsod Hill Country). Such an example is the recultivation of the iron ore mine in Rudabánya, the maintenance of waste heaps from mining (which have developed their own ecological systems; Szabó et al. 2004), slag tailings related to energy production (Image 12.7), or the environmental stabilization of abandoned industrial areas. The other group of conflicts refer to activities that are related to the present use of the surface. For example, a small change in the composition of water infiltrating into karst areas may cause the redissolution of dripstones (Image 12.8), indicating the high sensitivity of this ecological system. We must also pay great attention to ensure the resupply of karst water bodies and maintain the water quality of springs, which are very sensitive indicators of the surface use of the catchment area. The third conflict may develop between the distinguished recreational features of the macro region and its precious vegetation. It is the task of the national parks to deal with this conflict.

**Image 12.8** Redissolution of drinstones in Baradla Cave in Aggtelek (own photo)

## References


Balla Z, Csongrádi J, Havas L, Korpás L (1981) A börzsönyi vulkanitok kora és a K/Ar kormeghatározások pontossága. Földtani Közlöny 111:307–324

Hajdú-Moharos J, Hevesi A (2000) A Kárpát-Pannon térség természeti tájbeosztása. In: Karátson D. (szerk.) Pannon enciklopédia. Kertek, Budapest, 274–285

Hámor G, Ravasz-Baranyai L, Balogh Kad, Árva-Soós E (1987) K/Ar dating of miocene acid and intermediate volcanic activity of Hungary. A Magyar Állami Földtani Intézet Évkönyve 62:149–154

Mauritz B (1910) AMátra hegység eruptív kőzetei. Math Term Tud Közl 30(4):133–247

MET1 (2009). http://www.met.hu/omsz.php?almenu_id=climate&pid=climate_Hw&mpx=0

Mezősi G (2008) Magyarország környezetföldrajza. JATEPress, Szeged

Pécsi M (1996) Geomorphological regions of Hungary. MTAFKI, Budapest

Péczely G (1987) Éghajlat. In: Ádám L. – Marosi S. – Szilárd J. (szerk.) Dunántúli-középhegység. A. Akadémiai Kiadó, Budapest, 243–281

Soó R (1964–1980) A magyar flóra és vegetáció rendszertani-növényföldrajzi kézikönyve. I–VI. Akadémiai Kiadó, Budapest

Stefanovits P (1981) Talajtan. Mezőgazdasági Kiadó, Budapest

Székely A (1989) Geomorpology of the intra-carpathian volcanic range. In: Geomorphological and

geoecological essays (Studies in Geography in Hungary 25) Akadémiai Kiadó, Budapest, pp 49–60
Székely A (1993) Die Geomorphologie der Innerkarpatischen Vulkanischen Gebirge in Nord-Ungarn. Annales Scien. Bud. Sectio Geographica 22–23:5–21
MT = Glatz F. szerk (2002) Magyar Tudománytár I. Föld, víz, levegő. Kossuth, Budapest
szerk Korpás L (1998) Magyarázó a Börzsöny és a Visegrádi-hegység földtani térképéhez. MÁFI, Budapest
Hortobágyi – Simon T szerk. (1981) Növényföldrajz, társulástan és ökológia. Tankönyvkiadó, Budapest
Balla Z (1978) A Magas-Börzsönyi paleovulkán rekonstrukciója. Földtani Közlöny 108:119–136
Balogh K (1964) A Bükk hegység földtani képződményei. A Magyar Állami Földtani Intézet Évkönyve 48:246–719
BNP (2009) http://www.bnpi.hu/?c=vedett
Borger H, Beck RK (2000) Landschaftentwicklung, ökologische Grundlagen und anthropogene Veranderungen im westlichen Aggtelek-Karst. Mitt. Verb. Dt. Höhlenund Karstforsch 46(1–2):15–24
Borsod (1986) 200 éves a borsodi szénbányászat. http://mek.oszk.hu/07500/07506/07506.pdf
Bulla B (1964) Magyarország természeti földrajza. Tankönyvkiadó, Budapest 424
C Baksa, Csillag J, Földessy J, Zelenka T (1981) A hypothesis about the tertiary volcanic activities of the Mátra mountains. NE Hungary. Acta Geologica Hungarica 24(2-4):337–349
Cholnoky J (1936) Magyarország földrajza. Franklin, Budapest
Császár G, Galácz A, Haas J, Hámor G, Kecskeméti T, Knauer J, Korpásné Hódi M, Krolopp E, Nagymarosy A, Szederkényi T (1998) A hazai földkéreg rétegtani tagolásának helyzete. Földtani Közlöny 128 (1):99–121
Csontos L (1999) A Bükk hegység szerkezetének főbb vonásai. Földtani Közlöny 129(4):611–651
DINPI (2009) http://www.dinpi.hu
Gy Gábris (1983) Characteristic features of drainage patterns in the volcanic mountains of Hungary. Annales Budapestiensis Sec. Geogr 18–19:21–31
Gyarmati P (1977) A Tokaji-hegység intermedier vulkanizmusa. Intermediate volcanism in the Tokaj Mountains, MÁFI, Budapest
Gyarmati P (2000) A Tokaji-hegység. In: Karátson (szerk.) Pannon enciklopédia. Kertek, Budapest, 354–357
Hámor G (2001) Magyarázó a Kárpát-medence miocén ősföldrajzi és fáciestérképéhez. MÁFI, Budapest
Hevesi A (1978) A Bükk szerkezet- és felszínfejlődésének vázlata. Földrajzi Értesítő 27(2):169–204
Hevesi A (1980) Adatok a Bükk hegység negyedidőszaki ősföldrajzi képéhez. Földtani Közlöny 110:540–550
Hevesi A (1986) A Bükk hegység felszínfejlődése és karsztja. Kandidátusi értekezés, kézirat, Budapest
Hevesi A (2000) A Bükk. In: Karátson D. (szerk.) Pannon enciklopédia. Kertek, Budapest, 337–344
Horváth A (2000) Hazai újholocén klíma- és környezetváltozások vizsgálata régészeti adatok segítségével. Földrajzi Közlemények 124(1–4):149–164
Horváth G, Munkácsy B, Pintér Z, Csiky J, Karancsi Z, Prakfalvi P (1997) A Medves. Földrajzi Értesítő 46(3–4):217–248
Jakucs L (1971) A karsztok morfogenetikája. Akadémiai Kiadó, Budapest
Jakucs L, Móga J (2000) A Gömör–Tornai-karszt. In: Karátson D. (szerk.) Pannon enciklopédia. Kertek, Budapest, 345–352
Jaskó S (1981) Üledékfelhalmozódás és kőszénképződés a neogénben. MÁFI, Budapest
Jugovics L (1940) Adatok a Somoskő és Rónabánya környéki bazaltelőfordulások ismeretéhez. A Magyar Királyi Földtani Intézet évi jelentése az 1933–35. évekről 4, 1511–1516
Karancsi Z (2002) Természetes és antropogén eredetű környezetváltozás a Medves-térség területén. Doktori (PhD) értekezés, kézirat, SZTE
Karátson D (2000) ABörzsöny. In: uő (szerk.) Pannon enciklopédia. Kertek, Budapest, 329–331
Karátson D (2007) Börzsönytől a Hargitáig. Typotex, Budapest
Kerényi A (1991) Talajerózió. Akadémiai Kiadó, Budapest
Kiss L (1980) A földrajzi nevek etimológiai szótára. Akadémiai Kiadó, Budapest
Kordos L (1984) Magyarország barlangjai. Gondolat, Budapest
Korpás L (1998) Paleokarst. MÁFI, Budapest
Korpás L (2000) A Visegrádi-hegység. In: Karátson D. (szerk.) Pannon enciklopédia. Kertek, Budapest, 327–328
Kubovics I, Pantó G (1970) Vulkanológiai vizsgálatok a Mátrában és a Börzsönyben. Akadémiai Kiadó, Budapest
Láng S (1953) Az Északi-középhegység geomorfológiai problémái. Földrajzi Értesítő 1:54–62
Láng S (1967) A Cserhát természeti földrajza. Akadémiai Kiadó, Budapest
Marosi S, Somogyi S (1990) Magyarország kistájainak katasztere. I-II, Magyar Tudományos Akadémia Földrajztudományi Kutatóintézet, Budapest
Martonné EK (2006) Magyarország tájföldrajza. KLTE, Debrecen
Mattányi Z (2006) Komplex tájértékelés egy Ipoly menti mintaterületen. Phd-értekezés, kézirat. Budapest
Mezősi G (1984) A Sajó-Bódva köz felszínfejlődése. Földrajzi Értesítő 33:181–206
Mezősi G (1985) A természeti környezet potenciáljának felmérése (Elmélet–Módszer–Gyakorlat 37). MTAFKI, Budapest
Nemerkényi A (1987) Alakmérési vizsgálatok a Kárpátok vulkáni vonulatának egykori kitörési központjain. Földrajzi Értesítő 36(3–4):273–279

Pécskay Z, Balogh K, Széky-Fux V, Gyarmati P (1986) Geochronological investigations on the neogene vulcanism of the Tokaj Mountains. Geol. Zbornik 37(5):635–655

Péczely G (1979) Éghajlattan. Nemzeti Tankönyvkiadó, Budapest

Noszky J (1940) Vándorlások a Palócföldön. Athenaeum, Budapest

Pécsi M (1959) A magyarországi Duna-völgy kialakulása és felszínalaktana. Akadémiai Kiadó, Budapest

Pinczés Z (1986) Periglaciális formák és üledékek térbeli rendje egy vulkanikus hegy lejtőjén. Földrajzi Értesítő 35(1):28–42

Pinczés Z (1998a) A Tokaji-hegység kialakulása és geomorfológiai értékei. Földrajzi Közlemények 46 (1–2):1–10

Pinczés Z (1998b) A Tokaji-hegység geomorfológiai nagyformái. Földrajzi Értesítő 47(3):379–395

Schweitzer F (1993) Domborzatformálódás a Pannóniai--medence belsejében a fiatal újkorban és a negyedidőszak határán. MTA, doktori értekezés, Budapest

Sütő L (2001) A felszín alatti bányászat domborzatra gyakorolt hatásai a kelet-borsodi szénmedencében. I. MFK, Szeged. www.geography.hu

Szabó J (1978) A Cserehát felszínfejlődésének fő vonásai. Földrajzi Közlemények 102(26):246–268

Szabó J (1995) Csuszamlásos folyamatok szerepe a magyarországi tájak fejlődésében. Kossuth, Debrecen

Szabó J (1996) A csuszamlásos folyamatok szerepe a magyarországi tájak geomorfológiai fejlődésében. Kossuth, Debrecen

Szabó M, Karátson D, Angyal Z (2004) A Salgótarján melletti salakmeddők vegetációjának jellemzői és felszínfejlődése. 2. Magyar Földrajzi Konferencia, Szeged. www.geography.hu

Szádeczky KE (1938) Geologie der rumpfungarländi schen Kleinen Tiefebene. Mitteilungen der berg- und hüttenmaschinen Abteilungen

Székely A (1983a) Verleichende vulkanische Mittelgebirgforschung in Ungarn. Ungarn-Deutschland Kolloquim, Hamburg. Trofenik, München, 207–246

Székely A (1983b) A pleisztocén periglaciális domborzatformálás Magyarországon. Földrajzi Értesítő 32 (3-4):389–398

Székely A (1997) Vulkánmorfológia. Eötvös, Budapest

szerk Kovács F (2005) Természeti erőforrások, ásványi nyersanyagok felhasználásának hatékony fejlesztési lehetőségei, energia- és környezetgazdálkodás. Kézirat, NFT, Budapest

szerk Tardy J (1996) Magyarországi települések védett természeti értékei. Mezőgazda, Budapest

Tóth G (1984) A Központi-Bükk és geomorfológiai körzetei. Földrajzi Értesítő 33(4):333–345

Tóth J, Almási I (2001) Interpretation of observed fluid potential patterns in a deep sedimentary basin under tectonic compression: Hungarian Great Plain, Pannonian Basin. Geofluids 1:1–27

Varga G (1975) A Mátra hegység földtana. MÁFI Évkönyv 57/1. Műszaki Könyvkiadó Budapest

Veres L (2002) Hutatelepülések. In: Baráz Cs. (szerk.) Hegyek, erdők, emberek. Bükki Nemzeti Park, Eger, 493–498

Z Ruszkiczay-Rüdiger, Dunai T, Fodor L, Bada G, Leél-Őssy Sz, Horváth E (2006) A negyedidőszaki függőleges kéregmozgások számszerűsítése a Duna völgyében a korábbi kronológiai adatok és új, kozmogén 3He kitettségi kor mérések alapján. Földtani Közlöny 135(3):373–402

Zámbó L (1998) Az Aggteleki-karszt felszínalaktani jellemzése. Földrajzi Értesítő 47(3):359–379

Zelenka T (1997) A nemesfémek kutatásának története a Kárpát-medencében. Földtani Kutatás 34:4–8

# Appendix A: Landscape Units of Hungary

*Macro-region*
**Meso-region**
*Group of micro-regions*
Micro-region

***Great Hungarian Plain***
**Danube Plain**
*Pest Plain*

1. Vác-Pest Danube Valley
2. Pest alluvial plain

*Csepel-Mohács Plain*

3. Csepel Plain
4. Solt plain
5. Kalocsai Sárköz
6. Tolnai-Sárköz
7. Mohács Island
8. Terraced Plain of Mohács

**Danube-Tisza Interfluve**

9. Gerje-Perje Plain
10. Pilis-Alpár Sand Ridge
11. Kiskunság Sand Ridge
12. Bugac Sand Ridge
13. Dorozsma-Majsa Sand Ridge
14. Kiskunság Loess Ridge

**Bácska Plain**

15. Illancs
16. Bácska Loess Plain

**Mezőföld**
*North Mezőföld*

17. Érd-Ercs Ridge
18. Váli-víz Plain

*Danube-Sárvíz Interfluve*

19. Central Mezőföld
20. Velence Basin
21. Sárrét
22. Sárvíz Valley
23. South Mezőföld

*West Mezőföld*

24. Enying Ridge
25. Kálóz-Igar Loess Ridges
26. Sió Valley



G. Mezősi, *The Physical Geography of Hungary*, Geography of the Physical Environment,
DOI 10.1007/978-3-319-45183-1

**Dráva Alluvial Plain**

27. Dráva Plain
28. Fekete-víz Plain
29. Nyárád-Harkány Plain

**Upper Tisza Region**

30. Bereg Plain
31. Szatmár Plain
32. Bodrogköz
33. Rétköz

**Middle Tisza Region**
*Floodplain of Middle Tisza*

34. Taktaköz
35. Borsod Floodplain
36. Heves Floodplain
37. Szolnok Floodplain
38. Jászság

*Nagykunság*

39. Tiszafüred-Kunhegyes Plain
40. Szolnok-Túr Plain
41. Tiszazug
42. Hortobágy

**Lower Tisza Region**

43. Marosszög
44. South Tisza Valley

**Northern Alluvial Fans**
*Tápió-Galga-Zagyva Region*

45. Hatvan Plain
46. Tápió Region

*Gyöngyös-Heves Region*

47. Gyöngyös Plain
48. Heves Plain

*Borsod-Zemplén Plain*

49. Borsodi-Mezőség
50. Sajó-Hernád Plain
51. Harangod

**Nyírség**
*Sand Region of Nyírség*

52. Central Nyírség
53. North-Eastern Nyírség
54. South-Eastern Nyírség
55. South Nyírség
56. West or Loess Nyírség

**Hajdúság**

57. Hajdúhát
58. South Hajdúság

**Berettyó-Körös Region**
*Berettyó Region*

59. Dévaványa Plain
60. Nagy-Sárrét
61. Berettyó-Kálló Interfluve
62. Érmellék Loess Ridge

*Körös Region*

63. Bihar Plain
64. Kis-Sárrét
65. Körös Plain

**Körös-Maros köze**
*Békés-Csanád Ridge*

66. Csanád Ridge
67. Békés Ridge

*Békés-Csongrád Plain*

68. Békés Plain
69. Csongrád Plain
70. Körösszög

***Little Hungarian Plain***
**Győr Basin**
*Szigetköz-Moson Plain*

71. Szigetköz
72. Moson Plain

*Fertő-Hanság Basin*

73. Fertő Basin
74. Hanság

*Rábaköz*

75. Kapuvár Plain
76. Csorna Plain

**Marcal Basin**

77. Marcal Valley
78. Kemenesalja
79. Pápa-Devecser Plain

**Komárom-Esztergom Plain**

80. Győr-Tata Terrace System
81. Igmánd-Kisbér Basin
82. Almás-Tát Danube Valley

***West Hungarian Border Region (Foothills of the Alps)***
**Lower Alps**
*Sopron Mountains*

83. Sopron Mountains
84. Fertőmellék Hills
85. Sopron Basin

*Kőszeg-Vas Mountains*

86. Kőszeg Mountains
87. Mount Vas and Kőszeg Foothills
88. Pinka Plain

*Upper-Őrség-Vas Ridge*

89. Upper-Őrség
90. Vas Ridge

**Sopron-Vas Plain**

91. Ikva Plain
92. Répce Plain
93. Gyöngyös Plain
94. Rába Terraced Plain
95. Rába Valley

**Kemeneshát**

96. Lower Kemeneshát
97. Upper Kemeneshát

**Zala Hills**
*Western Zala Hills*

98. Upper Zala Valley
99. Kerka Region (Hetés)
100. Central Zala Hills (Göcsej)

*Eastern Zala Hills*

101. Egerszeg-Letenye Hills
102. Principális Valley
103. Zalaapáti Ridge
104. Lower Zala Valley
105. Zalavári Ridge

*Mura Plain*

106. Mura Plain (left side)

***Transdanubian Hills***

**Balaton Basin**

107. Kis-Balaton Basin
108. Nagyberek
109. Somogy Plain
110. Balaton
111. Riviéra of Balaton
112. Tapolca Basin
113. Riviéra of Keszthely

**Outer Somogy**

114. Western Outer Somogy
115. Eastern Outer Somogy
116. Southern Outer Somogy

**Inner Somogy**

117. Marcali Ridge
118. Eastern Inner Somogy
119. Western Inner Somogy
120. Central Dráva Valley

**Mecsek and Tolna-Baranya Hills**

*Mecsek Region*

121. Mecsek Mountains
122. Baranya Ridge

*Tolna Hills*

123. Völgység
124. Tolna Ridge
125. Szekszárd Hills

*Baranya Hills*

126. Pécs Plain
127. Geresd Hills
128. Villány Mountain
129. South Baranya Hills

*Zselic*

130. North Zselic
131. South Zselic

***Transdanubian Mountains***

**Bakony Region**

*Keszthely Mountains*

132. Tátika Group
133. Keszthely Plateau

*Balaton Uplands*

134. Badacsony-Gulács Group
135. Balaton Uplands and Basins
136. Vilonya Mountains

*South Bakony*

137. Veszprém-Nagyvázsony Basin
138. Kab-hegy-Agár-tető Group
139. Sümeg-Tapolca Ridge
140. Devecseri-Bakonyalja

*North Bakony*

141. High Bakony
142. Bakony Basins
143. East Bakony
144. Veszprém-Devecser Graben

*Bakonyalja*

145. Pápai-Bakonyalja
146. Pannonhalma Hills
147. Súri-Bakonyalja

**Vértes-Velence Mountains**
*Vértesalja Hills*

148. Bársonyos
149. Által-ér Valley
150. Mór Graben

*Vértes Mountains*

151. Vértes Plateau
152. Vértes Foothills
153. Gánt Basin

*Velence Mountain and its surroundings*

154. Zámoly Basin
155. Söréd Ridge
156. Lovasberény Ridge
157. Velence Mountain

**Dunazug Mountains**
*Gerecse Region*

158. West Gerecse
159. Central Gerecse
160. East Gerecse
161. Gerecse Basins

*Bicske-Zsámbék Basin*

162. Etyek Hills
163. Zsámbék Basin

*Buda Mountains*

164. Buda Mountains
165. Tétény Plateau
166. Budaörs and Budakeszi Basins

*Pilis Mountains*

167. Pilis Mountains
168. Pilis Basins

***North Hungarian Mountains***
**Visegrád Mountains**

169. Danube Bend
170. Visegrád Mountains

**Börzsöny Mountains**

171. Central Börzsöny
172. Börzsöny Basins
173. Börzsöny Outer Mountains

**Cserhát Region**
*West Cserhát*

174. Kosd Hills
175. Nézsa-Csővár Hills

*East Cserhát*

176. Central Cserhát
177. Galga Valley
178. Ecskend Hills
179. Cserhátalja

*North Cserhát*

180. Terény Hills
181. Szécsény Hills

*Karancsság*

182. Karancs
183. Litke-Etes Hills

*Gödöllő-Monor Hills*

184. Gödöllő Hills
185. Monor-Irsa Hills

**Mátra Region**
*Central Mátra*

186. High Mátra
187. West Mátra
188. South Mátra

*Foothills of the Mátra (Mátraalja)*

189. East Mátraalja
190. West Mátraalja
191. Mátralába
192. Parád-Recsk Basin

**Bükk Region**
*Central Bükk*

193. Bükk Plateau
194. North Bükk
195. South Bükk

*Southern Foothills of the Bükk (Bükkalja)*

196. Tárkány Basin
197. Egri-Bükkalja
198. Miskolci-Bükkalja

*Northern Foothills of the Bükk (Bükklába)*

199. Tardona Hills
200. Uppony Hills

**Aggtelek-Rudabánya Mountains**
*Aggtelek Karst*

201. Aggtelek Mountains
202. Alsó-hegy

*Rudabányai-Szalonna Mountains*

203. Rudabánya Mountains
204. Szalonna Mountains
205. Bódva Valley
206. Torna Hills

**Tokaj-Zemplén Mountain Region**
*Zemplén Mountains*

207. Central Zemplén
208. Abaúji-Hegyalja

*Foothills of Tokaj-Hegyalja*

209. Mount Tokaj
210. Szerencs Hills
211. Hegyalja

*Hegyköz*

212. Hegyköz Hills
213. Vitány Blocks

**North Hungarian Basins**
*Ipoly Valley and Nógrád Basin*

214. Lower Ipoly Valley
215. Central Ipoly Valley
216. Nógrád Basin

*Upper Zagyva-Tarna Hills*

217. Zagyva Valley
218. Medves Region
219. Upper Tarna Hills
220. Tarna Valley

*Gömör-Heves Hills*

221. Ózd-Egercsehi Basin
222. Pétervásár Hills

*Borsod Basin*

223. Sajó Valley
224. Putnok Hills

*Cserehát*

225. Szendrő Mountains
226. Rakaca Basin
227. East Cserehát
228. West Cserehát

*Hernád Valley Basin*

229. Hernád Valley
230. Szerencsköz

# Appendix B: The Landscape Types of Hungary

The **fourteen landscape types** listed in the National Atlas of Hungary (MNA) can be classified into **four larger groups** (MNA 1989).

## I. Moderately continental, plain landscapes utilized mostly by agriculture

1. Floodplain with high groundwater level and hydromorphic soils, cultural steppe
2. Floodplain mostly with medium groundwater level and meadow-steppe soils, cultural steppe
3. Loess plain in a basin with steppe soil, cultural steppe
4. Blown sand on alluvial plain, cultural steppe with an orchard-vineyard forest mosaic, medium and low groundwater level
5. Cultural steppe of an alluvial plain in basin margin, with a dense water network and mosaic-like turkey oak forest remains, with steppe and forest soils

Eighteen subtypes within the above-mentioned five main plain landscape types are distinguished on the basis of the differences in water household, soil types and relief.

## II. Erosional hills predominantly with agricultural and sylvicultural, and locally with an industrial character

6. Foothill surfaces and hills segmented by erosional and derasional valleys, cultural steppe with a mosaic of orchard-vineyard-turkey oak forest and low groundwater level (under moderately warm and dry submediterranean climatic influence in the Transdanubian Hills, under moderately cool and dry subcontinental climatic influence in the North Hungarian Mountains).
7. Independent hills segmented by erosional and derasional valleys, cultural steppe with vineyards and larger mixed forest remains, mostly with low groundwater level (the different climatic influences depend on the situation within the Carpathian Basin: moderately warm and moderately dry submediterranean, moderately warm and moderately humid subatlantic and subalpine).
8. Smaller basins and hills enclosed by mountains, cultural steppe with turkey oak-oak forest remains and low groundwater level (the different climatic influences—subcontinental and subatlantic—depend on the location within the Carpathian Basin).



G. Mezősi, *The Physical Geography of Hungary*, Geography of the Physical Environment,
DOI 10.1007/978-3-319-45183-1

The fourteen hilly landscape types are distinguished on the basis of their climatic features, soil-forming rock, soils and segmentation by valleys.

## III. Mountains, forested landscape types

9. Low mountains, prevalence of the subcontinental climatic influence, turkey oak (and hornbeam) forests (lower than 650 m asl).
10. Low mountains, presence of subatlantic and submediterranean climatic influences, turkey oak and hornbeam forests.
11. Low mountains with forests mainly under subatlantic climatic influence.
12. Cooler and more humid beech-forested landscape type in mountains in North Hungary

Nine further subtypes are distinguished within the mountain, forested landscape types, mainly on the basis of dominant rock type and soil conditions.

## IV. Specific landscape types

13. Large valleys within a hilly or mountainous landscape
14. Lacustrine landscape or landscape of lake surroundings

## Reference

MNA (1989) Magyarország Nemzeti Atlasza. Carthographia, Budapest

# Index





G. Mezősi, *The Physical Geography of Hungary*, Geography of the Physical Environment,
DOI 10.1007/978-3-319-45183-1

## E



## F

MIX
Papier aus verantwortungsvollen Quellen
Paper from responsible sources
FSC® C105338

If you have any concerns about our products,
you can contact us on
**ProductSafety@springernature.com**

In case Publisher is established outside the EU,
the EU authorized representative is:
**Springer Nature Customer Service Center GmbH**
**Europaplatz 3, 69115 Heidelberg, Germany**

Printed by Libri Plureos GmbH
in Hamburg, Germany